Advances in Wastewater Research

Series Editor

Maulin P. Shah, Environmental Microbiology Consultant, Ankleshwar, Gujarat, India

The Advances in Wastewater Research highlights the prospective future development of wastewater treatment techniques, research and field applications. The series also focuses on the limitations and challenges associated with some generally accepted wastewater strategies and evaluates the possible applications of these corrective strategies to eliminate toxic pollutants from the environment through integrated technologies in wastewater treatment. The series covers the research areas including, but not limited to, toxic pollutants removal from wastewater, methods and procedures, materials, management of wastewater, valorisation possibilities of waste resulting from the removal of toxic pollutants from wastewater, wastewater recycling, environmental impact, and wastewater policies. The series contribute to remove, reduce, or neutralize water contaminants that threaten human health, and ecosystem productivity and integrity. The series includes peer-reviewed monographs, edited volumes, textbooks, and conference proceedings.

Souravi Bardhan · Maulin P. Shah
Editors

Applications of Hydrogels in Modern Wastewater Treatment

Springer

Editors
Souravi Bardhan
Department of Environmental Science
Netaji Nagar College for Women
Kolkata, West Bengal, India

Maulin P. Shah
Department of Research Impact
and Outcome Research, Research
and Development Cell
Lovely Professional University
Phagwara, Punjab, India

ISSN 3004-9725 ISSN 3004-9733 (electronic)
Advances in Wastewater Research
ISBN 978-981-95-1597-4 ISBN 978-981-95-1598-1 (eBook)
https://doi.org/10.1007/978-981-95-1598-1

This Springer imprint is published by the registered company Springer Nature Singapore Pte Ltd.
The registered company address is: 152 Beach Road, #21-01/04 Gateway East, Singapore 189721, Singapore

Contents

Chapter 1
Introduction and History of Hydrogels

Raje Sengar and Swati Dubey

Abstract Hydrogels are classified as soft materials manufactured by combining artificial or natural polymers to produce a 3-dimensional crosslinked porous network assembly. They can hold and retain a substantial quantity of water or biological juices, keeping their structural integrity. They are typically composed of hydrophilic polymer chains that are interconnected by chemical or physical cross-linking. Their tunable mechanical properties, adaptability, biocompatibility, biodegradability, and high-water absorption capacity make them highly attractive for various applications, including biosensors, medicine, cosmetics, and wastewater treatment. Hydrogels, due to their hydrophilic nature, are ideal for structural and functional modifications. These modifications enable the engineering of stimuli-sensitive materials that react to exterior factors like temperature, pH, magnetic or electric field, ionic strength, and ultrasound, etc., making them ideal for advanced technical applications especially in heavy metal ion removal. The goal of this chapter is to deliver a summary of the fundamentals and uses of hydrogels specifically in wastewater remedies. In this chapter, we go over the introduction, followed by a timeline of the history of hydrogels. Furthermore, this chapter includes the elemental aspects, key advancements, and numerous applications of hydrogels in various fields, with emphasis on wastewater treatment supported by a comprehensive literature review. Additionally, the chapter also discusses the challenges faced in their development and implementation, along with the emerging trends in hydrogel research that could shape the next generation of hydrogel-based technologies.

Keywords Hydrogel · Natural polymer · Synthetic polymers · History of hydrogels · Crosslinking · Applications of hydrogels

---

R. Sengar (✉) · S. Dubey
Department of Applied Chemistry, Samrat Ashok Technological Institute, Vidisha, Madhya Pradesh, India
e-mail: raje2183@gmail.com

S. Bardhan and M. P. Shah (eds.), *Applications of Hydrogels in Modern Wastewater Treatment*, Advances in Wastewater Research,
https://doi.org/10.1007/978-981-95-1598-1_1

Abbreviations

AA	Acrylic Acid
AAm	Acrylamide
Arg-UPEAs	Hydrogel using arginine-based poly (ester amides)
CN-cl-PL(AA)NHG	Chitosan-crosslinked-poly (alginic acid) nanohydrogel
Cr (VI)	Hexavalent chromium
DMSA	Dexamethasone acetate
HEMA	Hydroxyethyl methacrylate
IPN	Interpenetrating polymer network
LCST	Lower critical solution temperature
PAA	Poly (acrylic acid)
PAM	Polyacrylamide
PCL	Poly(ε-caprolactone)
PEG	Polyethylene glycol
PGLA	Poly (glycolic-co-lactic acid)
PHEMA	Polyhydroxyethyl methacrylate
PHPMA	N-2-hydroxypropyl methacrylamide
Pluronic-DA	Pluronic diacrylate
PNIPAAm	Poly(N-isopropylacrylamide)
PVA	Polyvinyl alcohol
semi-IPN	Semi-interpenetrating polymer network
UCST	Upper critical solution temperature

1.1 Introduction

1.1.1 Background

Hydrogels are 3-dimensional interconnected polymeric structures that can soak and hold a significant volume of water or biofluids without any alteration in their structural integrity. They are typically composed of hydrophilic polymer chains that are physically or chemically interlinked, as shown in Fig. 1.1.

They are distinguished by their capacity for high-water content, comprising 20% to over 90% of their weight, which provides them with mechanical properties and environmental responsiveness comparable to natural tissue, such as skin and cartilage. Their ability to absorb water stems from the presence of hydrophilic functional groups such as $-OH$, $-CONH_2$, $-SO_3H$, and $-COOH$ groups (Bashir et al. 2020) on the polymer backbone, while the cross-linked network ensures resistance to dissolution. Hydrogels are ideal for structural and functional modifications due to their hydrophilic nature. These modifications enable the engineering of stimuli-sensitive materials that respond to extrinsic factors like temperature, pH, magnetic or electric

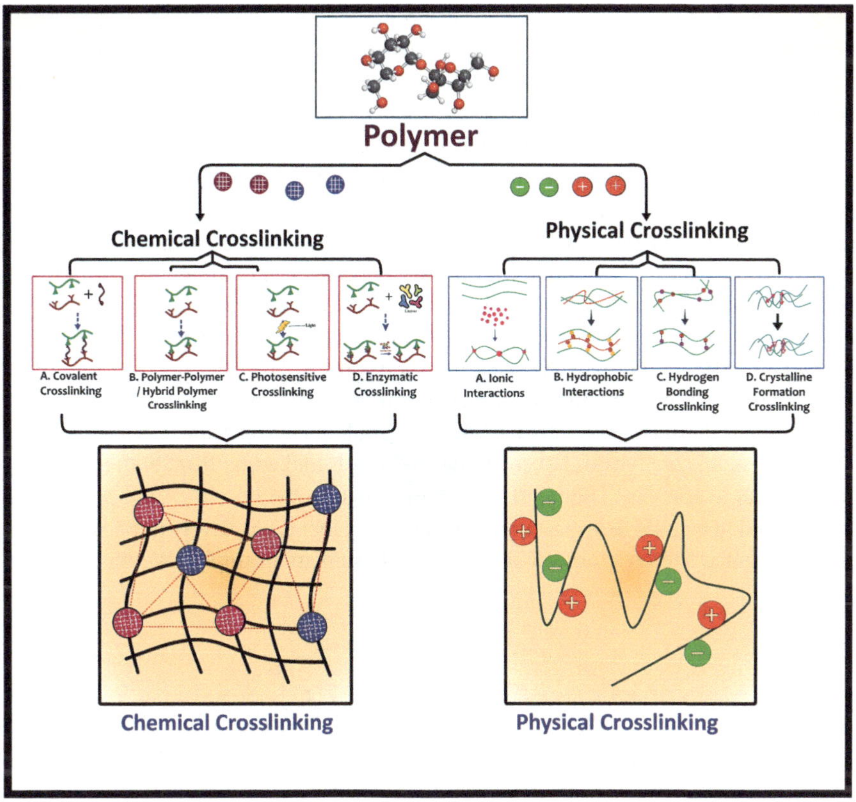

Fig. 1.1 Crosslinking methods employed in hydrogels synthesis

fields, ultrasound, and ionic strength, etc., making them ideal for advanced technical applications.

Hydrogels have been widely used across diverse fields due to their versatile and tunable properties. The most defining characteristics of hydrogels, including biodegradability, biocompatibility, hydrophilicity, superabsorbency, viscoelasticity, softness, and fluffiness, make them an irreplaceable candidate for a wide variety of cutting-edge applications (Hu et al. 2023) in diverse sectors such as biosensors, tissue engineering, biomedical engineering, drug delivery systems, food technology, agriculture engineering, waste water treatment, and more.

In biomedicine, hydrogels have revolutionized drug delivery systems, offering controlled release mechanisms and enhanced biocompatibility. They are widely used as scaffolds in tissue engineering, which imitate cytoplasm, promoting cell growth and regeneration. Additionally, hydrogels find applications in wound care, contact lenses, and as carriers for bioactive molecules in precision medicine. Hydrogels have also become pivotal in energy storage and electronics, particularly in the development of flexible sensors, supercapacitors, and bioelectronic devices. They are also used as smart materials for advanced technologies like soft robotics and wearable devices. In cosmetics and personal care, hydrogels are key ingredients in moisturizers, facial masks, and hygiene products, offering enhanced hydration and skin-soothing properties.

In environmental science, hydrogels contribute significantly to sustainable practices. They show wide application in wastewater treatment to remove pollutants and as superabsorbent materials in agriculture to improve soil water retention, reducing water consumption and promoting efficient irrigation practices (Zhang et al. 2022). Water is essential for various processes, including the hydrological cycle, industry, agriculture, and domestic use. However, freshwater availability is limited and increasingly threatened by contamination from marine dumping, oil spills, industrial discharge, and sewage waste. Figure 1.2 provides an overview of the many contaminants found in wastewater. The most prevalent contaminant in polluted water among them is heavy metals, which degrade the sustainable environment.

Heavy metals are harmful even at minimal levels and come from both natural and anthropogenic sources. The rapid growth of industries, economies, and populations has resulted in heavy metal poisoning of water, which has negatively impacted human health globally. Mercury, arsenic, lead, and cadmium are examples of common heavy

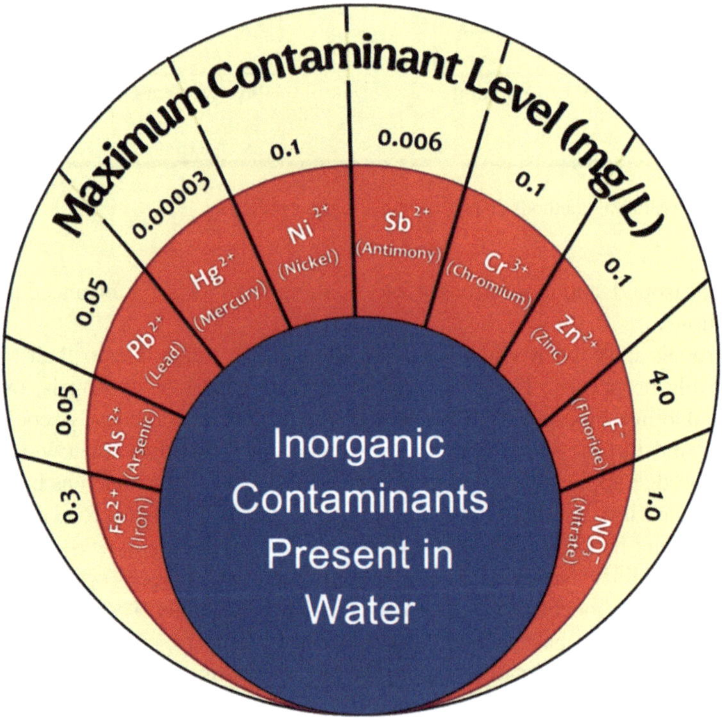

Fig. 1.2 Common inorganic contaminants detrimental to ecosystems and human health

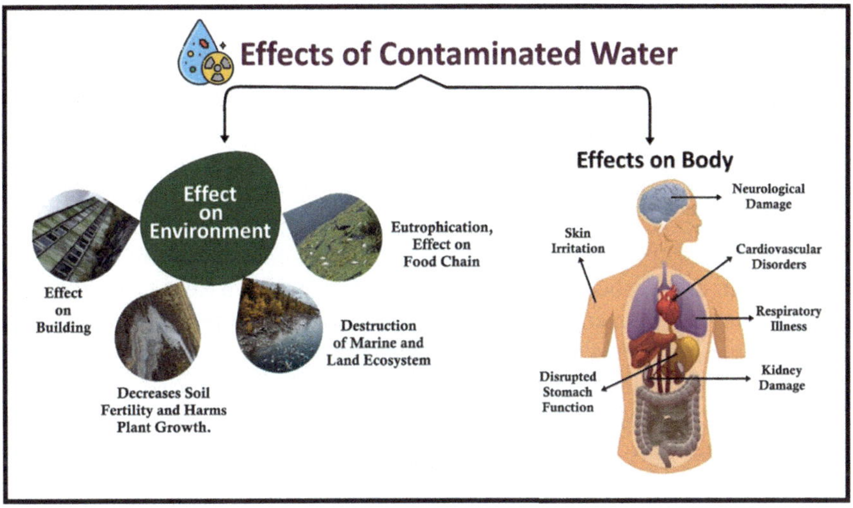

Fig. 1.3 Effect of contaminated drinking water

metals that are detrimental to ecosystems and human health (Fig. 1.2) since they are not biodegradable and their long-term exposure from drinking water as depicted in Fig. 1.3, causes cancer (kidney, bladder, skin, and lungs etc.), chest pain, organ damage, and neurological disorders.

Thus, water conservation and purification are critical challenges of this century. Effective treatment technologies are needed to reduce pollution, along with efforts to minimize pollutant emissions. Adsorption is a simple and economical method for removing low-concentration pollutants from large water bodies and hazardous substances from wastewater. Various solid adsorbents, like activated charcoal, modified clays, zeolites, and ion-exchange resins, are being developed for water treatment, with efficiency depending on adsorbent performance (Akhtar et al. 2024). Three-dimensional organic polymer hydrogels have gained prominence as efficient adsorbent materials for removing various contaminants, including heavy metal ions, synthetic dyes, and organic pollutants. Their superior adsorption capacity, structural adaptability, and user-friendly characteristics make them particularly valuable for environmental remediation applications. Recent advancements focus on designing hydrogel adsorbents with tailored functionalities, leveraging their physicochemical properties and polymer networks for improved efficiency.

The importance of hydrogels lies in their versatility, tunability, and capacity to bridge the gap between biology, chemistry, and engineering. As research advances, hydrogels are poised to drive innovation across industries, making them a vital component of modern scientific and technological progress. This chapter will explore the various developmental stages of hydrogels over the years, highlighting their evolution and advancements. Additionally, it will cover the properties and diverse applications of hydrogels.

1.2 History of Hydrogels

The concept of hydrogels was originally defined in 1894 by Lee and colleagues, who characterized these materials as a distinct category of colloidal systems. Although, as per today's standards, the described substance was not precisely a hydrogel, it instead was an inorganic salt-based gel. Since then, its definition has undergone a significant transformation. The history of hydrogels begins in the late nineteenth century, featuring a series of significant advancements and key milestones in their discovery, development, and application across various fields, which have transformed them into the materials we recognize today. Thomas Graham, in the nineteenth century, led groundbreaking studies on gels, developing silica gel through sol–gel chemistry and contributing significantly to the understanding of gel formation (Ungureanu et al. 2024). Initially, hydrogels were simple materials, but over time, they have developed into highly engineered substances integral to cutting-edge technologies. The first hydrogel developed for biomedical implantation was reported in 1949, which is commercially known as Ivalon®. This was synthesized by interlinking formaldehyde with poly (vinyl alcohol) (Kirschner and Anseth 2013). In the realm of contact lenses, the first reported lenses were made from poly (methyl methacrylate) (PMMA), a rigid, bioinert material originally used for airplane cockpit canopies during World War II.

In 1960, Wichterle and Lim manufactured PHEMA, the first documented hydrogel, which proved transformative for the contact lens industry. This breakthrough not only enhanced the comfort and functionality of lenses but also catalyzed the rapid growth of the hydrogel market, establishing it as a cornerstone of modern material science (Nguyen et al. 2015). The history of hydrogels, dating back to the early twentieth century, is marked by notable milestones that have shaped their discovery, development, and applications across various fields, as illustrated in Fig. 1.4. The history of hydrogels is divided into different generations showcasing significant advancements in design and functionality. Evolving from basic chemically crosslinked networks to advanced responsive and smart hydrogels, these innovations have broadened their applications across various fields. These are as follows.

1.2.1 The First-Generation Hydrogels (1960–1980)

The word hydrogel itself was coined in the 1960s, but the concept of water-absorbing polymers dates back much earlier. In the 1930s, researchers began to explore the properties of hydrogels more extensively. The development of synthetic polymers, especially those that could absorb water and form gel-like structures, paved the way for further exploration of their potential. The introduction of PVA and PEG hydrogels marked an important milestone in the understanding and manipulation of hydrogels. A key turning point in the history of hydrogels came after World War II. The development of new synthetic polymers in the postwar period allowed for the advancement

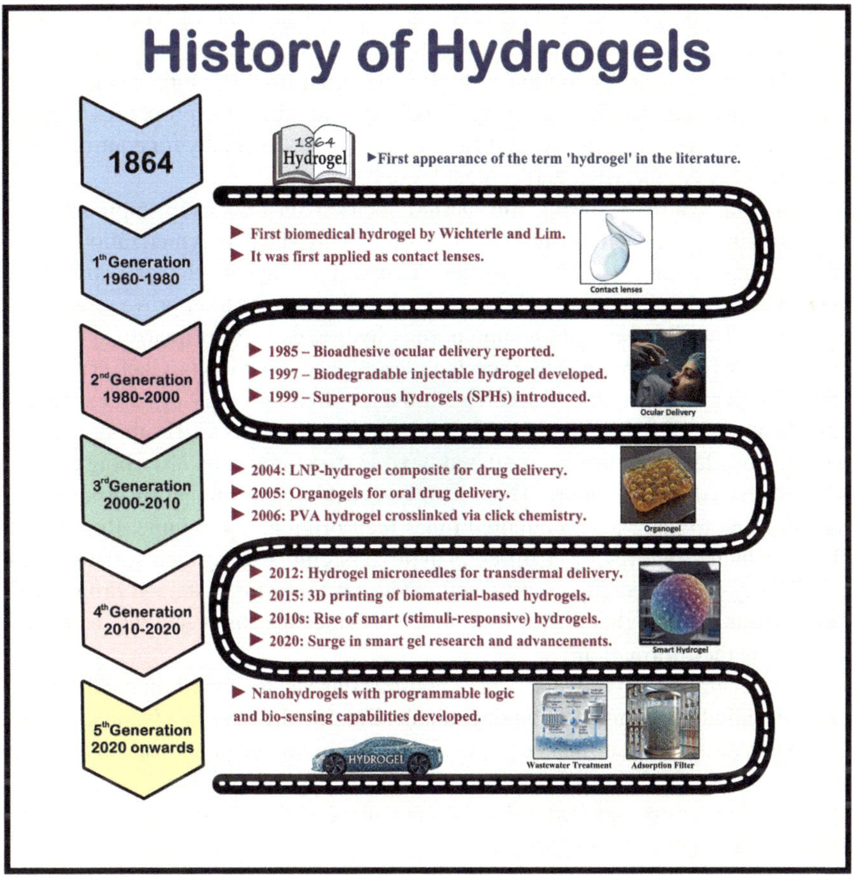

Fig. 1.4 Timeline presenting important developments in the history of hydrogel research

of hydrogels from basic materials to more complex, functional substances. In the 1950s, scientists began to synthesize cross-linked polymer networks, which led to the formation of robust hydrogels. These new materials exhibited significant water retention properties and could be produced with higher precision.

First, first-generation hydrogels (also known as hydrophilic synthetic polymers) are divided into three main classes according to their composition and method of synthesis. The first group includes alkene monomer-derived synthetic polymers using radical polymerization, such as poly(acrylamide) (PAM) and poly(2-hydroxyethyl methacrylate) (PHEMA)—materials whose biomedical significance persists decades after they were first synthesized. The second category includes hydrophilic polymer networks with covalent cross-links, i.e., polyvinyl alcohol (PVA) and polyethylene glycol (PEG), which have vast uses in regenerative medicine. The last category consists of natural polysaccharide-based systems, mainly cellulose hydrogels, with controlled release characteristics in pharmaceuticals (Ciolacu et al. 2020).

In 1960, hydrogels made their first major leap in the field of biomedical applications. Researchers developed hydrogels for wound dressings and other medical uses. This marked the beginning of hydrogels being used in more specialized and practical applications. The capability of hydrogels to mimic the properties of natural tissue opened doors to their use in the biomedical field, especially for controlled drug release, tissue engineering, and contact lenses.

Hydrogel synthesis during this period focused on two main approaches: (1) polymerizing water-soluble monomers using free-radical polymerization and (2) crosslinking existing synthetic polymers. Crosslinking agents played a dual role by both linking polymer chains and propagating radical chains to improve crosslinking efficiency. The first-generation superporous hydrogels were primarily synthesized using hydrophilic monomers like acrylamide and acrylic acid salts. Chen et al. (1999) created superporous hydrogels which show fast-swelling behavior and exceptional absorbent properties. In their study, they demonstrate that dried SPHs, though initially hard and brittle, transform into soft and flexible structures upon absorbing moisture due to their hydrophilic nature. These hydrogels exhibit exceptional absorbent potential, swelling to more than a hundred times their original dry volume. Roorda et al. (1986) in their article reviewed different hydrogels, emphasizing their soft, tissue-like consistency, high biocompatibility, and extensive study for sustained drug release. Their research highlights the intricate water-polymer structures and the interactions underlying their formation.

First-generation hydrogels were non-responsive to environmental stimuli, mechanically fragile, and non-biodegradable, limiting their use in advanced biomedical applications. Their lack of specificity for biomolecule or cell interactions further restricted their functionality in complex biological systems. First-generation hydrogel limitations spurred the development of more responsive, durable, and bioactive next-generation hydrogels.

1.2.2 The Second-Generation Hydrogels (1990–2000)

The limitations of traditional hydrogels provoked the development of advanced responsive hydrogels, which signified the introduction of second-generation hydrogel technology. These smart materials have dynamic responses to environmental stimuli such as thermal changes, pH levels, and biochemical concentrations. Their adaptive nature triggers controlled functions like polymerization regulation, site-specific therapeutic delivery, and dynamic pore architecture creation. These smart hydrogels, which came into existence when responsive biomaterials were in vogue, respond with three main classification types according to their mechanism of activation.

1.2.2.1 Temperature-Responsive Hydrogels

Temperature or thermal-responsive hydrogels experience a volume phase transition from sol to gel state at a critical temperature, namely LCST and UCST, from low to high temperatures. The most commonly used temperature-sensitive hydrogels include polylactide, PEG-polyester block copolymers, PGLA, PCL, PNIPAAm, and degradable hydrolytic hydrogels such as PHPMA. Among various temperature-sensitive polymers, PNIPAAm is a notable polymer for exhibiting a lower critical solution temperature (LCST) in aqueous solutions. In aqueous solutions, crosslinked poly (NIPAAm) creates a thermally reversible hydrogel with an LCST of approximately 33 °C (Mathur et al. 1996). Wu and his team (Wu et al. 1992) worked on the synthesis of macroporous hydrogels possessing enhanced swelling ratios at temperatures below the LCST. In another study, Okano and his team (Okano et al. 1990) published research on thermo-sensitive gels for medical applications, concentrating on thermoresponsive pulsatile drug release systems using PNIPAAm gels with an LCST. They suggested that the temperature for switching drug release on and off could be controlled by modifying the hydrophobic component in PNIPAAm gels. In a similar study, Katono et al. (1991) developed IPNs of poly(AAm-co-BMA) and poly(PAAc) for temperature-sensitive drug release. Drug release is activated at higher temperatures, while cooling induces pulsatile release followed by an "off" state, driven by surface layer shrinking and dense polymer layer formation.

1.2.2.2 pH-Sensitive Hydrogels

pH-responsive hydrogels can hydrolyze at high or low pH environments, respectively. pH-sensitive polymers possess acidic (like carboxylic or sulfonic acids) or basic groups (such as ammonium salts) that react to environmental pH changes by taking or releasing protons.

Chen et al. (1997) developed a novel natural semi-IPN composed of chitosan and silk fibroin. They studied its pH and ion sensitivity, exploring its potential application in chemomechanical systems like artificial muscles. Similarly, Allcock and Ambrosio (1996) developed a novel class of pH-sensitive hydrogels based on polyphosphazenes. These water-soluble polymers were cross-linked using gamma irradiation, forming hydrogels upon water absorption. The hydrogels exhibited greater swelling in basic buffers compared to acidic ones, with higher ionic side group content further enhancing their pH sensitivity.

1.2.2.3 Biomolecule-Sensitive Hydrogels

Biologically responsive hydrogels alter their properties upon recognizing specific biological targets like nutrients, enzymes, antibodies, receptors, or cells. These molecular recognition events alter intermolecular interactions, triggering large-scale changes like swelling, shrinking, or transitions from a solution to a gel state.

Hydrogels exhibit conformational changes in response to specific biomolecule concentrations, such as glucose oxidase-based hydrogels for insulin delivery. Glucose enters the hydrogel, where it is converted into gluconic acid by glucose oxidase, resulting in a lowered pH of the surrounding environment. This triggers protonation of the hydrogel's amine groups, increasing lysis, enabling the release of insulin from the matrix, and creating an auto-regulated insulin release system (Bajpai et al. 2013). A similar type of study was proposed by Kost et al. (1985) in their article. Their study reports the development of glucose-sensitive membranes that swell when in contact with glucose, potentially useful for insulin delivery or glucose monitoring. The membranes, synthesized via radiation-induced polymerization, are hydrogels with 60–90% water content, responsive to pH and glucose levels. Glucose oxidation by glucose oxidase generates gluconic acid, triggering swelling and permeability changes. Variations in formulations highlight the potential for creating membranes suitable for artificial pancreas applications. Peppas (1997) described several bio-responsive hydrogels and their applications in their review article. They noted that advances have been made in water-swollen, crosslinked biomedical polymers for drug, protein, protein conjugates, enzymes, site-specific targeted delivery, etc.

1.2.3 The Third-Generation Hydrogels (2000–2010)

During the 1990s, hydrogels saw significant growth in both academic research and commercial applications. As soon as hydrogel-based contact lenses were introduced in the market, they brought about a revolution in the field of optometry and eye care. This revolution was due to the high-water content present in these hydrogel-based lenses, which maintains better permeability and moisture in the eyes, therefore making them more comfortable than traditional hard lenses. Simultaneously, researchers focused on improving the mechanical properties and biocompatibility of hydrogels. The introduction of novel polymerization methods, such as radical polymerization and ionic gelation, enabled the production of hydrogels with better control over their physical properties, such as swelling behavior, mechanical strength, and degradation rates. Additionally, hydrogels became important materials in agriculture for controlled water release and soil conditioning. The hydrogels synthesized during this era are called third-generation hydrogels.

The key feature of this generation of hydrogels was the cross-linking method. Cross-linking is used to alter the physical and chemical properties of materials derived from polymers and biopolymers by creating physical or chemical bonds between polymer compounds and cross-linking agents. Bi and Liang (2016) provided an extensive review of various research approaches for the chemical synthesis of interlinked hydrogels in situ and highlighted their diverse applications in the biomedical sector. They highlight the application of various cross-linking agents, such as glyoxal, glutaraldehyde, different dialdehydes like starch, chitosan, alginate, and squaric acid used to chemically modify polymer and biopolymer materials.

1.2.4 The Smart Hydrogels or Fourth-Generation Hydrogels (2010–2020)

The 1990s and 2000s saw hydrogels gaining prominence in the field of biomedicine. The evolution of polymer science has enabled the creation of intelligent hydrogel systems capable of dynamic environmental responsiveness. Recent years have witnessed significant scientific focus on these adaptive materials, which demonstrate tunable properties in response to external triggers. These stimuli may include: pH, temperature, light, magnetic fields, or electrical signals, etc., enabling controlled drug release, ensuring precise drug delivery, enhancing treatment effectiveness and safety, as shown in Fig. 1.5. Therefore, this enables the progress of advanced multifunctional materials for applications like biosensors, wearable devices, and targeted drug delivery. Additionally, controlled drug release in response to stimuli creates new possibilities for treating diseases such as cancer, diabetes, and tissue regeneration (Puccetti et al. 2024).

The field of tissue engineering and regenerative medicine has driven remarkable progress in hydrogel technology, establishing these materials as fundamental components for biomedical applications. Their unique properties make them indispensable for wound healing, cell culture, tissue regeneration, etc. Pishavar et al. (2021) examined the advancements of multifunctional, self-healing hydrogels in cancer therapy, tissue engineering, and regenerative medicine. These regenerative medicines investigate the impact of materials on cellular functions to repair tissues and enhance cancer therapies. They discussed that hydrogels gained attention for their ability to copy or mimic the extracellular matrix, making them valuable in tissue engineering and drug delivery. With high drug-loading capacity and sustained release, they are ideal for therapeutic applications. Recently, self-healing supramolecular hydrogels, which are biocompatible, tissue-mimicking, and injectable, have shown promise in preclinical trials. These hydrogels allow for localized drug delivery, addressing surgical complications, inflammation, and infections. Similarly, Sikdar and his colleagues (Sikdar et al. 2021) discuss recent progress in smart hydrogel synthesis, their responsiveness, and methods for improving their functional properties for various applications. They highlight that smart materials can detect and react to environmental changes, with smart hydrogels responding to stimuli such as pH, temperature, light, and electric fields. Additionally, they note the rapid expansion of research in this field, driven by the promising applications of smart hydrogels in biomedical, industrial, and healthcare sectors.

1.2.5 Hydrogels in the Twenty-First Century—The Fifth-Generation Hydrogels (From 2020 Onwards)

Over recent years, hydrogels have been investigated as a promising material for multiple applications. In the twenty-first century, hydrogels have been increasingly

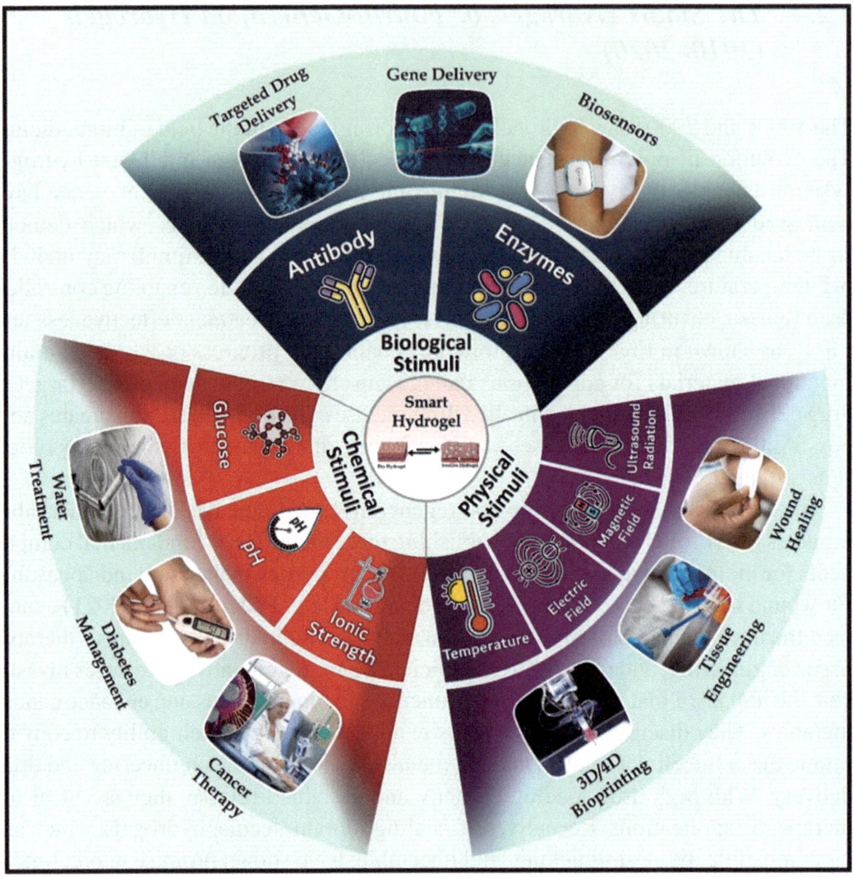

Fig. 1.5 Types of environmental sensitivity to develop smart hydrogel

developed into nano-scale forms, creating a multifunctional framework that enables the production of hybrid systems. These systems leverage the combined properties of hydrogels and nanomaterials, resulting in innovative materials with enhanced functionalities (Saha et al. 2023). These hybrid systems are being synthesized through various physicochemical methods and structural modifications, including the incorporation of different crosslinkers, alterations in polymer network architectures, integration of natural and synthetic polymers, addition of inorganic materials, and the development of homogeneous networks and stereocomplex structures.

In 2003, Gong et al. utilized free radical polymerization to develop the first novel double-network hydrogels, which exhibited exceptionally high mechanical strength, achieving a compressive stress of 17.2 MPa and with strain (92%) (Gong et al. 2003). Machado et al. (2024) in their review remarked that hydrogel scaffolds with drug-loaded nanostructures enhance bioavailability at tumor sites, showing strong efficacy against skin cancer. They also highlighted that hydrogels support solid nanoparticles in photothermal therapy while promoting wound healing and cancer cell death. Additionally, intravenously administered nanogels effectively combat metastasis. Sharma and colleagues (Sharma et al. 2024) highlight the unique properties of nanohydrogels, such as size, biocompatibility, and tunable drug release, making them valuable in biomedicine. They are widely used in tissue engineering, drug delivery, and diagnostics, with a focus on targeted therapeutic release. Their controlled-release capabilities are particularly promising for cancer treatment, enhancing efficacy while minimizing harm to healthy tissues, improved mechanical strength, and controlled drug release.

In another study, Naseer (2024) highlighted the critical challenges of water availability and quality, stressing the need for advanced treatment technologies. She discussed how untreated wastewater from industrialization harms the environment and human health. Her study emphasizes the efficiency of nanocomposites combined with polymers in wastewater treatment. She concludes that bio-adsorbents and nanocomposites have great potential to improve viability and effectiveness in the field.

1.3 Key Advancements in Hydrogel Development

It is well known that hydrogels have an exceptional ability to retain water. Their high-water content, flexible structure, and porous nature allow them to closely resemble living tissues. However, this was not always the case in their early stages of discovery. Over the years, significant advancements have been made to improve their properties, enhancing their efficiency in a variety of applications, such as in drug delivery, sensors, tissue engineering, environmental and industrial fields. Some key advancements include:

1.3.1 Biocompatibility and Biodegradability

Hydrogels used in medical and pharmaceutical applications need to be non-toxic and able to degrade safely within the body without causing harmful side effects. Therefore, one of the most important areas of hydrogel development has been in improving biocompatibility and biodegradability. Therefore, a key focus in hydrogel development was the enhancement of biocompatibility and biodegradability. To meet these requirements, researchers concentrated on developing hydrogels with these essential properties.

Pan and his co-workers (Pan et al. 2020) have engineered an innovative polysaccharide hydrogel system with injectable and biodegradable properties, characterized by an interconnected porous network. This advanced biomaterial combines nano-hydroxyapatite with self-repairing hydrogel components, demonstrating stem cell migration, supporting effective irregular bone filling. Similarly, chitosan-based hydrogel bio-scaffold exhibited enhanced biodegradability, antibacterial properties, uniform porosity, pH, and sensitivity. They reported that in vitro tests demonstrated over 90% inhibition of Staphylococcus aureus. And so, the designed hydrogel effectively prevents bacterial spread and inflammation while enhancing dental implant success rates.

Using a hybrid hydrogel platform, Wu et al. (2014) offer a unique drug encapsulating approach for prolonged and regulated drug release. They used arginine-based poly (ester amides) (Arg-UPEAs), alginate, and pluronic diacrylate (Pluronic-DA) via UV photo-crosslinking to create a temperature-responsive, environmentally friendly, and nontoxic hydrogel. The hydrogels were thoroughly characterized for swelling behavior, biodegradation, mechanical properties, and morphology. The in vitro studies confirmed their biocompatibility, supporting cell attachment and proliferation. They encapsulated drugs like insulin into synthesized nanoparticles and analyzed the release behavior using techniques such as HPLC and LC–MS, confirming prolonged drug release over several weeks.

1.3.2 Synthesis Techniques

The development of novel polymerization methods, such as click chemistry and self-assembly techniques, has allowed for more specific control over hydrogel structures. This has led to the creation of hydrogels with specific properties, such as enhanced mechanical strength, making them stronger and more elastic, more efficient drug release profiles, water purification, etc. The swelling behavior of hydrogel matrices significantly governs pharmaceutical compound permeation by regulating fluid dynamics and solute transport through their three-dimensional polymeric architectures. The mesh size determines drug-polymer interactions and controls drug diffusion. Different polymerization techniques, such as free radical, cationic, or anionic methods (Saha et al. 2023), can be used to achieve varying mesh sizes. So, as another advancement, researchers started focusing on synthesizing hydrogel systems by using various crosslinking methods to enhance the swelling capacity, controlled drug release, and better mechanical strength.

Advancements during 2010–2020 in hydrogel systems, occurred, focusing on their use in encapsulating living cells and controlled drug release. This period explored various cross-linking methods for hydrogel synthesis, including physical and chemical crosslinking, crystallization, free radical polymerization, high-energy radiation, and enzyme-mediated crosslinking, highlighting their applications in novel biomedical fields. Hydrogels are developed through chemical cross-linking, physical cross-linking (via temperature or pH changes), and polymer blending. These methods enable the development of hydrogels with tailored properties for applications in tissue engineering, biomedicine, and sensing. Tan et al. (2009) developed injectable composite hydrogels using N-succinyl chitosan functionalized with hyaluronic acid via a Schiff base mechanism. They found that increasing N-succinyl chitosan content improved the compressive modulus, a key factor in cartilage tissue engineering.

Multiple studies have highlighted the widespread use of cellulose and alginate in developing biocompatible hydrogels, owing to their cost-effectiveness, biodegradability, hydrophilicity, and non-toxicity. They are suitable applicants for water treatment, biomedical fields, etc., due to the presence of rich hydrophilic functional groups. The combination of these polysaccharides enhances hydrogel properties, including mechanical strength and adsorption effectiveness. Based on these studies, Rana and his colleagues (Rana et al. 2024) review recent advancements in cellulose/nanocellulose-alginate-based hydrogels, focusing on their applications in dye and heavy metal adsorption, wound healing, drug delivery, and tissue engineering, as well as the influence of additives on their performance.

1.3.3 Smart Hydrogels

The main challenge of drug delivery systems, especially in the case of chemotherapeutic drugs, is the high toxicity of antitumor drugs, which cause severe side effects and limit dosage levels. These drugs particularly harm rapidly dividing cells and vital organs like the liver, kidneys, and hematopoietic system. Therefore, as the population ages, there is an increasing demand for smart biomaterials that support the repair, replacement, and regeneration of both hard and soft tissues. Since then, smart hydrogels, also known as responsive hydrogels, have become a major focus of research. This system is termed smart because it can sense external stimuli. These materials can change their shape, volume, or other properties in response to external stimuli, enabling specific functions like controlled release of encapsulated drugs. For example, temperature-sensitive hydrogels are widely utilized in medicine delivery systems to release their therapeutic agents when a specific temperature threshold is reached.

The advancement of smart hydrogels enables targeted drug delivery, significantly reducing side effects by localizing the drug within cancerous tissue, thereby enhancing treatment efficacy while minimizing systemic toxicity. Recent studies highlight the growing importance of smart hydrogels in pharmaceutical applications due to their unique characteristics. As noted by Bordbar-Khiabani and Gasik (2022), these intelligent polymer networks have become particularly valuable for controlled drug delivery owing to their durability, physicochemical properties, and tissue compatibility. These hydrogels can alter their wettability, swelling capacity, physical properties, and molecular permeability in response to external stimuli such as pH, temperature, electric and magnetic fields, light, and biomolecule concentration, enabling the controlled release of encapsulated drugs. Hu et al. (2023) reported that smart hydrogels are advanced biomaterials with applications in diagnostics and therapy. These hydrogels react to physical, chemical, and biological stimuli, triggering structural changes such as volume shifts, phase transitions, and color changes. They offer high swelling capacity, flexibility, and biocompatibility, resembling human tissues. Their ability to provide controlled, on-demand responses makes them highly promising for clinical applications.

1.3.4 Hybrid Hydrogels

While hydrogels have natural properties that are used in various fields but more work was needed on their mechanical strength, stability, release behavior, bio-degradation, etc. Natural hydrogels face challenges such as poor stability, rapid release, and weak mechanical strength. While in synthetic hydrogels, structural irregularities and network defects, such as phase-separated regions, entangles, and hanging chain ends were observed by different researchers. These flaws were affecting their mechanical properties, diffusion rates, and the biological activity of embedded molecules. Hybrid hydrogels are synthesized by mixing natural and synthetic polymers. Due to the blending of two different origin materials, they can provide improved properties like water absorption, mechanical strength, compatibility with nature, and thus better performance in different sectors. And so, researchers started focusing on the synthesis of hybrid hydrogels to address individual challenges associated with natural and synthetic polymers and to leverage the application of blending materials.

Weerasundara et al. (2021) explore the use of hydrogel-based materials for water purification, highlighting recent advancements, challenges, and future prospects. Polymer and graphene hydrogels show promise in removing inorganic and organic contaminants and can be combined with various adsorbents to enhance efficiency. Their three-dimensional structure offers benefits like easy handling, large-scale application potential, and reusability. However, most studies remain at the laboratory scale, requiring further research for industrial applications and broader contaminant removal. This review provides a comprehensive analysis of these aspects. Likewise, in another review, Rigoletto and colleagues (Rigoletto et al. 2022) examine recent approaches for water purification that utilize hybrid materials combining techniques such as adsorption, photocatalysis, and enzymatic catalysis. The review also discussed that to improve efficiency, reduce costs, and minimize environmental impact, various reusable materials have been developed and tested on natural and spiked water matrices. The review also discusses their properties, effectiveness, and limitations.

Recent research by Tummino and colleagues (2020) investigated sustainable approaches for environmental remediation through the development of novel biocomposite hydrogels. Their work focused on fabricating crosslinked polymeric films using marine-derived sodium alginate (brown algae extract), crustacean-sourced chitosan, or eco-friendly crosslinking agents from composted organic waste. The study evaluated the adsorption efficacy of these renewable materials against three representative water contaminants: crystal violet (triphenylmethane dye), Rhodamine B (xanthene dye), and Orange II (azo dye). The study demonstrates the effectiveness of these hydrogels due to their diverse active functionalities, structural properties, and physicochemical characteristics. The toxic nature of the targeted dyes highlights the importance of developing sustainable removal solutions.

In another study, El-saied and El-Fawal (2021) present a sustainable approach to contaminant removal using superabsorbent nanocomposite hydrogels. These materials were synthesized by grafting AA and AAm onto green starch, incorporating

$NiFe_2O_4$ and TiO_2 nanoparticles. Characterized through various analytical techniques, the hydrogels were tested for the adsorption and photodegradation of Cr (VI) and Congo red dye under visible light. Results showed that $NiFe_2O_4$/superabsorbent nanocomposite hydrogels exhibited superior photocatalytic efficiency compared to TiO_2/superabsorbent nanocomposite hydrogels and conventional hydrogels. The composites demonstrated high reusability over five cycles and could be easily removed by filtration, making them promising green polymers for water purification.

1.4 Applications of Hydrogels

Hydrogels have become indispensable functional materials due to their unique combination of properties, including environmental degradation capability, biological compatibility, superior hydration capacity, and stimulus-responsive behavior. These characteristics enable diverse applications across several domains, like (Fig. 1.6):

- Biomedical sector (regenerative medicine scaffolds, advanced wound care dressings, ophthalmic devices)
- Therapeutic delivery (controlled-release drug carriers, targeted medication systems)
- Environmental remediation (heavy metal sequestration in water purification, pollutant adsorption matrices)
- Agricultural innovations (soil conditioners for drought mitigation, nutrient delivery platforms, smart fertilizer formulations)

1.4.1 Biomedical Applications

Hydrogels are extensively utilized in the biomedical field, such as drug delivery systems, wound care, and tissue engineering. Their biocompatibility, water retention, and ability to deliver active compounds make them ideal for these applications. They can be designed to give prolonged release of drugs, enhancing therapeutic efficacy. Active hydrogels are highly valued as wound dressings due to their ability to protect wounds, absorb exudate, retain moisture, inhibit microbial growth, and accelerate healing. They also seal wounds effectively, preventing infections and promoting cell proliferation (Bag et al. 2023). Key features include biocompatibility and ease of application, ensuring patient compliance. These qualities make hydrogels a promising option for wound care, driving ongoing research and advancements in their design. A wide range of hydrogel-based wound dressings is currently available in the market. Among the most commonly used are those made from alginate (such as *Algisite* and *Kaltostat*), cellulose derivatives (like *Comfeel Plus Contour Dressing* and *Gentell CMC Fiber Dressing*), and collagen or gelatin-based materials

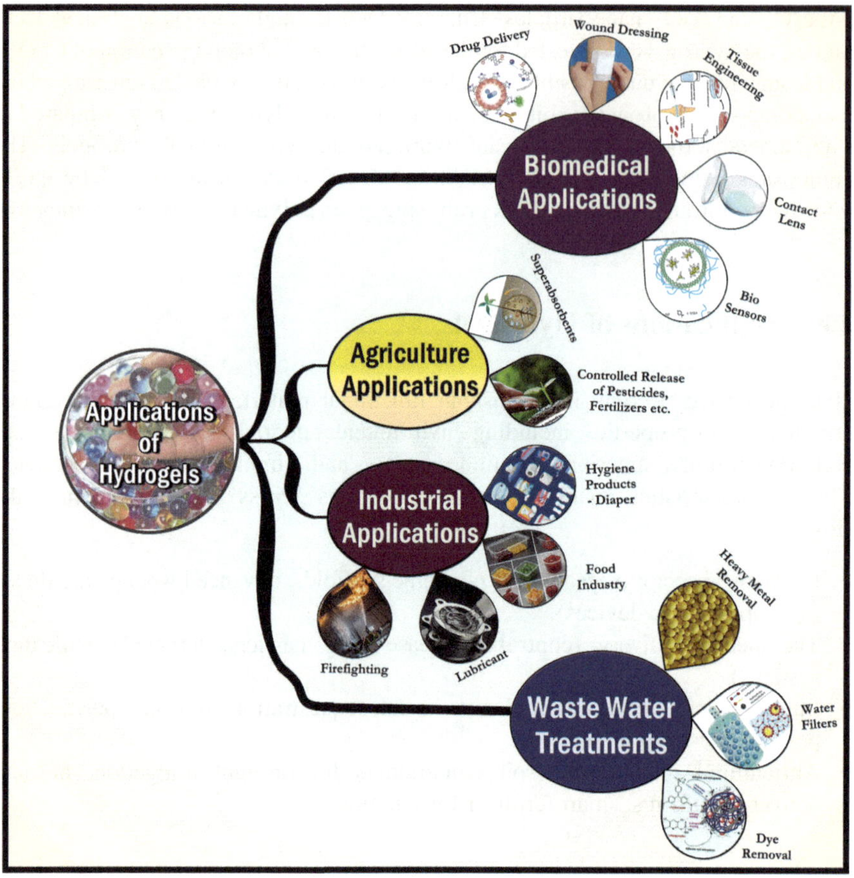

Fig. 1.6 Applications of hydrogels in various fields

(for example, *Biobrane* and *CollaSorb*). These products are typically offered in a variety of pre-cut sizes by multiple manufacturers to suit different wound care needs.

In addition to wound care, hydrogels play a crucial role in the production of contact lenses. Their excellent oxygen permeability, biocompatibility, and optical clarity make them particularly suitable for this application, especially in the management of eye conditions. Contact lenses made from hydrogels are commonly used for correcting vision, therapeutic purposes, or cosmetic enhancement. The soft lenses are generally composed of water-rich polymers, including hydroxyethyl methacrylate (HEMA), silicone, or polyvinyl alcohol (PVA) hydrogels. In the field of silicone contact lenses, Acuity 200 is one of the latest developments showing the maximum oxygen permeability in the market (Kalangara et al. 2023). PVA-based hydrogels are a different variety of contact lens material that is very biocompatible and hydrophilic. These lenses have outstanding wettability qualities, and there are several adjustments that may be made. Furthermore, the characteristics of contact lenses are altered by testing polymers like chitosan or hyaluronic acid. Hydrogels are also used as a lubricating material for dry mucosal tissues, joints, and medical procedures like catheter insertion, etc. They help prevent friction-related degradation and cell damage in mechanically stressed conditions. Water provides lubrication through hydrogen bonding, while polymers in hydrogels trap and retain water, enhancing their lubricating properties (Fuchs et al. 2020).

1.4.2 Drug Delivery Systems

Drug delivery systems are another field where hydrogels are extensively used. Ideally, drug delivery systems should efficiently achieve, maintain, and regulate the desired drug release, ensuring a therapeutic dose reaches the targeted area in the body. Their widespread use in drug delivery systems is attributed to their outstanding features, such as a porous architecture, biocompatibility, easy tunability, and excellent permeability to solutes. The porous structure of hydrogels acts as a supportive matrix for drug incorporation, enabling the encapsulation of therapeutic agents within their water-swollen network. The medication then subsequently diffuses into the immediate aqueous medium to allow regulated release at the intended spot. Directed drug distribution in the human body at specific places decreases undesirable systemic consequences and, ultimately, improves medical therapies. Hence, for achieving controlled and extended drug release, researchers are working on designing novel drug delivery systems based on natural polymers (chitosan, alginate, fibrin, gelatin, etc.), or synthetic polymers (polysaccharides, proteins, nucleic acids, etc. and or a combination of both, such as alginate/PVA, PVA/chitosan/gelatin, etc. The invention and engineering of specific hydrogels as long-term drug delivery vehicles for conditions like diabetes, inflammatory arthritis, inflammatory bowel disease, colorectal cancer, and atopic dermatitis has been the subject of numerous papers that demonstrate creative ideas.

Gull et al. (2019) in their research article discussed that although it is usual practice to provide painkillers and anti-inflammatory medications in a systematic manner, regulated drug distribution provides long-lasting therapeutic advantages. In this work, chitosan and polyvinyl pyrrolidone are combined with tetraethyl orthosilicate, a cleavable crosslinker, to create multi-responsive hydrogels. The hydrogels' cytotoxicity, biological degradation, durability, swelling behavior, and antibacterial activity were all examined. Their efficacy for long-term medication delivery was validated by drug release tests conducted in different solutions. The findings demonstrate the effective creation of non-toxic hydrogels that may find use in regulated medication delivery systems.

Hydrogels serve their purpose for regulated release of water-soluble medicines, but their hydrophilic characteristics make them difficult to use with weakly soluble compounds. To overcome this, hydrophobic medications can be placed in hydrogels for long-term release after being encapsulated in micro- or nanoparticles, such as

PLGA particles or PEG-PLA micelles. Soft contact lens materials, such as pHEMA-based hydrogels, have employed this hybrid system to prolong drug release.

Another example of a hydrogel's application is painless transdermal drug delivery. Blanzat and Labie (2023) has discussed the usage of hydrogels as novel therapeutic vehicles for carrying active ingredients and penetrating the epidermal barrier. This article covers the construction of hydrogels, physicochemical characteristics to optimize drug loading and release using hydrogels. They explore the usage of hydrogels, which were combined with various nano-carriers such as vesicles, capsules, and polymeric or inorganic nanoparticles, to improve skin absorption and drug safety while also addressing potential side effects that may limit their use. They improve transdermal drug administration by extending the formulation's longevity in the dermis and improving drug absorption.

1.4.3 Agricultural Applications

Hydrogels are widely used in agriculture for various applications, including improving crop production through their superabsorbent properties and controlled release of fertilizers and pesticides. They also serve as carriers for minerals, pesticides, and growth stimulants, integrating into the soil to enhance plant growth. Hydrogels are a useful tool in agriculture since they can hold moisture and progressively release bio-preparations, insecticides, and fertilizers. They greatly improve the preservation of soil water and lessen the need for frequent irrigation. Hydrogels act as soil conditioners that enhance the soil's physicochemical characteristics, hence increasing its fertility and improving soil properties by acting as water reservoirs, reducing osmotic pressure, and enhancing water availability for plants. They reduce crop's production cost while fostering root progress, seed growth, and production. They are widely used in hydroponics, dryland farming, and horticulture, hence preventing soil erosion and enabling the gradual release of fertilizers for sustained benefits Vedovello and colleagues (2024) review the challenges of increasing agricultural demand, soil degradation, and water scarcity due to reliance on pesticides and fertilizers. They highlight polymeric hydrogels as promising solutions for improving water efficiency and nutrient release. The authors discuss hydrogel production methods, cross-linking techniques, and their role in enhancing soil moisture retention and plant growth. This review also states that ongoing research continues to refine hydrogel formulations for sustainable agricultural applications. Similarly, Zhu et al. (2024) discussed agro-waste-derived superabsorbent hydrogels in their article and suggested that these types of hydrogels are essential for supporting ecology-friendly practices and sustainable agriculture. These hydrogels help reduce water consumption, lower carbon footprints, minimize waste, and enhance crop yields. The development of superabsorbent hydrogels using raw agricultural waste products, including cellulose, hemicellulose, and lignin, that are taken from agricultural wastes, has been the focus of recent studies. In their review, they highlighted different physical and chemical crosslinking methods to synthesize hydrogels and explored the way of using superabsorbent hydrogels in improving soil characteristics, particularly by increasing water retention and facilitating the gradual release of fertilizers to support sustainable agriculture.

1.4.4 Industrial Applications

Hydrogels are utilized in commercial settings in water purification systems, sanitation products, and superabsorbent items like diapers. They are used in a variety of industries, including as hygiene goods, agriculture, pharmaceuticals, separation technologies, textiles, wastewater treatment, electrical applications, soft actuators, construction, water-absorbing rubbers, dewatering, and fire-extinguishing gels.

Omidian and coworkers (2024) discussed about nature of cellulose-based superabsorbent hydrogels. They explored the fact that superabsorbent hydrogels are altering various sectors like agriculture, medicine, environmental management, and construction. Due to their biodegradability, exceptional water-holding capacity, and versatile nature, they are well-suited for improving soil fertility, enhancing water retention, and supporting various industrial uses, making them an environmentally friendly option in the construction sector. Hu et al. (2024) reviewed the synthesis, tribological behavior, and applications of hydrogels as lubricants in industrial and biomimetic settings. They discussed various lubricant types, highlighting the environmental concerns of mineral oil-based lubricants and the persistent challenges of synthetic alternatives, such as leakage and creeping. Solid lubricants were introduced to address these issues, particularly in extreme environments like aerospace. The review also explored literature on stimuli-responsive hydrogels, such as PNIPAM microgels, which exhibit temperature-sensitive lubrication due to phase transitions. Below the LCST, they remain swollen with low friction, while above the LCST, dehydration causes shrinkage and increased friction.

1.4.5 Environmental Applications

Despite being abundant, only a small fraction of Earth's water is freshwater, and even less is suitable for use. Freshwater supplies are being depleted due to pollution and rising water demand, creating a shortage. Heavy metals and dyes are the main contaminants found in wastewater, which is becoming more contaminated due to industrialization, urbanization, and inadequate waste management. Heavy metals like arsenic and chromium are extremely poisonous and carcinogenic, while dyes block sunlight, lowering dissolved oxygen and endangering human health. And so, effective industrial wastewater treatment is essential due to the significant effects they have on the environment and human health.

Hydrogels are being investigated for environmental uses, such as in water purification, waste treatment, and energy storage systems. Hydrogels, due to their porous crosslinked structure, have gained lots of attention for wastewater treatment. Their ability to absorb and retain water makes them useful in controlling water flow in environmental systems. Hydrophilic functional groups of hydrogels easily attract and bind to dyes and heavy metals, enabling efficient adsorption and retention within their three-dimensional network structure. This makes hydrogels extremely effective for removing these contaminants from water. According to recent research, certain carbohydrates may almost entirely remove colors and heavy metals from wastewater. Using hydrogels based on carboxylated cellulose nanofibrils, Zhou et al. (2014) were able to remove almost 90% of Pb^{2+} in four hours, with an effective adsorption rate of 171 mg/g.

Hexavalent chromium (Cr (VI)) is highly soluble and toxic, causing health issues like diarrhoea, liver infections, and plant damage. It is also mutagenic and oncogenic, penetrating cell membranes and disrupting DNA replication by binding to phosphate groups, ultimately leading to cell death. Also, due to its high solubility, it easily enters the bloodstream and exerts toxic effects. Hence, its removal from water is necessary. Keeping this fact in mind, Sharma and coworkers (Sharma et al. 2017) in their study synthesized chitosan-crosslinked-poly (alginic acid) nanohydrogel (CN-cl-PL(AA)NHG) by the co-polymerization method. This nanohydrogel proved to be an efficient adsorbent for the removal of Chromium (VI) metal ion from water. The nanohydrogel was examined using FTIR, SEM, and TEM, and revealed a smooth surface with particles ranging in size from 30–80 nm. Adsorption factors such as rotation time, pH, temperature, metallic ion quantity, and adsorbent dosage were investigated and modified, revealing that the chitosan-crosslinked-poly (alginic acid) nanohydrogel effectively removed Chromium (VI) from aqueous solutions. The process adhered to pseudo-second-order kinetic behavior and was found to be spontaneous and chemically driven, as evidenced by the results of thermodynamic analysis. Singh et al. (2023) developed a nano-adsorbent filter with gamma-aluminum oxide nanoparticles for removing arsenic and fluoride from water. The propylene filter, embedded with 140–150 g of low-cost gamma-alumina nanoparticles, achieved adsorption capacities of 20–25 mg/g fluoride and 25–30 mg/g arsenic, demonstrating its effectiveness in water treatment. Hence, the designed technique that uses nanoadsorbent in a sediment-removing filter candle offers a home defluoridation device that can treat drinking and ground water contaminated with arsenic and fluoride without using power.

1.5 Future Directions and Challenges

The present review emphasizes the development of next-generation hydrogel materials that exhibit multifunctionality, including the ability to respond to multiple external stimuli or to integrate with other components, such as nanomaterials. The future of hydrogels lies in creating smart, responsive systems for personalized medicine, environmental sustainability, and biotechnological applications. Despite the progress, several challenges remain, including:

i. Improving the mechanical strength to avoid collapsing during the high-temperature water treatment process.
ii. Stability of hydrogels to avoid disintegration during the water treatment process.
iii. Enhancing their scalability for industrial use.
iv. Addressing issues related to their long-term biodegradability in the human body.
v. In membrane technology, clogging, pore size management, and degradation must be managed since they impact the membranes' permeability and selectivity. As a result, they must be taken into account to ensure high water flux and effective removal of particular contaminants.
vi. Production cost is high, thus these materials are not economically feasible, hence the need to develop sustainable and economic technologies for their development.

The potential for hydrogels in the future is vast. From advanced drug delivery systems to regenerative medicine and environmental applications, hydrogels are poised to play an increasingly important role in addressing global challenges. The continued development of smart hydrogels will likely revolutionize fields ranging from healthcare to environmental management.

1.6 Conclusion

Hydrogels have progressed from mere water-absorbing materials to advanced, multifunctional polymers with applications in many fields. As polymer chemistry and synthesis methods continue to improve, the future of hydrogels looks bright, particularly with the emergence of smart materials and novel biomedical breakthroughs.

This chapter highlights advancements in hydrogels for various applications during the past 70 years, beginning with the ground-breaking work of Wichterle and Lim in the 1960s. This chapter discusses the sequential development of hydrogels over the past 70 years, categorized into various generations. As per the discussion, the initial progression involved the transition from simple cross-linked networks to advanced double-network composites. In later decades, the development advanced further with the emergence of stimuli-responsive hydrogels, smart hybrid polymers, and, ultimately, the development of nanohydrogels is being observed in the current era. The next section of the chapter discusses significant advancements in hydrogel development from the first generation to the present, which were based on property enhancements and improved efficiency across various applications. The main breakthroughs for the development of crosslinked, biocompatible, biodegradable, stimuli-sensitive, intelligent, and hybrid hydrogels occurred periodically. Finally, the chapter discussed hydrogel applications across various sectors, with a key focussing on the application of hydrogels in wastewater treatment and heavy metal ion removal.

In conclusion, hydrogels are versatile soft materials with a three-dimensional crosslinked network capable of high-water retention while maintaining structural integrity. Their tunable properties, biocompatibility, and biodegradability make them valuable in diverse fields, including medicine, biosensors, cosmetics, and wastewater treatment. The hydrophilic nature of these materials enables structural modifications to create environmentally-responsive systems, making them particularly effective for advanced applications such as heavy metal ion removal, drug delivery, and agricultural applications. The ongoing development of hydrogels will continue to contribute to innovations in medicine, biotechnology, and environmental science, such as in water purification, waste treatment, and energy storage systems, shaping the future of material science.

References

Akhtar MS, Ali S, Zaman W (2024) Innovative adsorbents for pollutant removal: exploring the latest research and applications. Molecules 29(18):4317

Allcock HR, Ambrosio AM (1996) Synthesis and characterization of pH-sensitive poly (organophosphazene) hydrogels. Biomaterials 17(23):2295–2302

Bag N, Bardhan S, Roy S, Roy J, Mondal D, Guo B, Das S (2023) Nanoparticle-mediated stimulus-responsive antibacterial therapy. Biomaterials Sci 11(6):1994–2019

Bajpai AK, Kankane S, Chouhan R, Goswami S (2013) 9 Biological stimulus-responsive. Polymeric Biomaterials: Med Pharmaceutical Appl 2(2):245

Bashir S, Hina M, Iqbal J, Rajpar AH, Mujtaba MA, Alghamdi NA, Wageh S, Ramesh K, Ramesh S (2020) Fundamental concepts of hydrogels: synthesis, properties, and their applications. Polymers 12(11):2702

Bi X, Liang A (2016) In situ-forming cross-linking hydrogel systems: chemistry and biomedical applications. Emerging Concepts Anal Appl Hydrogels 86:541–547

Bordbar-Khiabani A, Gasik M (2022) Smart hydrogels for advanced drug delivery systems. Int J Mol Sci 23(7):3665

Chen J, Park H, Park K (1999) Synthesis of superporous hydrogels: hydrogels with fast swelling and superabsorbent properties. Inc. J Biomed Mater Res 44:53–62

Chen X, Li W, Zhong W, Lu Y, Yu T (1997) PH sensitivity and ion sensitivity of hydrogels based on complex-forming chitosan/silk fibroin interpenetrating polymer network. J Appl Polym Sci 65(11):2257–2262

Ciolacu DE, Nicu R, Ciolacu F (2020) Cellulose-based hydrogels as sustained drug-delivery systems. Materials (Basel, Switzerland) 13(22):5270

El-saied HAA, El-Fawal EM (2021) Green superabsorbent nanocomposite hydrogels for high-efficiency adsorption and photo-degradation/reduction of toxic pollutants from waste water. Polym Testing 97:107134

Fuchs S, Shariati K, Ma M (2020) Specialty tough hydrogels and their biomedical applications. Adv Healthcare Mater 9(2):e1901396

Gong JP, Katsuyama Y, Kurokawa T, Osada Y (2003) Double-network hydrogels with extremely high mechanical strength. Adv Mater 15(14):1155–1158

Gull N, Khan SM, Zahid Butt MT, Khalid S, Shafiq M, Islam A, Asim S, Hafeez S, Khan RU (2019) In vitro study of chitosan-based multi-responsive hydrogels as drug release vehicles: a preclinical study. RSC Adv 9(53):31078–31091

Hu B, Gao J, Lu Y, Wang Y (2023) Applications of degradable hydrogels in novel approaches to disease treatment and new modes of drug delivery. Pharmaceutics 15(10):2370

Hu L, Yang Y, Yu W, Xu L (2024) Hydrogels for lubrication: synthesis, properties, mechanism, and challenges. Lubricants 12(6):186

Kalangara J, Kuruvilla M, Sarantopoulos KD (2023) Treatment of ocular pain not responsive to traditional dry eye disease treatments. In: Dry eye disease, pp 221–228. Elsevier
Katono H, Sanui K, Ogata N, Okano T, Sakurai Y (1991) Drug release off behavior and deswelling kinetics of thermo-responsive IPNs composed of poly (acrylamide-co-butyl methacrylate) and poly (acrylic acid). Polym J 23(10):1179–1189
Kirschner CM, Anseth KS (2013) Hydrogels in healthcare: from static to dynamic material microenvironments. Acta Mater 61(3):931–944
Kost J, Horbett TA, Ratner BD, Singh M (1985) Glucose-sensitive membranes containing glucose oxidase: activity, swelling, and permeability studies. J Biomed Mater Res 19(9):1117–1133
Labie H, Blanzat M (2023) Hydrogels for dermal and transdermal drug delivery. Biomaterials Sci 11(12):4073–4093
Machado RL, Gomes AC, Marques EF (2024) Hydrogels as versatile colloidal platforms to combat skin cancer–physicochemical features, strategies and advances. J Molecular Liquids, 126453
Mathur AM, Moorjani SK, Scranton AB (1996) Methods for synthesis of hydrogel networks: a review. J Macromolecular Sci Part C: Polymer Rev 36(2):405–430
Naseer A (2024) Role of nanocomposites and nano adsorbents for heavy metals removal and dyes. an overview. Desalination and Water Treatment, 100662
Nguyen QV, Park JH, Lee DS (2015) Injectable polymeric hydrogels for the delivery of therapeutic agents: a review. Eur Polymer J 72:602–619
Okano TY, Bae YH, Jacobs H, Kim SW (1990) Thermally on-off switching polymers for drug permeation and release. J Control Release 11(1–3):255–265
Omidian H, Akhzarmehr A, Chowdhury SD (2024) Advancements in cellulose-based superabsorbent hydrogels: sustainable solutions across industries. Gels 10(3):174
Pan Y, Zhao Y, Kuang R, Liu H, Sun D, Mao T,... Li J (2020) Injectable hydrogel-loaded nanohydroxyapatite that improves bone regeneration and alveolar ridge promotion. Mater Sci Eng C 116:111158
Peppas NA (1997) Hydrogels and drug delivery. Curr Opin Colloid Interface Sci 2(5):531–537
Pishavar E, Khosravi F, Naserifar M, Rezvani Ghomi E, Luo H, Zavan B,... Ramakrishna S (2021) Multifunctional and self-healable intelligent hydrogels for cancer drug delivery and promoting tissue regeneration in vivo. Polymers 13(16):2680
Puccetti M, Pariano M, Schoubben A, Giovagnoli S, Ricci M (2024) Biologics, theranostics, and personalized medicine in drug delivery systems. Pharmacol Res 201:107086
Rana AK, Gupta VK, Hart P, Thakur VK (2024) Cellulose-alginate hydrogels and their nanocomposites for water remediation and biomedical applications. Environ Res 243:117889
Rigoletto M, Calza P, Gaggero E, Laurenti E (2022) Hybrid materials for the removal of emerging pollutants in water: classification, synthesis, and properties. Chem Eng J Adv 10:100252
Roorda WE, Boddé HE, De Boer AG, Bouwstra JA, Junginer HE (1986) Synthetic hydrogels as drug delivery systems. Pharmaceutisch Weekblad. Scientific Edition 8(3):165–189
Saha A, Bardhan S, Roy S, Dutta S, Das S (2023) Biopolymeric hydrogels: a new era in combating heavy metal pollution in industrial wastewater. Membranes for water treatment and remediation, 209–226
Sharma G, Naushad M, Ala'a H, Kumar A, Khan MR, Kalia S, Sharma A (2017) Fabrication and characterization of chitosan-crosslinked-poly (alginic acid) nanohydrogel for adsorptive removal of Cr (VI) metal ion from aqueous medium. Int J Biol Macromolecules 95:484–493
Sharma P, Ghosh T, Sengupta N, Patra P (2024) Exploring nanohydrogels: crafting innovations for biomedical advancements. Chem Select 9(28):e202401396
Sikdar P, Uddin MM, Dip TM, Islam S, Hoque MS, Dhar AK, Wu S (2021) Recent advances in the synthesis of smart hydrogels. Mater Adv 2(14):4532–4573
Singh I, Singh A, Dubey S, Tomar AS, Arya P, Shrivastava A (2023) Nanoadsorbent based user-friendly household filter for the purification of fluoride and arsenic contaminated drinking water. US011766641B2

Tan H, Chu CR, Payne KA, Marra KG (2009) Injectable in situ forming biodegradable chitosan-hyaluronic acid based hydrogels for cartilage tissue engineering. Biomaterials 30(13):2499–2506

Tummino ML, Magnacca G, Cimino D, Laurenti E, Nisticò R (2020) The innovation comes from the sea: chitosan and alginate hybrid gels and films as sustainable materials for wastewater remediation. Int J Mol Sci 21(2):550

Ungureanu C, Răileanu S, Zgârian R, Tihan G, Burnei C (2024) State-of-the-art advances and current applications of gel-based membranes. Gels 10(1):39

Vedovello P, Sanches LV, da Silva Teodoro G, Majaron VF, Bortoletto-Santos R, Ribeiro C, Putti FF (2024) An overview of polymeric hydrogel applications for sustainable agriculture. Agriculture 14(6):840

Weerasundara L, Gabriele B, Figoli A, Ok YS, Bundschuh J (2021) Hydrogels: novel materials for contaminant removal in water—a review. Crit Rev Environ Sci Technol 51(17):1970–2014

Wu J, Zhao X, Wu D, Chu CC (2014) Development of a biocompatible and biodegradable hybrid hydrogel platform for sustained release of ionic drugs. J Mater Chem B 2(38):6660–6668

Wu XS, Hoffman AS, Yager P (1992) Synthesis and characterization of thermally reversible macroporous poly(N-isopropylacrylamide) hydrogels. J Polym Sci A 30:2121–2129

Zhang Y, Tian X, Zhang Q, Xie H, Wang B, Feng Y (2022) Hydrochar-embedded carboxymethyl cellulose-g-poly (acrylic acid) hydrogel as stable soil water retention and nutrient release agent for plant growth. J Bioresources Bioproducts 7(2):116–127

Zhou Y, Fu S, Zhang L, Zhan H, Levit MV (2014) Use of carboxylated cellulose nanofibrils-filled magnetic chitosan hydrogel beads as adsorbents for Pb (II). Carbohyd Polym 101:75–82

Zhu J, Zhang Z, Wen Y, Song X, Tan WK, Ong CN, Li J (2024) Recent advances in superabsorbent hydrogels derived from agro waste materials for sustainable agriculture: a review. J Agric Food Chem 72(41):22399–22419

Chapter 2
Fundamentals of Hydrogels

Janki Trivedi, Chaitali Patel, Nirav Bhavsar, and Rishit Soni

Abstract Hydrogels are complex three-dimensional networks of crosslinked polymer chains that can hold large quantities of water, and they are quite versatile materials with applications in many fields. Hydrogel synthesis techniques range from chemical to physical methods. In chemical synthesis, network structure and properties can be precisely manipulated via crosslinking reactions. In contrast, physical synthesis techniques, which include non-covalent interactions or self-assembly methods, provide advantages such as simplicity and reversibility. Describing synthesis methods is fundamental to allowing for tailored hydrogels for applications by modulating, for example, crosslinking density, the composition of the polymers, and environmental conditions. The architecture of hydrogels is a key factor in determining their characteristics and functionality. Due to its tunable properties, hydrogels are utilized for wound healing, tissue engineering, controlled drug delivery, and biosensing. Hydrogels are a potential candidate for wastewater treatment because of their ability to adsorb the contaminants. It is also used for controlled release of treatment agents and bioremediation processes. Beyond biomedicine, hydrogels serve as carriers for agricultural chemicals, adsorbents for environmental pollutants, thickeners in food formulations, and matrices for controlled release of fragrances and cosmetics.

Keywords Hydrogels · Polymerization · Synthesis · Structure · Properties · Applications

J. Trivedi (✉) · C. Patel · N. Bhavsar
Shri Alpesh N. Patel Post Graduate Institute of Science and Research, Anand, Gujarat, India
e-mail: jankitrivedi6@gmail.com

R. Soni
Shri P. M. Patel Institute of Biosciences, Anand, Gujarat, India

S. Bardhan and M. P. Shah (eds.), *Applications of Hydrogels in Modern Wastewater Treatment*, Advances in Wastewater Research,
https://doi.org/10.1007/978-981-95-1598-1_2

List of Abbreviations

3D	Three-Dimensional
AIBN	2,2′-Azobisisobutyronitrile
Cd	Cadmium
DNA	Deoxyribonucleic Acid
GA	Glutaraldehyde
Hg	Mercury
IPN	Interpenetrating Polymer Network
PAA	Polyacrylic Acid
PAM	Polyacrylamide
PCL	Polycaprolactone
PEG	Polyethylene Glycol
PLA	Polylactic Acid
PLGA	Poly (Lactic-Co-Glycolic Acid)
PVA	Polyvinyl Alcohol
PVP	Polyvinyl Pyrrolidone
TEGDMA	Tetraethyleneglycol Dimethacrylate
UV	Ultraviolet
VSNP	Vinyl Hybrid Silica Nanoparticles

2.1 Introduction

Hydrogels are intriguing three-dimensional networks of intertwined polymer chains that are renowned for their exceptional ability to hold and absorb water. This special property makes hydrogels incredibly adaptable for a wide range of applications across multiple industries by enabling them to expand and retain large amounts of moisture without disintegrating. Their crosslinked networks, which are created by physical or chemical processes and give them a stable and flexible structure, are what give them their water-attracting qualities (Bustamante-Torres et al. 2021). This chapter will focus on exploring different types of hydrogels, their production processes, and the polymerization strategies used in this chapter, offering a thorough grasp that is necessary for cutting-edge research and innovative applications. Over the span of fifty years, hydrogels have seen an incredible transformation, going from basic materials to complex compounds with a wide range of applications. When Czech chemists Otto Wichterle and Drahoslav Lím created cross-linked networks of water-attracting polymers for the first time in the 1960s. Soft contact lenses, which were produced from hydrogel for biomedical use, paved the way for the widespread use of hydrogels in pharmaceutical and medical applications (Wichterle and Lím 1960). Research on hydrogels exploded in the 1970s and 1980s, resulting in advancements in wound dressings, medication delivery systems, and other medical equipment. Hydrogels were perfect for these applications because of their exceptional capacity to absorb and hold water while preserving their structural integrity.

With the introduction of "smart" hydrogels in the 1990s, notable breakthroughs were made. The presence of particular ions or molecules, pH levels, temperature variations, and other environmental cues might cause these materials to alter their physical characteristics, such as swelling or solubility (Peppas and Hilt 2008). The development of advanced biosensors, tissue engineering, and targeted drug delivery has all benefited from their revolutionary potential.

Hydrogels' functions have expanded significantly in recent years, including improving food technology, acting as scaffolds for three-dimensional cell cultures in regenerative medicine, and acting as efficient water-retaining agents in agriculture. An intriguing advancement is the creation of nanocomposite hydrogels, which use nanoparticles to improve their functional and mechanical qualities, increasing their applicability in various environments (Hwang et al. 2021).

2.1.1 Key Features of Hydrogel

The ideal hydrogel should have the following properties (represented in Fig. 2.1).

2.2 Types of Hydrogels

Hydrogels can be divided into various types (as shown in Fig. 2.2) on the basis of their chemical makeup, physical structure, and how they respond to changes in their environment.

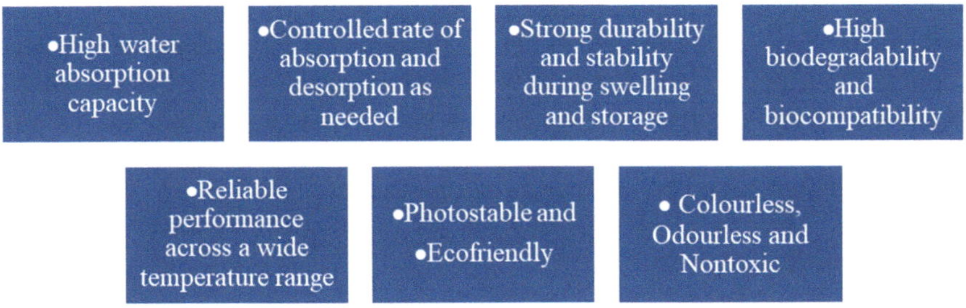

Fig. 2.1 Ideal properties of hydrogel

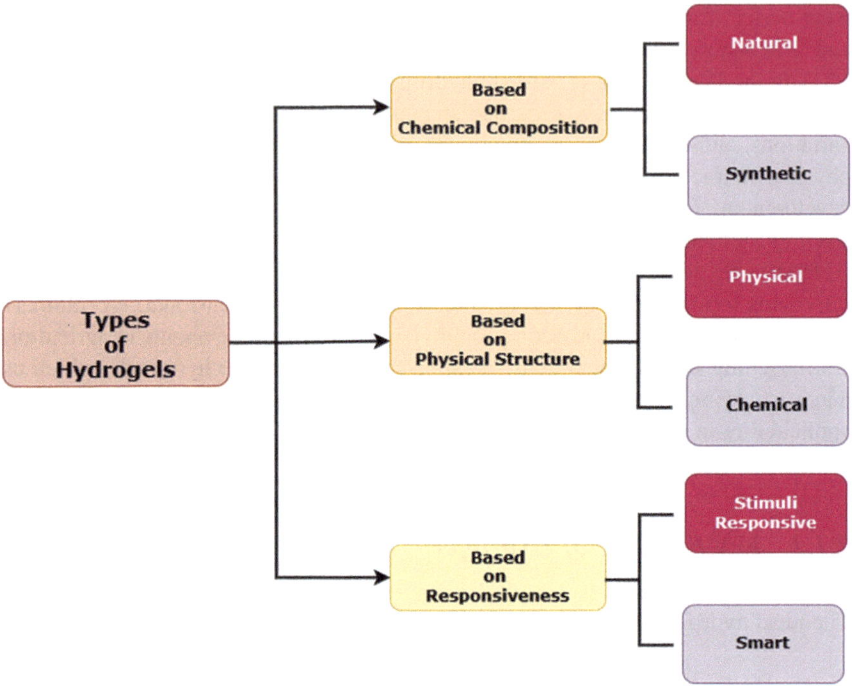

Fig. 2.2 Flowchart on the types of hydrogels

2.2.1 Based on Chemical Composition

2.2.1.1 Natural Hydrogels

Natural hydrogels are derived from proteins like collagen and gelatin as well as polysaccharides like cellulose, chitosan, and alginate. These materials are ideal for biological applications because they are both biocompatible and biodegradable. Hydrogels based on polysaccharides, such as chitosan and alginate, are well-known for their capacity to gel and their compatibility with living tissues. Collagen, gelatin, and other proteins improve the hydrogels' biological activity and cell-interaction capabilities, which makes them beneficial for tissue scaffolding and cell culture (Zhao et al. 2023b).

2.2.1.2 Synthetic Hydrogels

Synthetic hydrogels are composed of polymers such as polyvinyl alcohol (PVA), polyacrylamide, and polyethylene glycol (PEG). These materials are appropriate for applications needing certain qualities like strength or responsiveness because they offer greater control over mechanical and chemical properties. PEG-based hydrogels are important because of their low immunogenicity and because of this, they are commonly used in drug delivery systems; while PVA hydrogels are appreciated for their mechanical durability and are used in various industrial applications (Ho et al. 2022).

2.2.2 Based on Physical Structure

2.2.2.1 Physical Hydrogels

Physical hydrogels are formed via non-covalent interactions such as hydrogen bonding, ionic interactions, or hydrophobic interactions. These connections enable adjustable gelation, allowing the hydrogels to modify and enable them to react to environmental changes (Ho et al. 2022). For example, temperature-sensitive hydrogels use hydrophobic interfaces to form or disrupt the gel networks, making them effective in drug delivery systems where they deliver their contents in response to a change in body temperature.

2.2.2.2 Chemical Hydrogels

Strong chemical connections between polymer chains are used to form the solid, hard structure of chemical hydrogels. There are several chemical ways to do this, including using agents like glutaraldehyde or starting the process with UV light. These hydrogels are well known for their strength and durability, making them a better alternative for maintaining structure in, tissue engineering scaffolds (Xue et al. 2022).

2.2.3 Based on Responsiveness

2.2.3.1 Stimuli-Responsive Hydrogels

These hydrogels change their characteristics when they encounter different environmental triggers, like shifts in pH, temperature, light, or electric fields (El-Husseiny et al. 2022). Some examples include those that respond to temperature changes, variations in pH, or light exposure.

2.2.3.2 Smart Hydrogels

These smart hydrogels have multiple response mechanisms to adapt their behaviour based on the environmental cues. For example, smart hydrogels can respond to temperature, pH, and light simultaneously. They are useful in sophisticated applications that need flexible and responsive actions (El-Husseiny et al. 2022).

2.3 Synthesis and Polymerization of Hydrogels

Hydrogels refer to unique polymeric materials that are able to absorb and hold significant amounts of water or other polar solvents in their three-dimensional cross-linked networks. It means—"A hydrogel is an organized three-dimensional (3D) crosslinked network of polymer which is capable of retaining a high amount of water or polar solvent" (Sabbagh et al. 2019). A hydrogel consists of water, by which it swells but does not dissolve, hence can maintain its shape. Based on how much water you want the hydrogel to hold and which way of binding your polymer chains within a gel, it largely defines its central structural unit.

Hydrogel synthesis involves various methods that can be classified into chemical and physical approaches. Each method has its advantages and limitations depending on the desired properties and applications of the hydrogel.

2.3.1 Natural Hydrogels

In order to create hydrogels, monomers having reactive functional groups are frequently polymerized and cross-linked to create a complex network. A cross-linker joins linear polymer chains to form a gel matrix in the widely used process of free-radical polymerization to create hydrogels (Peppas and Bures 2000). According to Kumar et al. (2000), ionic gelation is an additional strategy in which network formation is facilitated by ionic interactions between polymers or between polymers and ions. Physical gelation techniques are also important; they use hydrogen bonds or van der Waals forces, which are non-covalent interactions, to form a hydrogel matrix (Hennink and van Nostrum 2012). Mechanical strength, swelling behaviour, and reactivity to environmental stimuli are among the features of the hydrogel that are influenced by the synthesis technique and monomer or polymer choices (Ahmed 2015).

Depending on the density of the hydrophilic group in the polymer, the hydrophilic polymers used to create hydrogel can absorb different amounts of polar material (Mantha et al. 2019). Major classes of polymers that are used to construct hydrogels are shown in Fig. 2.3. The wide variety of chemical structures seen in natural polymers results in a classification system with multiple unique categories. Fundamental biological polymers like DNA and nucleic acid are among them, as are polyamides—collagen being a key example—and polysaccharides like chitin, chitosan, starch, cellulose, and different gums like alginate and carrageenan. Polyphenols, such as lignin, as well as organic and inorganic polyesters and polyanhydrides, such as polyphosphazene and polysebacic acid, are further categories.

To improve the structural stability and its ability to eliminate heavy metals, chitosan, for example, has amino groups that can chelate metal ions. The sulfate groups that are prevalent in carrageenan, a sulphated polysaccharide, interact with cationic dyes to facilitate their adsorption. In the presence of divalent cations, alginate gels form a porous structure that can capture organic contaminants. When these biopolymers are combined in hydrogel form, their overall adsorption capacity and selectivity are increased, making them effective and eco-friendly for water purification solutions (Chai et al. 2017).

Natural polymers, such as polysaccharides and proteins, are often used to prepare hydrogels as they have inherent biocompatibility due to the cellular recognition or even integration provided by the abovementioned attribution (Buwalda et al. 2014).

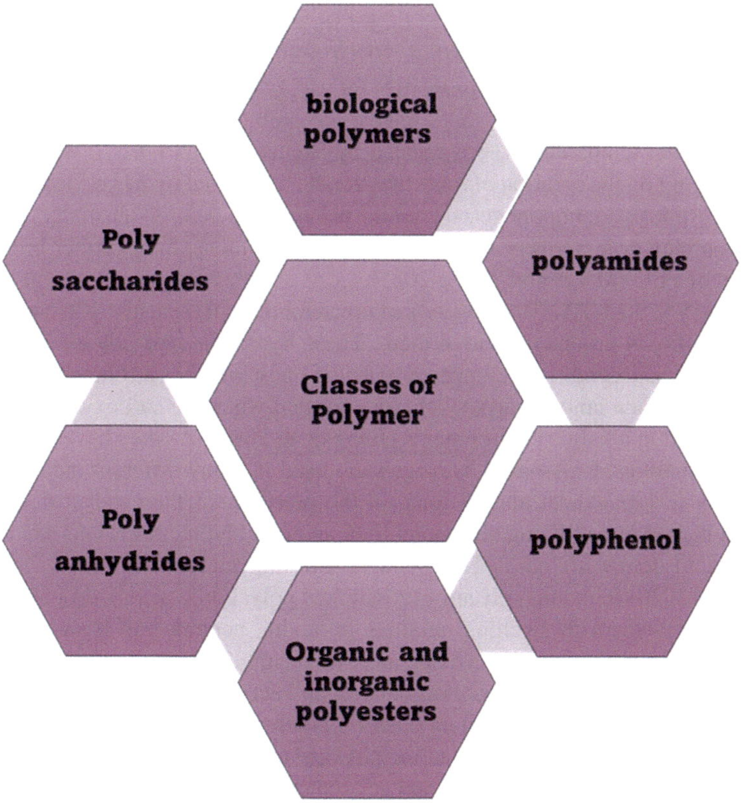

Fig. 2.3 Types of polymers used for hydrogel preparation

These hydrogels (e.g., chitosan and other polysaccharide-based based) generally exhibit poor mechanical properties (Kumar and Muthu 2015). Different sorts of improvement methodologies are utilized to manage these inadequacies. For example, to improve the mechanical strength and resilience of natural polymer-based hydrogels, they are crosslinked (Burdick et al. 2001; Schwach-Abdellaoui et al. 1998) and grafted with comonomers, as well as polyvinyl alcohol (PVA) is often blended with this hydrogel to improve both the mechanical properties and softness (Zhang et al. 2017).

2.3.2 Synthetic and Semi-synthetic Hydrogel

Synthetic hydrogels incorporate polymers that allow precise control over mechanical properties. Among the commonly used synthetic polymers are polycaprolactone, like poly lactic acid (PLA), polyvinyl pyrrolidone (PVP), polyvinyl alcohol (PVA), and poly ethylene glycol (PEG) (Smith and Doe 2019). Due to its great water solubility, biocompatibility, and biodegradability, PEG stands out among these (Lee and Park 2018). Moreover, it is very useful for conjugating proteins, peptides, and other medications (Brown and White 2020). PEG is also widely utilized to alter natural polymers, altering their properties for use. For example, PEG-graft-chitosan, a thermoresponsive hydrogel that congeals at ambient temperature but becomes liquid at lower temperatures, was created by Bhattarai et al. (2010). With research concentrating on the creation of multifunctional PEG-based hydrogels that integrate stimuli-responsive components, including pH or temperature sensitivity, enabling more complex drug delivery systems, recent developments have further broadened the uses of PEG (Miller and Zhang 2021).

Polyacrylic acid (PAA) hydrogels are known for their pH sensitivity because of the anionic carboxyl groups present in them. These hydrogels typically exhibit greater swelling in basic conditions compared to acidic ones, which is attributed to the electrostatic repulsion among carboxylate ions formed when the carboxylic groups are deprotonated in alkaline environments (Jiang et al. 2019). As they have exceptional adhesive qualities, PAA-based hydrogels are used in many different industries, but especially in biomedical applications. In the presence of the catalyst dextranase, hydrogels prepared utilizing ultraviolet light and glycidyl methacrylate-dextran and acrylic acid demonstrated a pH-dependent swelling mechanism at pH 7.4. Also, Lee et al. (2020) found that glycerol-crosslinked polyacrylic acid hydrogels showed 100% improvement in swelling qualities in acidic, neutral, and basic pH ranges when prepared with benzoyl peroxide and Novozym 435 as the crosslinking agent. Additionally, Nho and colleagues produced polyacrylic acid hydrogels by irradiating acrylic acid solutions with an electron beam at a maximum dose of 75 kilograys (kGy). They then investigated the physical characteristics of these hydrogels, including adhesion, swelling ratio, and gel content (Nho et al. 2017).

Kou et al. (2022) created slabs and cylinders of poly 2-hydroxyethyl methacrylate-co-methacrylic acid hydrogels using Tetraethyleneglycol Dimethacrylate (TEGDMA) as the crosslinking agent, and the hydrogel slabs were created using a redox initiator system including sodium metabisulfite and ammonium persulfate. The initiator and crosslinking agent for the hydrogel cylinders were 2,2′-azobisisobutyronitrile (AIBN) and TEGDMA, respectively. Numerous fields have made use of hydrogels based on derivatives of polyacrylamide. Hydrogels made of polyacrylamide are neutral, hydrophilic, and have physico-chemical properties of great importance that make them valuable for use as biomaterials or in the immobilization of cells.

Polyacrylamide-based hydrogels are neutral in nature, they are hydrophilic and exhibit valuable physical–chemical properties, making them suitable for biomaterials, whole cell and enzyme immobilization, heavy metal ion removal, bioremediation, wastewater treatment, and bioseparation. Saraydin et al. (2013) synthesized polyacrylamide hydrogels using chemical-free radical initiation and gamma-ray irradiation. Different swelling ratios (255–1450%) were seen in the gels, depending on the concentration, crosslinking agent, and synthesis process. Vinyl hybrid silica nanoparticles (VSNPs) were used as crosslinking agents to improve the hydrogel's dynamic polymerization and stress dissipation. The sol–gel method was used to create these nanoparticles from vinyl-triethoxysilane. The combined impacts of the VSNPs and the poly(acrylamide) network resulted in an enhanced VSNP-PAM hydrogel.

Gamma rays, which are well-known for their ability to effectively crosslink and improve the structural integrity of hydrogels, have been used to create gelatin hydrogels (Liu et al. 2020). Lejardi et al. (2013), for instance, created hydrogel blends of chitosan, glycolic acid, and modified polyvinyl alcohol. These mixtures showed promise for efficiently encapsulating and delivering anti-cancer medications when employed as hydrogel microspheres. Similarly, acrylamide was grafted onto polyvinyl alcohol and crosslinked with chitosan using glutaraldehyde to generate a unique pH-sensitive interpenetrating network (IPN). According to Wang et al. (2018), this IPN hydrogel demonstrated encouraging outcomes for the regulated release of antimicrobial medications.

The poor mechanical strength of naturally occurring polymers limits their utility in biological applications. On the other hand, synthetic polymers have a number of benefits, including higher mechanical strength, cost-effectiveness, and the flexibility to manufacture particular qualities suited to applications (Smith et al. 2020; Jones et al. 2018). Nevertheless, due to their production of non-biodegradable solid waste, synthetic polymers are not environmentally beneficial (Doe et al. 2017). As per Lee et al. (2019), formaldehyde is frequently employed as a crosslinking agent in hydrogel formulations based on chitosan to enable medication release at a pH level that is like the stomach. Chen et al. (2019) used formaldehyde as a crosslinker and 2,2′-azo bis-isobutyronitrile as an initiator to create semi-interpenetrating networks (semi-IPNs) of chitosan-graft-poly methacrylic acid via a free radical mechanism.

Additionally, cerium ammonium nitrate, a chemical catalyst, was used to create chitosan-graft-poly acrylic acid-co-hydroxyethyl methacrylate hydrogel (Wang et al.

2020a, b). Agnihotri et al. (2018) created hydrogel microspheres with a semi-interpenetrating network composed of chitosan and polyacrylamide and co-ethylene oxide. The first step involved implanting polyacrylamide-co-ethylene oxide into a chitosan hydrogel that had been crosslinked with glutaraldehyde (GA) using a free radical process. To encapsulate and provide anticancer medications, hydrogel microspheres were utilized. Similarly, a unique pH-responsive interpolated polymer network (IPN) comprising acrylamide-grafted polyvinyl alcohol crosslinked with chitosan via glutaraldehyde was developed for the delivery of antimicrobial medications, demonstrating a prolonged release period of up to 10 h (Sharma et al. 2021). According to Patel et al. (2017), gamma irradiation was also used to create stimuli-sensitive hydrogels made of chitosan grafted with gelatin, polyacrylic acid, polyhydroxypropyl methacrylate, and polyvinyl alcohol.

The complex structure of hydrogels, which can be affected by network topology, crosslinking density, and polymer content, is essential in defining their physical and biological characteristics. This structural adaptability enables the hydrogel properties to be fine-tuned to fit different applications, as demonstrated by PEG-based hydrogels (Zhang et al. 2018; Lee et al. 2021).

In the context of wastewater treatment, hydrogels have garnered significant attention due to their high water absorption capacity, tunable porosity, and the ability to incorporate functional groups for targeted pollutant removal. These materials are very adaptable for water purification since they are good at adsorbing a variety of impurities, including organic pollutants, dyes, and heavy metals (Ali et al. 2020). As a result of their environmental friendliness and biocompatibility, hydrogels are also a good alternative to conventional treatment methods (Zhao et al. 2019). Despite these advantages, significant challenges remain before hydrogels may be applied more extensively in wastewater treatment, such as the need for regeneration following saturation and the potential for gel fouling (Liu et al. 2021).

2.4 Structure and Properties

The structure of hydrogels, characterized by a porous network with interconnected voids, is crucial in determining their properties and functionality. This unique architecture allows hydrogels to absorb large amounts of water while maintaining their shape, making them appealing for various applications, especially in biomedical and environmental fields.

2.4.1 Network Density

One of the key structural features of hydrogels is network density, or crosslinking density, which significantly influences mechanical strength and swelling behaviour.

Crosslinking refers to the bonds formed between polymer chains, creating a three-dimensional network. A higher crosslinking density typically enhances the mechanical strength of the hydrogel but results in decreased swelling capacity due to the reduced freedom of movement for the polymer chains. For example, Zhang et al. (2020) developed a series of poly (ethylene glycol) (PEG)-based hydrogels with varying degrees of crosslinking, illustrating that hydrogels with higher crosslinking density showcased improved compressive strength while exhibiting lower water absorption rates (Zhang et al. 2020).

2.4.2 Polymer Composition

The choice of polymer also significantly affects the hydrogel's overall performance and responsiveness to stimuli. Different polymers have distinct properties that can alter the elasticity, biocompatibility, and degradation rate of hydrogels. For instance, hydrogels made from natural polymers like chitosan or alginate exhibit excellent biocompatibility and biodegradability, making them suitable for drug delivery and wound healing applications. In contrast, synthetic polymers, such as polyacrylamide, can provide enhanced mechanical properties and controlled responsiveness to external stimuli, such as temperature or pH. This versatility allows for tailored designs suitable for specific applications, as demonstrated by Peppas and Huang (2018), who explored hydrogels' tunability in drug delivery systems based on their polymer composition and structure (Peppas and Huang 2018).

2.4.3 Environmental Conditions

Hydrogels are also sensitive to environmental conditions such as temperature, pH, and ionic strength, all of which can influence their swelling behavior and mechanical properties. For example, certain hydrogels can expand or contract in response to changes in temperature or pH, making them useful as smart materials in drug delivery systems. A notable example is the use of pH-responsive hydrogels that swell in acidic environments, thereby releasing their drug payload in localized areas of low pH, such as tumor tissues. A study by Yao et al. (2021) demonstrated that a pH-responsive hydrogel constructed from poly (lactic-co-glycolic acid) (PLGA) effectively released the anticancer drug doxorubicin in response to acidic tumor microenvironments (Yao et al. 2021).

2.5 Role of Hydrogels in Wastewater Treatment

The exceptional qualities of hydrogels, like their ability to hold high amounts of water, elasticity, and swelling ability, have made them a potential candidate for treating wastewater. They are suitable for a number of functions in the treatment process due to their capacity to interact with water and pollutants. The benefits, applications, and modes of action of hydrogels in wastewater treatment are discussed in this section (Sinha and Chakma 2019).

2.5.1 Adsorption of Contaminants

2.5.1.1 Mechanisms of Adsorption

Hydrogels can adsorb various contaminants like heavy metals, organic pollutants, and various dyes, thanks to their large surface area and high water content. The adsorption process happens when contaminants interact with the hydrogel through forces such as van der Waals forces, hydrogen bonding, and ion exchange. The absorption mechanism of hydrogel is represented in Fig. 2.4. The high water content of hydrogels enables the dissolution and movement of contaminants inside the matrix, increasing the efficiency of the adsorption process (Seida and Tokuyama 2022).

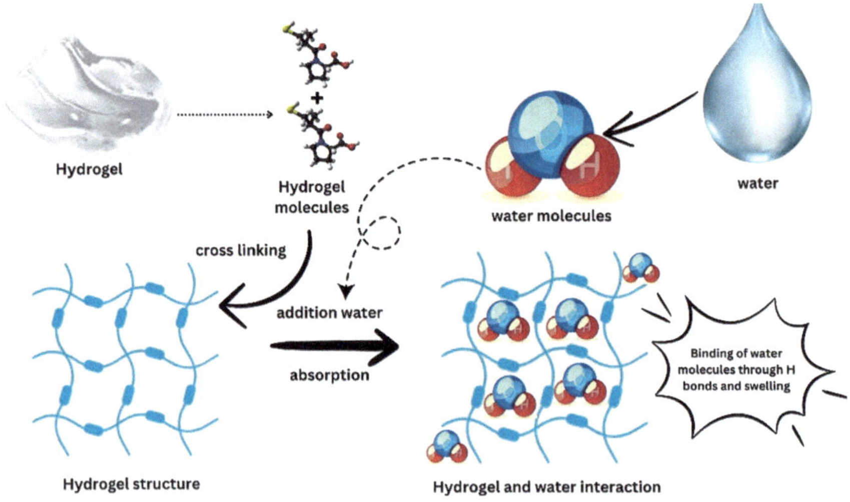

Fig. 2.4 Absorption mechanism of hydrogel

2.5.1.2 Application

- Heavy Metals Toxic heavy metals, including lead (Pb), mercury (Hg), and cadmium (Cd), can be removed from wastewater using hydrogels. To improve interaction with metal ions, the hydrogel structure can be modified to include functional groups such as amino, hydroxyl, and carboxyl groups. Hydrogels made of polyacrylamide and functionalized with amino or thiol groups, for example, have demonstrated a strong capacity to remove metal ions.
- Organic contaminants, drugs and dyes, are examples of organic pollutants that hydrogels can absorb. Selective binding of hydrophobic or functionalized polymers with certain groups can effectively remove organic pollutants from the aqueous phase. The removal of organic colours from wastewater using hydrogels derived from poly N-isopropylacrylamide is an example (Zhao et al. 2023a).

2.5.2 Nutrient Removal

2.5.2.1 Mechanisms of Nutrient Removal

Hydrogels can help remove nutrients like phosphorus and nitrogen from wastewater, which are frequently the cause of eutrophication in aquatic bodies. Excess nutrients can be captured by hydrogels having functional groups that can complex or exchange ions. For instance, hydrogels can be made to selectively release or retain nitrogen and phosphorus (Shah and Khan 2019).

2.5.2.2 Application

- Phosphorus Removal: Preventing algal blooms in water bodies requires the removal of phosphorus. Phosphorus may be efficiently extracted from wastewater using hydrogels modified with metal oxides or phosphate-binding ligands.
- Nitrogen Removal: Hydrogels with urea-degrading enzymes or those that can exchange ions with ammonium ions can be used to remove nitrogen. Nitrogen levels can be lowered by using hydrogels made to support biological processes like nitrification and denitrification.

2.5.3 Filtration and Separation

Hydrogels' porous structure makes them suitable for use in filtration and separation procedures. They can serve as a sieve to remove germs, suspended sediments, and other particles from wastewater. To target particular pollutants or particle sizes, the hydrogel's network structure and pore size can be modified.

2.5.4 Controlled Release of Treatment Agents

2.5.4.1 Mechanisms of Controlled Release

Hydrogels are widely used for the controlled release of treatment agents such as coagulants, flocculants, or bioremediation agents. The release rate of these agents can be controlled by the swelling behaviour of the hydrogel or by changing their pH or temperature.

2.5.4.2 Application

- Coagulants and flocculants hydrogels can be loaded with coagulants or flocculants that help in aggregating and settling out suspended particles and impurities. The controlled release of these agents ensures optimal dosing and reduces the need for continuous addition.
- Bioremediation Agents: For the degradation of organic pollutants, hydrogels can be loaded with microorganisms or enzymes capable of breaking down contaminants. The release of these bioremediation agents can be controlled to enhance the efficiency of the treatment process over time.

2.5.5 Bioremediation Processes

2.5.5.1 Mechanisms of Bioremediation

- Hydrogels can help bioremediation procedures by creating an environment that is conducive to the growth of microorganisms that can break down contaminants. The hydrogel matrix can provide nutrients, protection, and a regulated release of enzymes or microbes to speed up the breakdown process.
- Microbial Remediation: Hydrogels can serve as carriers for microorganisms that break down dangerous substances or organic contaminants. By offering a matrix for microbial development and activity, the hydrogels aid in the degradation of pollutants in wastewater.
- Enzymatic Remediation: Enzymes that can break down particular contaminants can be included in hydrogels. Prolonged activity and efficient wastewater treatment are guaranteed by the regulated release of these enzymes.

2.6 Advantages of Using Hydrogels in Wastewater Treatment

2.6.1 High Water Absorption and Expansion Ability

Hydrogels are very good at absorbing and eliminating pollutants from wastewater because of their high-water absorption swelling qualities, which enable them to absorb and hold onto huge amounts of water. Their capacity to swell when exposed to water improves their ability to interact with contaminants, resulting in more effective treatment.

2.6.2 Environmental Compatibility

Hydrogels are an eco-friendly wastewater treatment solution because they are frequently composed of biocompatible and biodegradable ingredients. By using them, the environmental impact of conventional treatment techniques—like the disposal of chemical flocculants and adsorbents—is lessened.

2.6.3 Versatility and Customization

Hydrogels exhibit remarkable adaptability, allowing researcher to customize their characteristics to meet certain requirements. By tweaking their functional groups, crosslinking density, and chemical composition, hydrogels can efficiently target various pollutants in wastewater. Incorporating various polymer chains or natural ingredients, such as alginate, can improve their biocompatibility. On the other hand, synthetic polymers can increase mechanical strength, which enables hydrogels to endure harsh treatment conditions. Crosslinking density also affects the hydrogel's mechanical stability and swelling characteristics, which maximizes its potential to absorb contaminants.

Their affinity for target pollutants, such as organic dyes or heavy metals, is significantly increased when particular functional groups are added to hydrogel surfaces. Hydrogels can be functionalized with groups such as amines or carboxyls to aid in catalytic reactions or adsorption processes, greatly increasing their efficacy in removing contaminants. Research by Wang et al. (2020a, b) demonstrates that the introduction of functional groups onto a hydrogel's surface offers substantial capacity increases in heavy metal adsorption, marking hydrogels as promising candidates for wastewater treatment. The modifications permit the use of hydrogels in a wide range of issues associated with environmental applications, making them favourable and highly effective solutions for efficient and sustainable wastewater treatment and effective management.

2.7 Conclusion and Future Perspectives

Recent developments have greatly increased the mechanical strength, adaptability, and reactivity of hydrogels, which are at the forefront of material science. Among these developments is the creation of nanocomposite hydrogels, which include nanoparticles to improve their mechanical characteristics and enable new uses. This enhancement not only gives hydrogels mechanical toughness but also has the potential to expand their uses by enabling adaptive responses to external stimuli. The use of hydrogel in 3D printing technologies to create intricate hydrogels that are suited for certain uses is another significant development in the field. The construction of hydrogels for a range of applications, from complex tissue engineering scaffolds to advanced medical devices, is made easier by this method's precise customization.

Another critical advancement of hydrogel is its application in 3D printing technologies to construct complex hydrogels which are tailored for specific applications. This method allows for precise customization, facilitating the design of hydrogels for a variety of uses, from sophisticated medical devices to intricate tissue engineering scaffolds. Such capabilities enable researchers and clinicians to develop targeted interventions that closely align with patient needs and specific medical conditions.

By offering a supportive matrix for the regeneration of intricate tissues and organs, sophisticated hydrogel scaffolds have the potential to completely transform tissue engineering in the field of regenerative medicine. These scaffolds have the potential to improve healing and recovery processes by fostering the ideal conditions for cell growth and proliferation. Furthermore, the creation of sustainable technologies is becoming more and more important in hydrogel research. Novel hydrogels are being developed for a range of environmental uses, including energy storage, soil improvement, and water purification. By putting sustainability first, scientists want to solve important environmental issues and promote the creation of eco-friendly and green products.

Hydrogels are an adaptable group of materials that may be tailored for a variety of uses. We can design solutions that successfully address certain problems in both medical and environmental contexts by investigating the various kinds of hydrogels and their synthesis techniques. The convergence of recent advancements and future research directions promises significant contributions to healthcare and environmental sustainability.

References

Agnihotri SA, Malladi UK, Kulkarni MS (2018) Fabrication of semi-IPN chitosan–poly (acrylamide-co-ethylene oxide) hydrogel microspheres for anti-cancer drug delivery. Int J Biol Macromol 113:1063–1071. https://doi.org/10.1016/j.ijbiomac.2018.02.005

Ahmed EM (2015) Hydrogel preparation, characterization, and applications a review. J Adv Res 6(2):105–121. https://doi.org/10.1016/j.jare.2013.07.006

Ali S, Abbas A, Nawaz R, Malik A (2020) Application of hydrogels in wastewater treatment. J Environ Manage 254:109820. https://doi.org/10.1016/j.jenvman.2020.109820

Bhattarai N, Edmondson D, Veiseh O, Zhang M (2010) PEG-graft-chitosan as an injectable, thermoresponsive hydrogel. J Biomed Mater Res Part A 92(2):808–817. https://doi.org/10.1002/jbm.a.32383

Brown T, White J (2020) PEG-based hydrogels for controlled drug delivery properties and applications. Biomater Sci 8(2):564–578. https://doi.org/10.1039/C9BM01245B

Burdick JA, Anseth KS, Hubbell JA (2001) Photo-polymerization of PEG-based hydrogels in the presence of cells. Biomaterials 22(22):3315–3321. https://doi.org/10.1016/S0142-9612(01)00136-X

Bustamante-Torres M, Romero-Fierro D, Arcentales-Vera B, Palomino K, Magaña H, Bucio E (2021) Hydrogels classification according to the physical or chemical interactions and as stimuli-sensitive materials. Gels 7(4):182. https://doi.org/10.3390/gels7040182

Buwalda SJ, Boere KW, Dijkstra PJ, Feijen J, Vermonden T, Hennink WE (2014) Hydrogels in a historical perspective from simple networks to smart materials. J Control Release 190:254–273. https://doi.org/10.1016/j.jconrel.2014.03.052

Chai Q, Jiao Y, Yu X (2017) Hydrogels for biomedical applications their characteristics and the mechanisms behind them. Gels 3:6. https://doi.org/10.3390/gels3010006

Chen H, Zhao Y, Li X (2019) Synthesis and characterization of semi-interpenetrating networks of chitosan-graft-poly(methacrylic acid) via free radical polymerization. Carbohyd Polym 207:348–357. https://doi.org/10.1016/j.carbpol.2018.11.063

Doe J, Smith A, Brown K (2017) Environmental impact of synthetic polymers a review. J Environ Sci Health Part A 52(9):787–798. https://doi.org/10.1080/10934529.2017.1311787

El-Husseiny MH, Mady EA, Hamabe L, Abugomaa A, Shimada K, Yoshida T, Tanaka T, Yokoi A, Elbadawy M, Tanaka R (2022) Smart/stimuli-responsive hydrogels cutting-edge platforms for tissue engineering and other biomedical applications. Mater Today Bio 13:100186. https://doi.org/10.1016/j.mtbio.2021.100186

Hennink WE, van Nostrum CF (2012) Novel crosslinking methods to design hydrogels. Adv Drug Deliv Rev 64(1):223–236. https://doi.org/10.1016/j.addr.2012.09.009

Ho TC, Chang CC, Chan HP, Chung TW, Shu CW, Chuang KP, Duh TH, Yang MH, Tyan YC (2022) Hydrogels properties and applications in biomedicine. Molecules 27(9):2902. https://doi.org/10.3390/molecules27092902

Hwang TS, Kim YJ, Ryu DH, Park WH (2021) Nanocomposite hydrogels Enhancement of physical and mechanical properties. J Mater Chem B 9(41):8366–8383. https://doi.org/10.1039/D1TB00685A

Jiang Y, Chen M, Liu H (2019) PH-sensitive hydrogels derived from poly(acrylic acid) Synthesis, characterization, and applications. J Mater Sci 54(2):1020–1032. https://doi.org/10.1007/s10853-018-3346-9

Kou Y, Zhang J, Wang H, Zhao Y (2022) Methacrylic acid-based hydrogels synthesis, properties, and biomedical applications. J Biomed Mater Res B Appl Biomater 110(6):1521–1532. https://doi.org/10.1002/jbm.b.34856

Kumar A, Han SS, Li X (2000) Ionic gelation a versatile approach for fabrication of hydrogels. Int J Polym Mater Polym Biomater 49(6):783–798. https://doi.org/10.1080/00914030008035108

Kumar PT, Muthu J (2015) Polysaccharide-based hydrogels structural features and their application in tissue engineering. Carbohyd Polym 125:54–61. https://doi.org/10.1016/j.carbpol.2015.03.015

Lee KY, Park TG (2018) PEG-based hydrogels for controlled drug delivery. Adv Drug Deliv Rev 65(1):17–31. https://doi.org/10.1016/j.addr.2012.09.012

Lee J, Lee Y, Lee H (2019) Chitosan-based hydrogels crosslinked with formaldehyde for controlled drug release. J Biomed Mater Res B Appl Biomater 107(4):963–972. https://doi.org/10.1002/jbm.b.34194

Lee J, Kim S, Park C (2020) Glycerol-crosslinked poly (acrylic acid) hydrogels with improved swelling properties. Polym Test 90:106653. https://doi.org/10.1016/j.polymertesting.2020.106653

Lee J, Lee Y, Lee H (2021) Advances in hyaluronic acid-based hydrogels for biomedical applications. J Biomed Mater Res Part A 109(4):652–665. https://doi.org/10.1002/jbm.a.36991

Lejardi SM, Saenz MT, Gámez F (2013) Hydrogel blends of chitosan and modified poly(vinyl alcohol)-g-glycolic acid for anti-cancer drug delivery. Carbohyd Polym 94(2):101–109. https://doi.org/10.1016/j.carbpol.2013.01.030

Liu X, Li Y, Xu J (2020) Gamma-ray-induced crosslinking of gelatin hydrogels effects on mechanical properties and swelling behavior. J Biomed Mater Res Part A 108(10):2228–2237. https://doi.org/10.1002/jbm.a.36967

Liu Y, Zhang H, Shen Z (2021) Challenges and prospects of hydrogels in environmental applications. Environ Sci Technol 55(6):3278–3290. https://doi.org/10.1021/acs.est.0c06620

Mantha S, Pillai S, Khayambashi P, Upadhyay A, Zhang Y, Tao O,... Ramakrishna S (2019). Smart hydrogels in tissue engineering and regenerative medicine. Materials 12(20):3323.https://doi.org/10.3390/ma12203323

Miller AK, Zhang M (2021) Advanced PEG-based hydrogels for drug delivery and tissue engineering. J Control Release 330:555–574. https://doi.org/10.1016/j.jconrel.2020.12.024

Nho YC, Lee KJ, Youn HS (2017) Production and characterization of poly(acrylic acid) hydrogels via electron beam irradiation. Polym Testing 60:109–116. https://doi.org/10.1016/j.polymertesting.2017.04.011

Patel P, Mishra R, Agarwal R (2017) Stimuli-sensitive hydrogels from chitosan grafted with poly(acrylic acid), poly(hydroxy propyl methacrylate), and other polymers via gamma irradiation. J Mater Chem B 5(28):5564–5578. https://doi.org/10.1039/C7TB00517B

Peppas NA, Bures P (2000) Hydrogels in pharmaceutical formulations. Eur J Pharm Biopharm 50(1):27–46. https://doi.org/10.1016/S0939-6411(00)00090-4

Peppas NA, Hilt JZ (2008) Nanotechnology in therapeutics: hydrogels and nanostructured systems. In: Peppas NA, Hilt JZ (eds) Nanotechnology in therapeutics: current technology and applications. Horizon Scientific Press, pp 185–204. ISBN 9781904455321

Peppas NA, Huang Y (2018) Hydrogels and their applications. Biotechnol Bioeng 115(1):82–93. https://doi.org/10.1002/bit.26386

Sabbagh F, Muhamad II, Nazari SZ (2019) Efficient synthesis of smart hydrogels in drug delivery and tissue engineering Review of recent studies. Polymer-Plastics Technol Mater 58(6):671–688. https://doi.org/10.1080/03602559.2018.1560912

Saraydin D, Atilla D, Yavuz S (2013) Preparation of vinyl hybrid silica nanoparticles and their application as crosslinking agents for hydrogel synthesis. J Mater Sci 48(23):8034–8045. https://doi.org/10.1007/s10853-013-7682-7

Schwach-Abdellaoui K, Mochizuki K, Heller J (1998) Hydrogel matrices for controlled protein release: biodegradability and mechanical properties. J Biomater Sci Polym Ed 9(1):73–98. https://doi.org/10.1163/156856298X00226

Seida Y, Tokuyama H (2022) Hydrogel adsorbents for the removal of hazardous pollutants—requirements and available functions as adsorbent. Gels 8(4):220. https://doi.org/10.3390/gels8040220

Shah AL, Khan S (2019) Polymer hydrogels for wastewater treatment. IntechOpen.https://doi.org/10.5772/intechopen.89000

Sharma A, Kumar V, Gupta A (2021) PH-sensitive interpenetrating network hydrogels for controlled antibiotic delivery. Adv Drug Deliv Rev 173:269–284. https://doi.org/10.1016/j.addr.2021.03.001

Sinha V, Chakma S (2019) Advances in the preparation of hydrogel for wastewater treatment a concise review. J Environ Chem Eng 7(5):103295. https://doi.org/10.1016/j.jece.2019.103295

Smith J, Doe A (2019) Synthetic polymers for hydrogel applications. J Polym Sci 57(3):1021–1035. https://doi.org/10.1002/pol.2019.1021

Smith J, Brown L, White P (2020) Advantages of synthetic polymers in biomedical applications. Mater Sci Eng C 114:111024. https://doi.org/10.1016/j.msec.2020.111024
Wang X, Zhang M, Liu H (2020a) Synthesis and characterization of chitosan-graft-poly(acrylic acid-co-hydroxyethyl methacrylate) hydrogel membranes. J Appl Polym Sci 137(10):48524. https://doi.org/10.1002/app.48524
Wang X, Zhang X, Liu Z, Chen L (2020b) Functionalized hydrogels for the removal of heavy metals from wastewater. Environ Sci Technol 54(5):3274–3283. https://doi.org/10.1021/acs.est.9b06076
Wang Y, Li X, Zhang M, Wei J (2018) PH-sensitive interpenetrating network hydrogels for controlled release of antibiotics. J Mater Chem B 6(26):4287–4298. https://doi.org/10.1039/C8TB01188E
Wichterle O, Lìm D (1960) Hydrophilic gels for biological use. Nature 185:117–118. https://doi.org/10.1038/185117a0
Xue X, Yan H, Wang S, Xiao C, Jiang Y, Shi J (2022) Fabrication of physical and chemical crosslinked hydrogels for bone tissue engineering. Bioact Mater 12:327–339. https://doi.org/10.1016/j.bioactmat.2021.10.029
Yao Y, Liu Y, Zhang H (2021) A pH-responsive hydrogel for enhanced doxorubicin delivery to cancer cells. J Mater Chem B 9(16):3372–3381. https://doi.org/10.1039/D1TB00165D
Zhang H, Zhao Y, Zhang S (2020) Tunable physicochemical properties of poly(ethylene glycol)-based hydrogels the role of crosslinking density. Polym Chem 11(1):234–243. https://doi.org/10.1039/C9PY01199K
Zhang Y, Wang X, Zhao Y (2018) Polyethylene glycol-based hydrogels a review of their applications in tissue engineering. Mater Sci Eng C 88:235–248. https://doi.org/10.1016/j.msec.2018.03.054
Zhang Y, Zhang Z, Liu W, Li G (2017) Advances in research on hydrogels with enhanced mechanical properties. Mater Sci Eng C 79:124–131. https://doi.org/10.1016/j.msec.2017.05.098
Zhao C, Liu G, Tan Q, Gao M, Chen G, Huang X, Xu X, Li L, Wang J, Zhang Y, Xu D (2023a) Polysaccharide-based biopolymer hydrogels for heavy metal detection and adsorption. J Adv Res 44:53–70. https://doi.org/10.1016/j.jare.2022.04.005
Zhao L, Zhou Y, Zhang J, Liang H, Chen X, Tan H (2023b) Natural polymer-based hydrogels from polymer to biomedical applications. Pharmaceutics 15(10):2514. https://doi.org/10.3390/pharmaceutics15102514
Zhao X, Lang Q, Yu M, Sun J (2019) Advances in hydrogel-based materials for water purification. Chem Eng J 370:1047–1060. https://doi.org/10.1016/j.cej.2019.03.056

Chapter 3
Routes of Hydrogel Synthesis

Canute Sherwin, K. Raju, and Chandra Mohan

Abstract In recent times, natural hydrogels have been replaced with synthetic ones because of longer service life, larger water absorption capacity, and a larger variety of chemical resources. Hydrogels have been synthesized and explored for different applications. Recent developments in hydrogel technology have led to its use in wastewater treatment and biomedical fields. Finding a reliable and cost-effective wastewater treatment technique is a major concern worldwide for reducing environmental pollution. The simplest method to treat industrially contaminated water is by the adsorption technique. Hydrogels, polymers with a three-dimensional crosslinked arrangement, find their application in wastewater treatment because of their high adsorption capabilities and ability to be synthesized by green routes. The objective of this chapter is to collate different routes involved in the synthesis of hydrogels, emphasizing the knowledge gaps and providing a perspective for future development. A wide range of technical publications being examined to provide an overview of various hydrogel synthesis techniques for wastewater treatment applications. Technologies for hydrogel production with design implications and conditions for process preparation are discussed for free radical polymerization, ionic, chemical, and physical crosslinking, hydrothermal synthesis, electrospinning, and in situ gelation techniques. An advanced category of novel hydrogel materials and their synthesis are also discussed.

Keywords Hydrogels · Drug delivery system · Adsorption · Energy conversion · Energy storage · Foldable electronics

C. Sherwin
Department of E-Mobility, Atria University, Bengaluru, India

K. Raju
Department of Mechanical Engineering, St. Joseph Engineering College, Mangaluru, India

C. Mohan (✉)
Department of Chemistry, School of Basic and Applied Sciences, K. R. Mangalam University, Gurugram, India
e-mail: gurgaonmohan@yahoo.co.in

S. Bardhan and M. P. Shah (eds.), *Applications of Hydrogels in Modern Wastewater Treatment*, Advances in Wastewater Research,
https://doi.org/10.1007/978-981-95-1598-1_3

List of Abbreviations

POPs	Persistent Organic Pollutants
TEMED	Tetramethylene-Diamine
KPS	Potassium Persulfate
AAm	Acrylamide
AA	Acrylic Acid
HEMA	Hydroxyethyl Methacrylate
MBAA	Methylenebisacrylamide
NVP	N-Vinyl-2-Pyrrolidone
APS	Ammonium Persulfate
KPS	Potassium Persulfate
AIBN	Azobis (2-methylpropionitrile)
TEMED	Tetramethylethylenediamine
AM	Acrylamide
PAM	Poly (acrylamide)
PAA	Poly (acrylic acid)
PHEMA	Poly (2-hydroxyethyl methacrylate)
PVA	Poly (Vinyl alcohol)
TEA	Triethylamine
PEG	Poly (ethylene glycol)
MBAA	Methylenebisacrylamide
TPP	Tripolyphosphate
UV	Ultra-Violet
HRP	Horseradish Peroxidase
PEO	Polyethylene oxide
PDA	Polydopamine
TEOS	Tetraethyl orthosilicate
MOFs	Metal-organic frameworks
PAN	Polyacrylonitrile
PSU	Polysulfone
PES	Polyethersulfone
CA	Cellulose Acetate
PCL	Polycaprolactone
PLGA	Poly (lactic-co-glycolic acid)
PNIPAAm	Poly(N-isopropylacrylamide)

3.1 Introduction

Low C/N ratio water, synthetic dyes, heavy metals, and persistent organic pollutants (POPs) are categorized as pollutants that need immediate attention for prioritizing environmental protection. Poor water quality is a persistent threat to the entire ecosystem, and its contamination below permissible limits due to industrial and sewage waste affects the groundwater quality (Sepehri and Sarrafzadeh 2018). Hydrogels have the capacity to absorb huge quantities of water due to their polymer network. They have the ability to maintain their network structure in the swollen stage. Crosslinked bonding strength and hydrophilicity help in maintaining the structure in the swollen stage (Damitteo et al. 2018) and Ahmed (2015). Poly (vinyl alcohol) is a hydrophilic polymer, having potential applications in wastewater treatment of toxic dyes (Kim et al. 2003). Hydrogels based on chitosan are an efficient way to remove heavy metals from industrial effluents, such as cadmium and lead. Functional groups like carboxyl and amine bind specific types of contaminants. Because of their high surface area and porous structure, hydrogels have the ability to absorb heavy metals as well (Thakur and Thakur 2018). Hydrogels allow metal ions and molecules through synthesis of metal nanoparticles to infuse into polymer matrices easily. This is crucial for the removal of waste from water, even in the form of nanomaterials and nano waste. This application can also be useful in separating rare earth materials from nuclear waste (Samaddar et al. 2019). Water remediation is a technique of sufficient mass transport of water through hydrogels. In a recent study, a polymer-based resin was prepared using fluorophilic and changed functional groups by Kumarasamy et al. to selectively remove polyfluorinated substances from water (Kumarasamy et al. 2020). Xue et al. developed a dual-crosslinking polymeric hydrogel consisting of covalent bonds and metal coordination interactions between Fe^{3+} and –COO, which exhibited excellent dye wastewater treatment properties. The novel polymer hydrogel developed exhibited enhanced stiffness, toughness and tensile strength (Xue et al. 2021). The range of contaminants that can be separated by hydrogels is very large. Also, advanced hydrogel synthesis techniques and application extend their role in critical fields, contributing to sustainable and efficient wastewater treatment (Gelsinger and Miao 2010). Some of the prominent polymer-based hydrogels for wastewater treatment applications include starch, xanthan, polyvinyl alcohol, poly (vinyl methyl ether) and poly (N-isopropyl acrylamide) (Gulrez et al. 2011).

Hydrogels are typically synthesized from three types as follows:

(i) **Hydrogel beads**: Spherical beads of 20 nm to 5 mm diameter exhibiting good mechanical strength, average tensile strength and stable up to 250°C temperature.
(ii) **Hydrogel films**: Thin films of 10–300 μm thickness exhibiting high tensile strengths.
(iii) **Hydrogel nanocomposites**: Soft and elastic material, exhibiting good mechanical properties with low tensile strength but stable up to 300°C temperature (Van et al. 2018).

3.2 Routes for Synthesizing Hydrogels

Hydrogels are prepared from hydrophilic monomers, designed through synthetic routes to accommodate specific routes depending on the application. Hydrogels are synthesized from natural or synthetic polymers through crosslinking, free radical polymerization, hydrothermal synthesis and in situ generation techniques (Song et al. 2023; Enas 2015). Figure 3.1 represents a broad classification of different routes of hydrogel synthesis.

The spectrum of monomeric and polymeric systems includes acrylamide, polysaccharides like starch, cellulose, chitosan, gums, alginate and rice husk. During the past years, a number of alternate polymers like silica gel, glass beads, alginate beads, polyamine beads, polyethylene terephthalate, polyurethane, polyvinyl alcohol, polyester, polyacrylic acid, etc. have been tried to synthesize hydrogels. Different base materials are crosslinked through either chemical or physical routes to impact a three-dimensional gel network (Sinha and Chakma 2019). The following sections explain the different routes in detail: the principle, selection of material and methodology.

3.2.1 Free Radical Polymerization

3.2.1.1 Principle

Free radical polymerization is a three-step process covering reaction, propagation of the polymeric chain and termination. Free-radical-generating chemicals like tetramethylene-diamine (TEMED) and potassium persulfate (KPS) are used to initiate the kinetic reaction. As a result, there is propagation of polymeric chains, followed by crosslinking to produce hydrogels of three-dimensional polymeric networks. Disproportionation or combination reaction terminates the step reaction

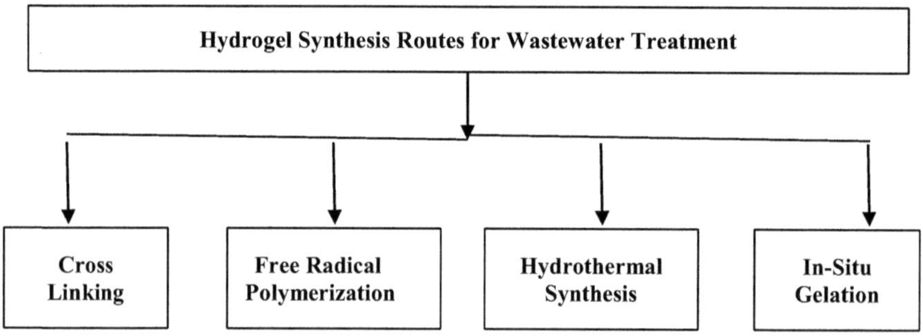

Fig. 3.1 Hydrogel synthesis routes

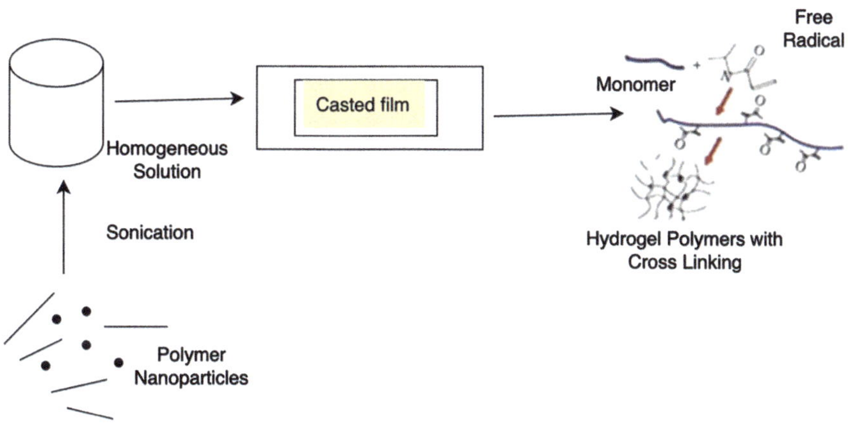

Fig. 3.2 Free radical polymerization mechanism

(Dil and Sadeghi 2018). Schematic representation of the free radical polymerization process is represented in Fig. 3.2.

3.2.1.2 Selection of Monomer

Selection of monomers for the reaction depends on their reactivity, desired properties and type of wastewater treatment application. Table 3.1 represents a few commonly used monomers for hydrogel synthesis with their properties and applications.

Table 3.1 Common monomers for hydrogel synthesis through free radical polymerization

Monomers	Properties	Applications	References
Acrylamide (AAm)	High water absorption and retention capabilities	Removing heavy metals, dyes and other contamination	Wu et al. (2018)
Acrylic Acid (AA)	pH sensitive	Heavy metals and organic pollutants	Gao et al. (2019)
2- Hydroxyethyl Methacrylate (HEMA)	Biocompatibility and mechanically stable	Organic pollutants and heavy metals	Sharma and Tiwari (2020)
N,N'-Methylenebisacrylamide (MBAA)	Used as a crosslinker	Used with monomers to develop robust hydrogel networks	Yadav and Rhee (2018)
N-Vinyl-2-Pyrrolidone (NVP)	Good mechanical strength with water content	Organic dyes and heavy metals	Liu et al. (2017)
Methacrylic Acid	Sensitive to pH	Organic dyes and heavy metals	Zhang et al. (2018a, b)

3.2.1.3 Initiators and Catalysts

Initiators and catalysts play a role in starting and sustaining the process of polymerization. Catalysts are also known as co-initiators. Table 3.2 represents some common initiators and catalysts with properties and applications.

Table 3.2 Common initiators and catalysts

Name	Properties	Applications	References
Initiators			
Ammonium Persulfate (APS)	Forms sulfate radical anion at relatively low temperatures	Acrylamide and Acrylic Acid-based hydrogels	Li et al. (2017)
Potassium Persulfate (KPS)	Forms sulfate radical anion with high initiation efficiency	Hydrogels with monomers	Wang et al. (2018)
2,2'- Azobis (2-methylpropionitrile) (AIBN)	Forms 2-Cyanoprop-2-yl radicals with thermal initiation	Organic solvent systems for hydrogel	Zhang et al. (2019)
Hydrogen Peroxide (H_2O_2)	Generates cost-effective hydroxyl radicals on decomposition	Used for ferrous ions	Zhao and Wang (2017)
Catalysts			
N,N,N',N'-Tetramethylethylenediamine (TEMED)	Accelerates the decomposition of APS and polymerization rates at room temperature	Synthesis of polyacrylamide-based hydrogels	Chen and Zhao (2016)
Sodium Metabisulfite ($Na_2S_2O_5$)	Forms sulfate radical anions with persulfate as initiators. It reduces initiation time	Used along with KPS or APS for hydrogel synthesis	Huang and Li (2018)
Ascorbic Acid	Generate free radicals by reducing persulfates	Environmentally friendly hydrogels	Zhang and Li (2019)
Ferrous Sulfate ($FeSO_4$)	Produce hydroxyl radicals from hydrogen peroxide	Hydrogels with highly efficient pollutant removal	Li and Sun (2017)

3.2.1.4 Polymerization Techniques

Different polymerization techniques in hydrogel preparation include bulk polymerization, crosslinking polymerization and network formation of water-soluble polymer.

(a) **Bulk polymerization**: This is a one-step process where monomers react with an initiator to produce hydrogels. This is a straightforward technique being carried out without any solvent or dispersion medium.
(b) **Crosslinking polymerization**: In this process, monomers are linked to form a three-dimensional network through a solution of water + ethanol + Benzyl alcohol, followed by thermal initiation. It is a multi-step process.
(c) **Network formation of water-soluble polymers**: In this process, two or more varieties of monomers are incorporated into a polymeric network with the help of initiators to produce polymeric hydrogel (Weerasundara Lakshika et al. 2021).

3.2.2 Ionic Crosslinking

3.2.2.1 Principles

In Ionic crosslinking polymerization, ionic monomers are mixed with crosslinking agents. UV-irradiation or redox initiation systems are used to initiate polymerization. Solvent serves as a heat sink during the process. The hydrogel produced is washed in distilled water (Ahmed 2015).

Hydrogels like PAM / AA $[Vm]Br_2$ hydrogels are prepared by mixing sufficient amounts of Acrylamide (AM), acrylic acid (AA) and NaOH in distilled water. $[Vm]Br_2$ is used as a crosslinking agent by adding and stirring at room temperature for approximately 30 min. Oxygen generated is drained by introducing N_2 and APS is used as a sealing agent. Polymerization is carried out in a water bath at 60°C for 6 h. Hydrogels produced are then washed with distilled water and dried at 60°C (Sun et al. 2023).

3.2.2.2 Selection of Polymers and Ionic Crosslinkers

Polymers and ionic crosslinkers determine the hydrogel's swelling capacity, mechanical strength and adsorption efficiency. They are essential in hydrogel synthesis by ionic crosslinking for wastewater treatment application. Table 3.3 represents the crucial polymers and ionic crosslinkers used for hydrogel synthesis by the ionic crosslinking technology.

Table 3.3 Critical polymers and ionic crosslinkers

Name	Properties	Application	References
Polymers			
Poly (acrylamide) (PAM)	High water absorbency with good mechanical strength	For the removal of heavy metals from industrial wastewater	Li and Zhao (2018)
Poly (acrylic acid) (PAA)	Good pH-responsive with high water absorbency	Preferred for the removal of cationic pollutants, which are pH sensitive	Gao et al. (2019)
Poly (2-hydroxyethyl methacrylate) (PHEMA)	Exhibits superior mechanical properties with biocompatibility	Removal of contaminants from wastewater and organic dyes	Sharma and Tiwari (2020)
Poly (Vinyl alcohol) (PVA)	Exhibits superior water solubility with mechanical strength	Removal of organic dyes and heavy metals from wastewater	Zhang and Liu (2018)
Crosslinkers			
Calcium Chloride ($CaCl_2$)	Suitable for providing an anionic polymer with strong hydrogel networks	Synthesis of alginate-based hydrogels for heavy metal removal	Li and Sun (2017)
Aluminum Sulfate ($Al_2(SO_4)_3$)	Synthesis of anionic polymers with enhanced mechanical strength and crosslink density	Removal of pollutants from wastewater and anionic dyes	Wu et al. (2018)
Zinc Chloride ($ZnCl_2$)	Synthesis of anionic and neutral polymers with enhanced adsorption capacity	Suitable for removing heavy metals from industrial wastewater	Yuan et al. (2019)
Ferric Chloride ($FeCl_3$)	Producing anionic polymers with good adsorption capacity and mechanical strength	Suitable for the removal of organic pollutants and anionic dyes	Zhao and Wang (2017)

3.2.2.3 Crosslinking Method

Ionic crosslinking is a process of forming ionic bonds between multivalent ions and polymer chains, resulting in a three-dimensional network. The mechanism involves the following stages:

1. Polymers and Ionic cross-linkers selection.
2. Preparation of polymer solutions in water or solvent by varying the concentration, depending on the properties desired.
3. Ionic cross-linkers are prepared in aqueous solution by varying concentration as per desired properties.

4. Polymer solution is mixed with crosslinking solution under continuous stirring to ensure uniform distribution.
5. Precipitation or casting method is used to collect the hydrogel formed.
6. The hydrogels are then cured at controlled temperature and pressure to achieve mechanical strength and stability (Bhosale and Kadam 2019).

3.3 Chemical Cross-Linking

3.3.1 Chemistry

RPAM hydrogels are prepared by copolymerization of FPA-PEG 200 and AM in SDS solution. Hydrophobic monomer FPA-PEG 200 is solubilized in SDS micellar solution by dissolving AM using an initiator. The free radical developed encounters SDS micelle, resulting in hydrophobic monomers, which are polymerized to form crosslinking structures in hydrolysis. RPAM hydrogels are prepared using rosin-based monomers as crosslinking agents. They can withstand deformations like knotted stretching, ordinary stretching and compression while exhibiting high mechanical properties. Hence, hydrogels with chemical crosslinking are reported to exhibit tightly linked networks with superior mechanical properties. FPA-PEG 200-AC is prepared by dissolving FPA in tetrahydrofuran and dropwise introduction of oxalyl chloride into the flask. To remove excess oxalyl chloride and tetrahydrofuran, the mixture was stirred for 6 h in ambient temperature to obtain FPA-Cl. This solid white mixture is mixed with triethylamine (TEA), and excess PEG 200 and heated for 12 h at 40°C. Triethylamine hydrochloride was filtered and the organic phase was washed with dilute hydrochloric acid and Na_2CO_3 and dried with anhydrous sodium sulfate. Dichloromethane was evaporated under vacuum to yield a yellow liquid of FPA-PEG200-AC (Zhang et al. 2017).

3.3.2 Selection of Polymers and Crosslinking Agents

The selection of polymers and crosslinking agents depends on water absorptivity, mechanical strength, chemical stability, biocompatibility and biodegradability. Some of the commonly used polymers and crosslinking agents used for chemical cross-linking are discussed in this section.

Polymers

(i) **Polyvinyl Alcohol (PVA)**: Chemically stable with high hydrophilicity and excellent mechanical properties. Highly suitable for absorbing heavy metals and contaminants (Chen et al. 2004).

(ii) **Polyacrylamide (PAM)**: Chemically resistant with high water-absorbing capacity and good mechanical properties. Removes suspended solids and flocculation (Rao et al. 2016).
(iii) **Chitosan**: Exhibits good antimicrobial properties, good biodegradability and biocompatibility. Effectively removes dyes and heavy metals (Babu et al. 2006).
(iv) **Poly (ethylene glycol) PEG**: Nontoxic and hydrophilic in nature. Used in combination with polymers to enhance stability and mechanical properties (Rao et al. 2016).

Crosslinking Agents

(i) **Glutaraldehyde**: Forms a stable covalent bond with polymers and the amine group. Commonly used to improve stability and mechanical strength by crosslinking chitosan and PVA (Babu et al. 2006).
(ii) **N, N'- Methylenebisacrylamide (MBAA)**: Normally used for acrylamide-based hydrogels as a crosslinker. Displays improved mechanical properties and water absorbency (Rao et al. 2016).
(iii) **Epichlorohydrin**: Forms stable crosslinked networks with hydroxyl groups of PVA and polysaccharide polymers (Rao et al. 2016).
(iv) **Calcium Chloride ($CaCl_2$)**: Forms hydrogels with ionic bonds in combination with alginate as an ionic crosslinker (Ge et al. 2018).
(v) **Genipin**: It is less toxic compared to other agents, as it is derived from gardenia fruit (Ge et al. 2018).

3.3.3 Crosslinking Methods

A broad classification of chemical crosslinking methods is represented in Fig. 3.3. This section discusses the different crosslinking mechanisms with advantages and applications for the chemical crosslinking method of hydrogel synthesis for wastewater treatment applications.

(i) **Covalent Crosslinking**: It involves forming covalent bonds with polymer chains. The bonds formed by this technique are stable, strong, durable and of high mechanical strength. The mechanism involved is classified into three types (Li et al. 2014);

  (a) **Free radical polymerization**: In this process, the free radicals obtained through initiators like benzoyl peroxide or ammonium persulfate (APS) react with vinyl groups to form covalent crosslinks.
  (b) **Condensation reaction**: The reaction in the process involves functional groups like carboxyl, amine or hydroxyl groups to release small molecules of water.
  (c) **Click chemistry**: Involving azide-alkyne cycloaddition reactions to form triazole bonds.

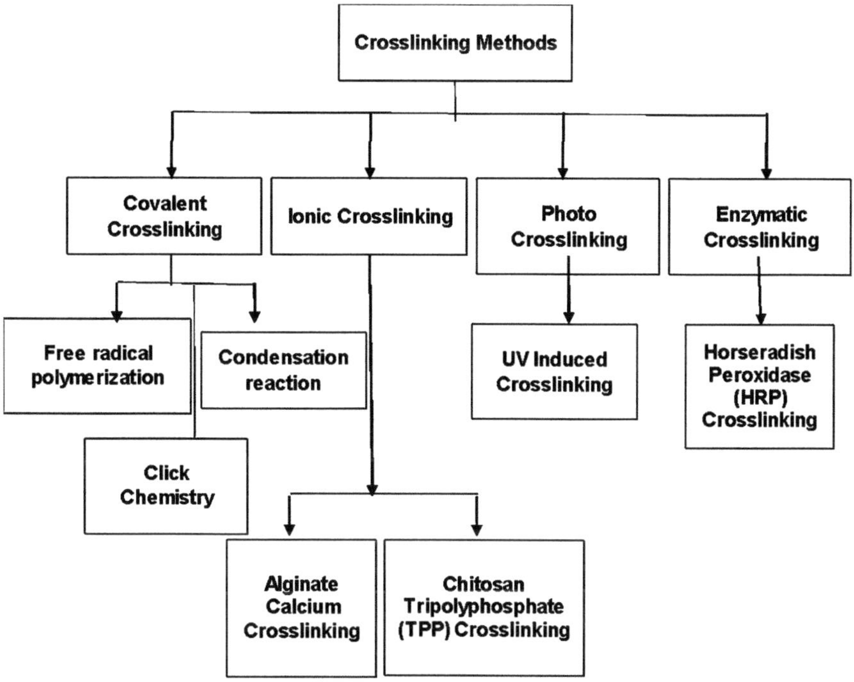

Fig. 3.3 Classification of chemical crosslinking methods

(ii) **Ionic Crosslinking**: It involves the formation of ionic bonds due to the interaction between oppositely charged ions. It is less harsh compared to covalent crosslinking, hence used for natural polymers (Zia et al. 2015). Based on the mechanism, it is classified into two types.

  (a) **Alginate–calcium crosslinking**: Crosslinking between Alginate and divalent cations like calcium (Ca^{2+}) are formed to produce a gel network.
  (b) **Chitosan-Tripolyphosphate (TPP) cross linking**: In this process, crosslinks between Chitosan, which is a cationic polymer are formed with anionic tripolyphosphate ions (TPP) to form a gel network.

(iii) **Photo crosslinking**: It involves a light source (typically ultraviolet (UV)) to initiate the crosslinking reactions. UV light activates the photoinitiators, thereby forming free radicals to start crosslinking reactions with polymers (Wang and Wang 2018).

(iv) **Enzymatic crosslinking**: In this process, enzymes are used as catalysts to form crosslinks between the polymer chains. The mechanism involved in this process is termed horseradish peroxidase (HRP) crosslinking, wherein HRP is used as a catalyst between the derivatives of aniline or phenol to form covalent crosslinks (Wang and Wang 2018).

3.4 Physical Crosslinking

3.4.1 Principle

Physical crosslinking is a process where polymer chains are linked together through non-covalent interactions rather than chemical bonds. This process creates a network structure that enhances the mechanical and thermal properties of the material without the need for chemical reactions. Physical crosslinking is often reversible and can be controlled by changing environmental conditions such as temperature, pH, or ionic strength.

3.4.2 Mechanisms of Physical Crosslinking

The various mechanisms of physical crosslinking are:

- **Hydrogen Bonding**: Hydrogen bonds are formed between polymer chains containing polar groups such as hydroxyl, carboxyl, or amine groups. These bonds are relatively weak and reversible, providing a means to form and break the crosslinked structure under different conditions. Polyvinyl alcohol (PVA) hydrogels are often crosslinked via hydrogen bonds formed between hydroxyl groups (Zhang et al. 2020a, b, c).
- **Ionic Interactions**: Ionic interactions occur between oppositely charged groups on polymer chains. These interactions are stronger than hydrogen bonds and provide more stable crosslinks. Alginate hydrogels are crosslinked using divalent cations such as calcium ions (Ca^2), which interact with the carboxyl groups on the alginate chains (Li et al. 2019).
- **Protein–Protein Interactions**: Proteins can be physically crosslinked through non-covalent interactions such as hydrogen bonding, electrostatic interactions, and van der Waals forces. These interactions are significant in natural biomaterials and engineered protein-based hydrogels. Gelatin hydrogels are formed through physical crosslinking of denatured collagen molecules via hydrogen bonds and hydrophobic interactions (Wang et al. 2021).
- **Hydrophobic Interactions**: Hydrophobic segments of polymer chains aggregate in an aqueous environment, leading to physical crosslinking. This mechanism is driven by the exclusion of water molecules from hydrophobic regions. Pluronic (PEO-PPO-PEO) block copolymers form hydrogels through the self-assembly of hydrophobic PPO blocks in water (Aida et al. 2020 and Cui et al. 2021).
- **Crystallization**: Crystalline regions within a polymer can act as physical crosslink points. These regions are formed when polymer chains arrange themselves in a regular, repeating pattern, providing strength and stability to the network. Polyethylene oxide (PEO) can form physically crosslinked networks through crystallization of PEO segments (Hsiao and Pochan 2023).

3.4.3 Selection of Polymer

The criterion for selection of polymer by physical cross linking includes hydrophilicity, mechanical strength, chemical stability, biocompatibility, non-toxicity and cost effectiveness. Some of the prominent polymers generated by physical cross linking include.

- **Poly (vinyl alcohol) (PVA)**: PVA is produced through hydrogen bonding and crystallization. It is highly biocompatible and hydrophilic in nature, used in membrane filter applications to separate inorganic and organic contaminants (Hua et al. 2017).
- **Poly (ethylene glycol) (PEG)**: Produced by physical crosslinking through hydrogen bonding and entanglement, PEG is nontoxic and hydrophilic in nature. It is used as a membrane filter for separating organic pollutants and heavy metals (Hua et al. 2017).
- **Chitosan**: It is produced through physical crosslinking using Ionic interaction and hydrogen bonding. It is a natural polymer, which is biodegradable and biocompatible. It is particularly used for removing heavy metals, microbes and dyes from water (Zhang et al. 2018a, b).
- **Agarose**: It is produced via. Hydrogen bonding with the gelation technique of physical crosslinking for purifying water. They form biocompatible gels with excellent mechanical properties (Zhang et al. 2018a, b).
- **Alginate**: They form gels through Ionic interaction with divalent cations. It is a natural polymer and used in membranes and beads for separating organic and heavy metal pollutants (Zhang et al. 2018a, b).

3.4.4 Types of Physical Crosslinking

The different types of physical crosslinking are:

- **Physical Adsorption**: Polymers can crosslink through physical adsorption of small molecules or ions onto the polymer chains. This can enhance mechanical properties and stability (Smith et al. 2022).
- **Reversible Covalent Bonds**: Certain reversible reactions, such as disulfide bonds (found in some biomedical polymers), can act as physical crosslinks. These bonds can break and reform under certain conditions, allowing for reprocessing or remodeling of the material (Zhang et al. 2020a, b, c).
- **Self-Assembly**: Block copolymers and certain supramolecular polymers can self-assemble into ordered structures, creating physical crosslinks between different blocks or segments of the polymer chains (Lehn et al. 2021).

These physical crosslinking methods are advantageous because they often allow for easier processing, recycling, or reshaping of materials compared to chemical crosslinking methods. They also offer tunable properties based on environmental conditions like temperature or pH.

3.4.5 Crosslinking Methods

Physical crosslinking methods are essential in the development of hydrogels, which are widely used in biomedical applications due to their tunable properties and biocompatibility. Recent studies have focused on improving the mechanical strength, stability, and functionality of these hydrogels through various physical crosslinking techniques.

- **Freeze–Thaw Cycling**: This method involves repeatedly freezing and thawing a hydrogel solution. During freezing, water crystallizes, leading to the aggregation of polymer chains into microcrystalline regions connected by physical entanglements and hydrogen bonds. Repeated cycles enhance the crystallinity and mechanical properties of the hydrogel. This technique is widely used in polyvinyl alcohol (PVA) hydrogels for applications such as wound dressings and flexible sensors (Zhong et al. 2024).
- **Salting-Out**: In this method, salts are added to the hydrogel solution to reduce the solubility of the polymer, causing it to precipitate and form a network structure through hydrophobic interactions. This technique is used to create hydrogels with specific mechanical and swelling properties suitable for drug delivery systems (Li et al. 2024).
- **Thermosensitive Hydrogels**: These hydrogels undergo sol–gel transition in response to temperature changes. For example, injectable thermosensitive hydrogels are used for localized drug delivery and tissue engineering. The physical crosslinking in these hydrogels is driven by temperature-induced phase separation, leading to the formation of a stable gel network at body temperature (Zhang et al. 2020a, b, c).
- **Hydrophobic and Electrostatic Interactions**: Hydrogels can be crosslinked through hydrophobic interactions or electrostatic interactions between charged polymer chains. These methods create soft, flexible, and self-healing hydrogels that are ideal for applications like electronic skin and drug delivery (Paiva et al. 2024).
- **Nanocomposite Hydrogels**: Incorporating nanoparticles into hydrogels can enhance their mechanical properties through physical interactions between the polymer matrix and the nanoparticles. This approach is used to develop hydrogels with improved toughness and functionality for various biomedical applications (Li et al. 2024).

Recent research highlights the combination of these physical crosslinking methods to create hydrogels with tailored properties for specific applications. For instance, combining freeze–thaw cycling with the addition of nanoparticles can produce hydrogels with enhanced mechanical strength and bioactivity, making them suitable for

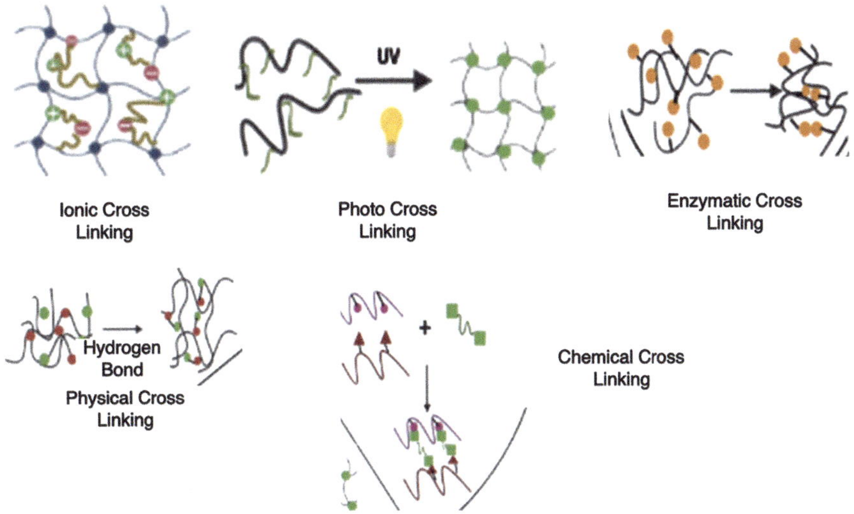

Fig. 3.4 Various forms of cross linking

complex medical applications such as tissue engineering and drug delivery (Paiva et al. 2024).

Schematic representation for various types of cross linking is displayed in Fig. 3.4.

3.5 Hydrothermal Synthesis

3.5.1 Principle

Hydrothermal synthesis is a method for crystallizing substances from high-temperature aqueous solutions at high vapor pressures. This process is essential for the formation of various advanced materials, including nanomaterials, zeolites, and single crystals. The technique is inspired by natural geological processes, simulating conditions found deep within the Earth's crust (Zhang et al. 2020a, b, c). The key Principles involved are:

- **High-Temperature and High-Pressure Environment**:

Temperature: Typically ranges from 100°C to 1000°C, depending on the substance being synthesized.

Pressure: Correspondingly high pressures, often several hundred megapascals (MPa), are required to maintain water in the liquid phase at these elevated temperatures (Corma et al. 2017a, b).

- **Aqueous Solution**: Water acts as a solvent, sometimes as a reactant, facilitating the dissolution of reactants and the subsequent nucleation and growth of crystals.

Autoclave: The reaction is carried out in a sealed vessel known as an autoclave, which can withstand the high pressure and temperature necessary for the reaction (Li et al. 2018a, b).

- **Crystallization**: The controlled environment allows for the slow and uniform growth of crystals. Parameters such as temperature, pressure, pH, and concentration of reactants can be precisely controlled to tailor the properties of the synthesized material (Li et al. 2018a, b).
- **Super saturation and Nucleation**: Super saturation is achieved by changing temperature or pressure, leading to nucleation. The rate of nucleation and growth is influenced by the concentration of the solute, temperature, and the presence of impurities or catalysts (Kumar et al. 2021).

The Mechanism involves dissolution of reactants in the aqueous phase, transportation of the dissolved species, nucleation and crystal growth.

3.5.2 Selection of Polymers and Precursors

Hydrothermal synthesis is a versatile technique used to produce a wide range of materials, including nanomaterials, composites, and biomaterials. The selection of polymers and precursors is crucial for tailoring the properties of the synthesized materials. The key considerations for the selection of polymers and precursors for hydrothermal synthesis include chemical stability, solubility, functional groups and decomposition temperature (Liu et al. 2021).

Commonly used polymers and precursors for hydrothermal synthesis include:

- **Polymers**:

  1. **Polyvinyl Alcohol (PVA)**: Used for its excellent solubility in water and ability to form strong hydrogen bonds, making it suitable for hydrothermal synthesis of composite materials (Liu et al. 2021).
  2. **Polyethylene Glycol (PEG)**: Acts as a stabilizer and template in the synthesis of nanomaterials due to its solubility and ability to control particle size (Wang et al. 2020).
  3. **Polydopamine (PDA)**: Known for its adhesive properties and ability to form uniform coatings, making it useful in the synthesis of functional nanomaterials (Liu et al. 2019).

- **Precursors**

  1. **Metal Salts**: Commonly used metal salts include nitrates, sulfates, and chlorides of metals like zinc, iron, and cobalt. These salts dissolve in water and

react to form metal oxides or hydroxides under hydrothermal conditions (Li et al. 2018a, b).

2. **Silica Precursors**: Tetraethyl orthosilicate (TEOS) is widely used in the synthesis of silica-based materials and zeolites due to its reactivity and ability to form a wide range of silicate structures (Corma et al. 2017a, b).
3. **Organic Precursors**: Organic compounds such as urea and citric acid are used as reducing agents or structure-directing agents in the hydrothermal synthesis of carbon-based materials and metal–organic frameworks (MOFs) (Wang and Wang 2021).

3.5.3 Hydrothermal Crosslinking Mechanism

Hydrothermal crosslinking is a process that involves the formation of covalent bonds between polymer chains under hydrothermal conditions (high temperature and pressure in the presence of water). This method is particularly useful for creating robust, stable materials with enhanced mechanical and thermal properties. The mechanism typically involves several stages, including polymer dissolution, chain mobility enhancement, crosslinking reaction, and subsequent formation of a three-dimensional network structure (Kumar et al. 2021). A schematic representation of the hydrothermal crosslinking mechanism is shown in Fig. 3.5.

The Key Steps in Hydrothermal Crosslinking are the following (Zhou and Zhang 2020)

1. **Polymer Dissolution and Swelling**: Under hydrothermal conditions, polymers can swell and partially dissolve in water, increasing the mobility of polymer chains and making reactive sites more accessible for crosslinking.
2. **Activation of Functional Groups**: High temperature and pressure can activate functional groups on polymer chains or incorporated crosslinking agents, facilitating the formation of covalent bonds.
3. **Crosslinking Reaction**: Crosslinking can occur through various mechanisms, such as condensation, addition, or radical reactions, depending on the types of

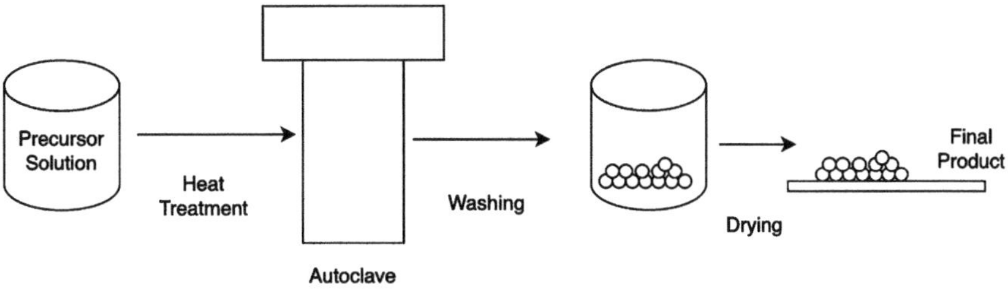

Fig. 3.5 Hydrothermal crosslinking mechanism

functional groups present. For instance, hydroxyl groups can undergo dehydration to form either bonds, or carboxylic groups can react with amines to form amide bonds.

4. **Network Formation**: As the crosslinking proceeds, a three-dimensional network structure forms, enhancing the material's mechanical strength, thermal stability, and chemical resistance.

3.6 Electrospinning

3.6.1 Overview

Electrospinning is a versatile and efficient technique used to produce nanofibers from a variety of polymers. The process involves the application of a high-voltage electric field to a polymer solution or melt, which results in the formation of thin fibers with diameters ranging from nanometers to micrometers. These nanofibers have numerous applications in areas such as biomedical engineering, filtration, textiles, and energy storage (Bhardwaj and Kundu 2010). The schematic representation of the electrospinning technique is displayed in Fig. 3.6.

Process of Electrospinning (Sundararaghavan et al. 2011).

- **Polymer Solution Preparation**: A polymer is dissolved in a suitable solvent to create a viscous solution.
- **High Voltage Application**: The polymer solution is loaded into a syringe equipped with a needle. A high voltage (typically in the range of 10–30 kV) is applied between the needle and a grounded collector.
- **Jet Formation and Fiber Spinning**: The electric field causes the polymer solution at the needle tip to form a Taylor cone. When the electrostatic force overcomes the

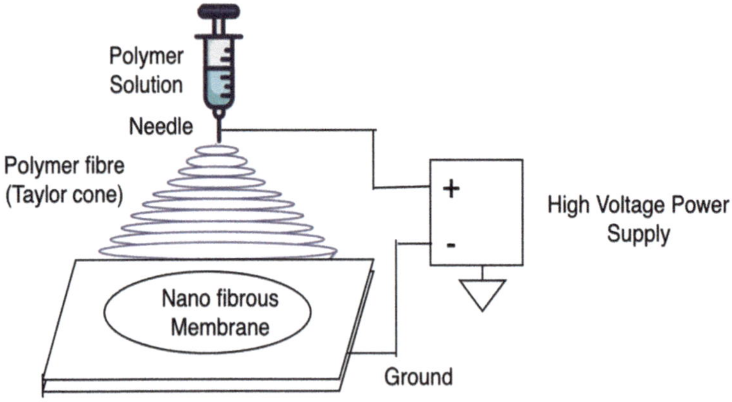

Fig. 3.6 Schematic diagram of the electrospinning process (Modified from Ikada 2006)

surface tension, a thin jet of the solution is ejected. The jet undergoes stretching and solvent evaporation, leading to the formation of solid nanofibers.
- **Fiber Deposition**: The nanofibers are collected on a grounded surface, such as a rotating drum or a flat plate. Electrospun nanofibers have high surface area-to-volume ratios, making them ideal for applications in air and water filtration.

3.6.2 Selection of Polymers

When selecting polymers for electrospinning, several factors must be considered, including the polymer's solubility, molecular weight, and the desired properties of the resulting nanofibers. Here are some commonly used polymers for electrospinning, along with references to relevant studies:

- **Polyvinyl Alcohol (PVA)**: Hydrophilic, biocompatible and biodegradable in nature, used for separating heavy metals, organic dyes and other contaminants from water (Sui et al. 2010).
- **Polyacrylonitrile (PAN)**: Highly resistant to chemicals with good mechanical strength and functionality. Used for adsorbing heavy metals, pathogens and organic compounds (Ma et al. 2011).
- **Polysulfone (PSU)**: It is specifically used for ultra and nanofiltration membranes for separating a wide variety of contaminants. It is known to be chemically and thermally stable with good mechanical strength (Mbareck et al. 2012).
- **Polyethersulfone (PES)**: Used for ultra and nanofiltration for removing organic compounds, microorganisms and heavy metals. They are highly resistant to chemicals, with good mechanical and thermal properties (Rajabzadeh et al. 2013).
- **Cellulose Acetate (CA)**: Specifically used for separating bacteria, heavy metals and organic pollutants from water. They are hydrophilic and biodegradable with good film-forming ability (Ma et al. 2011).

3.6.3 Preparing Electrospinning Solution

Preparing an electrospinning solution involves dissolving a polymer in a suitable solvent to create a homogenous mixture that can be electrospun into nanofibers.

Steps to Prepare an Electrospinning Solution

- **Selection of Polymer and Solvent**: Choose a polymer that can form nanofibers, such as polyvinyl alcohol (PVA), polycaprolactone (PCL), poly (lactic-co-glycolic acid) (PLGA), or cellulose acetate. Select a solvent that can dissolve the polymer completely. Common solvents include water, ethanol, dimethylformamide (DMF), and chloroform (Leila Tayebi et al. 2024).
- **Dissolution of Polymer**: Measure the desired amount of polymer and solvent based on the target concentration (usually ranging from 5 to 20% w/v, depending

on the polymer). Add the polymer to the solvent gradually while stirring continuously to avoid clumping.

- **Stirring and Heating**: Stir the mixture at room temperature or slightly elevated temperatures (if necessary) until the polymer is fully dissolved. This process can take several hours to overnight, depending on the polymer and solvent. Heating should be done carefully to avoid solvent evaporation or degradation of the polymer (Muqeet et al. 2019).
- **Filtration**: Filter the solution to remove any undissolved particles or impurities. This step ensures a homogeneous and smooth electrospinning process.
- **Viscosity and Conductivity Adjustment**: Adjust the viscosity and conductivity of the solution if necessary. These parameters can significantly affect the electrospinning process and the quality of the nanofibers. Additives like surfactants or salts can be used to modify these properties (Muqeet et al. 2019).

3.6.4 Crosslinking of Nanofibers

Crosslinking of nanofibers is an essential process in enhancing the properties and applications of nanofibrous materials. Recent studies highlight various innovative methods and applications for crosslinked nanofibers.

- **Photo Crosslinking for Food Packaging**: Recent advancements have explored the use of photo crosslinking for developing eco-friendly food packaging materials. For instance, nanofibers containing *Melissa officinalis* extract were photo crosslinked to create active food packaging that is an alternative to traditional plastic packaging. These nanofibers not only improve the mechanical properties but also impart antioxidant and antimicrobial properties, making them suitable for preserving food quality (Kim et al. 2005).
- **Ionic Crosslinking of Cellulose Nanofibers**: Another approach is ionic crosslinking, which significantly enhances the mechanical stability of cellulose nanofibers, making them suitable for dynamic adsorption applications. This method is particularly beneficial for creating materials with high mechanical strength and durability, which are essential for environmental and filtration applications (Golizadeh et al. 2019).
- **Siloxane-Based Crosslinking for Polymers**: Crosslinking lignin and cellulose with polymers using siloxane compounds has shown promising results. This method improves the compatibility of these materials, allowing for their use in various applications such as 3D printing and the development of sustainable composites. Siloxane-based crosslinking provides enhanced thermal and mechanical stability, making the materials more robust for industrial applications (Sundararaghavan et al. 2011).
- **Bioactive Packaging Materials**: Incorporating bioactive compounds into electrospun nanofibers is another innovative area. For example, nanofibers loaded with bioactive compounds like eugenol and encapsulated with chitosan/PEO have been developed for food packaging applications. These materials not only protect the

food from microbial contamination but also extend its shelf life (Zhang et al. 2005).

These recent advancements in the crosslinking of nanofibers demonstrate the potential for creating high-performance, sustainable materials for various industrial, environmental, and biomedical applications. The continued development and optimization of these methods are crucial for expanding the practical uses of nanofibrous materials.

3.7 In Situ Gelation

In situ gelation is a process where a liquid precursor transforms into a gel upon application to a specific site, typically within a biological system. This transformation can be triggered by various stimuli such as changes in temperature, pH, ionic strength, or the presence of specific ions or molecules. The step-by-step process of in situ gelation is displayed in Fig. 3.7.

3.7.1 Mechanisms

- **Temperature-Sensitive Systems**: These systems gel upon reaching body temperature. Polymers such as Pluronic® F127 and poly(N-isopropylacrylamide) (PNIPAAm) are commonly used (Liu and Zeng 2023).
- **pH-Sensitive Systems**: Gelling occurs in response to pH changes in the environment. Polymers like chitosan and poly (acrylic acid) are examples (Wang and Zhang 2023; Chen and Huang 2022).
- **Ionic Crosslinking**: Certain polymers gel in the presence of specific ions. For example, alginate gels in the presence of calcium ions (Zhao and Li 2023).
- **Enzymatic Reactions**: Enzymes present at the target site can induce gelation of specific polymer solutions (Lee and Park 2023).

3.7.2 Selection of Precursors

The selection of precursors for the in situ gelation process is important to ensure the rate of gel formation, stability for different wastewater conditions and contaminant affinity. The critical criteria involved in selection include; gelation time, stability, affinity to contaminants, biodegradability, non-toxicity and cost-effectiveness. Some of the prominent precursors for in situ gelation include chitosan, alginate, polyvinyl alcohol (PVA), polyacrylamide (PAM) and polyethylene glycol (PEG). The gelation process can be optimized by the selection of the right precursor, leading to efficient

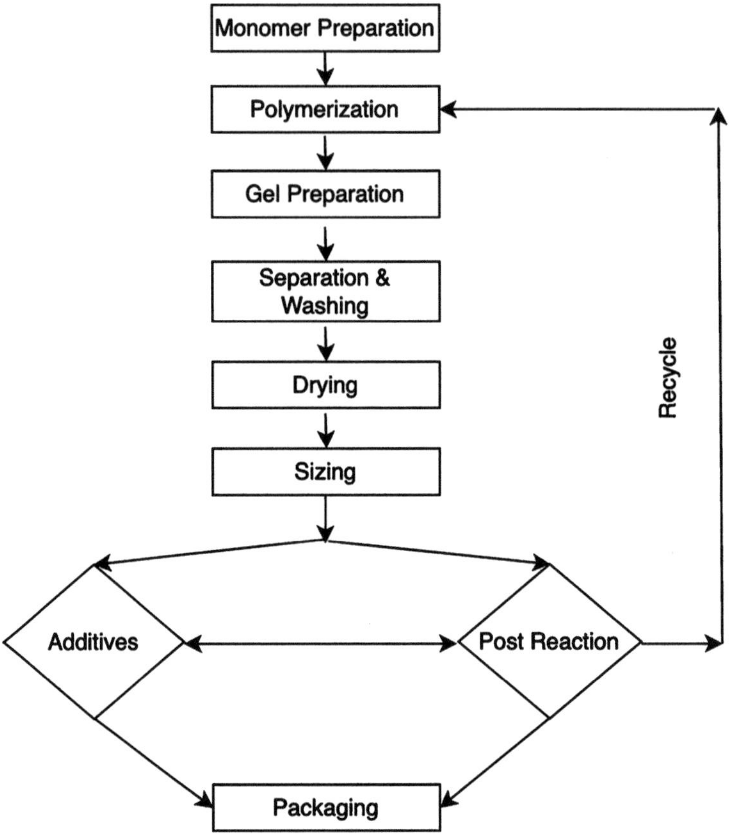

Fig. 3.7 Flowchart of in situ gelation process

wastewater treatment. The precursors used for in situ gelation processes are ideal for removing organic pollutants, heavy metals and dyes from wastewater (Zhang et al. 2022; Luo et al. 2021).

3.7.3 Gelation Parameters

The gelation parameters of in-situ forming hydrogels are critical factors that determine their performance and suitability for wastewater treatment applications. These parameters include gelation time, gel strength, swelling ratio, gelation temperature, crosslinking density and pH Sensitivity. Understanding and controlling these parameters allows for the design of hydrogels tailored to different wastewater treatment

Table 3.4 Key parameters involved in in situ gelation with importance, control methods and examples

Parameter	Importance	Control methods	Ideal hydrogel
Gelation time	Rapid gelation reduces the treatment time and prevents dispersion of contaminants	Adjusting pH, concentration of precursor and temperature	Chitosan-based hydrogels with glutaraldehyde crosslinkers
Gel strength	To withstand high shear force and flow of wastewater systems	Varying concentration and type of crosslinkers and molecular weight of polymers	Alginate gels crosslinked with calcium ions
Swelling ratio	To enhance the gels ability to capture and immobilize contaminants	Modifying the crosslinking density and polymer composition	Polyvinyl alcohol (PVA) exhibits high swelling ratios based on crosslinking degree
Crosslinking density	Influences mechanical properties and swelling nature of gel	Varying the type and concentration of crosslinking agents	Enhanced crosslinking density in polyacrylamide (PAM)
pH Sensitivity	pH of wastewater systems vary	Incorporate polymers and crosslinkers in response to the pH	Hydrogels prepared using acrylic acid are sensitive to pH
Temperature sensitivity	For application in environments with changing temperatures	Using thermosensitive polymers	Poly(N-isopropylacrylamide) (PNIPAAm) gels

applications (Tang and Hu 2021). Table 3.4 represents the key parameters with their importance, control methods and ideal hydrogels.

The methods discussed for hydrogel synthesis for water purification have distinct advantages and limitations regarding mechanical strength, stability, environmental impact, reusability, and scalability. Table 3.5 provides a comparative analysis of these methods.

Table 3.5 Comparison of different hydrogel synthesis techniques for water purification

Synthesis method	Mechanical strength	Stability and application	Environmental impact	Reusability	Scalability
Cross Linking (Physical/ Chemical)	Moderate to high Adsorption of heavy metals, dyes, and pollutants	High	Moderate	Moderate to High	High
Free radical polymerization	Moderate to high Selected Pollutant Removal	Moderate to High	Moderate	Moderate	High
Hydrothermal synthesis	High Nanocomposite hydrogels for heavy metals and dyes	High	High	High	Moderate to High
In situ gelation	Low to Moderate Smart hydrogels for real-time pollutant removal	Moderate	High	High	High

3.8 Conclusion

This chapter provides a comprehensive report on different hydrogel synthesis routes for wastewater treatment applications. Special emphasis was provided on the principles, methodologies and application of different routes. The various techniques discussed include free radical polymerization, ionic crosslinking, chemical crosslinking, physical crosslinking, hydrothermal synthesis, electrospinning and in-situ gelation. These diverse techniques were discussed with respect to their requirements, unique principles and applications. The optimum selection of polymers, crosslinking agents, initiators, catalysts and monomers is critical in hydrogel synthesis for wastewater treatment applications. Despite significant developments, challenges still remain for scaling up these methods to industrial applications with consistency in the properties of the hydrogel produced and environmental sustainability. Due to high water absorption capacity, capacity to tune porosity and ability to incorporate functional groups makes hydrogels are a promising material for water purification. The various synthesis techniques enable hydrogel development with tailored properties for ion exchange, adsorption and degradation. Development of nano-composite hydrogels and materials responsive to stimuli further enhances the efficiency by targeting organic pollutants, heavy metals, and microbial contaminants. Chemically crosslinked hydrogels display a high mechanical stability, whereas physically cross linked hydrogels offer eco-friendly and reusable properties.

Despite these advancements, scalability, cost-effectiveness and performance in a real-world environment remain a challenge. Future research focuses on optimizing hydrogel synthesis techniques for large-scale applications, improving biodegradability and developing multi-functional hydrogels to counter heavy metals, microbial pathogens and organic pollutants. Innovative hybrid solutions by integrating hydrogel technology with existing adsorption and filtering systems enhance efficiency and sustainability in water purification (Enas 2015).

References

Academic Press (2006) Ikada, Y. Tissue Engineering: fundamentals and applications. Oxford, United Kingdom. Academic Press

Ahmed EM (2015) Hydrogel: preparation, characterization, and applications: a review. J Adv Res 2015(6):105–121

Aida T, Meijer EW, Stupp SI (2020) Functional supramolecular polymers. Science 369(6509):eaav0634

Babu BV, Suresh Kumar SR, Anuradha G (2006) Adsorption of phenol by chitosan-based polymer. Chem Eng J 116(2):77–84

Bhardwaj N, Kundu SC (2010) Electrospinning: a fascinating fiber fabrication technique. Biotechnol Adv 28(3):325–347. https://doi.org/10.1016/j.biotechadv.2010.01.004

Bhosale RR, Kadam KS (2019) Carrageenan-based hydrogels for removal of dyes from aqueous solutions. Carbohyd Polym 207:351–361

Chen G, Zhao Y (2016) Role of TEMED in the free radical polymerization of hydrogels for environmental applications. Polym Chem 7(14):2397–2404

Chen JP, Chang KH, Wang MC (2004) Preparation and properties of polyvinyl alcohol–chitosan blend hydrogels for the adsorption of heavy metal ions. J Appl Polym Sci 91:2277–2284

Chen R, Huang Y (2022) PH-sensitive in-situ forming gels: mechanisms and applications in drug delivery. Mater Sci Eng C 134:112690. https://doi.org/10.1016/j.msec.2022.112690

Corma A et al (2017a) Hydrothermal synthesis of zeolites: mechanism and crystallization process. Chem Rev 118(12):6634–6658. https://doi.org/10.1021/acs.chemrev.7b00574

Corma A et al (2017b) Mechanisms of zeolite formation under hydrothermal conditions. Nat Rev Chem 1:0097. https://doi.org/10.1038/s41570-017-0097

Cui J, Yuan J, Zhang L (2021) Ionic interactions in polymer networks: recent advances and perspectives. Adv Mater 33(14):2005683

Damitteo R, Darling NJ, Segura T (2018) In situ forming injectable hydrogels for drug delivery and wound repair. Adv Drug Deliv Rev 127:167–184

Dil NN, Sadeghi M (2018) Free radical synthesis of nanosilver/gelatin-poly (acrylic acid) nanocomposite hydrogels employed for antibacterial activity and removal of Cu (II) metal ions. J Hazard Mater 351:38–53

Enas MA (2015) Hydrogel: preparation, characterization, and applications: a review. J Adv Res 6:105–121

Gao J, Cheng Y, Shi Z, Wang X (2019) Acrylic acid-based hydrogels with enhanced adsorption capacity for heavy metal ions and dyes. J Appl Polym Sci 136(20):47589

Ge SH, Liu HH, Chen JR (2018) lginate-based hydrogels: a review of their applications in wastewater treatment. Water Res 129:52–66

Gelsinger C, Miao JH (2010) Hydrogels in regenerative medicine. Mater Sci Eng R Rep 70(3–6):92–117

Golizadeh M et al (2019) Evaluation of cellular attachment and proliferation on different surface charged functional cellulose electrospun nanofibers. Carbohyd Polym 207:796–805. https://doi.org/10.1016/j.carbpol.2018.12.028
Hsiao BS, Pochan DJ (2023) Crystallization and physical crosslinking in polymers. Annu Rev Phys Chem 74:63–85
Hua K, Zhang Y, Zhang Z, Du Y (2017) Physical crosslinking of hydrogels: principles and applications. Prog Polym Sci 73:60–94
Huang Y, Li X (2018) Effect of sodium metabisulfite on the polymerization of hydrogels for wastewater treatment. J Environ Chem Eng 6(4):4572–4579
In Tech, Gulrez Syed KH, Saphwan A-A, Phillips Glyn O (2011) Hydrogels: methods of preparation, characterisation and applications, progress in molecular and environmental bioengineering–from analysis and modeling to technology applications, edited by prof. Angelo Carpi Tech 2011:117–150
Kim JH et al (2005) Electrospinning of poly(vinyl alcohol) mats with silver nanoparticles: optical properties and antibacterial activity. Macromol Res 13(5):467–473. https://doi.org/10.1007/BF03219034
Kim SJ, Park SJ, Kim SI (2003) Swelling behavior of interpenetrating polymer network hydrogels composed of poly (vinyl alcohol) and chitosan. React Funct Polym 55:53–59
Kumar S et al (2021) Hydrothermal synthesis of bioactive hydroxyapatite for bone tissue engineering. J Biomed Mater Res B Appl Biomater 109(3):336–347. https://doi.org/10.1002/jbm.b.34677
Kumarasamy E, Manning IM, Collins LB, Coronell O, Leibfarth FA (2020) Ionic fluorogels for remediation of per- and Polyfluorinated Alkyl substances from water. ACS Cent Sci 6:487−492
Lee Y, Park K (2023) Enzyme-responsive hydrogels for drug delivery applications. Prog Polym Sci 136:101575. https://doi.org/10.1016/j.progpolymsci.2023.101575
Lehn JM, Sauvage JP, Stoddart JF (2021) Supramolecular polymers: from molecular information to shape-controlled materials. Nat Rev Chem 5(4):326–341
Leila T, Mahboubi A, Bayat F, Moayeri-Jolandan S, Haeri A (2024) Photo-cross-linked nanofbers containing Melissa OFCINALIS extract as a novel active food packaging: an eco-friendly alternative for plastic packaging. J Polymers Environ 32:4385–4404. https://doi.org/10.1007/s10924-024-03209-5
Li D et al (2018a) Hydrothermal synthesis of ZnO nanorods using zinc nitrate and hexamethylenetetramine. J Alloy Compd 750:220–225. https://doi.org/10.1016/j.jallcom.2018.04.130
Li J et al (2019) Ionic and hydrogen bonding crosslinked polysaccharide-based hydrogels for biomedical applications. Carbohyd Polym 206:29–45. https://doi.org/10.1016/j.carbpol.2018.10.100
Li L, Chen S, Zhang Y, Deng H (2014) Water treatment by new-generation polymeric hydrogels: preparation, characterization, and applications. Prog Polym Sci 39(7):1230–1266
Li S, Sun J (2017) Use of ferrous sulfate in hydrogel polymerization for wastewater treatment. J Hazard Mater 330:47–55
Li X et al (2018b) Hydrothermal synthesis of high-quality single crystals for optical applications. Mater Today Phys 5:21–31. https://doi.org/10.1016/j.mtphys.2018.05.001
Li X, Li Y, Zhang X et al (2024) Cross-linking methods of the silk protein hydrogel in oral and Craniomaxillofacial tissue regeneration. Tissue Eng Regen Med 21:529–544. https://doi.org/10.1007/s13770-023-00624-y
Li X, Xu Z, Zhao J (2017) Ammonium persulfate-initiated polymerization for hydrogel synthesis in wastewater treatment applications. J Polym Sci Part A: Polym Chem 55(14):2239–2248
Li Y, Zhao J (2018) Poly(acrylamide)-based hydrogels for efficient heavy metal removal. J Environ Manage 228:395–403
Liu C, Zeng W (2023) Development of thermosensitive in-situ forming gels for local drug delivery. J Control Release 342:25–36. https://doi.org/10.1016/j.jconrel.2023.01.045
Liu X et al (2019) Polydopamine-coated nanomaterials for biomedical applications. Chem Soc Rev 48(10):2887–2904. https://doi.org/10.1039/C8CS00783K

Liu Y et al (2021) Hydrothermal synthesis of PVA-based nanocomposites for biomedical applications. J Biomed Mater Res Part A 109(6):1234–1245. https://doi.org/10.1002/jbm.a.37029

Liu Z, Ge H, Ma R, Wang Y (2017) NVP-based hydrogels for efficient removal of heavy metals and dyes from wastewater. J Environ Manage 202:244–252

Luo X, Chen H, Li Y (2021) Functional hydrogels for wastewater treatment: FABRICATION, application, and future perspectives. Environ Sci Technol 55(21):14490–14507. https://doi.org/10.1021/acs.est.1c03906

Ma H, Burger C, Hsiao BS, Chu B (2011) Ultra-fine cellulose nanofibers: new nano-scale materials for water purification. J Mater Chem 21(21):7507–7510

Mbareck C, Riedl B, Rodrigue D (2012) Electrospun nanofibrous membranes based on polysulfone-functionalized carbon nanotubes for water filtration. ACS Appl Mater Interfaces 4(2):1110–1117

Muqeet M, Qureshi UA, Mahar RB et al (2019) Ionic cross-linking of cellulose nanofibers: an approach to enhance mechanical stability for dynamic adsorption. Environ Sci Pollut Res 26:28842–28851. https://doi.org/10.1007/s11356-019-06076-z

Paiva MTP, Kishima JOF, Silva JBMD et al (2024) Crosslinking methods in polysaccharide-based hydrogels for drug delivery systems. Biomed Mater Dev 2:288–306. https://doi.org/10.1007/s44174-023-00118-4

Rajabzadeh S, Maruyama T, Sotani T, Matsuyama H (2013) Preparation of high performance PES hollow fiber UF membrane via thermally induced phase separation (TIPS) method. J Membr Sci 432:50–59

Rao RV, Gosavi SN, Gadkari SR (2016) Polyacrylamide-based hydrogels for water treatment applications. J Water Process Eng 10:1–10

Samaddar P, Kumar S, Kim K-H (2019) Polym Rev 2019(59):418

Sepehri A, Sarrafzadeh MH (2018) Effect of nitrifiers community on fouling mitigation and nitrification efficiency in a membrane bioreactor. Chem Eng Process Process Intensif 128:10–18

Sharma P, Tiwari S (2020) HEMA-based hydrogels for environmental remediation: an overview. J Polym Res 27(11):295

Sinha V, Chakma S (2019) Advances in the preparation of hydrogel for wastewater treatment: a concise review. J Environ Chem Eng 7(5):1–63

Smith AB, Jones CD, Wang X (2022) Physical adsorption in polymer materials: fundamentals and applications. J Polym Sci Part B: Polym Phys 60(7):721–736

Song M, Wang J, He J, Kan D, Chen K, Lu J (2023) Synthesis of hydrogels and their progress in environmental remediation and antimicrobial application. J Gels 9(16):1–24

Sui X, Shao C, Liu Y, Li X, Wang H (2010) Electrospun PVA/TEOS hybrid nanofibrous membranes for nanofiltration. J Membr Sci 349(1–2):267–274

Sun J, Jin Z, Wang J, Wang H, Zhang Q, Gao H, Jin Z, Zhang J, Wang Z (2023) Application of ionic liquid crosslinked hydrogel for removing heavy metal ions from water: different concentration ranges with different adsorption mechanisms. J Polymers 15(13):1–16

Sundararaghavan HG et al (2011) Biomimetic hydroxyapatite-containing composite nanofibers for bone tissue engineering. J Biomed Mater Res Part A 96(2):268–275. https://doi.org/10.1002/jbm.a.32989

Tang S, Hu J (2021) Fast in situ gelation of chitosan hydrogels via catechol-modified chitosan and an oxidizing agent. Carbohyd Polym 253:117179. https://doi.org/10.1016/j.carbpol.2020.117179

Van T, Vinh P, Duckshin L, Chul Y (2018) Hydrogel applications for adsorption of contaminants in water and wastewater treatment. J Environ Sci Pollut Res 25(25):24569–24599

Wang J, Wang S (2018) Preparation of biopolymer-based hydrogels for water treatment and their applications. Carbohyd Polym 195:71–85

Wang S, Wang Y (2021) Hydrothermal synthesis of MOFs using organic precursors. Coord Chem Rev 434:213764. https://doi.org/10.1016/j.ccr.2020.213764

Wang S, Zhang Q (2023) PH-responsive hydrogels for controlled drug delivery: recent advances and future perspectives. Carbohyd Polym 291:119523. https://doi.org/10.1016/j.carbpol.2023.119523

Wang X et al (2020) PEG-directed synthesis of metal oxide nanoparticles with controlled size and morphology. Adv Mater 32(15):1907061. https://doi.org/10.1002/adma.201907061
Wang Y et al (2021) Protein-based physical hydrogels for biomedical applications. Mater Sci Eng C 120:111789. https://doi.org/10.1016/j.msec.2020.111789
Wang Y, Wang W, Pan J (2018) Potassium persulfate as an initiator in the synthesis of hydrogels for wastewater treatment. Polym Bull 75(2):555–567
Weerasundara L, Bartolo G, Alberto F, Sik OY, Jochen B (2021) Hydrogels: novel materials for contaminant removal in water—a review. J Critical Rev Environ Sci Technol 51(17):1970–2014
Wiley-Scrivener (2018) Thakur VK, Thakur MK (Eds) Handbook of polymers for pharmaceutical technologies: biodegradable polymers
Wu T, Li H, Wu Z, Ding L (2018) Acrylamide-based hydrogel for efficient dye removal from wastewater. Colloids Surf A 553:128–135
Xue S, Ye Z, Tang Q, Wang, Yu (2021) A facile route to dual-crosslinking polymeric hydrogels with enhanced mechanical property. J Polymer Eng 41(8):690–694
Yadav M, Rhee KY (2018) MBAA crosslinked hydrogels for the removal of heavy metals from wastewater. J Ind Eng Chem 61:74–83
Yuan W, Liu X, Zhang J (2019) Zinc chloride crosslinked hydrogels for adsorption of heavy metals from wastewater. Colloids Surf B 182:110366
Zhang H, Liu Q, Zhang Y (2022) Recent advancements in hydrogel-based materials for wastewater treatment. J Environ Chem Eng 10(4):108152
Zhang H, Wang X, Wang X, Chang X, Wang Z (2018a) Biopolymer-based hydrogels in water treatment: an overview. Environ Sci Technol 52(9):5218–5230
Zhang H, Zhao Y, Zhang S (2018b) Preparation of MAA hydrogels for adsorption of organic pollutants from wastewater. Polym Bull 75(6):2229–2241
Zhang H, Huang X, Jiang J, Shang S, Song Z (2017) Hydrogels with high mechanical strength cross-linked by a rosin-based crosslinking agent. J RSC Adv 7(61):42541–42548
Zhang J et al (2019) Nickel cobalt oxide nanostructures for high-performance super capacitors: hydrothermal synthesis and electrochemical properties. J Power Sources 414:490–499. https://doi.org/10.1016/j.jpowsour.2018.11.033
Zhang L, Li Y (2019) Ascorbic acid as a co-initiator for hydrogel synthesis in wastewater treatment applications. Polym Degrad Stab 165:70–78
Zhang L, Liu M (2018) Poly(vinyl alcohol) hydrogels for wastewater treatment: a review. J Environ Chem Eng 6(3):3884–3893
Zhang X et al (2020a) Hydrothermal synthesis and applications of functional nanomaterials: a review. J Nanomater. https://doi.org/10.1155/2020/1234567
Zhang X et al (2020b) Temperature-responsive physical hydrogels for biomedical applications. Adv Drug Deliv Rev 160:78–100. https://doi.org/10.1016/j.addr.2020.09.014
Zhang Y et al (2005) Electrospun nanofibrous scaffold for vascular tissue engineering. Biomaterials 26(31):6176–6186. https://doi.org/10.1016/j.biomaterials.2005.03.001
Zhang Y, Huang Y, Yang J (2020c) Reversible covalent bonds in polymer networks: a comprehensive review. Chem Rev 120(23):13044–13122
Zhao X, Li Y (2023) Ionic cross-linked hydrogels for biomedical applications. Polym Rev 63(2):153–179. https://doi.org/10.1080/15583724.2023.2174579
Zhao Y, Wang Y (2017) Ferric chloride crosslinked hydrogels for efficient dye removal. J Appl Polym Sci 134(37):45267
Zhong Y, Lin Q, Yu H, Shao L, Cui X, Pang Q, Zhu Y, Hou R (2024) Construction methods and biomedical applications of PVA based hydrogels. Front Chem 12:1376799. https://doi.org/10.3389/fchem.2024.1376799
Zhou J, Zhang L (2020) Hydrothermal crosslinking of cellulose-based materials for advanced applications. J Mater Chem A 8(15):7413–7422. https://doi.org/10.1039/D0TA00953G
Zia KM, Zuber M, Ali M, Mujahid M (2015) Alginate-based polyurethanes: a review of recent advances and perspectives. Int J Biol Macromol 79:377–387

Chapter 4
Carbon Nanostructures Doped Hydrogels for Wastewater Treatment

Souravi Bardhan and Dipak Kr Chanda

Abstract The increasing requirement for ever newer and more efficient materials used to cleanse highly polluted waters brought carbon nanostructure-doped hydrogels into prominence as new remediation tools. Hydrogel matrices contain dispersed carbon nanostructures like carbon nanotubes, carbon dots, graphene, and fullerene to equip and improve their physicochemical properties. Hybrid materials have shown great adsorption capacities, surface areas, good mechanical strength, and enhanced chemical stability, and hence, they are suitable for the removal of various heavy metals, dyes, and organic contaminants from wastewater. This chapter outlines a thorough overview of properties and synthesis routes of carbon nanostructures and hydrogels, followed by a detailed discussion of the fabrication methods of carbon nanostructure-doped hydrogels. Improved contaminant removal efficiency will be evaluated based on recent advances and real-world opportunities in wastewater treatment. The chapter will also discuss the challenges that need to be overcome for scalable adoption, regeneration, environmental impact, and cost-effectiveness.

Keywords Carbon nanostructures · Hydrogels · Wastewater treatment · Adsorption · Carbon nanotubes · Graphene

Abbreviations

0D	Zero-dimensional
1D	One-dimensional

S. Bardhan
Department of Physics, Jadavpur University, Kolkata, India

Department of Environmental Science, Netaji Nagar College for Women, University of Calcutta, Kolkata, India

D. K. Chanda (✉)
Graphene Division and Head, Ferro Refratech, Adityapur Industrial Area, Jamshedpur, India
e-mail: dipak58c@gmail.com

S. Bardhan and M. P. Shah (eds.), *Applications of Hydrogels in Modern Wastewater Treatment*, Advances in Wastewater Research,
https://doi.org/10.1007/978-981-95-1598-1_4

2D	Two-dimensional
3D	Three-dimensional
BR	Bengal Rose
CC	Chromium complex
CDs	Carbon dots
CEL	Cellulose
CNTs	Carbon nanotubes
CQD	Carbon quantum dots
CS	Chitosan
g-C-EN-GO	GO with ethylenediamine and grafted with acrylamide
GNP	Graphene nanoplatelets
GNR	Graphene nanoribbons
GO	Graphene oxide
GQD	Graphene quantum dots
HGB	Hydrogel Blank (Control)
HGNT	Gelatin-coated MWCNT Hybrid Hydrogel
L-cys	L-cysteine
MB	Methylene blue
MWCNT	Multi-walled carbon nanotubes
NSAIDs	Nonsteroidal anti-inflammatory drugs
PANI	Polyaniline
PPy	Polypyrrole
PVA	Polyvinyl alcohol
RD	Rhodamine B
rGO	Reduced graphene oxide
ROS	Reactive oxygen species
SWCNT	Single-walled carbon nanotubes
UV	Ultra violet

4.1 Introduction

Rapid industrialization and urbanization have accelerated the pollution of water sources by industrial effluent, agricultural runoff, and emerging pollutants that threaten environmental sustainability and human health. Scenario differs when it comes to dealing with coagulation, filtration, or biological treatment because these techniques are ineffective when it comes to filtering out heavy metals, organic pollutants, dyes, and drugs (Saravanan et al. 2021). Scientists have been developing next-generation enhanced quality materials and advanced filters with higher adsorption, catalytic, and filtration abilities for tackling the ordeals. One of the candidates is hydrogels, which are 3D polymeric networks capable of holding much water and with different chemical properties, and can adsorb a wide range of complicated pollutants and wash them out (Hwang et al. 2021). But these conventional hydrogels have their shortcomings: low mechanical strength, poor adsorbing ability, and slower response speed, which places a barrier between them and large-scale application (Jayakumar et al. 2020). To overcome these limitations, embedding carbon nanostructures into hydrogel networks has emerged as a promising method to increase hydrogel's efficiency in wastewater treatment. These carbon nanostructures comprise graphene, carbon dots (CDs), carbon nanotubes (CNTs), and fullerenes, each with remarkable physicochemical properties such as a large surface area, superior electrical conductivity, and controlled surface functionalities (Georgakilas et al. 2015). The addition of these nanomaterials to hydrogel networks increases the mechanical strength, adsorption capacity, and catalytic activity of the hydrogel system. For example, graphene hydrogels have greater organic pollutant adsorption capacity through strong π-π interactions, whereas hydrogels containing CNTs have greater mechanical hardness and pollutant binding capacity due to their extremely long structure and functionalizable surface groups (Dong et al. 2018). CDs, in contrast, introduce photoreactive functions to the hydrogels, which can degrade pollutants upon light irradiation (Jiang et al. 2024). The interaction of hydrogels and carbon nanostructures provides several ways of removal, such as electrostatic interaction, hydrogen bonding, ion exchange, and size exclusion. In addition, these hybrid materials can be designed for the multifunctional application of photocatalytic degradation, electrochemical sensing, and controlled release of pollutants. Hence, carbon nanostructure-doped hydrogels remain a good prospect for sustainable and efficient wastewater treatment methods.

The last few years have also seen some material engineering advances that hold the promise of hybrid hydrogels in tackling some of the urgent environmental challenges. Hybrid polymer matrices of CNT and CQD composites have also been found to possess enhanced UV photocatalytic degradation and dye removal (Phang and Tan 2019). Hydrogels take advantage of their highly porous nature, which enables them to adsorb water and contaminants. The conductive and reactivity properties of carbon nanostructures facilitate easy charge transfer and catalysis (Zhai et al. 2015). The synergy leads to enhanced removal efficiency of a vast array of contaminants, thus making the feasibility of such materials for bulk wastewater removal treatments. There are, nonetheless, some issues with the optimization of synthesis processes, cost-effectiveness, and environmental sustainability over long periods of the carbon nanostructure-doped hydrogels. The reproducibility and scale-up of materials are influenced by the heterogeneity of feedstock, fabrication processes, and post-fabrication modifications. Furthermore, a thorough understanding of the environmental fate and potential toxicity of carbon-based nanomaterials in the aquatic ecosystem is imperative for their safe deployment (Jatoi et al. 2025). Overcoming these challenges will require transdisciplinary collaborations to realize optimal hydrogel-nanostructure interactions, develop standardized manufacturing protocols, and integrate these materials into existing wastewater treatment plants. This chapter deals with the synthesis, characterization, and application of carbon nanostructure-doped hydrogels in wastewater treatment. The merits and limitations of these materials are discussed, along with future perspectives. Using nanomaterial research with the flexibility afforded by the hydrogels, cleaner water is feasible with these hybrids.

4.2 Carbon Nanostructure-Doped Hydrogels

4.2.1 Carbon Nanostructure Types

Carbon, a nonmetal, forms the framework of organic chemistry and is crucial to the chemistry of life. The unique tetravalency of carbon allows the formation of four covalent bonds, thereby facilitating the creation of multiple structures, including highly complex rings and very long chains via catenation. It is this versatile bonding that brings carbon into more than 95% of all known chemical compounds. Diamonds, graphite, and fullerenes are all allotropes of carbon that exhibit unique chemical and physical characteristics. The hardness and optical clarity of diamonds are well known, while that, very well understood, is for the excellent electrical conductivity of graphite. Due to the versatility of bond formations with other elements, carbon forms giant groups of compounds industrially as well as biologically important (Slepičková Kasálková et al. 2021). This showed the properties of carbon, which formed nanostructures that have since piqued the curiosity of materials scientists to grow. Carbon has carbon nanostructures such as graphene and graphene oxide (GO), carbon nanotubes (CNTs), carbon dots (CDs), and countless other allotropes of carbon (Fig. 4.1), which stand out for their excellent electrical conductivities, strength, and potential application in technology and medicine. Understanding these nanostructures will open new doors of development toward futuristic technologies in various domains.

4.2.1.1 Graphene and Graphene Oxide (GO)

Graphene and its different derivatives—such as graphene quantum dots (GQD), graphene nanoplatelets (GNP), graphene nanoribbons (GNR), graphene oxide (GO), and reduced graphene oxide (rGO)—are significant in wastewater treatment and the elimination of impurities. Due to their superior physical and chemical properties, these materials are very valuable building blocks in hydrogel systems. Graphene, a one-atom-thick crystal of carbon atoms in a two-dimensional lattice, was initially separated and defined in 2004 by Professors Andre Geim and Kostya Novoselov at the University of Manchester, for which they received the Nobel Prize for Physics in 2010 (Deshmukh and Singh 2011).

Such nanostructures have a high surface area because of their two-dimensional morphology, which allows for effective adsorption of trace levels of impurities like heavy metals—lead, cadmium, and mercury and organic dyes like methylene blue and rhodamine from aqueous mediums (Saravanan et al. 2022). For example, electrostatic interactions, hydrogen bonding, and covalent bonds between the GO sheets and chitosan polymer chains lead to the formation of graphene oxide-chitosan hydrogel composites. Under near-neutral pH, chitosan amino groups protonate, allowing them to interact with negatively charged GO functional groups. The forces build stable

Fig. 4.1 Various types of carbon nanostructures based on their dimensionality

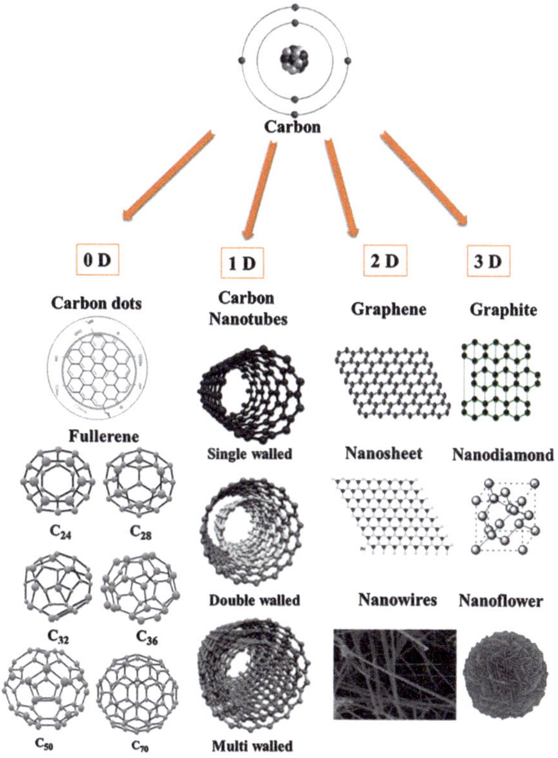

GO-chitosan hydrogels with great mechanical strength and thermal stability. In addition, incorporating chitosan-GO nanosheets enhances the surface area, pore size, and overall pore volume of the hydrogels, greatly improving their adsorption efficiency (Zhang et al. 2017).

Further, the oxygen-functional groups of GO, i.e., hydroxyl, carboxyl, and epoxy groups, enhance the strong interaction with the pollutants, enhancing contaminant attachment and removal. The functionalization also renders GO hydrophilic, allowing for uniform dispersion in hydrogel matrices, which enhances the adsorption capacity and mechanical strength of the hydrogels (Georgakilas et al. 2012). Studies have shown that GO-based hydrogels are stable and can be recycled efficiently under changing environmental conditions, validating their sustainability in wastewater treatment. Additionally, the addition of graphene or GO in hydrogels gives them responsiveness to environmental stimuli such as temperature and pH, making the targeted contaminant removed selectively.

4.2.1.2 Carbon Nanotubes (CNTs)

CNTs are diverse in high aspect ratios and tubular sizes and are used for performance enhancement in hydrogels. They have such a large surface area and with immense van der Waals forces that they adsorb nearly any type of organic impurity, including hydrocarbons, pesticides, and industrial dyes. CNTs, in particular, have main advantages in the production of highly effective sorbent material due to their hollow and layered structure that facilitates enhanced binding capacity with organic molecules through π–π stacking (Adewunmi et al. 2016).

Conversely, the use of pristine CNTs in adsorption applications is greatly limited by the lack of polar functional groups, leading to weak interaction with hydrophilic compounds. Therefore, the significance of both CNT and its polymer counterparts towards efficient wastewater remediation, with surface modification of CNT facilitating the improvement of the eco- and biocompatibility of the composite sorbent. They have, in particular, the following effects: by incorporating diverse chemical groups as functionalization, they can effectively disperse in hydrogel matrices, thus solving the aggregation problem that generally limits their efficiency. They display enhanced adsorption kinetics since they enhance interaction sites offered by the nanotube surface. They also provide superior mechanical and thermal stability, hence enhancing the longevity of composite materials during application in hostile environmental conditions (Venkataraman et al. 2019). Their conductivity potential holds promise for advancement in the area of electro-responsive hydrogels, hence making future applications more extensive in sensing and catalysis. According to Patil et al. (2021), CNT-based hydrogels can eliminate over 90% of the pathogenic bacteria and complicated organic pollutants in actual wastewater scenarios, demonstrating their practicality.

4.2.1.3 Carbon Dots (CDs)

Another newcomer to the carbon nanostructure family is carbon dot (CD), a zero-dimensional, small in size, with excellent photoluminescence, and easy surface functionalization (Himaja et al. 2015). These properties facilitate photocatalytic degradation and the removal of heavy metals. CDs, upon application in hydrogels, serve as photosensitizers in converting light energy to reactive oxygen species (ROS) for the degradation of organic contaminants like phenols and dyes. Moreover, their dense surface functional groups promote heavy metal chelation, like arsenic and chromium. CD-hydrogel composites are also anti-microbial, which may prevent biofouling in water treatment. The incorporation of CDs also makes the hydrogel UV or visible light-sensitive, enabling effective degradation of contaminants under environmentally friendly conditions (Saputra et al. 2024). In addition, CDs improve the recyclability of hydrogel systems by maintaining stable performance over multiple adsorption and desorption cycles. Improvements in CD synthesis, such as green and cost-effective methods, improve their scalability to water treatment processes.

4.2.1.4 Fullerenes

Fullerenes are carbon nanostructures resembling spheres. They exhibit high electronic properties and are very reactive. The same parameter is effective for fullerenes in wastewater treatment, as they can form stable complexes with pollutants (Georgakilas et al. 2015). Fullerenes can be incorporated into hydrogels by either dispersion, covalent functionalization, or host–guest interactions.

Since fullerene contains a hydrophobic part, surface modification with polar functional groups is required to account for its compatibility with hydrogel matrices. Fullerene-doped hydrogels show a strong affinity for organic pollutants, notably for hydrophobic contaminants such as pharmaceuticals and pesticides. They also have photocatalytic activity for pollutant degradation. Fullerenes in hydrogels improve the stability and reuse of these materials, suitable candidates for long-term applications.

4.2.2 Synthesis and Fabrication of Carbon Nanostructures Doped Hydrogels

The hybridization of carbon nanostructures with hydrogels is intriguing as they are capable of strengthening the mechanical, adsorption, and catalytic properties of the polymeric networks Carbon-based nanomaterials have demonstrated significant promise in wastewater treatment applications, largely owing to their unique physicochemical properties—such as high surface area, superior electrical conductivity, and adjustable surface chemistry—which make them highly suitable for integration into hydrogel frameworks. Synthesis and preparation of hydrogel through carbon-nanostructure-doping have divergent methods of implementation to support the homogenous distribution of chosen nanomaterials into the hydrogel structure without interference in the structural integrity of either entity or that of each other's functional ability (Anjali et al. 2019). Also important is that the synthesis protocol should be significant enough to impact the performance of the hydrogel for the purification of contaminants, as well as global stability within the aqueous environment. Various methods have been adopted for embedding carbon nanostructures into hydrogels, which give them varying features based on the intended use. Some of the most widely used techniques include in situ polymerization, physical blending, covalent cross-linking, and surface modification, which are discussed herein.

4.2.2.1 In Situ Polymerization

In situ polymerization is among the most commonly employed methods adopted for the embedding of carbon nanostructures into hydrogels. The procedure involves suspending the carbon nanostructures in a monomer solution and then initiating the polymerization reaction that eventually yields a cross-linked network hydrogel with the incorporated nanomaterials. Polymerization can be initiated through heating, chemical initiators, or photochemistry, based on the selection of a monomer mixture. One of the most prominent advantages is that the extensive intermolecular interactions between the polymer chains and the carbon nanostructures find application in enhancing the mechanical stability and strength of the obtained hydrogel (Karchoubi et al. 2024). Additionally, the method makes it easier for the nanomaterials to disperse uniformly throughout the hydrogel, which ultimately provides the maximum interphase for both impurity adsorption and catalytic action.

4.2.2.2 Physical Blending

Physical blending is a more straightforward route to the doped carbon nanostructure-hydrogel synthesis. Existing hydrogel networks are immersed in a solution containing carbon nanostructures such that nanomaterials are physically mixed within the hydrogel framework. The process does not involve chemical bonding between hydrogel and nanomaterials, a benefit that maintains the intrinsic properties of the two materials (Wang et al. 2024). On the other hand, however, the major shortcoming realized with the physical mixing process is the leaching of the nanomaterial after a very long period of time, potentially making the long-term efficiency and stability of a hydrogel as a wastewater treatment unit decline. Surface modification techniques are therefore applied to enhance the miscibility of the nanomaterials with the hydrogel matrix (Mahlangu et al. 2024).

4.2.2.3 Covalent Cross-Linking

Covalent cross-linking involves the formation of a chemical bond between hydrogel polymer chains and carbon nanostructures for effective embedding of the nanomaterials within the hydrogel network. It makes use of functionalized carbon nanostructures like carboxylated CNT or oxygenated graphene that can react with the polymer precursors to form stable covalent bonds (Dannert et al. 2019). The covalent bonding not only enhances the mechanical stability but also stabilizes against harsh environmental exposures, thereby making it more attractive to be used for long-term applications in wastewater treatment. Yet this synthesis route can possibly create other chemical alterations, which on the one hand can introduce the additional benefit of cost but on the other hand, add complexity to mass production.

4.2.2.4 Surface Functionalization

One particularly successful method for enhancing the compatibility and performance of hydrogels containing carbon nanostructures is surface functionalization. This complex process involves adding various functional groups, such as hydroxyl, amine, or carboxyl groups, to the surfaces of the carbon nanostructures (Liu and

Speranza 2019). These functional groups play a crucial role in improving the interaction between the carbon nanostructures and the hydrogel matrix, thereby promoting enhanced material synergy and integration. Surface functionalization improves not just the chemical and mechanical aspects of hydrogels, but also their overall functionality for various applications, ranging from biomedical to environmental, by creating a better interface. Ultimately, this approach produces complex hydrogel systems with distinct features that meet specific performance criteria. Functionalization can be achieved through methods such as chemical oxidation, plasma treatment, or polymer grafting. In addition to enhancing the aqueous dispersibility of carbon nanostructures (preventing aggregation and maximizing reactivity toward contaminants), surface functionalization can also introduce functionalities like increased hydrophilicity, charge selectivity, or catalytic activity, further enhancing the performance of the hydrogel in wastewater treatment.

4.2.3 Influence of Carbon Nanostructures on Hydrogel Properties

In formulating hydrogels, when carbon nanostructures are incorporated, changes occur generally to the chemical, physical, and functional properties; thereafter, the hydrogels become more robust with a view to environmental application. The properties become requisite to enhance pollution removal efficiency, mechanical strengthening, and hydrogel catalytic activity.

4.2.3.1 Changes in Mechanical Strength

For wastewater treatment applications, conventional hydrogels are severely limited by their intrinsically low mechanical strength, which jeopardizes long-term functionality and durability in dynamic industrial settings. Low crosslinking density and inadequate resistance to recurrent mechanical stresses during adsorption–desorption cycles are the causes of this structural fragility. In order to overcome these obstacles, scientists have used the nanoscale reinforcement properties of carbon-based nanofillers, such as graphene and CNTs, to incorporate them into hydrogel networks (Ho et al. 2022). Because of synergistic interactions like π-π stacking between aromatic polymer chains and graphene sheets, as well as hydrogen bonding that stabilizes the three-dimensional matrix, graphene-modified hydrogels have significantly higher tensile strength and elasticity. For example, adding graphene oxide (GO) to polyvinyl alcohol/sodium alginate hydrogels resulted in a 38% increase in compressive strength while preserving a high level of contaminant removal efficiency. Similar to this, CNTs' tubular nanostructure offers high aspect-ratio reinforcement, distributing mechanical loads throughout the hydrogel network and enhancing resistance to deformation under shear forces found in wastewater flow. These nanocomposite hydrogels are suitable for large-scale applications because they exhibit up to 97% contaminant removal efficiency with maintained structural integrity over several treatment cycles (Van et al. 2023). Weak mechanical strength- perhaps the most basic limitation of conventional hydrogels is a compromise in their functioning for long-term use in wastewater treatment plants. By incorporating nano-fillers into a hydrogel network, carbon nanostructures strengthen their mechanical properties, including tensile strength, elasticity, and toughness. For instance, graphene-based hydrogels exhibit improved structural integrity from the synergistic interaction of π-π stacking and hydrogen bonding between the graphene layers and polymer chains. Similarly, CNTs provide an avenue of high aspect ratio reinforcement, increasing hydrogel resistance to deformation and mechanical stresses.

4.2.3.2 Enhanced Adsorption Ability

The increased surface area and porosity characteristic of carbon nanostructures considerably improve the adsorption capacity of the hydrogels toward contaminants. Because of the high surface area and presence of various functional groups to interact strongly with the targeted pollutants, graphene and CNT-hydrogels absorb heavy metals and organic pollutants efficiently. Also, π-π interaction between carbon nanostructures and aromatic impurities improves selective adsorption of organic pollutants and thus maximizes the hydrogel efficiency for wastewater treatment.

4.2.3.3 Enhanced Catalytic and Photocatalytic Properties

Some carbon nanostructures, such as functionalized graphene and carbon dots, exhibit catalytic and photocatalytic characteristics useful for degrading organic pollutants. This will allow for optimizing hydrogel performance in wastewater treatment. For example, doped carbon dot-hydrogels could act as visible-light photocatalysts, enhancing dye, drug, and organic pollutant decomposition in wastewater (Zhang et al. 2018). Besides that, these carbon nanostructures promote electron-transfer reactions and enhance the redox potential of the hydrogel in degrading complex pollutants through redox reactions, thus holding great promise for developing advanced oxidation technologies for water treatment.

4.3 Applications of Carbon Nanostructures-Doped Hydrogels in Contaminant Removal from Wastewater

4.3.1 Heavy Metal Removal

Carbon-nanostructure-doped hydrogels attracted considerable interest due to their high selectivity and efficiency in adsorbing toxic heavy metal ions such as Pb^{2+}, Hg^{2+}, and As^{3+}. Doped functional groups enhance the hydrogel's chelation ability, binding metal ions through complexation. Functionalized graphene oxide (GO) hydrogels amplify this performance due to their high surface area, oxygen-containing functional groups, and superior metal-chelation properties. Studies have reported removal efficiencies exceeding 90% for lead and mercury ions under optimal conditions. This capability makes these hydrogels highly relevant in addressing contamination in industrial wastewater and mining effluents (Darban et al. 2022). Moreover, their reusability after desorption and regeneration cycles highlights their economic viability and environmental sustainability. The removal of heavy metals using carbon nanostructures happens differently due to the interior mechanisms of adsorption that depend on the order of the metal's affinity to the metal ions and the surface functionalization. For instance, in the case of CNT, the mechanism involves the release of protons (H^+) from the surface as a result of CNT surface oxidation, which is then followed by the binding of divalent metal ions onto the oxidized surface (García-Betancourt et al. 2020).

Fryczkowska et al. 2017 demonstrated improved adsorption of Fe^{3+} from industrial wastewater using 0.1% w/w GO-doped cellulose (CEL) (CEL/GO composite). The microporous structured composite can remove nearly 92% of ferric ions from wastewater. In another study (Yakout et al. 2017), GO was cross-linked with cellulose extracted from the halophytic trees (mangrove trees) or *Avicennia marina* and modified by ethylenediamine and grafted with acrylamide (g-C-EN-GO) for metal sorption from hard water samples. The g-C-EN-GO showed maximum adsorption capacity of 46.39 and 186.48 $mg.g^{-1}$ for the removal of Cu^{2+} and Pb^{2+}, respectively, at pH 7 and at 25 ± 1 °C. Moreover, the current quest for eco-friendly hydrogels led to the development of various green-synthesized sorbents for heavy metal removal. For instance, Bessa and team (Bessa et al. 2020) developed a GO-chitosan green nanocomposite for the remediation of Hg^{2+} in different water samples (tap water, river water, and seawater). It demonstrated a 97%, 81.4%, 13%, and 7% removal percentage for 10 mg L^{-1} spiking of ultrapure water, tap water, river, and seawater, respectively (Fig. 4.2). In the case of seawater, the lower affinity towards the sorbent was due to stable chloro-complexes formation of Hg^{2+}. Similarly, GO and iron(III) oxide hydroxide incorporated in chitosan effectively improves hydrogel and biosorption stability, along with enhancing the amount of functional groups such as -NH, -OH, C = C, -CH, -CO, -COOH, and Fe–O that promote adsorption. The abundant amino (-NH) and hydroxyl (-OH) functional groups provided effective adsorption and removal capacity of 84 and 95% for Pd^{2+} and Cd^{2+} at a contact time of ~50 min (Parastar et al. 2021).

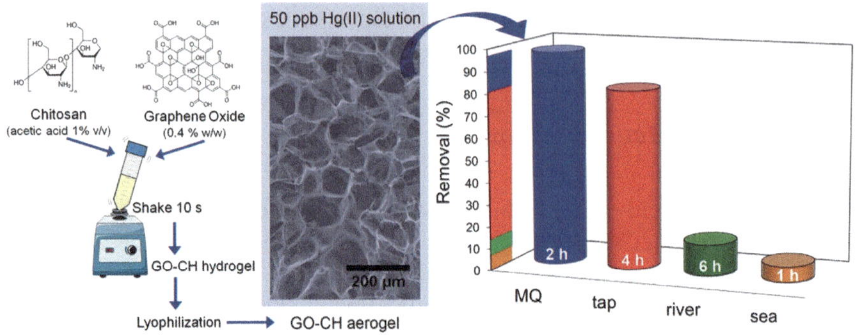

Fig. 4.2 Schematic illustration of the synthesis of graphene oxide-chitosan hydrogel and consecutive removal of heavy metal (Hg^{2+}) from various water samples (Bessa et al. 2020). Open Access

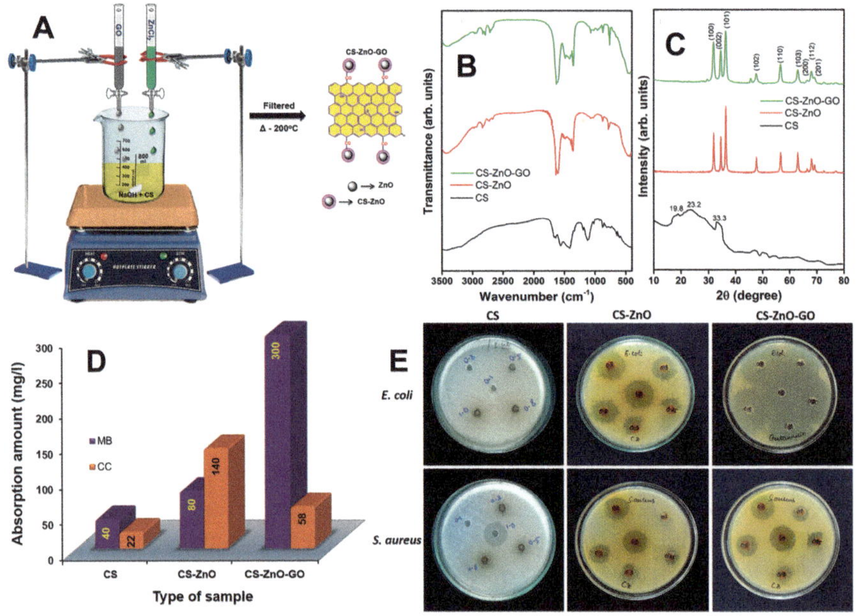

Fig. 4.3 **a** Schematic illustration of chitosan–ZnO–graphene oxide synthesis; **b** FTIR and **c** XRD spectra of chitosan (CS), CS–ZnO and CS–ZnO–GO hybrid structures, **d** A comparison of the adsorption capacities for methylene blue (MB) and chromium complex (CC) dyes using chitosan (CS), CS–ZnO, and CS–ZnO–GO demonstrates a marked improvement in dye uptake with the incorporation of graphene oxide (GO). Additionally, the antibacterial activity of the materials was evaluated against a Gram-positive (*S. aureus*) and Gram-negative bacteria (*E. coli*) (Sanmugam et al. 2017). Open Access

Other than such heavy metal pollutants, the extensive use of nuclear energy is resulting in excessive discharge and leaching of uranium into the water, mainly in hexavalent U(VI) ions form, which is of great danger to ecological balance and environmental security because it is highly chemically and radiologically toxic. Thus, for the U(VI) removal and recovery from aqueous solution, Liao et al. 2019 developed a highly porous phosphate-functionalized graphene hydrogel electrode material that could achieve kinetic equilibrium within 21 min, and its highest adsorption capacity was as high as 545.7 mg g^{-1} at 1.2 V and pH 5.0. In addition, the hydrogel electrode had high selectivity for U(VI) with other competing metal ions, with high regeneration capacity, demonstrating only an acceptable decrease of 7.5% in removal ratio after six cycles.

4.3.2 Dye Removal

Various types of dyes are daily released into the environment due to production activities in diverse industries, mainly textiles, leather, paper and pulp, plastic and paints, cosmetics, pharmaceuticals, and electroplating form an integral portion of the polluters that create water pollution. Currently, more than 100,000 various forms of dyes are commercially manufactured, resulting in immense issues. Aside from potential toxicities, even though a minute proportion of the dye is blended with terrestrial water resources, it causes significant concerns, such as blocking the sun's light directly, influencing photosynthesis processes, and lowering the concentration of water-soluble oxygen, which leads to damage in aquatic life (Alsukaibi 2022). Thus, removing dyes from wastewater before they are discharged into natural water bodies is critically important to prevent environmental contamination. Among various treatment methods, adsorption is considered one of the most effective techniques for removing pollutants from wastewater. Its widespread use is largely attributed to its cost-effectiveness, requiring low capital investment and offering economical operational costs compared to many physical, chemical, and biological alternatives (Kahya and Erim 2022). These hydrogels are particularly advantageous in treating effluents from textile, paper, and leather industries, which often contain high concentrations of persistent dyes. Their regeneration potential, achieved by simple washing with solvents, underscores their applicability for long-term use in water treatment processes.

The surface densities of the dyes can be positive (cationic or basic dyes, such as methylene blue, crystal violet, etc.) or may have negative surface charge densities (anionic or acidic dyes, such as Congo red, eosin, etc.). Biopolymers, such as chitosan (poly-(1 $\rightarrow$ 4)-2-amino-2-deoxy-b-d-glucose) are positively charged due to the protonation of amine groups in their backbones and have a tendency to adsorb negatively charged dyes. Moreover, the incorporation of nanostructures, such as the hydrophilic GO, which contains a large surface area with numerous oxygen-containing groups like epoxides and carboxylic acids on its surface, enhances the dye adsorption capacity, along with fortifying mechanical strength. Thus, the removal of industrial dyes such as methylene blue and methyl orange has been achieved effectively using hydrogels doped with GO and carbon nanotubes (CNTs). These materials interact with dye molecules via π-π stacking and hydrogen bonding, allowing for efficient adsorption (Yuan et al. 2016). The large surface area and tunable porosity of GO and CNT-based hydrogels further enhance their adsorption capacity, with reports of adsorption capacities reaching over 500 mg/g in certain configurations (Zhang et al. 2022).

Several studies have highlighted the use of electric fields to improve the efficiency of sorbent materials. For instance, Cirillo et al. (2021) designed a hybrid hydrogel composed of gelatin-coated multi-walled carbon nanotubes (MWCNTs), referred to as HGNT, and assessed its effectiveness in adsorbing three different organic dyes with varying electrical properties: Methylene Blue (MB), Rhodamine B (RD), and Bengal Rose (BR). To thoroughly characterize the adsorption process, they applied various empirical models to analyze the adsorption mechanisms, kinetics, and thermodynamic parameters (shown in Fig. 4.4). These included adsorption isotherms such as Langmuir, Freundlich, Redlich–Peterson, Sips, Dubinin–Radushkevich, and Temkin, alongside kinetic models like pseudo-first-order, pseudo-second-order, Avrami, fractional power, intraparticle diffusion, and Elovich. Notably, the MWCNTs' electro-responsive nature allowed modulation of both sorption capacity and rate when an external voltage was applied. Specifically, applying a negative polarization to the hybrid hydrogel resulted in increased adsorption for cationic dyes and decreased uptake for anionic or neutral species. Thermodynamics, isothermal and kinetic experiments demonstrated favorable sorption processes and the use of an external 12 V electric field was found capable of affecting the quantity of adsorbed dyes: stronger interactions with cationic MB dye as sorption potential increase to 33.45 mg g^{-1} at 12 V from 19.72 mg g^{-1} at 0 V, while reduces to 13.06 mg g^{-1} (12 V) from 28.93 mg g^{-1} (0 V) and 15.85 mg g^{-1} (12 V) from 36.75 mg g^{-1} (0 V) in cases of RD and BR, respectively.

4.3.3 Oil–Water Separation

Carbon-based hydrogels have emerged as promising candidates for oil–water separation due to their high hydrophobicity and oleophilic properties. These hydrogels are engineered to absorb oils selectively while repelling water, making them ideal for cleaning up oil spills and industrial effluents. Additionally, their porous structure allows for high absorption capacities, often exceeding 20 times their weight in oil (Kang et al. 2020). Unlike traditional separation methods, these hydrogels can be reused multiple times without significant loss of efficiency, as they retain their structural integrity after repeated absorption–desorption cycles. This feature, combined with their eco-friendly synthesis, positions them as sustainable solutions for managing oil-contaminated water bodies.

Yun et al. 2024 prepared a regenerated cellulose hydrogel with hydrophobic alkylated GO-coating for oil adsorption. The hydrogel was modified with

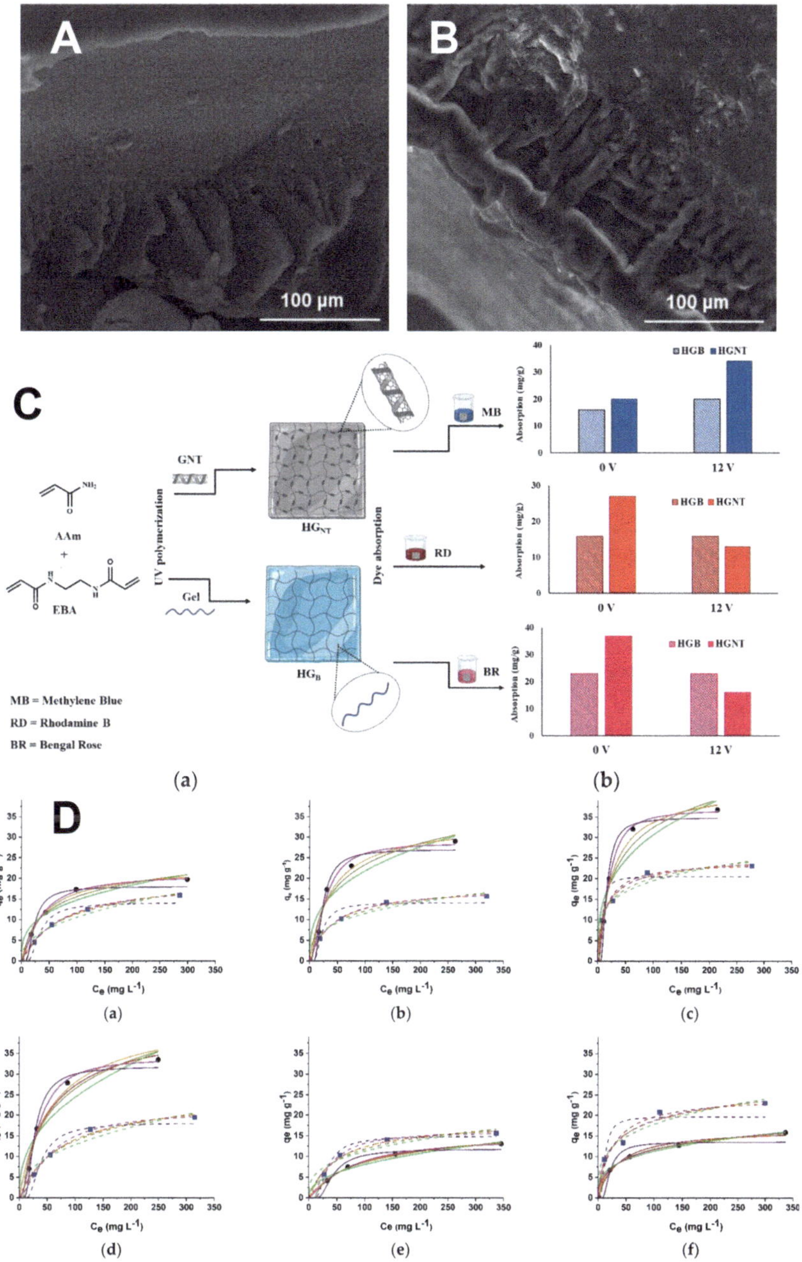
A
100 µm
B
100 µm
C
AAm
+
EBA
UV polymerization
GNT
Gel
HG$_{NT}$
HG$_B$
Dye absorption
MB
RD
BR
MB = Methylene Blue
RD = Rhodamine B
BR = Bengal Rose
HGB HGNT
Absorption (mg/g)
0 V
12 V
(a)
(b)
D
q_e (mg g^{-1})
C_e (mg L^{-1})
(a)
(b)
(c)
(d)
(e)
(f)

◂**Fig. 4.4** SEM pictures of **a** Hydrogel Blank (Control) (HGB) and **b** Gelatin-coated MWCNT Hybrid Hydrogel (HGNT) samples, indicating the uniform distribution of MWCNT in the polymer network of HGNT; **c** Schematic illustration of synthesis of hybrid hydrogels and (b) dye adsorption performance at various voltage conditions; **d** Adsorption isotherms including Langmuir (red), Freundlich (green), Redlich–Peterson (orange), Sips (magenta), Dubinin–Radushkevich (purple), and Temkin (dark yellow) were applied to model the adsorption behavior of Methylene Blue (MB) [(a,d)], Rhodamine B (RD) [(b,e)], and Bengal Rose (BR) [(c,f)] dyes using HGNT (represented by solid lines) and HGB (represented by dashed lines). The data were evaluated under two conditions: 0 V (a–c) and 12 V (d–f). Experimental results for HGNT and HGB are indicated by circles and squares, respectively. (Cirillo et al. 2021) Open Access

polyethyleneimine and di-epoxide crosslinked to improve the elastic recovery potential for the squeezing process. The hydrogel demonstrated an oil adsorption capacity of >84% and high recovery through the squeezing process even after ten consecutive reuses. Ghasemi et al. 2019 demonstrated the real-life application on the adsorption of oil spill from Caspian sea water. The team developed super hydrophobic, graphene-chitosan (GC) nanocomposite absorbent that showed maximum sorption of 48.2 (gr oil/gr GC) from the seawater sample, making it a promising material for oil spill combating.

4.3.4 Pharmaceutical Degradation

Pharmaceutically active compounds, which mainly comprise nonsteroidal anti-inflammatory drugs (NSAIDs) such as diclofenac, ibuprofen, naproxen, antidepressants like diazepam, meprobamate, antibiotics like amoxicillin, ciprofloxacin, or erythromycin, anticonvulsants, hormones, etc. which enter the water through various routes like industries, hospitals, homes, aquaculture, or feedstock plants. Consumption of such pharmaceutical contaminants in drinking water, even at low concentrations or long-term exposure, has various adverse effects, including the emerging issues of the development of antibiotic resistance (Fortunato and Mba 2022). Ma et al. 2020 combined sodium alginate with GO and CNTs with the help of L-cysteine (L-cys) and H_2O_2 to prepare triple-network composite hydrogels (CNTs/L-cys@GO/SA), with high efficiency removal of antibiotics, such as ciprofloxacin. The three-dimensional multi-network structure enhanced the surface and space, which improved the adsorption capacity. Under a weak acidity environment, it demonstrated adsorption capacity of 181 mg g^{-1} and 200 mg g^{-1} at moderately low temperatures of 25 °C and 15 °C, respectively.

Currently, the application of hydrogel thin films for contaminant removal is also gaining the limelight. For example, in a recent report, Kumar and Barakat 2024 developed GO-chitosan-based polyaniline (PANI) nanoparticles-doped positively charged hydrogel thin film (CS/GO/PANI) that acted as the ibuprofen scavenger from the aqueous solution. Non-electrostatic forces like π–π bonding, hydrogen bonding, etc., were responsible for the adsorption process and obeyed Elovich and Redlich-Peterson models. Moreover, for practical applications and cost-effectiveness, recovery and multiple use with effective potential are of utmost significance. Even after five cycles, the film showed effective decontamination potential, as ethanol demonstrated high regeneration potential.

Photocatalytic hydrogels, incorporating materials like carbon dots (CDs) and GO, have demonstrated significant potential in degrading organic pollutants under sunlight. These hydrogels leverage the photocatalytic properties of CDs and GO, which generate ROS under light irradiation. These ROS effectively break down organic contaminants, such as phenols and pesticides, into less harmful compounds like water and carbon dioxide. The process is energy-efficient, as it utilizes solar energy, and environmentally benign, as it avoids secondary pollution. Studies have shown degradation efficiencies exceeding 85% for several pollutants within a few hours of exposure (Pirilä et al. 2015). This innovative approach integrates water purification with renewable energy utilization, providing a sustainable pathway for tackling organic pollution in wastewater.

4.4 Challenges and Future Directions

These carbon nanostructure-doped hydrogels contain materials with gigantic potential in drug delivery, tissue engineering, and sensor technology. However, several issues must be overcome before they become everyday materials (Zhang et al. 2021, 2025).

4.4.1 Scalability

Industrially available carbon nanostructure-doped hydrogels reveal numerous drawbacks due to high prices and complex synthesis processes. Material inputs in the preparation of these hydrogels pricey endeavour in any sense- involve graphene oxide, carbon nanotubes, or fullerenes. Step by step to prepare these hydrogels involves the functionalization of carbon nanostructures, cross-linking with the polymer matrix, and optimization of gelation conditions. This is not possible, however, in bulk preparations, because some will need specialized equipment or careful control of reaction conditions, which again is more expensive. The carbon nanostructures in the hydrogel matrix also need to be dispersed and stabilized, which introduces yet another level of complexity to bulk production. If scaling issues are to be addressed, innovation thus has to be infused into synthesis processes (e.g., reducing steps or streamlining materials utilized) and into creating manufacturing technologies that are cost-effective and capable of applications.

4.4.2 Regeneration

One of the most salient shortcomings of carbon nanostructure-doped hydrogels is the regeneration and recyclability of the materials. Reusability or regeneration of the hydrogel after the first time of use is critical for any long-term usage, such as drug delivery and tissue engineering. Nonetheless, at the moment, numerous carbon nanostructure-doped hydrogels are experiencing irreversible degradation during repetitive loading or under aggressive conditions owing to the lack of stability of the carbon nanostructures in some environments and their dynamic or repetitive bad interaction with the polymer matrix. Also, due to the repetitive application of mechanical stress or stimulus to the hydrogel, its network structure may be damaged, ultimately damaging its overall functionality. Regenerative hydrogels should be engineered with characteristics like self-healing capabilities, mechanical robustness, and stability. Presently, studies are concentrated on reversible cross-linking, supramolecular interactions, and hybrid hydrogels with stimuli-responsive characteristics for better reusability of carbon nanostructure-doped hydrogels. However, there is still a lot that needs to be accomplished before they can be utilized on a greater scale.

4.4.3 Biodegradability

The limitations of carbon nanostructure-doped hydrogels concerning mechanical properties and functionalization have proven to be incredibly useful, but in comparison, biodegradability and long-term environmental impact are unsatisfactorily investigated. In any biomaterial use, particularly in the case of in vivo drug delivery or tissue engineering, one most critical characteristics a hydrogel should possess.

4.5 Conclusion

Carbon nanostructure incorporation into hydrogels is a novel method of wastewater treatment improvement by combining the best of both worlds: distinctive carbon material properties and the hydrogel's three-dimensional porous structure. The chapter thus provides the vast potential of carbon nanostructure-doped hydrogels against a variety of water pollutants such as heavy metals, organic dyes, pharmaceuticals, and microplastics. The high surface areas, controllable surface chemistry, and excellent mechanical strength, salient properties of carbon nanostructures such as graphene oxide, carbon nanotubes, and carbon quantum dots, are designed to enhance hydrogels' adsorption ability, selectivity, and recyclability. In addition, there is also the advantage of good water retention and swelling behavior under hydrophilic conditions to promote pollutant adsorption and removal.

Despite their great potential and strong promise, much has to be done before carbon nanostructure-doped hydrogels can be utilized on a large scale in wastewater treatment. First, the synthesis of such materials is often complex and energy-intensive, a factor that limits their cost viability and scalability. The second one concerns the long-term stability and recyclability of such hydrogels in actual wastewater systems, which could be subjected to fouling, degradation, or loss of adsorption capacity over several cycles of operation. Third, as already pointed out, the environmental effect of carbon nanostructures-their toxicity and persistence in the environment-must be extensively studied for sustainable and safe use. Subsequent studies should pursue lower-cost, more scalable synthesis routes to engineer carbon nanostructure-doped hydrogels. Recyclable and biodegradable hydrogels would both alleviate concerns over long-term persistence in the environment and produce waste. Higher end characterization, such as in situ spectroscopy and computational modeling, can potentially provide even greater insight into adsorption mechanisms and kinetics that can inform the design of much more effective and selective materials. Smartness, i.e., stimulus-sensitive behavior or healing capability, might also function as an improvement on the quality and longevity of such hydrogels in wastewater's dynamic state.

Thus, hydrogels doped with carbon nanostructures represent one of the most promising material candidates for the treatment of wastewater, with significant efficiency in contaminant removal and multifunctional properties. But scaling-up, stability, and environmental concerns issues still exist; nevertheless, the research and advancements associated with these will offer an avenue to breach these obstacles. This idea has the potential to go a long way in realizing clean and sustainable water resources worldwide through the inherent attributes of carbon nanostructures and hydrogels. Thus, future work should be directed toward devising cost-effective, environmentally friendly, and large-scale solutions for the future commercialization of such materials for water treatment purposes.

Table 4.1 Various carbon nanostructure doped hydrogels for the removal of pollutants from wastewater

Carbon nanostructure doped hydrogel	Pollutant removed	Removal capacity and performance	References
Fe_3O_4-graphene oxide-chitosan nanohybrid	Hg^{2+}	84.96% removal in 8 min; 339.82 $mg{\cdot}g^{-1}$	Bulin et al. (2023)
Fe_3O_4-graphene oxide-chitosan magnetic adsorbent	Zn^{2+}	96.73% removal in 38 min; 386.92 $mg{\cdot}g^{-1}$	Bulin et al. (2023)
Magnetic chitosan-functionalized 3D graphene-$NiFe_2O_4$ nanocomposite	Pb^{2+}	100% removal (pH 8.5, 36 mg adsorbent, 26 $mg{\cdot}L^{-1}$ Pb^{2+}, 18 min)	Nasiri et al. (2018)
Chitosan-assisted Fe-Al double-layered hydroxide/ reduced graphene oxide (FAH-rGO/CS)	As(V)	96.73% removal; 167.79 $mg{\cdot}g^{-1}$	Priya et al. (2020)
3D chitosan-carbon nanotube-PVA nanocomposite hydrogel	Cr^{6+}	217.4 $mg{\cdot}g^{-1}$ (Langmuir model, pH 1.5)	Eldeeb et al. (2020)
Poly(vinyl alcohol)-multiwalled carbon nanotube (MWCNT) hydrogel	Pb^{2+}	80% removal (1000 $\mu g{\cdot}L^{-1}$ Pb^{2+}, pH 3.0, 8 h)	Abraham et al. (2012)
Magnetic polypyrrole/ chitosan/graphene oxide (M-PPy/CS/GO)	Ponceau 4R dye; E. coli and Fusarium	85% removal (210 min, pH 2, 0.01 g adsorbent, 2.88 ppm dye); stable for 4 reuse cycles	Salahuddin et al. (2020)
ZnO/functionalized CNT/ chitosan	Reactive Blue 4 (RB-4)	93.54% degradation (photocatalysis)	Humayoun et al. (2023)
Polypyrrole-carbon nanotube (PPy-CNT) 3D porous hydrogel anode	Congo red	94.35% removal (50 $mg{\cdot}L^{-1}$, 48 h, 200 Ω, 480 mV); 99.55% with activated sludge	Qi et al. (2022)
CNT-CS-dopamine-coated PVDF membrane	Anionic dye wastewater; oil–water emulsions	99.4% (emulsions); 98.8% (dye wastewater)	Zhao et al. (2024)
Co_3O_4/g-C_3N_4/multiwall carbon nanotube hydrogel	Rhodamine B	87% removal (3 ppm, 120 min, visible light)	Shafiee et al. (2019)
Alginate/carbon nanotube-carbon dot-magnesium fluorohydroxyapatite hydrogel beads	Methylene blue	94% removal (120 mg adsorbent, pH ~ 4.2, 48 h)	Mallakpour et al. (2021)
Carbon nanotube/chitosan hydrogel	Acid Red 73 (azo dye)	92.23% removal; 101.07 $mg{\cdot}g^{-1}$ (70 mg adsorbent, pH 5)	Wei et al. (2023)

(continued)

Table 4.1 (continued)

Carbon nanostructure doped hydrogel	Pollutant removed	Removal capacity and performance	References
Sodium alginate-immobilized β-cyclodextrin/MWCNT hybrid hydrogel	Perfluorooctanesulfonic acid (PFOS)	91.6% removal (1 g adsorbent, 10 $mg{\cdot}L^{-1}$ PFOS, pH 3, 45 min)	Zakaria et al. (2023)
Carbon dot cross-linked polyvinylpyrrolidone hybrid hydrogel	Malachite Green, Crystal Violet, Eosin Y; *S. aureus*, *E. coli*	Equilibrium capacities: 5.75 $mg{\cdot}g^{-1}$ (Malachite Green), 5.3 $mg{\cdot}g^{-1}$ (Crystal Violet), 8.75 $mg{\cdot}g^{-1}$ (Eosin Y)	Nayak et al. (2020)
Magnetic Fe/porous carbon/sodium alginate hydrogel	Norfloxacin (NOR)	Complete degradation (5 $mg{\cdot}L^{-1}$ NOR, 2 $mL{\cdot}min^{-1}$ flow rate)	Liu et al. (2022)
Graphene hydrogel-metal (GH-M; M: Co, Ni, Cu) nanocomposites	Rifampicin (RIF)	90.47% (GH-Cu), 92.60% (GH-Co), 93.69% (GH-Ni); 30 $mA{\cdot}cm^{-2}$, pH 5.0, 30 $mg{\cdot}L^{-1}$ RIF, 90 min	Ebrankhahan et al. (2022)
3D porous graphene hydrogel	Ciprofloxacin	235.6 $mg{\cdot}g^{-1}$ adsorption capacity	Ma et al. (2015)
Carbon dots/polyethylene glycol diglycidyl ether/ polypropylene glycol diglycidyl ether hydrogel	PFOS	2.82 $mg{\cdot}g^{-1}$ adsorption capacity	Wang et al. (2022)

References

Abraham TN, Kumar R, Misra RK, Jain SK (2012) Poly (vinyl alcohol)-based MWCNT hydrogel for lead ion removal from contaminated water. J Appl Polym Sci 125(S1):E670–E674

Adewunmi AA, Ismail S, Sultan AS (2016) Carbon nanotubes (CNTs) nanocomposite hydrogels developed for various applications: a critical review. J Inorg Organomet Polym Mater 26:717–737

Alsukaibi AK (2022) Various approaches for the detoxification of toxic dyes in wastewater. Processes 10(10):1968

Anjali J, Jose VK, Lee JM (2019) Carbon-based hydrogels: synthesis and their recent energy applications. J Mater Chem A 7(26):15491–15518

Bessa A, Gonçalves G, Henriques B, Domingues EM, Pereira E, Marques PA (2020) Green graphene–chitosan sorbent materials for mercury water remediation. Nanomaterials 10(8):1474

Bulin C (2023) Adsorption mechanism and removal efficiency of magnetic graphene oxide-chitosan hybrid on aqueous Zn (II). Int J Biol Macromol 241:124588

Bulin C, Zheng R, Song J, Bao J, Xin G, Zhang B (2023) Magnetic graphene oxide-chitosan nanohybrid for efficient removal of aqueous Hg (Π) and the interaction mechanism. J Mol Liq 370:121050

Cirillo G, Curcio M, Madeo LF, Iemma F, De Filpo G, Hampel S, Nicoletta FP (2021) Carbon nanotubes hybrid hydrogels for environmental remediation: evaluation of adsorption efficiency under electric field. Molecules 26(22):7001

Dannert C, Stokke BT, Dias RS (2019) Nanoparticle-hydrogel composites: from molecular interactions to macroscopic behavior. Polymers 11(2):275
Darban Z, Shahabuddin S, Gaur R, Ahmad I, Sridewi N (2022) Hydrogel-based adsorbent material for the effective removal of heavy metals from wastewater: a comprehensive review. Gels 8(5):263
Deshmukh MM, Singh V (2011) Graphene—an exciting two-dimensional material for science and technology. Resonance 16:238–253
Dong C, Lu J, Qiu B, Shen B, Xing M, Zhang J (2018) Developing stretchable and graphene-oxide-based hydrogel for the removal of organic pollutants and metal ions. Appl Catal B 222:146–156
Ebratkhahan M, Zarei M, Akpinar IZ, Metin Ö (2022) One-pot synthesis of graphene hydrogel/M (M: Cu Co, Ni) nanocomposites as cathodes for electrochemical removal of rifampicin from polluted water. Environ Res 214:113789
Eldeeb TM, El Nemr A, Khedr MH, El-Dek SI, Imam NG (2020) Novel three-dimensional chitosan-carbon nanotube–PVA nanocomposite hydrogel for removal of Cr6+ from wastewater. Desalin Water Treat 184:163–177
Fortunato A, Mba M (2022) A peptide-based hydrogel for adsorption of dyes and pharmaceuticals in water remediation. Gels 8(10):672
Fryczkowska B, Wyszomirski M, Puzoń M (2017) Obtaining and application of new cellulose-and graphene oxide-based adsorbents for treatment of industrial waste containing heavy metals. J Ecol Eng 18(6):43–52
García-Betancourt ML, Jiménez SIR, González-Hodges A, Salazar ZEN, Escalante-García IL, Aparicio JR (2020) Low dimensional nanostructures: measurement and remediation technologies applied to trace heavy metals in water. In: Trace metals in the environment-new approaches and recent advances. IntechOpen
Georgakilas V, Otyepka M, Bourlinos AB, Chandra V, Kim N, Kemp KC, Kim KS (2012) Functionalization of graphene: covalent and non-covalent approaches, derivatives and applications. Chem Rev 112(11):6156–6214
Georgakilas V, Perman JA, Tucek J, Zboril R (2015) Broad family of carbon nanoallotropes: classification, chemistry, and applications of fullerenes, carbon dots, nanotubes, graphene, nanodiamonds, and combined superstructures. Chem Rev 115(11):4744–4822
Ghasemi O, Mehrdadi N, Baghdadi M, Aminzadeh B, Ghaseminejad A (2020) Spilled oil absorption from Caspian sea water by graphene/chitosan nano composite. Energy Sources, Part A: Recover, Util, Environ Eff 42(23):2856–2872
Himaja AL, Karthik PS, Singh SP (2015) Carbon dots: the newest member of the carbon nanomaterials family. Chem Rec 15(3):595–615
Ho TC, Chang CC, Chan HP, Chung TW, Shu CW, Chuang KP, Tyan YC (2022) Hydrogels: properties and applications in biomedicine. Molecules 27(9):2902
Humayoun UB, Mehmood F, Hassan Y, Rasheed A, Dastgeer G, Anwar A, Yoon D (2023) Harnessing bio-immobilized ZnO/CNT/chitosan ternary composite fabric for enhanced photodegradation of a commercial reactive dye. Molecules 28(18):6461
Hwang TS, Kim YJ, Ryu DH, Park WH (2021) Nanocomposite hydrogels Enhancement of physical and mechanical properties. J Mater Chem B 9(41):8366–8383
Jatoi AS, Ahmed J, Bhutto AA, Jeyapaul AS (2025) Recent advances and future perspectives of carbon-based nanomaterials for environmental remediation. Braz J Chem Eng 42(1):151–169
Jayakumar A, Jose VK, Lee JM (2020) Hydrogels for medical and environmental applications. Small Methods 4(3):1900735
Jiang M, Wang Y, Li J, Gao X (2024) Review of carbon dot–hydrogel composite material as a future water-environmental regulator. Int J Biol Macromol 131850
Kahya N, Erim FB (2022) Graphene oxide/chitosan-based composite materials as adsorbents in dye removal. Chem Eng Commun 209(12):1711–1726
Kang W, Cui Y, Qin L, Yang Y, Zhao Z, Wang X, Liu X (2020) A novel robust adsorbent for efficient oil/water separation: Magnetic carbon nanospheres/graphene composite

Karchoubi F, Ghotli RA, Pahlevani H, Salehi MB (2024) New insights into nanocomposite hydrogels; a review on recent advances in characteristics and applications. Adv Ind Eng Polym Res 7(1):54–78

Kumar R, Barakat MA (2024) Flexible multifunctional chitosan/graphene oxide/polyaniline hydrogel thin films for adsorption of ibuprofen from aqueous solution. Cellulose 31(7):4347–4366

Liao Y, Wang M, Chen D (2019) Electrosorption of uranium (VI) by highly porous phosphate-functionalized graphene hydrogel. Appl Surf Sci 484:83–96

Liu D, Gu W, Zhou W, Xu Y, He W, Liu L, Liu Y (2022) Magnetic Fe/carbon/sodium alginate hydrogels for efficient degradation of norfloxacin in simulated wastewater. J Clean Prod 369:133239

Liu W, Speranza G (2019) Functionalization of carbon nanomaterials for biomedical applications. C5(4):72

Ma J, Yang M, Yu F, Zheng J (2015) Water-enhanced removal of ciprofloxacin from water by porous graphene hydrogel. Sci Rep 5:13578

Ma J, Jiang Z, Cao J, Yu F (2020) Enhanced adsorption for the removal of antibiotics by carbon nanotubes/graphene oxide/sodium alginate triple-network nanocomposite hydrogels in aqueous solutions. Chemosphere 242:125188

Mahlangu OT, Motsa MM, Richards H, Mamba BB, George MJ, Nthunya LN (2024) The impact of nanoparticle leach on sustainable performance of the membranes–a critical review. Environ Nanotechnol, Monit Manag 100984

Mallakpour S, Behranvand V, Mallakpour F (2021) Adsorptive performance of alginate/carbon nanotube-carbon dot-magnesium fluorohydroxyapatite hydrogel for methylene blue-contaminated water. J Environ Chem Eng 9(2):105170

Nasiri R, Arsalani N, Panahian Y (2018) One-pot synthesis of novel magnetic three-dimensional graphene/chitosan/nickel ferrite nanocomposite for lead ions removal from aqueous solution: RSM modelling design. J Clean Prod 201:507–515

Nayak S, Prasad SR, Mandal D, Das P (2020) Carbon dot cross-linked polyvinylpyrrolidone hybrid hydrogel for simultaneous dye adsorption, photodegradation and bacterial elimination from waste water. J Hazard Mater 392:122287

Parastar M, Sheshmani S, Shokrollahzadeh S (2021) Cross-linked chitosan into graphene oxide-iron (III) oxide hydroxide as nano-biosorbent for Pd (II) and Cd (II) removal. Int J Biol Macromol 166:229–237

Patil TV, Patel DK, Dutta SD, Ganguly K, Randhawa A, Lim KT (2021) Carbon nanotubes-based hydrogels for bacterial eradiation and wound-healing applications. Appl Sci 11(20):9550

Phang SJ, Tan LL (2019) Recent advances in carbon quantum dot (CQD)-based two dimensional materials for photocatalytic applications. Catal Sci Technol 9(21):5882–5905

Pirilä M, Saouabe M, Ojala S, Rathnayake B, Drault F, Valtanen A, Keiski RL (2015) Photocatalytic degradation of organic pollutants in wastewater. Top Catal 58:1085–1099

Priya VN, Rajkumar M, Magesh G, Mobika J, Sibi SL (2020) Chitosan assisted Fe-Al double layered hydroxide/reduced graphene oxide composites for As (V) removal. Mater Chem Phys 251:123108

Qi L, Xu H, Tang R, Liu L, Chen Y, Wen Q (2022) Study on power generation and Congo red decolorization of 3D conductive PPy-CNT hydrogel in bioelectrochemical system. Int J Hydrogen Energy 47(37):16568–16579

Salahuddin NA, EL-Daly HA, El Sharkawy RG, Nasr BT (2020) Nano-hybrid based on polypyrrole/chitosan/grapheneoxide magnetite decoration for dual function in water remediation and its application to form fashionable colored product. Adv Powder Technol 31(4):1587–1596

Sanmugam A, Vikraman D, Park HJ, Kim HS (2017) One-pot facile methodology to synthesize chitosan-ZnO-graphene oxide hybrid composites for better dye adsorption and antibacterial activity. Nanomaterials 7(11):363

Saputra AMA, Piliang AFR, Goei R, HTS RR, Gea S (2024) Synthesis, properties, and utilization of carbon quantum dots as photocatalysts on degradation of organic dyes: a mini review. Catal Commun 187:106914

Saravanan A, Kumar PS, Jeevanantham S, Karishma S, Tajsabreen B, Yaashikaa PR, Reshma B (2021) Effective water/wastewater treatment methodologies for toxic pollutants removal: Processes and applications towards sustainable development. Chemosphere 280:130595

Saravanan A, Kumar PS, Srinivasan S, Jeevanantham S, Vishnu M, Amith KV, Vo DVN (2022) Insights on synthesis and applications of graphene-based materials in wastewater treatment: a review. Chemosphere 298:134284

Shafiee MRM, Parhizkar J, Radfar S (2019) Removal of Rhodamine B by g-C3N4/Co3O4/MWCNT composite stabilized in hydrogel via the synergy of adsorption and photocatalysis under visible light. J Mater Sci: Mater Electron 30(13):12475–12486

Slepičková Kasálková N, Slepička P, Švorčík V (2021) Carbon nanostructures, nanolayers, and their composites. Nanomaterials 11(9):2368

Van TN, Quang TD, Hung TT, Xuan QC, Van TH, Kim H, Ahn D (2023) Enhancing mechanical properties of polyvinyl alcohol/sodium alginate gel beads by graphene oxide for the aerobic sludge immobilization in wastewater treatment. Environ Eng Res 28(5)

Venkataraman A, Amadi EV, Chen Y, Papadopoulos C (2019) Carbon nanotube assembly and integration for applications. Nanoscale Res Lett 14:1–47

Wang WR, Chen PY, Deng J, Chen Y, Liu HJ (2022) Carbon-dot hydrogels as superior carbonaceous adsorbents for removing perfluorooctane sulfonate from water. Chem Eng J 435:135021

Wang R, Cheng C, Wang H, Wang D (2024) Swollen hydrogel nanotechnology: advanced applications of the rudimentary swelling properties of hydrogels. ChemPhysMater

Wei J, Yan L, Zhang Z, Hu B, Gui W, Cui Y (2023) Carbon nanotube/Chitosan hydrogel for adsorption of acid red 73 in aqueous and soil environments. BMC Chemistry 17(1):104

Yakout AA, El-Sokkary RH, Shreadah MA, Hamid OGA (2017) Cross-linked graphene oxide sheets via modified extracted cellulose with high metal adsorption. Carbohyd Polym 172:20–27

Yuan H, Meng LY, Park SJ (2016) A review: synthesis and applications of graphene/chitosan nanocomposites. Carbon Letters 17(1):11–17

Yun H, Jung S, Choi J, Cho SY, Kwak HW (2024) Highly efficient sorbent utilizing regenerated cellulose as an eco-friendly template for humic acid removal and oil–water separation processes. Sep Purif Technol 341:126812

Zakaria AF, Kamaruzaman S, Abdul Rahman N, Yahaya N (2023) Sodium alginate immobilized β-Cyclodextrin/Multi-walled carbon nanotubes as hybrid hydrogel adsorbent for perfluorinated compounds removal. J Polym Environ 31(5):1895–1914

Zhai Y, Zhu Z, Dong S (2015) Carbon-based nanostructures for advanced catalysis. ChemCatChem 7(18):2806–2815

Zhang Y, Zhang M, Jiang H, Shi J, Li F, Xia Y, Li H (2017) Bio-inspired layered chitosan/graphene oxide nanocomposite hydrogels with high strength and pH-driven shape memory effect. Carbohyd Polym 177:116–125

Zhang X, An W, Li Y, Hu J, Gao H, Cui W (2018) Efficient photo-catalytic hydrogen production performance and stability of a three-dimensional porous CdS NPs-graphene hydrogel. Int J Hydrogen Energy 43(21):9902–9913

Zhang K, Luo X, Yang L, Chang Z, Luo S (2021) Progress toward hydrogels in removing heavy metals from water: problems and solutions—a review. Acs EsandT Water 1(5):1098–1116

Zhang Y, Luo J, Zhang H, Li T, Xu H, Sun Y, Gao B (2022) Synthesis and adsorption performance of three-dimensional gels assembled by carbon nanomaterials for heavy metal removal from water: a review. Sci Total Environ 852:158201

Zhang Z, Lu Y, Zhao Y, Cui L, Xu C, Wu S (2025) Current developments in chitosan-based hydrogels for water and wastewater treatment: a comprehensive review. ChemistrySelect 10(6):e202404061

Zhao J, Liu H, Qi Y, Wang R, Lv Z, Yu Y, Sun S (2024) Construction of robust DA@ CNTs/CS hydrogel coatings on hydrophobic PVDF membrane by deposition of self-assembly for efficient separation of oil-water emulsion and dye wastewater. Sep Purif Technol 333:125956

Chapter 5
Treatment of Contaminated Water by Carbon Dot Hydrogel Composite

Zinnia Chowdhury, Debdutta Mandal, and Niladri Shekhar Samanta

Abstract Humanity's ability to progress sustainably is gravely challenged by water pollution and scarcity, so it's critical to create materials or a method that can effectively and precisely detect contaminants, clean up aquatic ecosystems, and use marine resources. A "win–win" tactic for preserving aquatic habitats is the composite of hydrogels with three-dimensional networks and photo-luminescent and hydrophilic carbon dots (CDs). This has shown that the previously suggested strategy is feasible. This chapter reviews the development of carbon dot hydrogel polymers for use in aquatic environments. The initial section throws light on the details of hydrogel preparation techniques, followed by the feasibility analysis of the carbon hydrogel composite. The study also threw light on the synthesis techniques of the carbon dot and carbon dot hydrogel composite. The development of CD hydrogel material research in the field of water treatment and water purification has also been examined in terms of the mechanisms' potential applications. The readers of this chapter will benefit from the insight into the CD hydrogel development and its utilities for future research.

Keywords Wastewater treatment · Carbon dots (CDs) · Hydrogel · Composite material · Adsorption

---

Z. Chowdhury · D. Mandal
Department of Chemical Engineering, Calcutta Institute of Technology, Uluberia, Howrah, India
e-mail: chowdhuryzin@gmail.com

D. Mandal
e-mail: debduttamandal23@gmail.com

D. Mandal
Department of Chemical Technology, University of Calcutta, Kolkata, India

N. S. Samanta (✉)
Department of Petroleum Engineering, Presidency University, Bangalore, Karnataka, India
e-mail: niladrisamanta91@gmail.com

S. Bardhan and M. P. Shah (eds.), *Applications of Hydrogels in Modern Wastewater Treatment*, Advances in Wastewater Research,
https://doi.org/10.1007/978-981-95-1598-1_5

Abbreviations

AA	Acryl acid
AM	Acrylamide
AMPS	2-Acrylamido-2-methylpropanesulfonic acid
BGQD	B-doped grapheme quantum dots
BW	Bound water
CDs	Carbon dots
CFHS	Based fluorescent hydrogel
CMC	Carboxymethyl cellulose
CNS	Cellulose nanocrystals
CS	Chitosan
DCPL	Down conversion PL
FW	Free water
GMA	Glycidyl methacrylated
GO	Grapheme oxide
GQDS	Graphene quantum dots
IW	Intermediate water
MBA	Methylene bis-acrylamide
MCC	Microcrystalline cellulose
NCDS	Nitrogen-doped carbon dots
NIPAM	N-isopropylacrylamide
ODA	Oxidized dextran
PAA	Polyacrylic acid
PAM	Polyacrylamide
PL	Photoluminescence
PLYS	Polylysine
PNIPAM	Poly-(N-isopropylacrylamide)
PPHS	Polyvinyl alcohol hydrogel
PVA	Polyvinyl alcohol
UCPL	Up conversion PL

5.1 Introduction

The survival of life on Earth and public health depend on having the potential availability of safer and cleaner drinking water. However, the alarming rise in industrialization and urbanization has led to widespread foulness of aquatic sources with enormous categories of water pollutants, which include heavy metals, pathogens, and organic compounds. It is vitally important to choose the best wastewater treatment technology to reduce the concentration of these contaminants in water. Conventional water treatment technologies were mainly categorized into three segments, namely, physical methods, chemical methods, and biological methods. Physical treatment methods mainly involved the removal of suspended solids through processes like sedimentation, filtration, and centrifugation (Ahmed et al. 2021). Coagulation, flocculation, and disinfection techniques were considered under chemical treatment methods and were mainly applied to agglomerate pollutants for their easy removal and to destroy harmful pathogens (Saravanan et al. 2021). Utilizing microorganisms' metabolic activity, biological treatment methods break down organic contaminants and transform nutrients into safe forms.

However, these conventional water treatment techniques frequently fall short of fully eliminating these impurities, which calls for the creation of sophisticated and effective wastewater treatment technology. In the recent era, nanotechnology has emerged as a promising frontier in the field of environmental remediation, offering innovative solutions to tackle water pollution. Carbon dots, or C-dots, have attracted a lot of attention among the several nanomaterials investigated for this purpose because of their special physicochemical characteristics, which include a large surface area, adjustable surface chemistry, superior photostability, and low toxicity. These characteristics make C-dots ideal candidates for diverse applications, ranging from sensing and imaging to catalysis and environmental remediation. On the other hand, the ability to absorb and hold onto high water content is possessed by hydrogels, which are three-dimensional structures made of hydrophilic polymers. Their porous structure and high water content facilitate the uptake and retention of contaminants from aqueous environments, thereby serving as effective adsorbents in water treatment applications. However, hydrogels alone often lack sufficient selectivity and efficiency for the removal of specific pollutants.

Carbon dot incorporation into hydrogel matrices represents a synergistic approach to overcome these limitations. In 2016, Hu et al. introduced carbon nanodots into a polyacrylamide network to synthesize cross-linked polyacrylamide (PAM) carbon nanodot hydrogel. It was shown in the study that the synergistic effect between carbon nanodots and polyacrylamide escalated the mechanical, swelling and recoverable properties of the fabricated hydrogel (Hu et al. 2016). Later in 2020, Nayak et al. cross-linked a hydrogel with carbon dots and fabricated a pH-responsive DNA—Carbon dots—Poly(vinylpyrrolidone) hybrid hydrogel. The study revealed that the incorporation of carbon dots has enhanced the hydrogel's mechanical characteristics and has also affected its photophysical characteristics.

By combining unique properties of both materials, carbon dot hydrogel composites can be tailored to target specific contaminants through adsorption, photocatalytic degradation, and other advanced mechanisms. The design flexibility of Cdot hydrogel composites allows for customization of their properties and performance characteristics, making them versatile tools for addressing various water quality challenges. Recently, in 2019, Singh et al., synthesized carbon dot induced hydrogel composites and applied them for various wastewater treatment applications, which include the eradication of heavy metal ions, detergents, organic matters, and dissolved salts. The study claimed that carbon dots induced hydrogel composites can be a promising technology in effectively treating various pollutants in water. A year later, Singh and Jelinek, (2020), synthesized C dot CMCPNIPAM hydrogel composites for applying it in solar-mediated water purification. The study suggested that the synthesized hydrogel composite showed enhanced mechanical stability, efficiently adsorbed water pollutants, and enhanced water swelling. They concluded that the synthesized hydrogel composite can be an effective technology in remediating water pollutants. In the same year, Issa et al. (2020), fabricated polyvinyl alcohol (PVA) and nitrogen-impregnated carbon dots composite film and applied it to remove cadmium from wastewater. The study revealed that the fabricated composite successfully removed 90.9% and 95.34% cadmium ions from portable water and tap water, respectively.

This chapter explores the recent advancements and challenges in the development of CDs-hydrogel composites for the treatment of contaminated water. It provides an extensive discussion on the fabrication techniques, their characterization, and application in environmental pollution remediation. Case studies and experimental findings from recent research efforts are presented to highlight the efficacy and potential of Cdot hydrogel composites in removing different classes of contaminants from water matrices. Furthermore, the chapter discusses the underlying mechanisms governing the pollutant removal processes and examines the factors influencing the performance and durability of Cdot hydrogel composites under different environmental conditions. The environmental sustainability and capability of these materials are also evaluated to assess their feasibility for large-scale deployment in water treatment infrastructures. Ultimately, the goal of this chapter is to give scientists, engineers, and students a thorough grasp of the potential and difficulties of carbon dot hydrogel composites as advanced materials for contaminated water restoration. By elucidating the synergistic interactions between carbon dots and hydrogels, this work seeks to inspire further innovations and advancements in the quest for sustainable water treatment solutions.

5.2 Hydrogel Preparation Technique

Hydrogels are a kind of polymer chain that is hydrophilic in nature. Though hydrophilic monomers are employed to synthesize hydrogels, hydrophobic monomers are also occasionally added during hydrogel synthesis to modify their properties for specific applications. Hydrogels' primary sources can be broadly categorized as natural sources and artificial (petrochemical) sources. The artificial source can be further classified into two categories: hydrogels derived from polypeptides (proteins) and those derived from polysaccharides. Most hydrogels are either petrochemically produced or wholly manufactured. Typically, acrylic monomers are used in their synthesis. The most widely used components that are used in the industries for the production of hydrogels are acrylamide (AM), acryl acid (AA), and their potassium or sodium analogs. The two widely used methods for the synthesis of acrylic hydrogel networks involve: (1) polymer crosslinking with a multifunctional crosslinker in water, and (2) concurrently carrying out crosslinking and polymerization with a crosslinker of polyvinyl (Ahmed 2015).

Polymeric hybrids are often made using one of two tried-and-true processes: (a) hydrophilic monomers and their polymerization, or (b) functionalization or modification of preexisting polymers (natural or synthetic). Most hydrogels are either petrochemically produced or wholly manufactured. They are constructed from acrylic monomers. When making hydrogel on an industrial scale, the most often used ingredients are the sodium or potassium equivalents of acrylamide (AM) and acryl acid (AA). Using a polyvinyl crosslinker for simultaneous crosslinking and polymerization and a polyfunctional crosslinker for crosslinking a Water-soluble prepolymer are the two most used techniques for producing acrylic hydrogel networks. The most popular and versatile technique for producing artificial hydrogels is the use of multifunctional vinyl monomers with free radicals (Zohourian, and Kabiri 2008).

5.2.1 Synthesis of Polyacrylic Acid (PAA) Hydrogel

Though many types of monomers are used to create hydrogels, acrylics are mostly applied. Potassium or sodium salts, along with polyacrylic acid (PAA), are among the primary components frequently employed in the industrial synthesis of hydrogels. Conversely, AA looks like a white liquid that smells like vinegar and has the mobility to change into its dimer, DAA. Maintaining a low DAA level is necessary to prevent deficiencies in the final product, such as yield reduction, soluble fraction loss, residual monomers, etc. Manufacturers employ appropriate methods for working with AA, for instance, timely order placement, delivery within time, repulsion of moisture, and controlling the storage temperature, to reduce possible issues brought on by AA's natural tendency to dimerize with time (Zohourian, and Kabiri 2008). As previously indicated, the free radical initiating polymerization of acrylic monomers is a common method used to synthesize hydrogel materials. Either a hydrocarbon media with well-dispersed monomes or an aqueous medium using solution polymerization are used to manufacture the resins. These various methods have been briefly described in the latter parts.

5.2.2 Synthesis of Hydrogel by Solution Polymerization Technique

Acrylic acid (AA) and its salts are usually polymerized by free radicals with a crosslinker in the hydrogel manufacturing process. Either before or after the polymerization process, the carboxylic acid groups in the final product undergo partial neutralization. The most frequent way to start chemical reactions is with redox systems, in which an oxidizing agent and a reducing agent combine, or by thermally breaking down free radical initiators like peroxides or azo compounds (Don et al. 2008).

Using a water-soluble crosslinker, like acrylamide (AA), methylene bisacrylamide (MBA) and/or its salts, can be solution polymerized in an aqueous solution with ease. The reactants, which usually range from 10 to 70%, dissolve in water at the proper concentrations. A schematic of a general solution polymerization procedure is shown in Fig. 5.1.

The flow diagram includes several components that are typically present in real-world free radical copolymerization reactor setups (Park et al. 2002). The steady addition of monomers A and B, as well as the solvent, chain transfer agent, and initiator, as shown in Fig. 5.1. Furthermore, in addition to the fresh feeds, an inhibitor may be added as an impurity. Following their combination (stream 1) with the recycle (stream 2), these feed streams are delivered to the reactor (stream 3), which is believed to be a jacketed well-mixed tank. Pumping a coolant into the jacket helps disperse the heat generated during the polymerization process. Once the unreacted monomers, initiator, chain transfer agent, and solvent have left the reactor, they go to the separator (stream 4), where any remaining polymer, initiator, and chain transfer agent are discharged. They go to the purge point (stream 8) with the solvent and unreacted monomers (stream 7), which is necessary to keep inserts from building up in the system and represents venting and other losses. Following the purge, the recycle

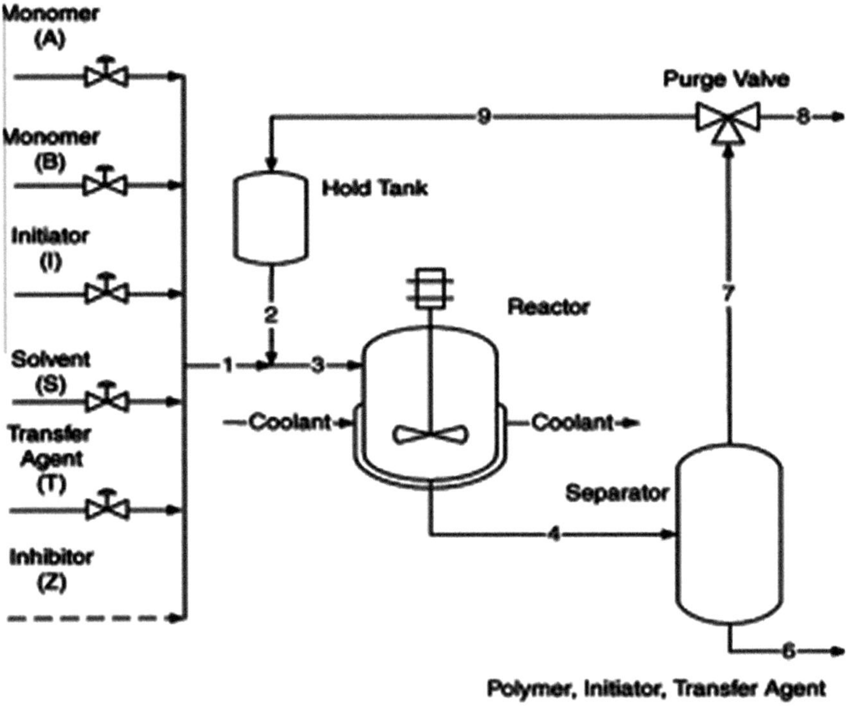

Fig. 5.1 Schematic diagram of solution polymerization with recycle stream. Reproduced with permission from Ahmed et al. (2015) ©Elsevier

hold tank stores the solvent and monomers (stream 9) and serves as a buffer to accommodate fluctuations in both the composition and flow rate of the recycled stream. Subsequently, the newly added feed is combined with the recycled stream effluent.

5.2.3 Synthesis of Hydrogel by Suspension Polymerization Technique

A very adaptable and versatile method for creating hydrogels with rapid absorption kinetics and a high swelling ability is inverse suspension (Hong et al. 2010). Water-soluble initiators are more efficient than those that can be dissolved in oil. Once the initiator dissolves in the aqueous dispersion medium, each particle functions as an individual microreactor for polymerization, as it contains all the necessary reactive components. The microspherical particles that are left over from the continuous organic phase can be easily separated either by filtration or by centrifugation. As soon as they dry, these particles or beads will create a powder that flows easily. The inverse suspension method exhibits additional benefits over the solution method, in addition to the distinctive flowing characteristics of these beads. Those mainly include better control over particle size distribution, reaction heat removal, and additional opportunities to modify particle structure or morphological change (Ma and Lee 2009). Using this technique, spherical hydrogel microparticles ranging in size from 1 μm to 1 mm are prepared. Fine monomer droplets are formed during suspension polymerization when the monomer solution is distributed in the non-solvent and stabilized with the addition of a stabilizer. Radicals produced when an initiator breaks down thermally start the polymerization process. The initiator, crosslinking agent, and monomers are then cleaned to remove them from the freshly formed microparticles. Figure 5.2 clearly depicts the block diagram and schematic diagram of hydrogel preparation by suspension polymerization technique.

5.2.4 Hydrogel Preparation and Process Optimization Using Grafted Starch

Natural polymer hydrogels are frequently nontoxic and biodegradable, especially those derived from polysaccharides. Numerous reports of technical work and research have been published. The two most important methods for enhancing starch's properties and broadening its uses are vinyl graft copolymerization and chemical modification. Free radicals are then produced on the starch granule's surface in order to produce starch graft copolymers, for instance, starch-g-polyvinyl alcohol, starch acrylonitrile, starch-g-polystyrene, and starch-g-methacrylonitrile, shown in Fig. 5.3. The corresponding vinyl monomers are then copolymerized with the starch granules.

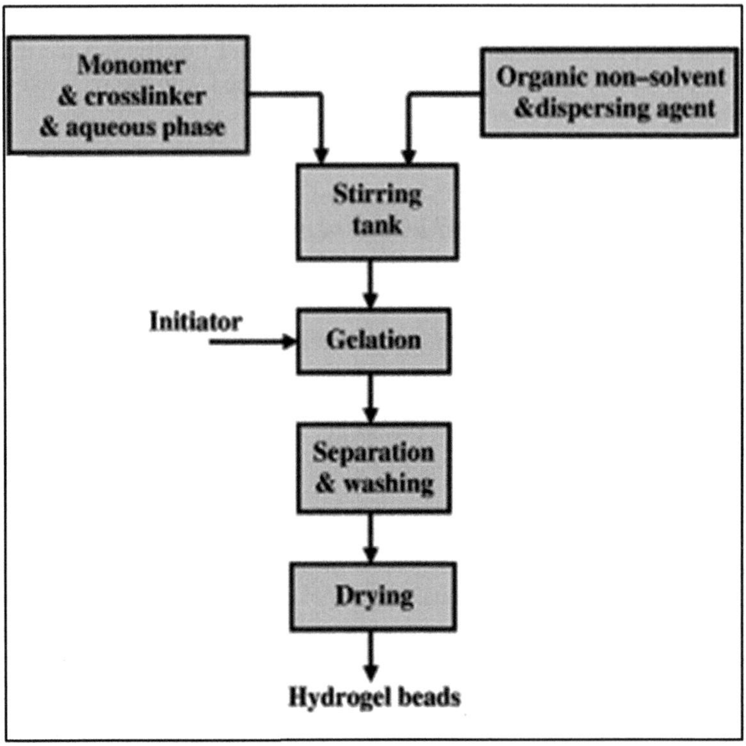

Fig. 5.2 Block diagram of suspension polymerization technique. Reproduced with permission from Ahmed et al. (2015). Elsevier

Because there is a nonbiodegradable portion of the polymer, these copolymers likewise have limited biodegradability (Talaat et al. 2008). It has been found in various research works that sugar-containing poly(acrylate) hydrogels have been prepared through the modification of natural polymers through hydrogel synthesis, such as biocatalytic synthesis. These authors discovered that they could completely eradicate the polyacrylamide hydrogels' relative brittleness and lessen the production of unwanted fine particles during wet milling by adding tiny amounts of agar. Acrylonitrile was grafted onto cassava starch by Raju and Raju (2001) using a polymerization process started by ceric ions. The effects of reactant concentrations and polymerization duration were examined by these authors. A manganese pyrophosphate redox-initiating method was used in the work of Qunyi and Ganwei (2005) to create superabsorbents by grafting 2-acrylamido-2-methylpropanesulfonic acid (AMPS) and acrylonitrile onto starch. The graft polymer's saponification time gradually decreased once AMPS was added. As a result, the overall time required to produce superabsorbents likewise shortens. Investigations were conducted to determine how the temperature and dosage of potassium hydroxide affected the superabsorbent's water absorbency.

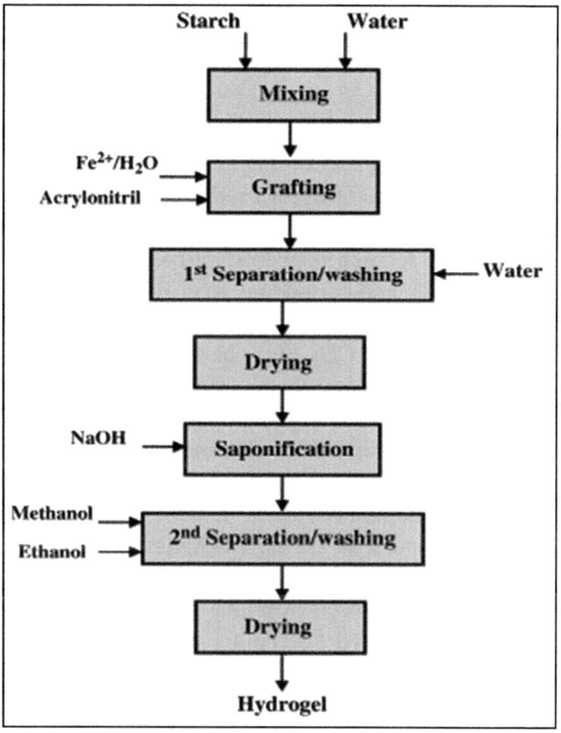

Fig. 5.3 Schematic flowchart of preparation of hydrogel based on grafted starch. Ahmed et al. (2015) Reproduced with permission from. Elsevier

5.2.5 Hydrogel Preparation by Irradiation Method

The synthesis of acrylamide hydrogel from aqueous solutions using γ-ray irradiation has been studied, revealing the influence of factors such as solution concentration, radiation dose, pH, and irradiation time through analysis of the resulting gels. Understanding the fundamental radiation chemistry is necessary for the design and optimization of radiation-based hydrogel production technologies that are effective, safe, and financially viable. This necessity has long been one of the primary driving forces behind research on polymer radiolysis in aqueous solutions. There are several ways to use the radiation approach to generate hydrogels: you can irradiate, polymer in an aqueous solution or a polymer monomer in solid form. There are certain disadvantages to the first approach, which involves irradiating hydrophilic polymers in a dry form (Francis et al. 2009). Specific synthesis methods like pressing or melting are required for sample preparation, and producing uniform macroscopic hydrogels may present certain difficulties. Furthermore, compared to irradiation in solution, obtaining a gel often needs significantly higher doses of ionizing radiation, and it might be challenging to fully exclude the oxygen (O_2) that could cause unintended side effects (Varshney 2007).

5.3 Strategies for the Synthesis of Carbon Dot Hydrogel Composite

The class of carbon-based nanoparticles known as carbon dots can be divided into multiple subgroups according to their morphological and crystalline characteristics. This structure can consist of a carbon skeleton of sp2 hybridization and sp3 hybridization, as well as nanoscale fluorescent carbon material (Deka et al. 2019; Bardhan et al. 2020). Carbon nanodots, carbon quantum dots (CQDS), graphene quantum dots (GQDS), and carbide polymer dots are the three main types of CDs that are separated according to the various surface group structures of the carbon nucleus (Xia et al. 2019). CDs often have a large number of amine, carboxyl, hydroxyl, and epoxide groups on their surface. The ability of these functional groups to form hydrogen bonds in water makes CQDs more soluble in water and makes it easier for them to hybridize with other materials (Zhao and Chung 2018; Roy et al. 2022). Hydrogel is a material made by crosslinking homopolymers or copolymers with a three-dimensional network structure. The crosslinking process can be physical or chemical (Xiao et al. 2022). The hydrogel possesses a potent ability to both adsorb and retain water from the soil. This allows it to efficiently boost soil fertility and maintain soil moisture levels by retaining and absorbing water inside the soil. The process of creating a CD-hydrogel composite involves evenly dispersing CDs throughout the hydrogel's internal structure and interacting with it to create a composite system. Strong water adsorption and retention capabilities allow the hydrogel to effectively block water loss and evaporation from the soil while also retaining and adsorbing water inside the soil.

5.3.1 Feasibility of the Carbon Dot Hydrogel Preparation Technique

The synthesis of CD-hydrogel composites is generally categorized into three processes: (i) Gelation of hydrogel into CDs solution, (ii) After gelation, premade CDs are implanted into a hydrogel, (iii) Quantum dots are generated in situ within a gel that has already been formed.

i. *Hydrogel creation in a CD solution:* CD-hydrogel composites are typically made by gelating hydrogel in a CD solution. This technique requires the inclusion of an initiator under the proper conditions after completely mixing the CD solution with the crosslinking matter. Zeng et al. in 2021 had synthesized a CS/CNCD composite hydrogel by mixing a suspension of cellulose nanocrystals (CNS) grafted with CDs with a 98 weight percent chitosan (CS) solution. Pentylene glycol was then added to cause a coupling reaction. The FTIR analysis method revealed that during the crosslinking reaction there is a formation of a novel –C = CN = C– structure. Polyethyleneimine (PEI) and organic molecules with either nitrogen oxygen as a functional group or oxygen as a functional group can form a strong hydrogen bond. By employing PEI to passivate the surface of the willow-based CDs, the degree of crosslinking and heat stability of synthesized PEICD/microcrystalline cellulose (MCC) hydrogels increases. The single-bonded OH groups and the hydrophilic component of the MCC hydroxide radicals shared a stronger hydrogen bond, which improved the photoluminescent characteristics of the CDs. It is possible to convert a solution into gel in some gel reaction systems without the need for additional polymer. Polylysine (PLYS)-CD was created by replacing the PLYS main amino group with an aldehyde in an oxidized dextran (ODA) (Yang et al. 2021). Schiff base reaction was used to create PLYSCD@ODA gel. It was found that the high concentration of ODA providing aldehyde groups in this reaction limited the movement of movement of polymer chains.

ii. *Premade CDs are implanted into a hydrogel following gelation*: Another method for creating CDs and hydrogel was to embed previously prepared CDs into the hydrogel following gelation. In this procedure, after a hydrogel has formed, prepared CDs are added chemically or mechanically. After three freeze–thaw cycles, pectin/polyvinyl alcohol hydrogel (PPHS) was immersed in glycidyl methacrylate (GMA) CD solution for an entire day to produce a high-toughness hydrogel (Jiang et al. 2024). An electrostatic bond existed between the carboxylic acid sites of the PPH and the GMACDS, in contrast to the PPH that was generated from a structure that had a tight polymer network and no ligand ions or compounds. In a similar vein, the hydrogel's secondary crosslinking caused by the addition of CDs resulted in toughening properties.

iii. Another method for creating a CD hydrogel composite from scratch, is the in situ synthesis of quantum dots in a perforated gel. Neither the hydrogel nor the CDs must be completely prepared prior to the synthesis in order to use this method. The CD-hydrogel composite was created using a one-step hydrothermal process. According to Chen et al. (2020), solid nitrogen-doped carbon dots (NCDS)/CNF gels were used to make a brown CD solution. The combination was heated for 24 h at 120–180 °C after being filled with CNF suspension and EDA in a hydrothermal reactor. In the NCDS/CNF gel structure, it was discovered that the NCDS and CNF cross-linked to create a strong binding by covalent and hydrogen bonding.

Thus, it created the most stable chemical hydrogel when its cross-linked structure gave it exceptional resistance to acids and alkalis. The production of N-doped spherical CQDS with a graphite lattice structure through amidation allowed the amide groups of EDA to be grafted onto the surface of the CDs. An ester link was created by the condensation and dehydration of the carboxyl groups on CA and the hydroxyl groups on CNF. The hydrothermal process caused epichlorohydrin to undergo ring-opening and react with the OH groups on CNF, resulting in the formation of ether bonds that improved the hydrogel's swelling capacity and chemical crosslinking density. Boron-doped graphene quantum dots (BGQD) anchored on a graphene hydrogel (GHBGQD) were created by Tam et al. (2019) by doping graphene oxide (GO) with boric acid and then adding glucose. The synthesized BGQD was gradually bonded to the graphene sheet of the GH utilizing $\pi\pi$ stacking. The rendered GHBGQD showed outstanding performance for hydrogen precipitation, oxygenolysis, and electrocatalytic oxygen reduction.

5.3.2 Advantages and Disadvantages of CD-Hydrogel Composite Synthesis Strategies

The synthesis technique of the CD-hydrogel composite has some advantages and disadvantages. In case of carbon dots having 3D porous structure and high values of surface area, transparency and ductile properties are some of the advantages of it. Besides, it has some disadvantages like low mechanical strength and also insufficient degrees of crosslinking properties. In the case of hydrogel synthesis, hydrogels are easily aggregated but have some disadvantages, like hydrogels are generally soft and lacking of mechanical strength needed for practical application especially in flow conditions or large-scale production systems (Zhang et al. 2022). The tendency of hydrogels to swell in aqueous solution may have an impact on the synthesis procedure. The Win–Win strategies model was designed by Jiang et al. (2024) to explain the advantages and disadvantages of CD-hydrogel composite synthesis strategies. Figure 5.4 shows the advantages of the synthesis strategies for CD-hydrogel composites.

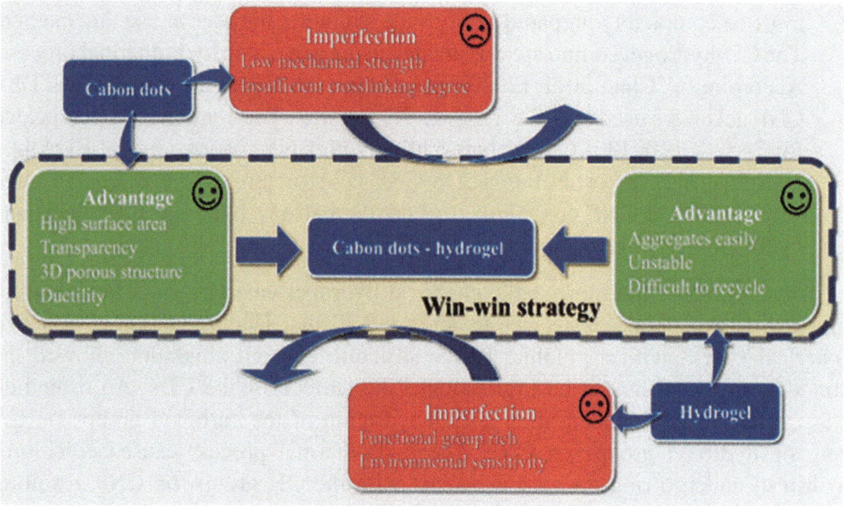

Fig. 5.4 Benefits of the CD-hydrogel composite production methods. Reprinted and permission was taken from Jiang et al. (2024). Elsevier

In many fields, carbon dot hydrogel composites outperform conventional treatments in many ways. They are superior to conventional drug delivery systems in biomedicine due to their biodegradability, biocompatibility, and tunable properties, which permit controlled release with fewer side effects. Carbon dots have better fluorescence, better photostability, and less toxicity than conventional imaging agents, making it possible to conduct diagnostics with greater precision and safety. These composites outperform conventional adsorbents in environmental remediation because they have a better adsorption capacity, larger surface area, and selectively remove pollutants, making water purification more effective and long-lasting (Bardhan et al. 2022). In addition, the synergistic combination of CDs and hydrogels enables the creation of multifunctional materials that are capable of simultaneously addressing multiple challenges, such as drug delivery and imaging or simultaneous pollutant removal and sensing, both of which are frequently impractical with conventional treatments that use only one component. Carbon dot hydrogel composites are a promising alternative to conventional treatment methods because of their versatility, ease of synthesis, and potential for cost-effectiveness.

5.4 Development of Carbon Dot Hydrogels

Several investigations have documented various techniques for creating CD-based hydrogels. Most concentrate on employing different crosslinking techniques and functionalized CDs. Thus, the kind of functionalized CDs and crosslinking techniques employed can classify the preparation techniques.

5.4.1 Functionalized CDs

The fabrication of CD-based hydrogels requires functionalized CDs as fundamental structural components. Functionalized CDs can easily build a stable hydrogel network by crosslinking with other units having many functional groups on their surface. Another essential element of CD hydrogels is polymers. Further encouraging the creation of hydrogel networks are the many functional groups along the covalent chains of the polymer that can crosslink with CDs, as depicted in Fig. 5.5 (Sui et al. 2020). It is important to talk about the hydrogels made from various functionalized CDs because the characteristics of CD hydrogels greatly depend on the functionalized CD that is employed.

5.4.1.1 CDs-Functionalized with an Amino Group

For the synthesis of CD hydrogels, amino-functionalized CDs ($CDsNH_2$) are one of the most crucial precursors. They are mostly made from nitrogen-rich substances

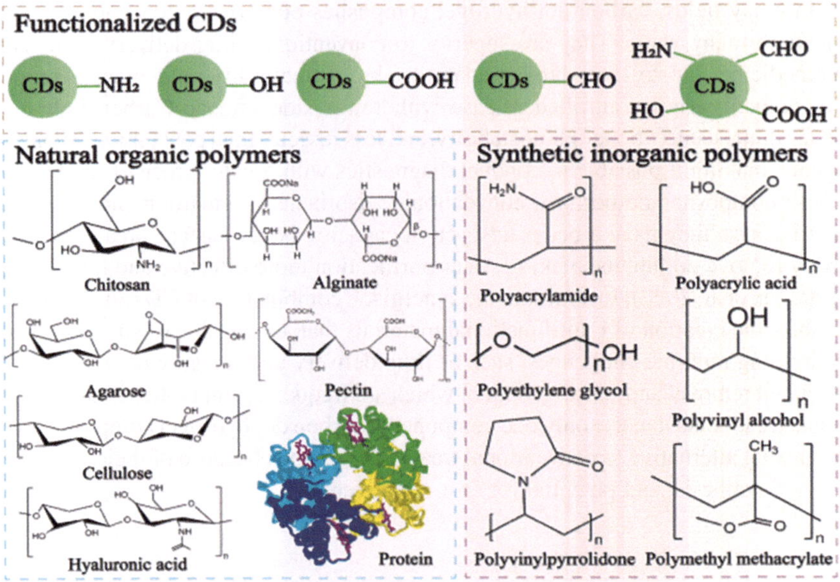

Fig. 5.5 Several functionalized carbon dots (CDs) and their polymers. Reproduced with permission from Cui et al. (2023). Elsevier

like ammonia, urea, and amino acids, which provide the nitrogen needed for amino functionalization. In order to bind the CDs through the creation of amide bonds, the amino groups on $CDsNH_2$ undergo acylation with carboxyl and aldehyde groups found on the polymer chains during hydrogel formation. A Schiff base reaction occurs when an amino group reacts with an aldehyde group. In a study, Yue et al. (2022) hydrothermally produced $CDsNH_2$ employing tetraphenyl-zinc-porphyrin and europium chloride monomers. The CDs next underwent a Schiff base reaction with aldehyde-functionalized hydroxyalanine, crosslinking at ambient temperature to generate a hydrogel, as shown in Fig. 6a. In this case, the CDs performed double functions: they were photosensitizers for photodynamic tumor therapy and crosslinkers for Schiff base reactions. The reaction between the carboxyl group and the amino group is also known as the peptide bond formation reaction. Using polytetrafluoroethylene, citric acid, and ethylenediamine precursors, Zhang et al. (2023) created $CDs\text{-}NH_2$ via hydrothermal synthesis. The amino groups on the CDs and the carboxyl groups on the nanocellulose chains crosslink to generate hydrogels. The improved sorption and photocatalysis of the CDs allowed the hydrogels to remove chromium with exceptional proficiency.

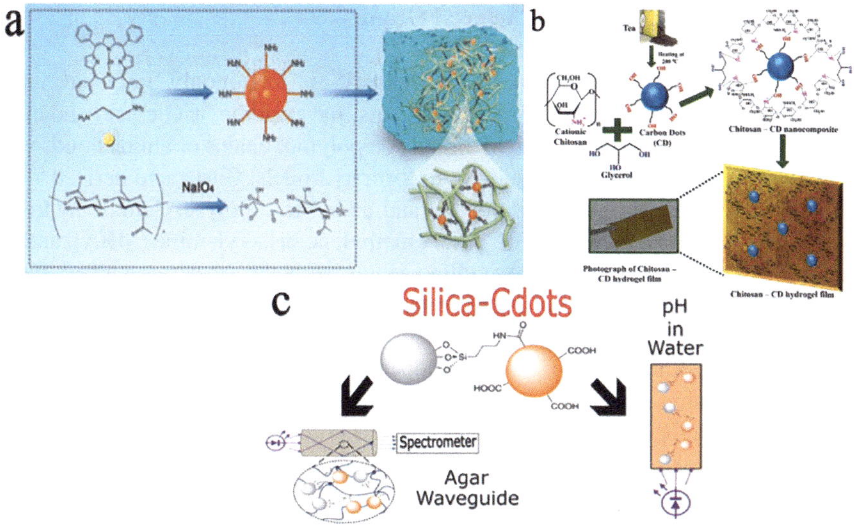

Fig. 5.6 **a** Synthesis of CDs-based hyaluronic acid hydrogel; **b** fabrication of chitosan CDs nanocomposite hydrogel; **c** synthesis of CD-agarose composite hydrogel. Reproduced with permission from Cui et al. (2023). Elsevier

5.4.1.2 CDs with Functionalized Hydroxyl Group

Hydroxyl-functionalized carbon dots (CDs-OH) are typically derived from precursors abundant in hydroxyl groups, including compounds like natural sugars, alcohols, and phenols. When CDs-OH react with the COOH groups on polymer chains, ester linkages are formed that attach to the CDs and cause them to crosslink into hydrogels (Fig. 6b). Konwar et al. (2015) synthesized hydroxyl-functionalized carbon dots (CDs-OH) from tea leaves and incorporated them into chitosan films through electrostatic attraction between the highly negatively charged CDs and the positively charged chitosan. The incorporation of CDs enhanced the ultraviolet (UV) shielding capability of the hydrogel nanocomposite films. Singh et al. (2020) used hydrothermal treatment to manufacture CDs-OH utilizing urea and citric acid as precursors. The combined CDs-OH, poly(N-isopropylacrylamide) (PNIPAM), and carboxymethyl cellulose (CMC) produce a hydrogel through free radical polymerization. The hydrogel system consists of PNIPAM, which serves as the CMC, the hydrogel's skeleton, which fixes the CD to the hydrogel's surface via the carboxylic acid functional group, and the CD, which acts as a photothermal agent to cause the hydrogel to undergo a phase shift that improves its mechanical stability. Polyphenols, abundant in tea, serve as a green synthetic precursor for CDs-OH.

5.4.1.3 CDs Modified with Carboxyl Group

Carboxyl-modified carbon dots (CDs-COOH) are commonly prepared using carboxylic acid containing starting constituents, for example, acetic acid and citric acid. Through ester linkages with hydroxyls on polymer chains or amide bonds with amines, CDs-COOH crosslink in order to form hydrogels. Citric acid derived CDs-COOH were made by Sun et al. (2021) and added to a polyacrylamide hydrogel along with polyvinyl alcohol (PVA), N, N′methylene bisacrylamide (MBA), and N-isopropylacrylamide (NIPAM). As gelling agents, N, N, N′, N′ tetramethylethylene-diamine and ammonium persulfate were added. The hydrogel exhibited a strong pH-responsive fluorescence due to the high quantum yield of 43.9% of CDs-COOH. In a different study, Amato et al. (2021) developed luminous Silica-dots by synthesizing CDs-COOH using citric acid as precursor material and coupling them to functionalized silica nanoparticles in order to achieve fluorescent Silica dots. These were subsequently cross-linked into agarose hydrogels, where the optical characteristics of the CDs allowed for pH sensing (Fig. 6c). To sum up, CDs-COOH are versatile cross-linkers that can be used to create luminous and responsive hydrogels because of their carboxyl-rich precursors.

5.5 Utilization of Carbon Dot Hydrogel Composite for Wastewater Treatment

For human society to flourish sustainably and maintain natural equilibrium, water purification is crucial. Water pollution and shortage can be addressed with carbon dot hydrogel via means of desalination, sorption, and photodegradation (Zhao et al. 2015).

5.5.1 Adsorption

Adsorption has established its noteworthy footsteps in water treatment for eliminating environmental contaminants, owing to its cost-effectiveness, high efficiency, and operational simplicity (Valadi et al. 2022; Park et al. 2020; Singh et al. 2021). Contaminants can be captured and retained on or within CD-based hydrogels via various mechanisms, including hydrogen bonding, electrostatic attraction, chelation, and complex formation. The large surface area-to-volume ratio of carbon dots (CDs) facilitates the adsorption of pollutants through mechanisms such as hydrogen bonding, electrostatic attraction, and π-π interactions. Additionally, the hydrogel matrix present on CDs also absorbs pollutants through diffusion and partitioning, which results in the removal of a wide range of noxious constituents. Using the NH_2 and OH surface residues for electrostatic attraction, or using another technique, first, contaminants are adsorbed on the hydrogel structure. The CD-hydrogel's porous structure allows contaminants to permeate from the exterior to the interior, where they are absorbed and rendered immobile. By altering the process used to synthesize the CD-hydrogel's utilized material, various functional groups, such as amine, hydroxyl groups, and carboxylic acid, can be loaded onto the material to increase its adsorption capacity. Using the EDC/NHs coupling approach, Yang et al. (2022) created a luminous hydrogel by grafting CDs onto carboxylate nanocellulose.

5.5.2 Photo-Catalytic Degradation

Carbon dots (CDs), owing to their exceptional photoluminescent characteristics—such as downconversion (DCPL) and upconversion photoluminescence (UCPL) improve the activity of semiconductor materials under visible and near-infrared light. By combining semiconductive materials with CD-hydrogel, new possibilities are opened up for various applications (Zhu et al. 2015). The superior electron transmission and storage properties of CDs are demonstrated. Likewise, the structural conjugation of the πsystem can expedite the transfer of electrons from semiconductors to their conducting network system. When CD-hydrogel is filled with semiconductive materials, its photocatalytic capacity increases. When CDs are exposed to sunlight, they transform visible light into energy that excites semiconductive materials, which in turn produces a whole that can accelerate the process of conduction and electron migration. The photodegradation of tetracycline hydrochloride was carried out using reduced graphene oxide (rGH) doped with CDs, obtained using the in situ growth approach (TCH) (Ma et al. 2021). It was found that adding more CDs to CDBVO/rGH and incorporating them increased the semiconductors' photocatalytic activity. Through the upconversion effect, CDs may absorb near-infrared light and transform it into visible light under illumination. The photogenerated charges can quickly traverse the surface of CDs and rGH due to rGH's large πconjugated structure and good electrical conductivity, which in turn generates the strong electron trapping capacity of CDs and the difference in conduction band energy levels between the three. The CDs in the hydrogel matrix also exhibit catalytic properties, enabling them to degrade pollutants such as organic dyes through photochemical reactions. Furthermore, the CDs can form complexes with metal ions such as Pb^{2+}, Hg^{2+}, and Cu^{2+} through coordination bonds, allowing for the removal of these toxic ions. Finally, the hydrogel matrix acts as a filter, excluding larger pollutants such as particulate matter and bacteria through size exclusion.

5.5.3 Desalination

The demand for water has risen due to the population expansion as well as degradation of the water environment due to pollution, and nowadays it is highly challenging for water security (Tian and Chen 2022). Since seawater is the primary component of the global water system, desalination plays a critical role in this regard. Solar-driven interfacial evaporation is one of the desalination technologies that is widely used since it is inexpensive and environmentally friendly (Wang et al. 2022). There are three main kinds of water molecules: bound water (BW), intermediate water (IW), and free water (FW) (Tang et al. 2022). Four nearby water molecules can establish a stable hydrogen bond with free water. By forming hydrogen bonds with nearby polar functional groups on the molecular chain of hydrophilic polymers, bound water was created. BW isn't a part of the extraction process. Differential scanning colorimetry was used by Singh et al. (2019) to calculate the heat uptake for the evaporation of water molecules from CD-hydrogels. When the hydrogel was doped with CD, its enthalpy dropped. The analysis of enthalpy and the swelling characteristics of hydrogel revealed that the water molecules that were physically connected to the hydrogel framework were substituted by the crosslinking process and the addition of CDs. Mohapatra 2022) reported that hybrid hydrogel modified CDs with Fe_3O_4 based on synthesized poly (vinyl alcohol) showed that the PPH could be adjusted to build a vertical structure that was aligned well and modulated the ratio of hydrogel raw materials. The IW/FW ratio increased as a result of this modulation procedure. Singh and Jelinek (2020) state that CD-hydrogel (Cdot-CMC-PNIPAM) was temperature-modified using poly (N-isopropyl acrylamide) (PNIPAM). CDs having the hydrogel and amide group, which enhance the water solubility of the gel and enable it to adsorb salt and pollutants for water purification. CDs use color contrast to provide an optical switching function. As the temperature rises, PNIPAM can reduce the amount of water molecules in the gel, facilitating desalination.

5.6 Future Outlook and Challenges

The synthesis techniques for CD-hydrogels are explained here. Additionally, the advancement of their research in the area of water quality detection and purification is examined. CD-hydrogel composites have a wide range of possible applications, but environmental detection and cleanup are two areas where they shine. This is due to their special qualities and the way they compensate for the shortcomings of their separate uses as composites. The creation and use of CD-hydrogel composites still present difficulties; therefore, this does not imply that the technology is sophisticated or dependable. First, because CDs are hydrophilic, they usually increase the degree of crosslinking in the hydrogel, improving its mechanical and swelling properties. Nevertheless, the CDs' optical qualities aren't always improved, as is frequently shown in investigations in which the composite's maximum PL quantum yield is less than that of the CDs alone due to the CDs' propensity to assemble inside the hydrogel's structure. Eventually, it could be how to reduce CDs' propensity to cluster during synthesis by making quantitative changes to the hydrogel's pore structure and surface functional group modifications to guarantee that the CDs are widely dispersed. Secondly, the properties of CD-hydrogel materials are still little understood by researchers, and there is now only one source of these materials, many of which are only useful for synthesizing the raw components needed to make CDs or hydrogels and not for the development of composite materials. The advancement of computer technology has made it possible to apply data-driven techniques like machine learning and computer simulation. This would make it easier to create and optimize the composition and properties of materials, enhancing the effectiveness and precision of research and development (Jiang et al. 2024). Lastly, unique identification methods are needed for CD-hydrogels in water purification applications in order to provide excellent selectivity and regenerative processing for recycling. It is possible to develop and combine smart-responsive CD-hydrogel composites with other processes, such as photocatalysis and thermocatalysis, to construct integrated water purification systems that can detect and degrade pollutants simultaneously. The implementation of systems can be achieved by employing CDs that exhibit photothermal effects or possess electron donor–acceptor capabilities.

5.7 Conclusion

In conclusion, a potential development in environmental remediation technology is the use of carbon dot hydrogel composites for the treatment of contaminated water. This innovative method provides a sustainable, environmentally acceptable solution in addition to increasing the effectiveness of pollutant removal through the synergistic effects of hydrogels and carbon dots. Together with hydrogels' water retention and biocompatibility, carbon dots' special qualities—such as their large surface area and adjustable surface chemistry—provide a flexible platform for dealing with a variety of pollutants. We may expect these composites to perform far better and be more applicable in real-world situations as research into their optimization and refinement proceeds. Ultimately, incorporating such advanced materials into water treatment procedures may be essential to protect our water supplies and foster a more wholesome environment for future generations.

References

Abdullah Issa M, Abidin Z (2020) Sustainable development of enhanced luminescence polymer-carbon dots composite film for rapid Cd2+ removal from wastewater. Molecules 25(15):3541

Ahmed EM (2015) Hydrogel: Preparation, characterization, and applications: a review. J Adv Res 6(2):105–121

Ahmed SF, Mofijur M, Nuzhat S, Chowdhury AT, Rafa N, Uddin MA, Show PL (2021) Recent developments in physical, biological, chemical, and hybrid treatment techniques for removing emerging contaminants from wastewater. J Hazard Mater 416:125912

Amato F, Soares MCP, Cabral TD, Fujiwara E, Cordeiro CMDB, Criado A, Bartoli JR (2021) Agarose-based fluorescent waveguide with embedded silica nanoparticle–carbon nanodot hybrids for pH sensing. ACS Applied Nano Materials 4(9):9738–9751

Bardhan S, Roy S, Chanda DK, Ghosh S, Mondal D, Das S, Das S (2020) Nitrogenous carbon dot decorated natural microcline: an ameliorative dual fluorometric probe for Fe^{3+} and Cr^{6+} detection. Dalton Trans 49(30):10554–10566

Bardhan S, Roy S, Das S (2022) Carbon dots: fundamental concepts and biomedical applications. Nanotechnol Biomed Appl 83–108

Chen X, Song Z, Li S, Thang NT, Gao X, Gong X, Guo M (2020) Facile one-pot synthesis of self-assembled nitrogen-doped carbon dots/cellulose nanofibril hydrogel with enhanced fluorescence and mechanical properties. Green Chem 22(10):3296–3308

Cui F, Xi L, Wang D, Ren L, Tan X, Li X, Li T (2023) Advanced in carbon dot-based hydrogels for antibacterial, detection and adsorption. Coord Chem Rev 497:215457

Deka MJ, Dutta P, Sarma S, Medhi OK, Talukdar NC, Chowdhury D (2019) Carbon dots derived from water hyacinth and their application as a sensor for pretilachlor. Heliyon 5(6)

Don TM, Huang ML, Chiu AC, Kuo KH, Chiu WY, Chiu LH (2008) Preparation of thermo-responsive acrylic hydrogels useful for the application in transdermal drug delivery systems. Mater Chem Phys 107(2–3):266–273

Francis S, Mitra D, Dhanawade BR, Varshney L, Sabharwal S (2009) Gamma radiation synthesis of rapid swelling superporous polyacrylamide hydrogels. Radiat Phys Chem 78(11):951–953

Hong J, Lee J, Rhym YM, Kim DH, Shim SE (2010) Polyelectrolyte-assisted synthesis of polystyrene microspheres by dispersion polymerization and the subsequent formation of silica shell. J Colloid Interface Sci 344(2):410–416

Hu M, Gu X, Hu Y, Wang T, Huang J, Wang C (2016) Low chemically cross-linked PAM/C-dot hydrogel with robustness and superstretchability in both as-prepared and swelling equilibrium states. Macromolecules 49(8):3174–3183

Jiang M, Wang Y, Li J, Gao X (2024) Review of carbon dot–hydrogel composite material as a future water-environmental regulator. Int J Biol Macromol 131850

Konwar A, Gogoi N, Majumdar G, Chowdhury D (2015) Green chitosan–carbon dots nanocomposite hydrogel film with superior properties. Carbohyd Polym 115:238–245

Ma Y, Lee P (2009) Investigation of suspension polymerization of hydrogel beads for drug delivery. Iran Polym J 18(4):307–313

Ma C, Xie Z, Seo WC, Din STU, Lee J, Kim Y, Yang W (2021) Carbon dot-coupled BiVO4/reduced graphene hydrogel for significant enhancement of photocatalytic activity: antibiotic degradation and CO2 reduction. Appl Surf Sci 565:150564

Mohana Raju K, Padmanabha Raju M (2001) Synthesis of novel superabsorbing copolymers for agricultural and horticultural applications. Polym Int 50(8):946–951

Mohapatra S (2022) Multifunctional bio-based photothermal hydrogel for highly efficient seawater desalination and contaminant adsorption. J Environ Chem Eng 10(6):108616

Nayak S, Prasad SR, Mandal D, Das P (2020) Hybrid DNA–Carbon Dot-Poly (vinylpyrrolidone) hydrogel with self-healing and shape memory properties for simultaneous trackable drug delivery and visible-light-induced antimicrobial photodynamic inactivation. ACS Appl Bio Mater 3(11):7865–7875

Park MJ, Hur SM, Rhee HK (2002) Online estimation and control of polymer quality in a copolymerization reactor. AIChE J 48(5):1013–1021

Park M, Wu S, Lopez IJ, Chang JY, Karanfil T, Snyder SA (2020) Adsorption of perfluoroalkyl substances (PFAS) in groundwater by granular activated carbons: roles of hydrophobicity of PFAS and carbon characteristics. Water Res 170:115364

Qunyi T, Ganwei Z (2005) Rapid synthesis of a superabsorbent from a saponified starch and acrylonitrile/AMPS graft copolymers. Carbohyd Polym 62(1):74–79

Roy S, Bardhan S, Hoque NA, Das S (2022) A Brief review on carbon-based nanosystems for detection of pollutants in industrial wastewater and living cells and their subsequent removal strategies. Nov Appl Carbon Based Nano-Mater 321–339

Saravanan A, Kumar PS, Jeevanantham S, Karishma S, Tajsabreen B, Yaashikaa PR, Reshma B (2021) Effective water/wastewater treatment methodologies for toxic pollutants removal: Processes and applications towards sustainable development. Chemosphere 280:130595

Singh S, Jelinek R (2020) Sunlight-activated phase transformation in carbon dot-hydrogel facilitates water purification and optical switching. ACS Appl Polym Mater 2(7):2810–2818

Singh S, Shauloff N, Jelinek R (2019) Solar-enabled water remediation via recyclable carbon dot/hydrogel composites. ACS Sustain Chem Eng 7(15):13186–13194

Singh N, Kumari S, Khan S (2021) Improved fluoride removal efficiency using novel defluoridation pencil. Mater Today Commun 28:102521

Sui B, Li Y, Yang B (2020) Nanocomposite hydrogels based on carbon dots and polymers. Chin Chem Lett 31(6):1443–1447

Sun L, Mo Z, Li Q, Zheng D, Qiu X, Pan X (2021) Facile synthesis and performance of pH/temperature dual-response hydrogel containing lignin-based carbon dots. Int J Biol Macromol 175:516–525

Talaat HA, Sorour MH, Aboulnour A, Shaalan HF, Ahmed EM, Awad AM, Ahmed M (2008) Development of a multi-component fertilizing hydrogel with relevant techno-economic indicators. Am J Agric Environ Sci 3(5):764–770

Tam TV, Kang SG, Kim MH, Lee SG, Hur SH, Chung JS, Choi WM (2019) Novel graphene hydrogel/B-doped graphene quantum dots composites as trifunctional electrocatalysts for Zn–air batteries and overall water splitting. Adv Energy Mater 9(26):1900945

Tang S, Gong J, Shi Y, Wen S, Zhao Q (2022) Spontaneous water-on-water spreading of polyelectrolyte membranes inspired by skin formation. Nat Commun 13(1):3227

Tian K, Chen Z (2022) What roles do smart sensors play in citizens' water use? from the perspective of household water-saving. Water Supply 22(3):3519–3525

Valadi FM, Shahsavari S, Akbarzadeh E, Gholami MR (2022) Preparation of new MOF-808/chitosan composite for Cr (VI) adsorption from aqueous solution: experimental and DFT study. Carbohyd Polym 288:119383

Varshney L (2007) Role of natural polysaccharides in radiation formation of PVA–hydrogel wound dressing. Nucl Instrum Methods Phys Res, Sect B 255(2):343–349

Wang Z, Huang H, Huang S, Lin P, Pan D, Wang H, Wang L (2022) Continuous and efficient purification of seawater using suspended photothermal nanocomposite fabrics with self-floatation. React Funct Polym 175:105270

Xia C, Zhu S, Feng T, Yang M, Yang B (2019) Evolution and synthesis of carbon dots: from carbon dots to carbonized polymer dots. Adv Sci 6(23):1901316

Xiao Z, Li Q, Liu H, Zhao Q, Niu Y, Zhao D (2022) Adhesion mechanism and application progress of hydrogels. Eur Polymer J 173:111277

Yang X, Li P, Tang W, Du S, Yu M, Lu H, Xing X (2021) A facile injectable carbon dot/oxidative polysaccharide hydrogel with potent self-healing and high antibacterial activity. Carbohyd Polym 251:117040

Yang J, Luo Z, Wang M (2022) Novel fluorescent nanocellulose hydrogel based on nanocellulose and carbon dots for detection and removal of heavy metal ions in water. Foods 11(11):1619

Yue J, Miao P, Li L, Yan R, Dong WF, Mei Q (2022) Injectable carbon dots-based hydrogel for combined photothermal therapy and photodynamic therapy of cancer. ACS Appl Mater Interfaces 14(44):49582–49591

Zeng H, Hu Z, Peng C, Deng L, Liu S (2021) Effective adsorption and sensitive detection of Cr (VI) by chitosan/cellulose nanocrystals grafted with carbon dots composite hydrogel. Polymers 13(21):3788

Zhang Z, Fu H, Li Z, Huang J, Xu Z, Lai Y, Zhang S (2022) Hydrogel materials for sustainable water resources harvesting and treatment: synthesis, mechanism and applications. Chem Eng J 439:135756

Zhang X, Peng J, Qi X, Huang Y, Qiao J, Guo Y, Wu Y (2023) Nanocellulose/carbon dots hydrogel as superior intensifier of ZnO/AgBr nanocomposite with adsorption and photocatalysis synergy for Cr (VI) removal. Int J Biol Macromol 233:123566

Zhao DL, Chung TS (2018) Applications of carbon quantum dots (CQDs) in membrane technologies: a review. Water Res 147:43–49

Zhao H, Jiang X, Dong Y, Chen W, Cui J (2015) Geographic information system-based optimization of sewage treatment facilities by evaluating pollution effects and governance demands. J Water Reuse Desalination 5(2):104–118

Zhu S, Song Y, Zhao X, Shao J, Zhang J, Yang B (2015) The photoluminescence mechanism in carbon dots (graphene quantum dots, carbon nanodots, and polymer dots): current state and future perspective. Nano Res 8:355–381

Zohourian MM, Kabiri K (2008) Superabsorbent polymer materials: a review. Iran Polym J 17(6):451–477

Chapter 6
Application of Hydrogels in Heavy Metal Sensing in Wastewater

Suvarna Yenduri and K. Naga Prashant

Abstract Hydrogels are adsorbent-type materials used for the treatment of wastewater especially for removing pollutants. Because they are persistent and harmful, heavy metal ions are among the pollutants that most worry the environment. Among the various benefits of hydrogels are their 3D network-like structure, higher water absorptivity and adjustable physicochemical characteristics. Diverse hydrogel-based methods have been investigated by researchers to deal with wastewater pollution by heavy metals. Radionuclides, dyes, and heavy metal ions have all been investigated for removal from wastewater and water using hydrogels. Heavy metals are enduring pollutants that are generated by human activity and industrial operations. Their eradication is essential to preserving both human health and aquatic ecosystems. Commonly utilized in the chemical sector, dyes can add to water contamination. Helping to lessen their effects are hydrogels. Because dyes break down slowly in nature, getting rid of them is hard. One possible answer is hydrogels. Hydrogel synthesis, characterization, and practical applications are all still being investigated by researchers. Interest areas are in the recycling and regeneration of hydrogels after treatment. Hydrogels have the potential to act as an absorbent for heavy metal removal in wastewater treatment. Their particular qualities qualify them to tackle environmental issues. Hydrogel performance optimization and practical application expansion will probably be the main topics of future study.

Keywords Hydrogels · Heavy metals · Water contamination · Metal sensing · Waste water treatment · Natural polymers · Biodegradable

S. Yenduri (✉) · K. Naga Prashant
Faculty of Pharmacy, Sri Adichunchanagiri College of Pharmacy, Adichunchanagiri University, Javarana Hally, Karnataka, India
e-mail: suvarna.ph52@gmail.com

S. Bardhan and M. P. Shah (eds.), *Applications of Hydrogels in Modern Wastewater Treatment*, Advances in Wastewater Research,
https://doi.org/10.1007/978-981-95-1598-1_6

Abbreviations

AIE	Aggregation-Induced Emission
CDs	Carbon Dots
CMC	Carboxylmethyl cellulose
CNC	Carboxylate Nano Cellulose
CNFs	Cellulose Nanofibrils
CNNPs	Carbon Nitride Nanoparticles
COF	Covalent Organic Framework
EDTA	Ethylene Diamine Tetraacetate
FNH	Fluorescence Nanocellulose Hydrogel
FTIR	Fourier-Transform Infra-Red
HB-PEGDA	Hyperbranched Poly (Poly (Ethylene Glycol) Diacrylate
HMIs	Heavy Metal Ions
KI	Potassium Iodide
LOD	Limit of Detection
NC	Nanocellulose
PDA-PPy	Dopamine polymer doped polypyrrole
PEI	Polyethylenimine
POCT	Point of Care Testing
QDs	Quantum Dots
SA-SH	Thiolate Sodium Alginate
SEM	Scanning Electron Microscopy
SPCE	Screen-Printed Carbon Electrodes
TpDq	Coumarin Augmented COF
UV	Ultra Violet
XRD	X-ray Diffraction

6.1 Introduction

The environment is the space that is inhabited or used by humans, animals, plants, and microbes. The land, the sea, with the atmospheric conditions of the earth form this entity. All four spheres of the earth system, like air, water, land, and living beings should coexist harmoniously. More compounds than in any other sector of the environment are known to be pollutants or environmental contaminants (Briffa et al. 2020). Many pollutants have seriously harmed the environment, including radioactive isotopes, inorganic particles, volatile organic contaminants, organometallic substances, gaseous pollutants, and nanoparticles (Masindi and Muedi 2018).

In this study, we examine strategies taken to reduce the environmental harm that heavy metals (HMs) produce. A special class of metals known as "heavy metals" that have relatively high atomic number, mass number, and density across the periodic table. Usually, these heavy metals take decades to biodegrade and thus remain in the environment for the long term. As these heavy metals are hazardous in trace amounts even untargeted living creatures are thought to be seriously threatened by toxic heavy metals (Das et al. 2023).

The issue of water contamination is a global concern that demands attention. Both man-made and natural processes are the main causes of metallic elements in wastewater. While there are many human causes, the natural effects include deterioration, including erosion of soil, precipitation drainage, lava flows from active volcanoes, and particle matter like aerosols. They consist of municipal and industrial waste effluents. Municipal effluent comes from discharges from homes and municipalities, whereas industrial effluent comes from activities like metal smelting and electroplating, tanning leather, even textiles, extraction of minerals, and nuclear power. Heavy metal ions, organic debris, and water pollutants are found in industrial and municipal effluents. Industries are the primary sources for the presence of heavy metal pollutants in water (Hegazi 2013).

Pollutants can traverse long distances in the form of surface waters, suspensions, and solutions in water. They can enter saltwater and oceans through temperature changes, variations in density, wind, and currents. By infiltrating the nutrient cycle through marine life, persistent contaminants can move to different environments and affect predators (Walker et al. 2005). A tendency for bioaccumulative and toxic properties of HMs in wastewater runoff makes them a potential hazard to the environment and living beings. Heavy metals are also nonbiodegradable. The presence of heavy metals in groundwater and soil is due to conventional agricultural practices such as irrigation using wastewater discharge (Barakat 2011).

Heavy metal ions have long been removed from the wastewater using various conventional techniques and technologies. These comprise, among other techniques, coagulation, solvent extraction, adsorption, oxidation-discount, ionic change, reverse osmosis, electrochemical treatment, evaporating recuperation, membrane era, and biological treatment. For treating these heavy metals, several techniques were being used; however, far too costly or sophisticated. The production of hazardous sludge, whose disposal is complicated and no longer environmentally friendly, is a significant drawback of these methods. For example, extracting metals at fewer than 100 μg per milligram is usually quite expensive. In contrast, a novel technique called bioremediation makes use of enzymes, microorganisms, healthy plants, and agricultural waste products (Odumbe et al. 2023).

Sustainable removal techniques are essential to preserving the health and balance of ecosystems. Both humans and apex predators are at risk for health problems due to heavy metal accumulation in creatures that go up the food chain. Heavy metal pollution can deteriorate water quality for drinking, farming, and recreational uses. Reducing energy use, utilizing renewable materials, and recovering precious metals for reuse are ways that sustainable removal technologies reduce their negative effects on the environment. Technologies for removal that are flexible are necessary to meet upcoming problems and protect the ecosystem and the health of society (Dhokpande et al. 2024).

6.1.1 Heavy Metal Contaminants

Cadmium, chromium, copper, nickel, zinc, arsenic, as well as lead, are the most frequently detected heavy metals in wastewater and create toxic effects on the environment and humans too (Matta and Gjyli 2016).

Lead and other heavy metals are impacted by both the environment and people in different ways. Lead does not bioaccumulate or grow in food chains, although it is a particle and has limited mobility and bioavailability. It has an impact on renal function, haemoglobin production, and the central nervous system. Lead is categorized by the International Agency for Research on Cancer as possibly harmful to human beings. Lead doesn't bioaccumulate in the majority of species, but it binds strongly to materials, including sediments, soil, and sewage sludge. It may build up in bones and in the biota by feeding on microscopic organisms like worms and mussels. Mercury is a special metal that may be hazardous to people and the environment due to its tendency to flow at room temperature. It may be converted into dimethylmercury and methylmercury, which are bioaccumulative and volatile and can travel great distances. Because organic mercury may reach quantities of up to 1 mg/litre, it is more harmful to aquatic life. Cadmium and its derivatives are bioavailable, bioaccumulative, and soluble in water. Although they are easily absorbed by molluscs and bacteria, they are not necessary for the existence of plants or animals. Cadmium is more readily absorbed in aquatic environments. The acute toxicity of the metal is species-specific and directly correlates with its free ionic concentration. Reduced development and larval mortality may result from prolonged exposure (Rajumon et al. 2020).

The biologically essential mineral chromium is present in plants, animals, and humans in both trivalent and hexavalent form. It is not as harmful as chromium hexavalent, which can lead to cancer and irritability. It is mostly because of human activity that this form of chromium is more hazardous to creatures in the water. Elevated levels have the potential to harm fish tissues and invertebrates, as well as increase fish susceptibility to infection. Chromium poisoning is more common in invertebrate animals, including insects, crabs, and polychaete worms. 0.05 mg/litre is the deadly concentration of chromium for several terrestrial and aquatic invertebrates. Copper can be found in food, drinking water, and the air, and humans absorb large amounts of it every day. Many wastewater sources, such as printed circuit board fabrication, electronics plating, paint manufacture, and printing processes, consist of copper. Nevertheless, too much copper might have negative health effects. Copper is bound to minerals and organic materials in the soil, which restricts its movement and affects both surface and groundwater. The naturally occurring element nickel, which is employed in many different sectors, may be harmful to people and animals in large doses. Ship cruise discharge wastewater contains nickel. Groundwater used for agriculture, silicon semiconductor production effluent, and electronic manufacturing wastewater all contain arsenic. Natural mobilization, mining operations, burning fossil fuels, arsenic pesticides, herbicides, agricultural desiccants, and animal feed are the main causes of environmental arsenic issues. There are three oxidation states of arsenic: −3, 0, +3, and +5, and each has a distinct hazardous impact Jaishankar et al. 2014).

6.1.2 Impact of Heavy Metals on Humans

In real terms, all chemicals are hazardous above and below their limits, including water, medicines, and even necessary components. Certain elements, nevertheless, including arsenic, lead, cadmium, and mercury, are classified as toxic since they are known to be harmful to living things at all concentrations and should not be ingested by the body, not even at ultra-trace levels (Baysal et al. 2013). Some of the heavy metals that show toxic effects on the health of humans are listed below in Table 6.1 (Gautam et al. 2014).

The order is Co < Al < Cr < Pb < Ni < Zn < Cu < Cd < Hg, in which some metals are found to be most hazardous to humans. The dosage, rate of release, and exposure time affect how hazardous heavy metals are to people. In the past few decades, mercury, cadmium, and lead have all attracted increased attention as heavy metals. 2 parts per billion, or ppb, of mercury is the maximum allowed by the US EPA in drinking water. As per WHO recommendations, Hg acceptable limits in wastewater and agricultural soils are 0.001 and 0.05 ppm, respectively (Ackermann et al. 1999).

Table 6.1 Effect of heavy metals on human health

Sl. no.	Heavy metal	Effect
1	Cadmium	Causes grave harm to human kidneys and bones. Acute impact on kids. Emphysema, anaemia, and pneumonia
2	Arsenic	Immunotoxin Reactive oxygen species production and lipid peroxidation cause genotoxicity, keratosis, hyperpigmentation, and melanosis in humans co-receptor expression modulation
3	Zinc	Anemia, stomach ache, phytotoxic and inability to coordinate muscles
4	Chromium	Gastrointestinal mucosal irritation, death in humans, and necrosis nephritis
5	Nickel	Eczema of the hands Phytotoxic Damaging fauna
6	Lead	Metabolic poisoning results from high dosages. Brain injury and hypertension, children's fatigue, irritability, anaemia, and behavioural changes
7	Copper	Corrosion and irritation of the mucosa, many aquatic animals suffer harm as a result of phytotoxicity. An irritated central nervous system is followed by depression
8	Mercury	Mutagenic consequences raise cholesterol and its toxicity

6.1.3 Sources of Heavy Metal Exposure

People may get exposed to cadmium by smoking cigarettes, eating contaminated food, working in primary metal industries, and being in cadmium-affected workplaces. According to Iranian research, canned fish samples showed levels of cadmium over the maximum permitted limit (MPL) because the pollutants that are rich in heavy metals were released into aquatic ecosystems. A limit of 5 ppb, or 0.005 ppm, has been imposed by the US EPA regarding cadmium (Cd) in water for human consumption. Human lead poisoning affects the nervous system, brain, skeleton, liver, heart, and kidneys. Some of the early symptoms of heavy metal poisoning are headache, irritation, dullness, loss of memory. Lead poisoning can cause anaemia and interference with haemoglobin production. Long-term exposure to low levels of lead by children can lead to cognitive impairment. The International Agency for Research on Cancer (IARC) states that "lead is a substance that may cause cancer in humans". According to the US EPA, the acceptable threshold level in water for consumption is 15 ppb of lead. WHO has recommended the permitted level of lead (Pb) in wastewater and agricultural soils, at levels of 0.1 and 0.01 ppm, respectively (Ayeni 2014).

Fertilizers and sewage are the two common sources of environmental chromium emissions. In addition to being highly soluble in water, hexavalent chromium compounds, other elements such as lead, strontium, calcium compounds are more toxic and also carcinogenic. Moreover, chromium-containing drugs have also been linked to slow-healing ulcers. It has also been demonstrated that chromate compounds are capable of causing DNA damage in cells. The WHO suggested 0.05 and 0.1 ppm acceptable values of Chromium (hexavalent) within wastewater and agricultural soils, respectively. Lethal doses of 6 to 40 mg/kg are associated with thallium toxicity. Anorexia, vomiting, bleeding from the gastrointestinal tract, abdominal discomfort, polyneuropathy, alopecia, failure of the renal system, skin erythema, convulsions, dysphoria, autonomic dysfunction, cardiotoxicity, and coma are among the symptoms of thallium poisoning. In China, 0.0001 ppm is the recommended acceptable level for thallium in drinking water. The literature that was accessed did not include the WHO's recommended acceptable levels for Tl in wastewater or agricultural soils (Han et al. 2023).

There are estimates that range from 1,50,000 to 1,80,000 metric tons per year for the quantity of nickel discharged into the environment globally. Humans are exposed to nickel through their diet, air, and water. Prior research has demonstrated that, in contrast to inhalation and cutaneous routes, the primary method of heavy metal exposure for the local population was consumption of dust polluted with nickel. An individual may have elevated amounts of nickel in their urine and tissues after being exposed to the metal. Exposure to nickel can cause adverse effects such as dermatitis, allergies, organ illnesses, and respiratory system cancer. The WHO has suggested acceptable levels of 0.02 and 0.05 ppm, respectively, for Ni in agricultural soils and wastewater (Kinuthia et al. 2020). Regulatory guidelines permissible values for a number of metallic elements in wastewater are listed below in Table 6.2.

Table 6.2 Regulatory guidelines for permissible levels of HMs in wastewater

S. no.	Heavy metals	WHO permissible limits (ppm)	USEPA permissible limits (ppm)
1	Mercury	0.001	0.05
2	Copper	0.05	1.3
3	Lead	0.01	0.0015
4	Chromium	0.05	0.1
5	Arsenic	0.01	0.05
6	Nickel	0.05	0.2
7	Cadmium	0.003	0.005
8	Thallium	–	–
9	Zinc	5	5

6.2 Various Sensing Mechanisms

Hydrogel-based sensors are gaining a lot of importance because of their flexibility, durability, adaptability, and biocompatibility properties. Additionally, it has the ability to give various levels of response on changing its physical or chemical properties, which makes it ideal for various sensing applications. Hydrogels can act as sensors through various sensing mechanisms and so play a role as mechanical sensors, electrochemical sensors, fluorescence sensors, optical sensors, chemical sensors, and biosensors. The addition of crosslinking agent is helpful for improving the biocompatibility of the sensor and also improving its mechanical and electrical properties (Zhu et al. 2023).

6.2.1 *Mechanical Sensors*

Mechanical sensors use capacitive sensing, piezo resistance, and piezoelectricity to sense changes in their environment. When subjected to mechanical stress, piezoresistive materials alter their electrical resistance, whereas piezoelectric materials generate an electrical charge. Two conductive surfaces that are separated by a dielectric material are used in capacitive sensors. Capacitance varies as pressure or force alters the distance. These changes can be used to calculate the applied force, pressure, or displacement. The dielectric constant, hydrogel thickness, electrode area, and electrode distance affect a capacitive sensor's capacitance. Two parallel plates with conductive electrodes and a thin hydrogel dielectric layer comprise the capacitive sensor. Sensor capacitance depends on the hydrogel layer dielectric constant and plate surface area. When a hydrogel-based capacitive sensor is subjected to variations in humidity, temperature, pH, or pressure, its water content varies, affecting its dielectric constant and capacitance (Villa et al. 2019).

To reduce external influences on sensor findings, the mechanical hydrogel sensor needs powerful anti-interference capabilities and stability. Inclusion of polyacrylamide/sodium alginate/carbon nanotube (PAM/SA/CNTs) causes a double-network hydrogel, which makes it strong and robust. The PAM/SA DN hydrogel network with CNTs generated hydrogels with controllable photomechanical deformations, such as item grabbing and bending. The addition of $CaCl_2$ to the hydrogel strain sensor made it sensitive to low temperatures as well. Further research on mechanical sensing materials and combination techniques is needed to develop hydrogels with improved sensitivity (Zhang et al. 2020).

6.2.2 Electrochemical Sensors

Electrochemical biosensors work by sensing the electrochemical characteristics of sensing components and their interactions with analytes. Interaction of the analyte with the sensing component causes an electrochemical reaction, which can be detected by amperometry, potentiometry, or voltammetry. Electrochemical biosensors use 'signal amplification strategies' using sensing components and transducer surfaces to improve their electrochemical characteristics and sensitivity. To ensure stability, durability, and convenience, sensors need a substrate with proper materials and dimensions. Electrochemical sensors can detect biomolecules, ions, and gases. It has wide applications in monitoring food and environmental samples and also in clinical diagnosis, such as blood glucose monitoring and Wilson's disease (Zhu et al. 2023).

6.2.3 Fluorescent Sensors

Hydrogel-based fluorescent sensors detect target analytes via sensing fluorescent molecules, hydrogel volume changes, and sensing interactions of molecules with the target analyte. Sensory molecules are usually embedded in the hydrogel matrix, which holds them. When activated by a certain wavelength, sensor molecules produce light at different wavelengths. When these sensor molecules touch the target analyte, their fluorescence changes, changing the light intensity or wavelength. Hydrogel swelling or shrinking changes the distance between sensor molecules, affecting fluorescence. Interaction with the target analyte changes the hydrogel's dimensions (Fig. 6.1). Hydrogel fluorescent sensors are used in environmental monitoring, food safety, and biomedical engineering (Wang et al. 2020).

The mechanical and electrical features of fluorescent sensors affect their sensitivity. Due to their high conductivity, fluorescent molecules have a shorter excitation lifespan, requiring careful design. Three processes are used to make colour-changing polymers. First, adding chromophores that change shape or state in response to stimuli. The second way involves manipulating ordered nanostructures to impact

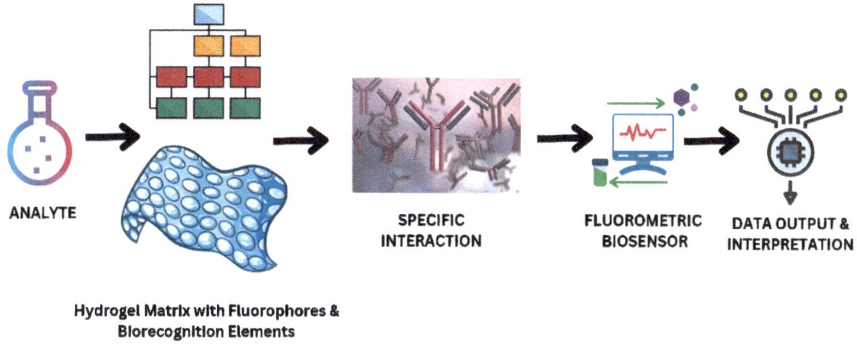

Fig. 6.1 Work flow of a typical fluorescent sensor

photon propagation to create 'structural' colour. The third technique generates strong luminescence employing metal centres of luminous lanthanide metals like Eu and Tb that dynamically coordinate with polymers (Weng et al. 2018).

6.2.4 *Biosensors*

Hydrogels can also act as biosensors by analysing biomolecules. These sensors contain enzymes, nucleic acids, antibodies, and aptamers and conduction components like electrochemical, optical, thermal, ion-sensitive, or piezoelectric. High sensitivity is necessary to detect biomolecules at low concentration, and it can be improved by altering pore shape and surface functionalization of cellulose-based hydrogels to recognize and collect target biomolecules. Biosensors can detect bacteria by embedding enzymes, antibodies, or nucleic acid probes in cellulose-based hydrogels (Gao et al. 2023).

Cellulose-based hydrogels are utilized in enzyme sensors to measure microorganism metabolic activity utilizing enzyme-catalyzed reactions. Hydrogels readily transport enzymes and bind to enzyme reaction products due to their high hydration and biocompatibility (Fig. 6.2). These hydrogel-based biosensors have applications in glucose sensing, antibody sensing, cell sensing, etc. (Zhang et al. 2025).

6.2.5 *Chemical Sensors*

Chemical sensors detect and measure the concentration, composition, and other chemical properties of substances in the environment. The target chemical reacts with a certain element in the sensor, which results in a readable signal. Metal ion sensing (Cr ions, copper ions, mercury ions), poisonous gas sensing, and compound

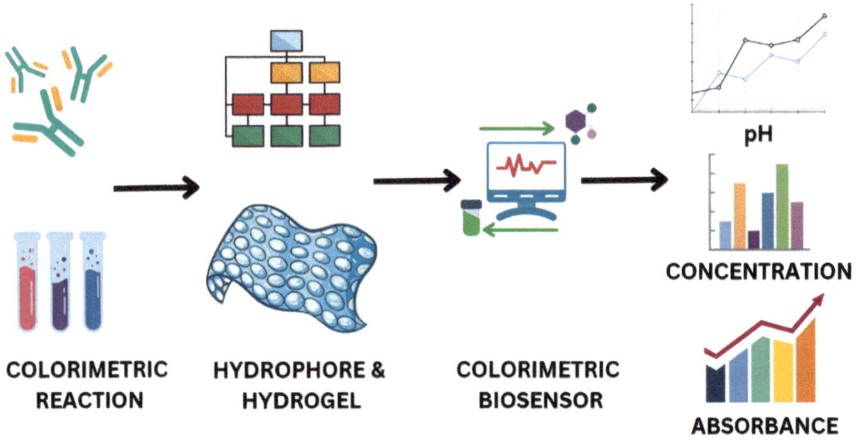

Fig. 6.2 Work flow of a typical biosensor sensor with colorimetric reaction

sensing (e.g., nitrite) are the three categories of chemical sensors. Cellulose-based hydrogels are good sensing components because of their great chemical reactivity and specificity. They can interact with particular molecules to detect and measure changes in target chemical compounds. These chemical sensors are mainly used in sensing of ions, toxic gases, nitrite, etc. (Zhang et al. 2025).

6.3 Hydrogel-Based Sensing Platforms for Detecting Heavy Metals in Water

Hydrogels are mainly used as sensing devices for heavy metal detection in various types of samples, especially in water. The role of hydrogels as sensors is discussed in detail below: Growing industrial activity has raised serious concerns about HMIs, including mercury, cadmium, and lead contaminating water supplies. Significant concentrations of these ions are dangerous and can lead to life-threatening illnesses in humans. The identification of HMIs within water is therefore crucial. Some traditional approaches have been rendered outdated by methodological drawbacks as technology has advanced, making room for a new generation of methods that make use of innovative sensors. Due to its unique physico-chemical properties, nanocellulose has recently attracted a lot of attention as a potential biocompatible material for the development of sensors. The various dimensions of nanocellulose-based sensors developed for HMI recognition are given particular focus in this study. Furthermore, several methods of sensing (such as optical and electrochemical), processes of sensing, and the functions of nanocellulose in these sensors are explored (Langari et al. 2022).

Gogoi et al. (2015) developed a hybrid solid detecting platform that utilizes an agar hydrogel sheet embedded in carbon dot particles and chitosan that allows for the identification of heavy metal ions. The rigid sensing platform can be constructed more easily by a simple electrostatic interaction between the ammonium group in the CD particles and the hydroxyl groups of agarose. The immersion of a hydrogel strip in solutions containing heavy metals such as lead, iron, manganese, chromium, copper causes variation in colour, specifically, Fe^{3+} turns brown, Pb^{2+} turns white, Mn^{2+} turns tan brown, Cr^{6+} turns yellow, and Cu^{2+} turns blue. These colour changes can be assessed in the reflectance spectrum of the UV–Visible range of the hydrogel strip, the corresponding metal ion's optical detection limit is found to be 1 pM for Cr^{6+}, 0.5 μM for Cu^{2+}, along with Fe^{3+}, Pb^{2+}, and Mn^{2+} to be 0.5 nM. Furthermore, it is shown that the hydrogel film may be used as an effective filtering membrane to separate the five heavy metal ions. The effective formation of colourful chelates with transition metals by chitosan is a critical essential property of this sensing platform. Consequently, the most optimal way to utilize the efficient hybrid hydrogel firm sensing platform is as an inexpensive, portable, colorimetric-optical heavy metal ion detector that can be used on-site and may be skilled in separating heavy metal ions. Additionally included are specifics of a potential mechanistic understanding of colorimetric analysis and ion separation (Gogoi et al. 2015).

Jingjing Guo et al. (2017) had developed a hydrogel waveguide doped with carbon dots (CDs) for the purpose of precisely identifying ions of heavy metals in aqueous solutions on-site. The efficient light confinement made possible by the waveguide's reflective index difference lowers light scattering loss. The innovative design may serve as a reliable and portable sensing platform for on-site heavy metal ion research and assessment. Based on carbon dots integrated nanocomposite hydrogels, a portable, dependable photonic system for real-time, selective, and on-site Hg^{2+} ion detection was presented. Through passive diffusion, Hg^{2+} ions infiltrated the hydrogel matrix, trapping functional CDs inside light-guiding hydrogels and quenching them. Total internal reflection at the hydrogel's contact with the surrounding aqueous medium allowed for the confinement of light. The waveguide for Hg^{2+} sensing was quantitatively characterized, showing a low limit of detection of 4 nM and a linear spectrum between 0 to 5 μM. Although this work focused on Hg^{2+} ions as the sensing target, by integrating other CDs functionalized with appropriate selectivity, the current design may also be utilized for the monitoring of other metal ions. The photonic gadget could offer a fresh and potentially effective method for measuring and analysing heavy metal ions within real-world settings (Guo et al. 2017).

Guo et al. (2019) created a novel fluorescent nano-cellulosic hydrogel in order to benefit from the exceptionally bright luminescence of carbon dot particles (CDs) and the exceptionally high ratio of aspect ratio of cellulose nanofibrils (CNFs). This hydrogel was shown to be both an efficient absorbent for heavy metal removal and can also play the role of a sensor due to its optical properties for heavy metal detection. Optical microscopy, SEM, XRD, FTIR, UV–vis, and luminescence spectroscopy were employed to assess this bright nano-cellulosic hydrogel's morphological features, chemical composition, and optical qualities. The study looked at how the nano-cellulosic hydrogel's adsorption effectiveness was affected by the initial concentration and contact time. The experimental results indicated that Fe^{3+}, Ba^{2+}, Pb^{2+}, and Cu^{2+} had maximum adsorption capacities of 769, 212, 2056, and 1246 mg g^{-1}, respectively. Taking into account a bright blue fluorescent quantum efficiency of 23.6%, the fluorescent nano-cellulosic hydrogel that was produced showed significant sensitivity for Fe^{3+} detection. On this hydrogel, the CNFs customized with CDs enhanced the adsorption-aggregator function of hydroxyl-induced heavy metal aggregation and enabled a fast apparent reaction with heavy metals as an optical sensor. This resulted in an increase in the heavy metals' adsorption capacity, as well as an improvement in the stability during the signal from fluorescence, in addition to the sensitivity used to count the number of heavy metals (Guo et al. 2019).

Li et al. (2020) developed a novel dual fluorescence hydrogel microsphere sensor that can identify Al^{3+} and Hg^{2+} concurrently and sensitively was developed as a consequence of an experiment designed to assess this strategy. The sensor, which is made of hyperbranched poly (poly (ethylene glycol) diacrylate) (HB-PEGDA) composite hydrogel microspheres and thiolate sodium alginate (SA-SH), was created utilizing a microfluidic device and a self-assembly technique. The incredibly densely packed microspheres are made up of two fluorescent probes. Considerable sensitivity for Al^{3+} and Hg^{2+}, as well as other metal ions, was shown by the functionalized hydrogel microspheres, with limits of detection 70 nM and 120 nM, respectively. Additionally, by eliminating Al^{3+} and Hg^{2+} with the addition of an EDTA and KI solution, the hydrogel microspheres may be easily removed from test samples and used again. The novel HB-PEGDA/SA-SH composite hydrogel microspheres, which include two encapsulated fluorescent probes, are easier to produce and more efficient than earlier sensors when it comes to detecting two distinct metal ions simultaneously. The optimized hydrogel-based microsphere sensor was successfully applied for environmental samples such as water and soil, and was found to produce accurate and precise results (Li et al. 2020).

Bhamore et al. (2020) employed agarose hydrogel Pb^{2+}, Fe^{3+}, and Mn^{2+} metal ion detection using films and glutathione (GSH)-capped Syzygium cumini carbon-dot (CDs) as an amalgamated solid probe. The foundation for creating a solid sensing platform is provided by the electrostatic interactions between agarose hydrogel and GSH-capped Syzygium cumini CDs. Good linearities were demonstrated by the hydrogel-based solid probe that was constructed, taking into account the observed concentration fluctuations of metallic ions between 0.005 and 0.075, 0.0075 and 0.1, and 0.0075 and 0.1 mM. Regarding Pb^{2+}, Fe^{3+}, and Mn^{2+} ions, the corresponding detection limits were 1.3, 2.5, and 2.1 μM (Bhamore et al. 2020).

Mao et al. (2021) created a semi-dry smart biosensing technology that performs well for detecting metal ions and is operable between typical wet and dry chemistry. The combination of sophisticated picture recognition with a stimuli-responsive hydrogel forms the foundation of this platform. Agarose serves as the hydrogel's matrix, and response elements are well-made fluorescent DNA probes. In a test sample, target metal ions have the Intelligent imaging is made possible by its capacity to penetrate through the hydrogel, which activates the DNA sensors, and produces fluorescent signals. This method allows for the easy and sensitive detection of metal ions, such as potassium (K^+) and mercury (Hg^{2+}), without the need for big gear or skilled personnel. For K^+ and Hg^{2+}, the detection limits were 0.34 mM and 5.6 nM, respectively. It is also possible to identify ions in plasma and lake water. Furthermore, the biosensing platform based on hydrogel exhibits strong durability, resistance to biological degradation, and selectivity. Another way to achieve high-throughput testing is to punch many test microwells into a single hydrogel piece. The concept and efficient application of this semi-dry chemistry-based method make up for the shortcomings associated with dry and wet chemistry and provide a workable way to carry out on-site testing. The biosensing platform in particular has a lot of potential for POCT analysis for disease diagnosis and pollution monitoring because of its benefits, which include the capacity to perform high-throughput tests employing an adaptable semi-dry hydrogel and easy operability without requiring large instruments and skilled personnel (Mao et al. 2021).

Zhong et al. (2021) successfully built a Pb (II) sensitive and selective electrochemical sensing platform built on an altered screen-printed carbon electrode (SPCE) called PDA-PPy/SPCE, or dopaminergic polymer doped polypyrrole hydrogel for the detection of lead. Consequently, the development of a high-performance technology for detecting lead ions (Pb (II)) in food and the environment is important. This study successfully built a Pb (II) sensitive and selective electrochemical sensing platform built on an altered screen-printed carbon electrode (SPCE) called PDA-PPy/SPCE, or dopaminergic polymer doped polypyrrole hydrogel. Dopamine hydrochloride, along with pyrrole, hybridized to form PDA-PPy hydrogel using a rapid and ecologically safe one-step production technique. The hydrogels PDA-PPy with excellent conductivity and selective adsorption capability, were combined to create a modified material that was used for electrodes to amplify signals. The results indicated that PDA-PPy/SPCE has an exceptional sensing capacity for Pb (II), with a high level of sensitivity of 0.17 μg L-1 μA^{-1} along with a small limit of detection of 0.15 $\mu g\ L^{-1}$. Furthermore, real sample analysis showed satisfactory recoveries, suggesting that it has potential applications (Zhong et al. 2021).

Jia et al. (2022) produced an innovative luminous hybrid polymer featuring dual-emission centres (Eu@Dye@TpDq hydrogel) on the basis of a covalent organic framework (COF). They created a dye (coumarin) augmented COF (TpDq) hybrid fluorescence material (Dye@TpDq) before the material formed as a hydrogel film. A Eu^{3+} ion was introduced to the material as an extra bright centre during the cross-linking procedure. Under a dual-emission fluorescent platform, it reacts to different heavy metal ions in two ways: it turns off its fluorescence in response to Cr (VI) and shows a ratio metric for the identification response to Cu (II). In addition to presenting a unique synthesis of double-emission COF-based hydrogels functionalized with Eu^{3+}, this work serves as inspiration for the creation of COF-based multifunctional analytical instruments (Jia et al. 2022).

Yang et al. (2022) developed a sensor aimed to detect and remove heavy metals (Fe^{3+} and Pb^{2+}) from aqueous solutions by developing a novel fluorescence nanocellulose hydrogel (FNH) constructed around carbon dots (CDs). A crucial basic ingredient in the manufacturing of food is water. Preserving the water's safety and purity is essential for the food sector. In this study, CDs were attached to the carboxylate nanocellulose (CNC) using the EDC/NHS coupling technique. FTIR analysis was performed to characterize the CNC, FNH, and nanocellulose (NC). Further investigation was done into how the adsorption environment affected FNH's ability to adsorb. Through CD grafting and carboxylation, nanocellulose's ability to adsorb Fe^{3+} and Pb^{2+} was significantly enhanced. Additionally, it was made possible for nanocellulose to respond quickly to Fe^{3+} as a sensor with optical properties, enabling the use of visual signals to determine the concentration of Fe^{3+}. A static adsorption experiment showed that the rate of elimination of Fe^{3+} and Pb^{2+} by FNH was greater than 69.4% and 98.2%, respectively, and the amount of adsorption ability approached 98.3 mg/g and 442.0 mg/g. FNH may have an acceptable level of detection (LOD) of as much as 62.5 mg/L. When it comes to detecting the concentration of Fe^{3+} in an aqueous solution, owing to the quenching effect of Fe^{3+} on fluorescence (Yang et al. 2022).

Wu et al. (2023) developed hydrogel-based AIE sensing sheets for fluorogenic mercury ion detection in water. A "turn-on" fluorescence sensing method was devised to detect Hg^{2+} in water ultrasensitively. TPE-4TA, an aqueous soluble aggregation-induced emission (AIE)-active probe, has a luminous tetraphenylene core and numerous anionic tetrazolate offspring. Molecularly dissolved in water, the probe remained black. However, mercury ions can spontaneously form infinite coordination polymer particles with numerous tetrazolate groups, causing a fluorogenic AIE response for in situ study. In addition, integrating AIE sensing molecules on a hydrophilic polyvinyl alcohol substrate created a hydrogel sheet that could detect Hg^{2+} in situ using a laser-induced fluorescence analysis setup in water. This method worked well in typical aquatic conditions with pH 4–7.5 and had great sensitivity (limit of detection down to 0.38 ppb/1.9 nM), selectivity, and quantification linearity. This research may produce a promising on-site environmental water analysis platform (Wu et al. 2023).

Chu et al. (2024) developed a novel fluorescent composite hydrogel N, P-CDs@CMC/PEI with a bio-based polymer matrix of carboxylmethyl cellulose (CMC), polyethylenimine (PEI) as a second interpenetrating network, and N, P-doped carbon dots (N, P-CDs) as a fluorescent probe to detect and capture HMIs in a simple one-step process to address the serious environmental pollution problems of toxic heavy metal ions in water Various characterization methods examined these composite hydrogels' shape, chemical structure, swelling ratio, mechanical strength, and fluorescence. The composite hydrogel exhibited selective fluorescence response to Hg(II) and Fe(III), with LOD values of 0.48 and 0.27 mg L^{-1}, respectively. The effects of adsorbent type, pH, HMIs concentration, and temperature on composite hydrogel adsorption were examined. The pseudo-second-order and Langmuir models explained CPH2 adsorption for Hg(II) and Cr(VI). Langmuir model calculated maximal adsorption capacities of 846.7 and 289.5 mg g^{-1} for the specified HMIs. Low concentrations of inorganic salts and organic acids had no influence on Hg(II) and Cr(VI) elimination, and the composite hydrogel showed high recyclability and stability after four cycles. Adsorption was caused by electrostatic attraction and coordinated covalent connections (Chu et al. 2024).

Li et al. (2024a, b) constructed cellulose-based hybrid hydrogel beads with carbon dots that adsorb and detect mercury ions in water. Heavy metal capture and wastewater treatment have extensively developed cellulose-based adsorbents. However, most powder adsorbents are too tiny to recycle and lack sensitive sensor moieties, making target detection problematic. Due to their superior fluorescence sensing capabilities, carbon dots (CDs) were proposed to be encapsulated in cellulosic hydrogel beads to detect and adsorb Hg (II) in water. Cellulose moulding also helped recycle it and decreased secondary pollutants from adsorbent degradation. The hydrogel beads' detection limit for Hg (II) was $8.8 \times 10 - 8$ M, below the WHO mercury effluent standard, demonstrating practicality in water treatment and detection. The maximal CB-50% Hg (II) adsorption capability was 290.70 mg/g. The hydrogel beads' sensitive and selective Hg (II) sensing capability was maintained after 2 months of storage due to the adsorbent materials' superior stability. After 2 weeks of water immersion, only 3.3% of CDs leaked, assuring Hg (II) assessment accuracy. After five regeneration cycles, the adsorbent retained over 80% of its initial adsorption capacity, demonstrating its durability and promise for sustainable environmental applications (Li et al. 2024a, b).

Guo et al. (2024) made a hydrogel-based ratiometric fluorescence sensor for portable mercury ion measurement utilizing C3N4-CdTe0.16S0.84 QDs nanocomposites. Portable detection of trace mercury ions (Hg^{2+}) in water is essential for mercury pollution prevention. This work generated graphitic carbon nitride nanoparticles (CNNPs) with green fluorescence as the self-calibration signal using a one-pot solvothermal approach. The oxidation–reduction approach produced CdTe0.16S0.84 QDs with red fluorescence as a response signal. Thus, CNNPs-CdTe0.16S0.84 QDs were used to create a ratiometric fluorescence sensor for trace Hg^{2+} visual detection. The linear range of the sensor was 0.05–2525 nM, with a detection limit of 0.009 nM. The sensor was more sensitive to Hg^{2+} than copper and silver. The Stern–Volmer equation and density functional theory were used to study the three metal ions' fluorescence quenching and sensitivity. The increased sensitivity to Hg^{2+} may be due to electron transport and electrostatic interactions with CdTe0.16S0.84 QDs. Based on this method, a smartphone-assisted CNNPs-CdTe0.16S0.84 QDs hydrogel-based fluorescent sensor was developed for trace Hg^{2+} measurement on-site. This innovative approach makes on-site water hazard monitoring easier (Guo et al. 2024). All the detection limits of heavy metals in various hydrogels were mentioned in Table 6.3.

6.4 Applications of Hydrogels in Heavy Metal Sensing

Heavy metal detection sensors based on hydrogel have been applied in a variety of real-world situations.

Table 6.3 Comparative data of detection limits of various hydrogel-based sensors for HM detection

S. no.	Hydrogel	Type of sensor	Detection metals	Detection limit
1	Chitosan	Optical sensor	Fe^{2+}, Cr^{6+}, Cu^{2+}, Fe^{3+}, Pb^{2+}, Mn^{2+}	1 pM for Cr^{6+}, 0.5 μM for Cu^{2+}, and 0.5 nM for Fe^{3+}, Pb^{2+}, and Mn^{2+}
2	PEG diacrylate	Fluorescent sensor	Hg^{2+}	4 nM
3	Cellulose	Fluorescent sensor	Fe^{3+}, Ba^{2+}, Cu^{2+} and Pb^{2+}	769, 212, 1246, 2056 mg g^{-1}
4	HB-PEGDA	Fluorescent sensor	Al^{3+}, Hg^{2+}	70 and 120 nM
5	Agarose	Visual sensor	Pb^{2+}, Fe^{3+}, and Mn^{2+}	1.3, 2.5 and 2.1 μM
6	Agarose	Biosensor	K^{+}, Hg^{2+}	0.34 mM and 5.6 nM
7	Dopamine polymer doped polypyrrole (PDA-PPy)	Electrochemical sensor	Pb^{2+}	0.15 μg L^{-1}
8	Coumarin modified COF (TpDq)	Fluorescent sensor	Cu^{2+}, Cr^{6+}	10^{-2}, 10^{-6} mM
9	Nanocellulose	Fluorescent sensor	Fe^{3+}	62.5 mg/L
10	PEG	Fluorescent sensor	Hg^{2+}	1.9 nM
11	CMC	Fluorescent sensor	Hg^{2+} and Fe^{3+}	For Hg^{2+} 0.48 and Fe^{3+} 0.27 mg L^{-1}
12	Cellulose	Fluorescent sensor	Hg^{2+}	$8.8 \times 10 - 8$ M
13	CNNPs	Fluorescent sensor	Hg^{2+}	0.009 nM

6.4.1 Wearable Sensors for Human Health Monitoring

Researchers have created resilient and flexible hydrogel-based sensors that are wearable on the body to track exposure to heavy metals. Heavy metal detection in real time is suitable for these sensors in perspiration or other body fluids because they attain superior mechanical durability and conductivity through the use of metal ion coordination. Constructed by combining different conductive fillers that provide electrical conductivity with traditional hydrogels that provide structural support, conductive hydrogels are inspired by chemical structure and synthetic methods of hydrogels (Wang et al. 2021).

6.4.2 Microfluidic Sensors for Water Quality Monitoring

Microfluidic fluorescence sensor arrays, which can detect numerous heavy metal ions concurrently in real-time, are a recent innovation. With the help of this technology, water quality may be quickly and accurately assessed by combining smartphone-based colour recognition with microfluidic chips. The microfluidic fluorescent sensor array developed here may satisfy the requirements, allowing in situ, in addition to quantitative determination of a range of heavy metal ions due to its low detection limit, fast reaction time, and lack of pretreatment requirements. It is an innovative method of engineering technology for multi-component environmental real-time sensing (Li et al. 2024a).

6.4.3 Underwater Sensors for Environmental Monitoring

Hydrogel sensors with a bioinspired design are able to function underwater and send data via Morse code. These hydrogels have outstanding mechanical qualities and anti-swelling characteristics thanks to their hierarchical structure, which mixes flexible and rigid components. Because of the involvement of dynamic ions (Br^-, Cl^-, and Zr^{4+}), the hydrogels evolved desirable sensing qualities that allowed for reliable signal output under a variety of frequencies and stresses. As a result, these hydrogels were modified to create underwater sensors that could send messages using Morse code. This bio-inspired design approach provided new insights into the creation of underwater sensors by achieving the ideal blend of mechanical properties, anti-swelling, and sensing capabilities in the hydrogels (Pan et al. 2024a).

6.4.4 Portable Heavy Metal Detectors

Heavy metals are being detected on-site in a variety of settings using portable sensors based on hydrogel. These sensors are useful for field applications like water, soil, and food product testing since they are simple to use and yield immediate results. Novel *in-situ* electrochemical sensors were developed for the detection of heavy metals, especially stripping voltammetry is preferably suitable because of its small size, easy installation, low cost, and multi-element detection capabilities. Screen-printed electrodes (SPEs) brought a revolution in the application of electrochemical sensors in portable devices due to their reproducibility, versatility, and low cost. Thus, the development of cost-effective electrochemical sensors using SPEs holds promise for meeting new regulations for the detection of heavy metals in drinking water. (García-Miranda Ferrari et al. 2020).

6.4.5 Water Quality Monitoring in Rivers and Lakes

Aquatic ecosystems are kept safe and healthy by the detection of heavy metal contamination in water bodies using hydrogel sensors. Placing sensors based on hydrogel in waterways to track levels of pollution from lead (Pb) and cadmium (Cd). Since most environmental heavy metal ions are multi-component, the demands of simultaneous multi-component detection cannot be satisfied by the single-to-single rapid detection techniques currently in use. Here, they have combined organic fluorescent switches with microfluidic sensing to create a microfluidic fluorescence sensor array that allows for the simultaneous and real-time detection of many heavy metal ions visually. Several "light-up" organic molecular probes were made, by a powerful ability to bind ions of heavy metals was developed, which led to the activation of probe fluorescence (Li et al. 2024b).

6.4.6 Industrial Effluent Monitoring

In order to monitor the heavy metal release from industrial effluents and ensure compliance with environmental standards, hydrogel sensors are utilized. Hydrogel-based sensors are used to continuously monitor Heavy metals are substances including chromium (Cr) and nickel (Ni), in industrial effluent. Hydrogel-based sensors frequently exhibit significant swelling in aquatic environments, which can lead to a decline in mechanical performance and signal distortion in the output, significantly limiting their wide range of applications. Using the principles of various dynamic interaction matching, an anti-swelling supramolecular hydrogel with robustness, quick self-recovery, and strain sensitivity was painstakingly engineered, taking ideas from the hierarchical, diverse cross-linking architecture found in biological tissues (Pan et al. 2024b).

6.4.7 Marine Environmental Monitoring

In order to safeguard marine life and ecosystems, hydrogel sensors are designed to detect heavy metal contamination in marine environments. To ensure the well-being of marine life, levels of heavy metals in coastal waterways are tracked using underwater hydrogel sensors (Pan et al. 2024b).

6.4.8 Monitoring Heavy Metal Concentrations in Industrial Effluents

Industry wastewater treatment systems incorporate hydrogel sensors to continually monitor heavy metal levels, such as chromium (Cr), lead (Pb), and cadmium (Cd). This aids businesses in adhering to environmental laws and preventing the release of hazardous chemicals into lakes and rivers. A study created hydrogel-based sensors to quantify and identify the heavy metals in wastewater from factories using particular chelating agents. These sensors can be integrated into automated monitoring systems and offer real-time data (Pan et al. 2024b).

6.4.9 Monitoring Heavy Metal Concentrations in Rivers

Hydrogel sensors are deployed in river systems in order to identify and track the existence of metals that are toxic metals, ensuring the safety and health of aquatic ecosystems and providing critical data for environmental protection agencies. Researchers have developed microfluidic hydrogel sensors capable of detecting lead (Pb) and cadmium (Cd) in river water. These sensors are highly sensitive and can be combined with smartphone technology for on-site analysis and immediate results. The sensor array developed here may meet the demands for in situ with quantitative detection of a range of heavy metal ions due to its low detection limit, fast reaction time, and lack of pretreatment. It is an innovative method of engineering technology for multi-component environmental real-time sensing (Li et al. 2024c).

6.5 Future Directions

Hydrogel is a sort of flexible platform with a range of biomedical uses, following logical arrangement and functional design. Material engineering is used to alter its physicochemical characteristics (such as pore size, stiffness, viscoelasticity, the microarchitecture, biodegradability, ligand presentation, and stimulus-responsive qualities, among others) along with the fate and cascades of cell signalling. Much innovative research has been conducted in the past few decades to investigate the interactions between cells and hydrogel matrices and determine the underlying processes, opening the door for the translation of hydrogel-based therapeutics from the laboratory to the clinic. We briefly reviewed the physicochemical characteristics of hydrogels that are and their manufacturing processes in this study. The detailed explanation and in-depth discussion that followed focused on the effects of various hydrogel characteristics affecting cell behaviours and signalling processes inside cells. The aforementioned behaviours or events encompassed integrin clustering, accumulation and activation of focal adhesion (FA) complex, rearrangement of the cytoskeleton, shuttling and activation of protein cyto-nuclei such as gene expression, restructuring of cellular compartments, Yes-associated protein (YAP), catenin, etc.), and other cell biology modification (e.g., propagating, emigration, proliferation, lineage commitment…). These served as the foundation for an outline and discussion of current developments in hydrogel technologies for in vitro as well as in vivo environments. These encompassed, among other things, bioimaging, biosensors, smart drug carriers, conductive wearable/implantable biodevices, and disease models. More crucially, the research and clinical translation possibilities of hydrogels were emphasized, along with a focus on the issues that still need to be resolved and the industry's prospects going forward. When taken as a whole, the thorough and comprehensive information in this review will illuminate the fundamental principles of design of novel biomedical hydrogels to comprehend and regulate cellular processes. These hydrogels can offer important clues for the development of future hydrogels and can be used for a variety of biomedical purposes (Cao et al. 2021).

Ouyang et al. (2022) A network of hydrophilic polymers arranged in three dimensions and heavily saturated with water makes up hydrogels, which are soft materials. In contrast to inflexible machinery, hydrogel-based robots include energy conversion processes that enable them to sense, adapt, alter, and respond in intricate surroundings. Regarding soft robotics and sensing applications, we address the developments and difficulties associated with hydrogel materials. Firstly, we present stimulus-responsive hydrogel sensors, which are able to convert external energy inputs from various stimuli into biological, optical, electrical, and geometric output signals. The development of hydrogel robot systems in recent times, which utilize their responsive qualities to accomplish a variety of locomotion models and purposes, is then thoroughly discussed. Researchers divide hydrogel robots into two primary categories based on the different driving forces and processes of locomotion: the active robot, which is driven by reactive materials, and the passive robot, which is shaped through stimuli-responsive swelling of soft hydrogels. We weigh the benefits and drawbacks of each approach and demonstrate how materials and structural designs might translate biomimetic concepts into technical capabilities. In the end, we offer a critical viewpoint on the major obstacles to functionality integration for hydrogel robotic systems as well as logical paths to advance hydrogel robots toward a variety of uses (Ouyang et al. 2022).

Zhang et al. (2016) Nanomaterials with distinctive optical characteristics have been created via recent developments within nanoscale science and engineering. Many fluorescent nanoprobes have been invented and created during the previous several years for the incredibly sensitive and selective imaging and detection of metallic ions in both in vitro and in vivo settings. We give a summary of the most recent advancements in the design and optical characteristics of many types of fluorescence nanoprobes based on carbon-based nanomaterials, semiconductor quantum dots, upconversion nanoparticles, and noble metal nanomaterials in this study. We also go into depth on how they are used in biomedical imaging of live cells and animals, as well as in the detection and measurement of metallic ions for monitoring the environment, food safety, and medical diagnostics (Zhang et al. 2016).

6.6 Conclusion

Hydrogel sensors have demonstrated significant promise in the realm of environmental monitoring for the detection and surveillance of heavy metal pollutants in diverse environments. These sensors are beneficial due to their exceptional sensitivity, specificity, and the ability to detect in real time. They have uses in aquatic environments, industrial facilities for treating wastewater, and can even be incorporated into wearable devices to monitor personal exposure. Nevertheless, there remain some obstacles and opportunities for enhancement in the advancement of hydrogel sensors for heavy metal detection. An obstacle that arises is the requirement for sensors that are capable of concurrently detecting various heavy metals, given that the majority of environmental samples consist of a combination of various metals. For this, the creation of sensor arrays or multiplexed detection systems would be necessary. Another obstacle is in optimizing the performance of the sensor, which encompasses enhancing its selectivity, stability, and also sensitivity. Enhancements in the design and production of hydrogel sensors, together with the integration of innovative materials and technologies, have the potential to tackle these difficulties. Moreover, by combining hydrogel sensors with microfluidics, nanomaterials, and wireless communication, their functionalities may be improved, and more sophisticated applications can be achieved. For instance, the integration of hydrogel sensors with microfluidic systems has the potential to facilitate automated and efficient analysis of heavy metals in intricate samples. Hydrogel sensors have demonstrated significant potential for detecting and monitoring heavy metal pollution in the environment. Through more study and advancement, these sensors have the potential to be vital in safeguarding the well-being and balance of ecosystems, while also bolstering initiatives aimed at preserving the environment.

References

Ackermann ROA, Dixon S, Richard Fitzgerald J, Hanrahan AD, Hughes DC, et al (1999) Pollution prevention and abatement handbook 1998: toward cleaner production. The World Region

Ayeni O (2014) Assessment of heavy metals in wastewater obtained from an industrial area in Ibadan, Nigeria Ocena tekih kovin v odpadnih vodah industrijske cone v Ibadanu v Nigeriji. In: Environmental science. https://api.semanticscholar.org/CorpusID:110299945

Barakat MA (2011) New trends in removing heavy metals from industrial wastewater. Arab J Chem 4(4):361–377. https://doi.org/10.1016/j.arabjc.2010.07.019

Baysal A, Ozbek N, Akm S (2013) Determination of trace metals in waste water and their removal processes. In: Waste water - treatment technologies and recent analytical developments. InTech. https://doi.org/10.5772/52025

Bhamore JR, Park TJ, Kailasa SK (2020) Glutathione-capped Syzygium cumini carbon dot-amalgamated agarose hydrogel film for naked-eye detection of heavy metal ions. J Anal Sci Technol 11(1):13. https://doi.org/10.1186/s40543-020-00208-8

Briffa J, Sinagra E, Blundell R (2020) Heavy metal pollution in the environment and their toxicological effects on humans. Heliyon 6(9):e04691. https://doi.org/10.1016/j.heliyon.2020.e04691

Cao H, Duan L, Zhang Y, Cao J, Zhang K (2021) Current hydrogel advances in physicochemical and biological response-driven biomedical application diversity. Signal Transduct Target Ther 6(1):426. https://doi.org/10.1038/s41392-021-00830-x

Chu Q, Liu Z, Feng F, Chen D, Qin J, Bai Y, Feng Y (2024) A novel bio-based fluorescent N, P-CDs@CMC/PEI composite hydrogel for sensitive detection and efficient capture of toxic heavy metal ions. J Hazard Mater 474:134757. https://doi.org/10.1016/j.jhazmat.2024.134757

Das S, Sultana KW, Ndhlala AR, Mondal M, Chandra I (2023) Heavy metal pollution in the environment and its impact on health: exploring green technology for remediation. Environ Health Insights 17. https://doi.org/10.1177/11786302231201259

Dhokpande SR, Deshmukh SM, Khandekar A, Sankhe A (2024) A review outlook on methods for removal of heavy metal ions from wastewater. Sep Purif Technol 350:127868. https://doi.org/10.1016/j.seppur.2024.127868

Gao ZF, Zhu H, Li Y, Yang X, Ren X, Wu D et al (2023) Revolutionizing biosensing with superwettability: designs, mechanisms, and applications. Nano Today 53:102008. https://doi.org/10.1016/j.nantod.2023.102008

García-Miranda Ferrari A, Carrington P, Rowley-Neale SJ, Banks CE (2020) Recent advances in portable heavy metal electrochemical sensing platforms. Environ Sci: Water Res Technol 6(10):2676–2690. https://doi.org/10.1039/D0EW00407C

Gautam R, Sharma S, Mahiya S, Chattopadhyaya M (2014) Heavy metals in water: presence, removal and safety

Gogoi N, Barooah M, Majumdar G, Chowdhury D (2015) Carbon dots rooted agarose hydrogel hybrid platform for optical detection and separation of heavy metal ions. ACS Appl Mater Interfaces 7(5):3058–3067. https://doi.org/10.1021/am506558d

Guo J, Zhou M, Yang C (2017) Fluorescent hydrogel waveguide for on-site detection of heavy metal ions. Sci Rep 7(1):7902. https://doi.org/10.1038/s41598-017-08353-8

Guo X, Xu D, Yuan H, Luo Q, Tang S, Liu L, Wu Y (2019) A novel fluorescent nanocellulosic hydrogel based on carbon dots for efficient adsorption and sensitive sensing in heavy metals. J Mater Chem A 7(47):27081–27088. https://doi.org/10.1039/C9TA11502A

Guo X, Xiao J, Zhang Y, Zhang Q, Yang J, Wei Y, et al (2024) Portable analysis of tract mercury ions by a hydrogel-based ratiometric fluorescence sensor using C3N4-CdTe0.16S0.84 QDs nanocomposites. SensS Actuators B: Chem 413:135846. https://doi.org/10.1016/j.snb.2024.135846

Han J, Zhang L, Ye B, Gao S, Yao X, Shi X (2023) The standards for drinking water quality of China (2022 Edition) will take effect. China CDC Wkly 5(13):297–300

Hegazi HA (2013) Removal of heavy metals from wastewater using agricultural and industrial wastes as adsorbents. HBRC J 9(3):276–282. https://doi.org/10.1016/j.hbrcj.2013.08.004

Jaishankar M, Tseten T, Anbalagan N, Mathew BB, Beeregowda KN (2014) Toxicity, mechanism and health effects of some heavy metals. Interdiscip Toxicol 7(2):60–72. https://doi.org/10.2478/intox-2014-0009

Jia Y, Wang J, Zhao L, Yan B (2022) A double responsive fluorescent platform for sensing heavy metal ions based on a dual-emitting fluorescent covalent organic framework hydrogel film. Dalton Trans 51(37):14352–14358. https://doi.org/10.1039/D2DT01810A

Kinuthia GK, Ngure V, Beti D, Lugalia R, Wangila A, Kamau L (2020) Levels of heavy metals in wastewater and soil samples from open drainage channels in Nairobi, Kenya: community health implication. Sci Rep 10(1):8434. https://doi.org/10.1038/s41598-020-65359-5

Langari MM, Antxustegi MM, Labidi J (2022) Nanocellulose-based sensing platforms for heavy metal ions detection: A comprehensive review. Chemosphere 302:134823. https://doi.org/10.1016/j.chemosphere.2022.134823

Li W, Jiang C, Lu S, Wang F, Zhang Z, Wei T et al (2020) A hydrogel microsphere-based sensor for dual and highly selective detection of Al3+ and Hg2+. Sens Actuators, B Chem 321:128490. https://doi.org/10.1016/j.snb.2020.128490

Li L, Xu S, Li X, Gao H, Yang L, Jiang C (2024a) Design of microfluidic fluorescent sensor arrays for real-time and synchronously visualized detection of multi-component heavy metal ions. Chem Eng J 493:152636. https://doi.org/10.1016/j.cej.2024.152636

Li M, Zhang P, Mao J, Li J, Zhang Y, Xu B et al (2024b) Construction of cellulose-based hybrid hydrogel beads containing carbon dots and their high performance in the adsorption and detection of mercury ions in water. J Environ Manage 359:121076. https://doi.org/10.1016/j.jenvman.2024.121076

Mao X, Mao D, Jiang J, Su B, Chen G, Zhu X (2021) A semi-dry chemistry hydrogel-based smart biosensing platform for on-site detection of metal ions. Lab Chip 21(1):154–162. https://doi.org/10.1039/D0LC00855A

Masindi V, Muedi KL (2018) Environmental contamination by heavy metals. In: Heavy metals. InTech. https://doi.org/10.5772/intechopen.76082

Matta G, Gjyli L (2016) Mercury, lead and arsenic: impact on environment and human health. Artic J Chem Pharm Sci. www.jchps.com

Odumbe E, Murunga S, Ndiiri J (2023) Heavy metals in wastewater effluent: causes, effects, and removal technologies. In: Trace metals in the environment. IntechOpen. https://doi.org/10.5772/intechopen.1001452

Ouyang Y, Huang G, Cui J, Zhu H, Yan G, Mei Y (2022) Advances and challenges of hydrogel materials for robotic and sensing applications. Chem Mater 34(21):9307–9328. https://doi.org/10.1021/acs.chemmater.2c01960

Pan S, Xiang C, Liu Z, Tong G, Zhang C, Zhu X (2024) Bio-inspired, robust, and anti-swelling hydrogel sensors for underwater information transmission. J Mater Chem C 12(24):8813–8824. https://doi.org/10.1039/D4TC00397G

Rajumon R, Aravind SP, Bhuvaneshwari S, Ranjitha J, Mohanraj P (2020) Removal of cadmium heavy metal ions from wastewater by electrosorption using modified activated carbon felt electrodes. Water Sci Technol 82(7):1430–1444. https://doi.org/10.2166/wst.2020.425

Villa SM, Mazzola VM, Santaniello T, Locatelli E, Maturi M, Migliorini L et al (2019) Soft piezoionic/piezoelectric nanocomposites based on Ionogel/$BaTiO_3$ nanoparticles for low frequency and directional discriminative pressure sensing. ACS Macro Lett 8(4):414–420. https://doi.org/10.1021/acsmacrolett.8b01011

Walker CH, Sibly RM, Sibly RM, Peakall DB (2005) Principles of ecotoxicology. CRC Press. https://doi.org/10.1201/9781439894224

Wang L, Xu T, Zhang X (2021) Multifunctional conductive hydrogel-based flexible wearable sensors. TrAC, Trends Anal Chem 134:116130. https://doi.org/10.1016/j.trac.2020.116130

Wang Y, Yu Y, Guo J, Zhang Z, Zhang X, Zhao Y (2020) Bio-inspired stretchable, adhesive, and conductive structural color film for visually flexible electronics. Adv Funct Mater 30(32). https://doi.org/10.1002/adfm.202000151

Weng G, Thanneeru S, He J (2018) Dynamic coordination of Eu–Iminodiacetate to control Fluorochromic response of polymer hydrogels to multistimuli. Adv Mater 30(11). https://doi.org/10.1002/adma.201706526

Wu S, Yang Y, Cheng Y, Wang S, Zhou Z, Zhang P, et al (2023). Fluorogenic detection of mercury ion in aqueous environment using hydrogel-based AIE sensing films. Aggregate 4(3). https://doi.org/10.1002/agt2.287

Yang J, Luo Z, Wang M (2022) Novel fluorescent Nanocellulose hydrogel based on Nanocellulose and carbon dots for detection and removal of heavy metal ions in water. Foods 11(11):1619. https://doi.org/10.3390/foods11111619

Zhang JingJing, Cheng F, Li J, Zhu J-J, Lu Y (2016) Fluorescent nanoprobes for sensing and imaging of metal ions: Recent advances and future perspectives. Nano Today 11(3):309–329. https://doi.org/10.1016/j.nantod.2016.05.010

Zhang J, Zeng L, Qiao Z, Wang J, Jiang X, Zhang YS, Yang H (2020) Functionalizing double-network hydrogels for applications in remote actuation and in low-temperature strain sensing. ACS Appl Mater Interfaces 12(27):30247–30258. https://doi.org/10.1021/acsami.0c10430

Zhang M, Xu T, Liu K, Zhu L, Miao C, Chen T, et al (2025) Modulation and mechanisms of cellulose-based hydrogels for flexible sensors. SusMat 5(1). https://doi.org/10.1002/sus2.255

Zhong J, Zhao H, Cheng Y, Feng T, Lan M, Zuo S (2021) A high-performance electrochemical sensor for the determination of Pb(II) based on conductive dopamine polymer doped polypyrrole hydrogel. J Electroanal Chem 902:115815. https://doi.org/10.1016/j.jelechem.2021.115815

Zhu J, Tao J, Yan W, Song W (2023) Pathways toward wearable and high-performance sensors based on hydrogels: toughening networks and conductive networks. Natl Sci Rev 10(9). https://doi.org/10.1093/nsr/nwad180

Chapter 7
Hydrogel Applications in Heavy Metal Adsorption from Wastewater

Zulnurain Khan, Bushra Mansoori, and Mohd Kamran Khan

Abstract Hydrogel consists of structural units that are either amphiphilic or hydrophilic, organized into interconnected macroscopic networks. Hydrogels are three-dimensional systems formed by chains of hydrophilic polymers. They hold water without undergoing dissociation. Polymer chain cross-linking may occur through chemical means (covalent bonding) or physical means (hydrogen bonding or hydrophobic interactions). PVA, PEG, and polyacrylic acid are prevalent examples of synthetic polymer-based gels. Toxic metal ions and many hazardous substances from several sectors perpetually contaminate the air, water, and soil. Numerous approaches have been extensively utilized to eliminate heavy metal ions. The techniques encompass membrane filtration (including nanofiltration, reverse osmosis, ultrafiltration, and electrodialysis), coagulation and flocculation, ion exchange, adsorption, and chemical precipitation (comprising sulfide, oxide, and chelating precipitation), among others. We evaluated and contrasted the efficacy of the hydrogels in removing Fe^{3+}, Cr^{3+}, and Hg^{2+} ions from solution through diffusion, coordination bond formation, and electrostatic interactions. The poly (AAm) hydrogel has a significant ability to eliminate heavy metal ions owing to its moderately acidic AAc groups. The hydrogel samples' capacity to release and recuperate heavy metal ions is equally remarkable. Hydrogels demonstrated a markedly enhanced compressive strength following the incorporation of heavy metal ions, hence broadening their potential applications across several industries.

Z. Khan
Department of Chemistry, Faculty of Science, Aligarh Muslim University, Aligarh, India

B. Mansoori
Department of Industrial Chemistry, Faculty of Science, Aligarh Muslim University, Aligarh, India

M. K. Khan (✉)
Interdisciplinary Nanotechnology Centre, Faculty of Engineering and Technology, Aligarh Muslim University, Aligarh, India
e-mail: kkhan6387@gmail.com

Department of Chemistry, King Fahd University of Petroleum and Minerals, Dhahran, Saudi Arabia

S. Bardhan and M. P. Shah (eds.), *Applications of Hydrogels in Modern Wastewater Treatment*, Advances in Wastewater Research,
https://doi.org/10.1007/978-981-95-1598-1_7

Keywords Collagen · PVA · Heavy metals · Hydrophilic · Diffusion

Abbreviations

PVA	Polyvinyl alcohol
PEG	Polyethylene glycol
PMMA	Poly-methacrylic acid
HEMA	Polyhydroxy ethyl methacrylate
PLA	Polylactic acid
PEO	Polyethylene oxide
PLL	Poly(L-lysine)
CMC	Carboxymethyl cellulose
ECH	Epichlorohydrin
CS	Chitosan
pH	Potential of hydrogen

7.1 Introduction

The hydrogel structure includes extensive hydrophilic or amphiphilic structural units, which preserve their shape while absorbing large amounts of water. The substance starts as a solution made of building blocks and soluble monomers, but becomes an insoluble hydrogel material through increased crosslinking processes. The crosslinking mechanism occurs in two ways, through hydrogen bonding as well as through covalent bonds, before the resulting strong structure allows water absorption while maintaining strength. Hydrogels derive their main characteristic from water absorption abilities, which make them excellent candidates for tissue compatibility applications. Hydrogels exist as naturally occurring proteins and DNA together with synthetic materials that include poly (2-hydroxyethyl methacrylate) and poly(N-isopropyl acrylamide), as well as biohybrid compounds (Peppas et al. 2006). Different compositions of hydrogels determine their water-absorption abilities, yet hydrogels need between 10-15% weight to display these properties. Design elements differentiate the degradation properties of hydrogels from those that keep their chemical composition intact. The basic structural characteristics determine the physical form which hydrogels will adopt (Khan et al. 2022). The formation of several defects in physical hydrogels occurs because of molecular entanglements along with hydrophobic sites, along with ionic regions, which also include structural imperfections such as chain loops and free ends (Hoffman 2002). Hydrogel synthesis depends on using solvent monomers or multifunctional polymers to increase the capacity to absorb and retain water (Bajpai 2001). Hydrogel features undergo direct modifications because of physical as well as chemical influence factors. The scientific definition of crosslinking describes the formation of polymer chain junction points that create network structures which enhance hydrogel resistance against degradation (Maitra and Shukla 2014) Interfaces Hydrogel water maintainability increases along with an increase in crosslinking density (Jones and Vaughan 2005). Aqua retention in hydrogels occurs through water absorption and maintenance because of amenable hydroxyl(–OH) groups, which form hydrogen bonds with water. This characteristic renders them appropriate for diverse applications, including:

- Compressed powder formulations, including oral pills and capsules
- Rigid structures, such as soft contact lenses
- Coatings present on pharmaceuticals, catheters, implants, or electrophoresis capillaries
- Microparticles utilized in bio-adhesive pharmacological administration or wound repair
- Encapsulated solids, exemplified by those utilized in osmotic pumps
- Membranes for transdermal pharmaceutical administration or electrophoresis matrices.

Due to their exceptional water retention and adaptability, hydrogels have become vital materials in biological, pharmacological, and industrial fields.

7.2 Categories

7.2.1 According to the Characteristics of the Polymer

Hydrogels exist in two types based on the used polymers, in addition to their manufacturing techniques.

7.2.1.1 Classification Based on Polymer Type

The hydrogel field splits into natural polymer hydrogels along with synthetic polymer hydrogels as its two fundamental groups (Hoffman 2002). The biological sources of natural hydrogels (Varghese et al. 2019) include alginate, pectin, carrageenan, chitosan, collagen, and poly-lysine, dextran, agarose, and pollutants. Many biomedical applications make use of such materials because they show high compatibility with biological systems. Synthetic hydrogels comprise polymers such as PVA, PEG, and PAA. These synthetic materials provide enhanced control over mechanical strength, degradation rate, and stability, rendering them advantageous for numerous industrial and medicinal applications.

They are additionally categorized based on surface charge.

- **Anionic hydrogels** are composed of negatively charged polymers, like alginic acid or alginate.

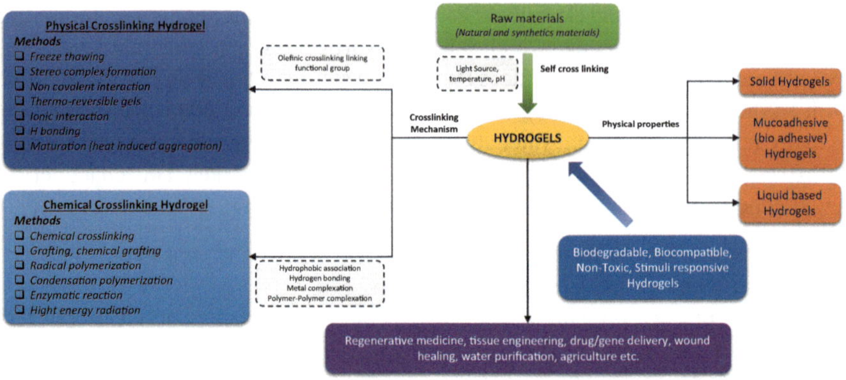

Fig. 7.1 Diagram illustrating the process of making hydrogels

- **Cationic hydrogels** contain positively charged polymers such as chitosan and poly-lysine, which are crucial for cellular contacts and adhesion.
- **Neutral hydrogels** comprise polymers including agarose, pullulan, and dextran. These neutral polymers can be combined with synthetic hydrogels and polysaccharides to improve particular features (Bahram et al. 2016).

Polysaccharide-based hydrogels are extensively utilized in pharmaceuticals owing to their biocompatibility and non-toxic characteristics. Substances such as alginate, chitosan, and hyaluronic acid have the capacity to retain substantial volumes of water, rendering them suitable for uses in medication administration, wound healing, and tissue engineering. Likewise, protein-based hydrogels, such as those derived from collagen and polylysine, have garnered interest because of their biological advantages (Zhu et al. 2019). These materials facilitate cellular proliferation, provide structural integrity, and can be tailored for many medical applications, especially in regenerative medicine. Their capacity to replicate natural tissues renders them essential in the development of sophisticated biological therapies (Fig. 7.1).

7.2.1.2 Classification Based on Preparation Method

According to their synthesis methodology, hydrogel synthesis involves one hydrophilic monomer, which receives crosslinking to produce a homopolymer product. Copolymer hydrogels use two or more different monomeric units to build the hydrogel, yet require at least one hydrophilic unit for water absorption that drives hydrogel expansion. Multipolymer hydrogels emerge from three or more co-monomers that collaborate to form an elaborate network structure. Interpenetrating Polymer Network (IPN) Hydrogels: These hydrogels consist of numerous polymer networks that are interlaced without chemical bonding. The procedure entails establishing an initial polymer network, followed by the incorporation of a secondary network that interlaces with the current framework to produce a unified hydrogel.

New strength occurs in chemical hydrogels due to covalent linkages that create structures with increased durability (Jaipan et al. 2017). Examples encompass gelatin-based and polymethacrylic acid (PMMA) hydrogels (Hussein 2017). Physical hydrogels remain intrinsically stable using basic forces, including hydrogen bonding and ionic forces, as well as molecular entanglements. Agar–agar (Xu et al. 2018), together with polyhydroxyethyl methacrylate (HEMA) (Kim et al. 2018), represent physical hydrogels. Researchers widely study hydrogels today because these materials offer flexible applications in two main areas: medicine delivery through capsules and tissue structure development. Researchers can guarantee specified functional properties by selecting appropriate polymers and appropriate crosslinking mechanisms for hydrogels.

7.3 Preparation

The methods for producing them involve different preparation processes, which shape all properties, including mesh size and morphology, as well as swelling and permeability. Physical or chemical crosslinking and grafting, together with radiation crosslinking, stand as common preparation methods (Fig. 7.2).

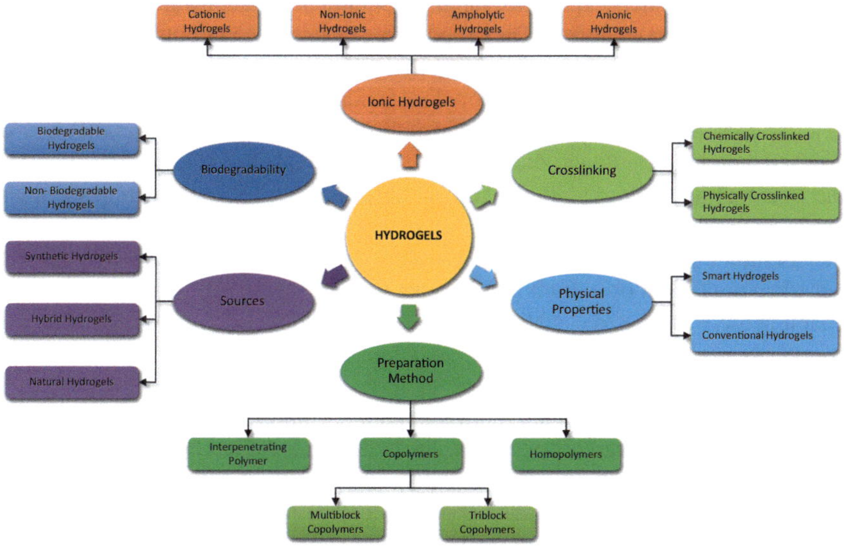

Fig. 7.2 Diagrammatic representation of the hydrogel classification system

7.3.1 Physiological Techniques

Applications of physical cross-linking procedures eliminate the necessity to use a crosslinker. The physical crosslinking of hydrogels finds its greatest importance within the food along with packaging industries. A review of different approaches for synthesizing physically cross-linked hydrogels appears in this document according to the literature research.

7.3.1.1 Through Thermal Regulation

Physically crosslinked hydrogels are formed by exposing polymer chains in semi-dilute aqueous solutions to successive heating and cooling cycles. These cycles transpire at temperatures exceeding and falling below the polymer's melting point, facilitating the transition of polymer chains through various structural configurations. During the preliminary phase, the heating and chilling process induces the polymer chains to assume a coiled shape. The subsequent phase involves the reiteration of cycles by the incorporation of salts like sodium (Na^+) or potassium (K^+), which promote the development of double helices (Gulrez et al. 2011). This leads to the creation of stable hydrogels. In certain instances, heating a polymer solution can initiate block copolymerization, resulting in hydrogel formation (Shoukat et al. 2020). These polymers experience rapid temperature cycling, resulting in transitions between coiled and gel forms. The introduction of salts such as potassium or sodium ions enhances gelation by promoting physical crosslinking. The heating process causes structural changes in polypropylene oxide solution, together with polyethylene oxide solution and polylactic acid polyethylene glycol hydrogels (Pan et al. 2020). Upon cooling down, the polymer chains establish junction zones that lead the hydrogel network to solidify. The initial random coil form of carrageenan also exists above its melting point before rearranging into stiff helical structures when the solution cools down. Prolonged double helical structures convert to a resistant gel when potassium or sodium ions interact with sulfonic (SO_3^-) groups, while also enabling these groups to establish repulsive effects.

A. **Complex Coacervation**

Complex coacervation is an alternative technique employed to create hydrogels via ionic interactions. Charged polymers, including polycations and polyanions, are combined in a solution under regulated pH and concentration parameters. The electrostatic attraction among these charged molecules results in a polymer network, which can exist in both soluble and insoluble states. Scientists refer to this new substance as complex coacervate, which functions as an ionic crosslinked hydrogel. The standard preparations in this field include xanthan gum as a polyanionic component along with chitosan (Malik et al. 2020) as a polycationic agent mixed with calcium alginate capsules that receive an alginate-poly(L-lysine) (PLL) shell. The use of coacervates stands prominent in biomedical applications through drug delivery systems and encapsulating methods, mainly due to their stable composition and suitable biological behavior.

B. **Ionic Interaction**

When polymeric materials interact with counterions that hold two or three valences, they undergo a process of crosslinking. Na^+-alginate gels form when negatively charged ions come near the gelling solution. The formation of hydrogels via ionic interaction crosslinking needs the inclusion of di- or trivalent counterion compounds. The crosslinking reaction occurs under neutral Ph (Funami et al. 2007) levels during temperature-neutral conditions. Crosslinking between polyanions and polycations leads to the creation of anionic polymers which form bonds with metal ions (Kuo and Ma, 2001).

7.3.1.2 Hydrogen Bonding

The synthesis method for this hydrogel depends on hydrogen bond crosslinking. The hydrogels form instantaneously when a hydrogen acceptor group receives the provided hydrogen donor. The mixture of tannic acid (TA) and poly(N-vinylpyrrolidone) (PVP) easily creates an agglomerated hydrogel named PVP–TA.

7.3.1.3 Crystallization

Freezing and thawing represent the method used for this operation. The generation of hydrogel coils through freezing followed by multiple thawing steps of polymer solutions makes up this process (Gupta et al. 2011). The interconnection of these coils produces an elastic hydrogel network. During freezing, ice forms in the amorphous area, but crystallite polymers keep expanding until they reach neighboring crystallite borders and create a porous structure when the solution thaws. The resulting mechanical properties of this crystallization method achieve satisfactory outcomes. These hydrogels possess superior mechanical strength, which outperforms other crosslinked materials because they transmit applied stress throughout their crystallite network. PG hydrogel production involves subjecting glycerol and PVA gels to freezing and thawing processes. Properties of mechanical strength developed through crystallization produce better results than different methods of crosslinking.

7.3.1.4 Chemical Method

Crosslinking processes take place when bifunctional polymer components react directly to form chains that result in a network structure. The resulting cross-linked polymer network features multiple functional groups (–NH2, –COOH, –OH) which interact with cross-linking agents based on adipic acid, glutaraldehyde, and di-hydrodiphenyl diamide. The crosslinked structure of hydrophilic polymers forms through condensation reactions that employ organic hydroxyl groups in aldehydes and oxalic acid, as well as maleic acid and di-isocarboxylic acid esters and dimethylurea, but also requires their pendant hydroxyl group as a crosslinker. The chemical components Epichlorohydrin (ECH) and glutaraldehyde produce hydrogels used extensively in biocompatibility research. When epichlorohydrin is cross-linked with carboxymethyl cellulose (CMC), the resulting product becomes the super-absorbent CMC/ECH hydrogel.

7.3.2 *By Chemical Cross-Linkers*

A cross-linking agent contains two or more reactive ends, which enable chemical interactions with proteins and primary amines and sulfhydryl groups and various other chemical substances. Three main cross-linking agents used are the chemicals citric acid and glutaraldehyde, and epichlorohydrin. The cross-linking agents with genipin accelerate the formation of cross-linked structures in hydrogels based on chitosan. The cross-linking agent function of citric acid exists in a hydrogel solution that contains both carboxymethylcellulose and poly (ethylene glycol). Hydrogels produced from gelatin/alginate need glutaraldehyde as their cross-linking agent. Chemical cross-linking represents an adjustable approach to improve hydrogel mechanical properties. The cross-linking agents expose serious environmental threats by harming how hydrogel systems interact with their physiological contents. Through electron beam and radiation methods, healthcare providers can avoid the negative effects of chemical cross-linking agents. The advantage of radiation cross-linking stems from its cleanliness, alongside energy efficiency and adjustable cross-linking degree, and its ability to eliminate all extraneous residues in the final production.

7.3.2.1 Grafting

The initial step starts through activation of the macromolecular chain backbones using a chemical process. Chitosan polymers are cross-linked with N,N'-methylene bisacrylamide because it enables acrylic acid grafting onto the chitosan backbone. The polymerization reaction, which produced CTS-g-PAA hydrogels, started by using potassium dihydrogen phosphate as the initial activator. The synthesis of CTS-g-PAA/cell hydrogels included cellulose as their monomeric material. The combination of thiourea formaldehyde resin and chitosan and cellulose structures produced the crosslinking agent (CTS/cell). The laboratory can generate the (CTS/cell)-g-PAA hydrogel through free radical polymerization of acrylic acid monomer to achieve cross-linking. The produced hydrogel possesses superior mechanical properties and better acid durability, together with higher water absorption and prolonged water retention capabilities. The research outcomes showed that the superabsorbent polymer effectively controlled soil nutrient delivery, thus improving its potential use in both agricultural and horticultural fields.

7.3.2.2 By Radiation Method

X-rays, alongside gamma rays and electrons, primarily serve as the producing agents in this technology, leading to hydrogel formation. Radiation-based hydrogel synthesis produces pure substances with minimal toxicity because it requires no chemical initiators. Hydrogels function as the main output of radiation technology, which stands out due to its straightforward nature and ecological sustainability, as well as its high operational effectiveness. Synthesis, along with sterilization, is combined into one operation, resulting in cost and production time reductions through this technology (Fig. 7.3).

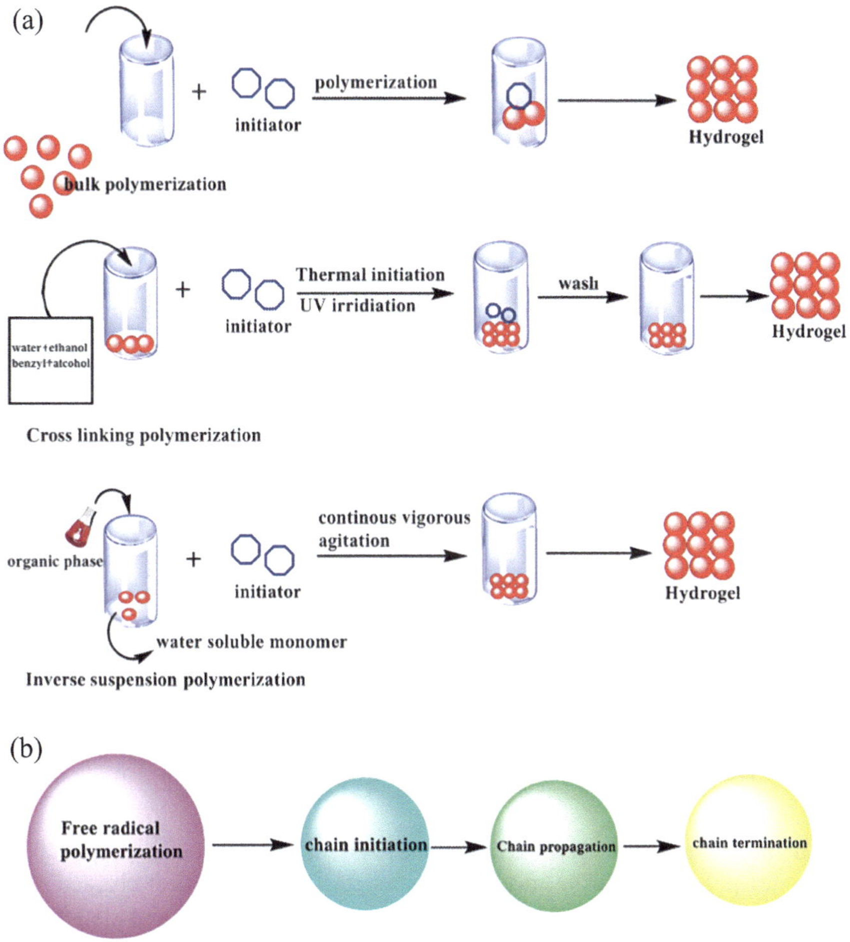

Fig. 7.3 **A** Various synthesis methods, such as inverse suspension polymerization of the polymerization hydrogel, bulk polymerization, and cross-linking polymerization; **B** The free radical polymerization reaction process in the solution

7.4 Characteristics of Hydrogels

7.4.1 *Elevated Hydrophilicity*

A key property of hydrogels is their capacity to absorb substantial quantities of water, attributable to the presence of hydrophilic functional groups in their polymeric structure. Their resistance to hydrolysis is attributable to cross-linking among polymer chains. This distinctive amalgamation renders hydrogels very advantageous in biomedical applications. Hydrophilic polymers, whether of natural or synthetic origin, facilitate the interaction of hydrogels with biological tissues, including mucous membranes and epithelial layers, hence improving their suitability for medical applications.

7.4.2 *Elevated Swelling Capacity*

Hydrogels are highly absorbent materials capable of holding substantial quantities of biological fluids while maintaining a soft, elastic, and viscoelastic nature. This characteristic is essential in regulating the immune response when employing hydrogels in biological applications. The extent of hydrogel swelling is determined by the specific crosslinker employed during its synthesis.

The swelling process consists of three stages:

- Primary Bound Water: Water molecules first infiltrate the hydrogel network and adhere to the polymer chains.
- Secondary Bound Water: The absorbed water increases further due to the relaxation of polymer chains.
- Free Water: Additional water is ultimately absorbed into the expanded hydrogel matrix, resulting in complete swelling.

Hydrogels, due to their capacity to retain water, are essential in wound healing, medication delivery, and tissue engineering applications.

7.4.3 *Mechanical Integrity*

For hydrogels to operate efficiently in medical and pharmaceutical applications, they must possess sufficient mechanical strength. The mechanical characteristics of hydrogels can be precisely modified by altering the degree of crosslinking. An increased degree of crosslinking produces a more robust and stiff gel network; nevertheless, excessive crosslinking may render the hydrogel brittle and diminish its flexibility. A reduced degree of crosslinking enhances elasticity, advantageous for applications necessitating adaptability and mobility. Hydrogel mechanical property evaluations require different testing methodologies that incorporate rheometry along with dynamic mechanical analysis, as well as compression testing and tensile analysis. Various tests help determine the perfect combination of strength with flexibility for different usage scenarios.

7.4.4 Biocompatibility

For hydrogels to be appropriate for the biomedical sector, they must be safe, non-toxic, and biocompatible. Most polymers intended for these uses must undergo stringent assessments to determine their toxicity both in vivo and in terms of cytotoxicity. Biocompatibility denotes a material's capacity to engage with the body without inducing harm. This encompasses two fundamental aspects: biosafety and bifunctionality.

- **Biosafety** denotes the lack of detrimental effects, including cytotoxicity, mutagenesis, or carcinogenesis. It also encompasses verifying that the body's response to the material is suitable—both locally, at the site of tissue interaction, and systemically, throughout the organism. The substance must not elicit any detrimental immunological responses or adverse side effects.
- **Bifunctionality** pertains to a material's capacity to execute its intended function. In tissue engineering, hydrogels must facilitate cell proliferation, adhesion, and regeneration while enhancing healing. This factor is crucial, particularly in applications where the material will be utilized for prolonged durations or incorporated into the body's healing mechanisms. It is essential to evaluate the potential hazards associated with leftover chemicals employed in the synthesis of hydrogels, including initiators, crosslinkers, solvents, and stabilizers. If these substances are not eliminated, they may permeate adjacent tissues and adversely affect the body. To mitigate this, purification techniques such as washing or dialysis are employed to eliminate any hazardous chemicals from the final gel. For hydrogels to be useful in biomedical applications, they must be safe and fulfill their intended function within the body, facilitating the healing process while preventing unwanted effects. People have allowed their careless waste disposal habits to create the global water contamination problem. Wastewater treatment currently faces a crucial challenge regarding finding budget-friendly as well as environmentally friendly methods for hazardous contaminant removal, including heavy metals and dyes, before water discharge occurs into natural reservoirs. Protecting our freshwater resources from depletion and contamination becomes vital in our attempt to safeguard them. Wastewater treatment employs adsorption as an affordable and basic technical method. Natural materials that absorb toxins show increasing industrial value because they are practical to obtain while being both inexpensive and friendly to nature. A range of activated carbon and clays together with natural zeolites and agricultural by-products is studied for testing their abilities to remove petroleum and heavy metals from wastewater.

Reno solution adsorption shows huge promise, yet scientists still face difficulties with process optimization for handling difficult water contaminants. The selection of materials needs to ensure that such substances do not create new environmental issues after extracting pollutants. Hydrogels stand as one of the most appealing groups of adsorbents.

7.4.5 Biodegradability

Biodegradable hydrogels decompose into innocuous byproducts via hydrolysis and enzymatic breakdown, enabling their absorption or excretion by the body. The degradation rate and process are contingent upon polymer composition, molecular weight, etc. Certain hydrogels undergo degradation through straightforward hydrolysis, but others necessitate enzymatic activity to dismantle intricate polymer architectures. Protein-based and polysaccharide-based hydrogels undergo degradation via enzymatic hydrolysis, mediated by particular enzymes, including proteases for protein hydrogels and glycosidases for polysaccharide hydrogels. Chitosan-based hydrogels undergo deterioration by enzymatic hydrolysis and oxidative mechanisms. Biodegradable hydrogels are especially advantageous in applications including controlled medication release, tissue engineering scaffolds, and wound dressings, as they decompose gradually without leaving detrimental residues in the body.

7.4.6 Physical and Chemical Properties

Despite significant progress, our comprehension of gel characteristics is still inadequate, complicating the effective design of novel gel-based systems. An essential element of this is comprehending the interaction of solute molecules with the gel, specifically their distribution between the gel phase and the adjacent liquid. The partitioning behavior is predominantly governed by two factors: size exclusion and molecular interactions, including attraction or repulsion. Size exclusion is contingent upon the gel's architecture and pore dimensions, whereas molecule interactions are influenced by variables such as charge, polarity, and chemical affinity.

7.4.6.1 Swelling

Hydrogels are polymeric networks that absorb and retain liquid, resulting in their expansion. The polymer structure retains the liquid, while the absorbed water functions as a selective filter, permitting specific molecules to traverse. The water absorption capacity of hydrogels ranges significantly, from 10–20% of their dry weight to several hundred times their mass. The interaction of water with a hydrogel dictates the movement of nutrients inward and the diffusion of biological products outward.

7.4.6.2 Mechanical Strength

The mechanical properties of a material may vary based on its composition and structure. The stiffness of a gel can be altered by modifying the degree of crosslinking, either by increasing or decreasing it by heating. The requisite stiffness of a complex prosthetic device, such as an intervertebral disc replacement, is contingent upon the type of cells being cultivated. A stiffer substance is required for osteoblast cell seeding than for adipocyte cultivation.

7.4.6.3 Porosity and Permeation

During hydrogel synthesis, phase separation frequently leads to the formation of holes or relatively small pores within the network. The tortuosity parameter significantly influences characteristics such as size and pore interconnectivity, which are essential in defining the material's structure. Nevertheless, obtaining precise measurements of these qualities is challenging. Porosity: It inherently denotes voids inside the substance's mass. Regulating porosity is essential primarily to enhance macromolecule release or to promote cellular mobility in hydrogel-based scaffolds. By adjusting porosity, hydrogel fits can be tailored for certain biomedical and technical purposes.

7.5 Utilization via Adsorption

Hydrogels represent advanced biomaterials through three-dimensional cross-linked fibers made from chains of highly hydrophilic polymers, which display exclusive physical and chemical characteristics along with substantial swelling ability and satisfactory biocompatibility, while providing extensive research fields for heavy metal elimination applications. Cellulose exists as the most abundant renewable resource that Earth contains. The preference for derivative-based hydrogel development surpasses straight cellulose (Chang and Zhang 2011) utilization since cellulose derivatives dissolve better in water. The commercial cellulose derivative CMC stands as one of the most highly used products because it dissolves easily in water. The preparation of CMC hydrogels mainly depends on crosslinking reactions using polycarboxylic acids as well as epichlorohydrin (ECH) and N, N′-methylene-bisacrylamide. Tests of heavy metal ions Pb^{2+}, Ni^{2+}, and Cu^{2+} adsorption capability (Bouyahmed et al. 2022a, b) for the hydrogels were carried out utilizing model wastewater solutions. The biopolymer extracted from chitosan (CS) has become a rising star (Alam et al. 2019) across several sectors, including environmental biotechnology, because it shows exceptional metal retention properties at a reasonable cost. The field makes use of numerous proprietary and unique nanoparticles, which work as encapsulants to boost hydrogel bead adsorption effectiveness in both hydrogel films and hydrogel nanocomposites (Khan et al. 2025). The pollution elimination prospects for hybrid hydrogels extend to a wide range of aquatic pollutants. The adsorption properties of cross-linked hydrogel polymers extend to various solvents while maintaining high adaptability through their special and specific nature.

This three-dimensional system represents hydrophilic groups that establish an efficient platform for adsorbing large amounts of dyes and metal ions from wastewater solutions. The precise chemical properties, alongside hydrophilicity and sensitivity, and utility characteristics of novel hydrogels make them more appealing options for wastewater treatment development. Heavy metal contamination within wastewater continues to increase because numerous industrial sectors are expanding, such as plating and electroplating, battery manufacturing, pesticides, mining, rayon production, tanning, fluidized bed bioreactors, textiles, metal smelting, petrochemical processes, paper production, and electrolysis. The released pollutants from these contaminants create substantial risks for the environment, together with human health hazards. Heavy metals such as lead, mercury, cadmium, arsenic, and chromium are non-biodegradable, leading to their accumulation in the environment over time and resulting in enduring damage. The effect of heavy metals on water supplies is particularly concerning. The influx of wastewater containing these metals into rivers, lakes, and oceans can severely impact aquatic ecosystems. Fish, flora, and various species assimilate these metals, resulting in toxicity and destabilizing entire ecosystems. Human exposure to contaminated water or seafood can lead to significant health complications, including organ damage, neurological disorders, and, in certain instances, cancer. This pollution disrupts the equilibrium of aquatic habitats, hindering the survival and prosperity of numerous species. In certain instances, it impairs the reproductive systems of aquatic organisms, resulting in population decreases. The contaminated water might become hazardous, unsuitable for consumption, bathing, or cultivation, exacerbating the issue. Confronting the issue of heavy metal contamination necessitates a comprehensive strategy. It encompasses the development of sophisticated treatment technologies, like ion exchange, reverse osmosis, and electrocoagulation, alongside the investigation of natural remedies such as bioremediation utilizing plants or microorganisms to absorb or neutralize pollutants. It is essential to implement more stringent controls on industry and promote cleaner alternatives that reduce the utilization of hazardous metals. Addressing this issue necessitates cooperation among industries, governments, and environmental organizations to safeguard both human health and the ecosystem. The escalating impact of effluent contamination on the atmosphere is raising concerns among researchers globally. Prior to entering water reservoirs, wastewater is challenging to filter. Consequently, the treatment of wastewater is essential prior to its discharge into natural water bodies. The treatment of wastewater is a relatively modern process. Hydrogels can serve as versatile platforms for the targeted detection of pollutant traces and the comprehensive purification of micro-polluted water. High capacity and rapid adsorption rates have garnered significant interest.

A primary advantage of hydrogels in water purification is their superabsorbent characteristics, facilitating water absorption and pollutant elimination. A multitude of hydrogels have been examined for their capacity to eliminate pollutants from water and wastewater in laboratory settings. The primary subjects of discourse were metalloids and heavy metals. Besides metals and metalloids, organic and inorganic colors, together with medicines, have garnered significant study for hydrogel clearance. Heavy metals are a critical category of inorganic pollutants that must be removed from aquatic environments.

7.6 Different Types of Contaminants

The hydrogel's hydrophobic characteristics and ability to retain increased water in its swelled state are ascribed to the polymer network structure, which also facilitates the rapid absorption of metal ions from aqueous solutions. Waste and byproducts contaminate water, rendering it discolored and unsuitable for human or animal consumption. Toxic compounds originating from dyes adversely affect ecosystems and human health. Industrial wastewater may contain a diverse array of pollutants at varying quantities (Shah 2024). Organic pollutants have garnered significant interest due to their frequently elevated toxicity and the challenges traditional approaches face in eliminating them. A category of organic contaminants referred to as phenols can impact all kinds of life, including humans and ecosystems.

Polymer hydrogel-based purification is constrained by mechanical strength for the aforementioned reason. The incorporation of GO and other inorganic species into the composite hydrogel formulation aims to enhance the mechanical strength of the resultant hydrogel, hence expanding its applicability. Besides enhancing mechanical qualities, graphene oxide sheets demonstrate excellent adsorption capabilities, facilitating the effective elimination of deleterious organic dyes from aqueous solutions. The practical requirements include filter media with substantial adsorption capacity, rapid kinetics for removal, cost-effectiveness, and recyclability in wastewater treatment. A porous Jute/Polyacrylic acid (Jute/PAA) hydrogel property was created to attain these. The Jute/PAA gel exhibits exceptional adsorption capacity for the extraction of heavy metal ions, specifically Cd^{2+} and Pb^{2+} from wastewater. Previous studies have demonstrated the successful manufacture of green carboxymethyl cellulose (CMC) hydrogel beads utilizing epichlorohydrin (ECH) as a cross-linking agent, achieved through the formation of an ether bond between ECH and CMC in a liquid wax suspension (Fig. 7.4).

7.7 Limitations of Adsorption Techniques and Future Perspectives

Adsorption treatment techniques exhibit strong development, yet they face several challenges that stop their widespread implementation. Current issues in wastewater treatment adsorption methods consist of three groups, including adsorption capacity,

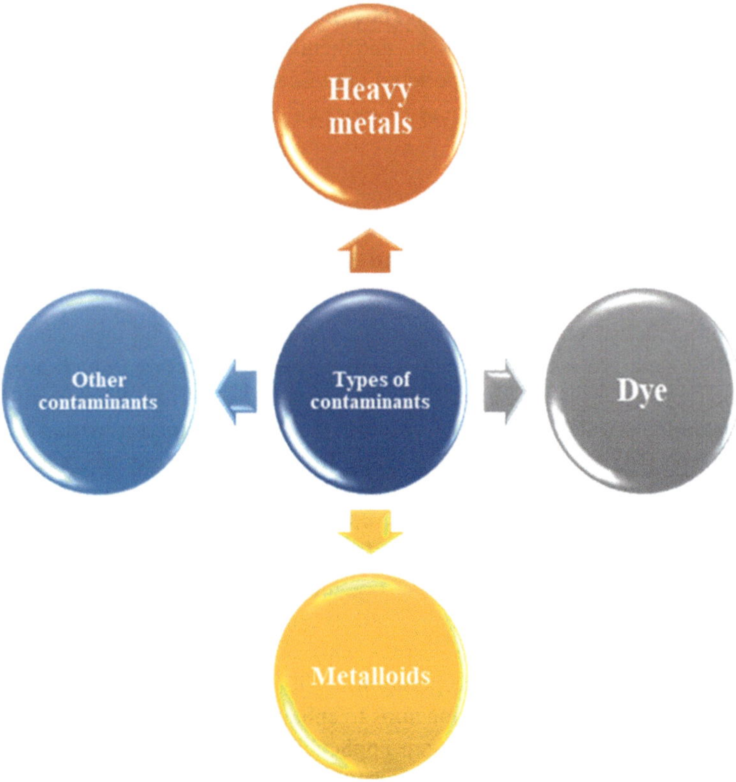

Fig. 7.4 Examples of types of contaminants

selectivity, and non-specific binding, and two other groups addressing degradability and stability of adsorbents, and sustainability of recovery and reusability, and economic feasibility, along with the specific details regarding these separate groups are presented throughout the following sub-sections.

7.7.1 Adsorption Capacity

Wastewater treatment through adsorption techniques requires an adsorption capacity to succeed in both effectiveness and efficiency. A range of variables affects the adsorption capacity according to Table 7.1. Most adsorbents used today demonstrate insufficient adsorption properties for handling wastewater contaminants that exist at high concentrations and present multiple kinds. Rising treatment efficiency requires large amounts of adsorbent materials, which increases operational costs and makes the process more complex. The adsorption capacity usually decreases when most adsorbents reach the saturation point, at which available binding sites on the adsorbent surface become fully occupied while the initial adsorbate concentration grows. It becomes essential to fabricate innovative adsorbents possessing better ability and selective binding affinity toward different contaminants present in wastewater.

7.7.2 *Future Perspective*

Active progress in wastewater treatment through adsorption technologies continues as researchers and technologists develop new methods of operation. Environmental engineering, along with materials science research, has led to the advancement from traditional to modern practices. Traditional procedures enabled the development of advanced techniques that use carbon nanotubes and metal–organic frameworks (MOFs) alongside graphene-based structures through their utilization of activated carbon and zeolites. This development shows how technical capabilities merge with cost-performance and sustainability goals alongside performance effectiveness in the evolution process.

Consequently, composite hydrogels exhibit superior potential as adsorbents in comparison to simple hydrogels. Chemical conjugation is an effective way for enhancing the adsorption capacity of hydrogels by including functional groups.

7.8 Conclusion

The review shows that hydrogel functional groups connect with contaminants with chemical and physical mechanisms. Compound hydrogels prove superior to basic hydrogels as adsorbents in cleanup processes. The adsorption capacity of hydrogels can be improved effectively through chemical conjugation methods that introduce functional groups. The food and agricultural wastewater treatment process requires careful evaluation of hydrogel safety since improper assessment might introduce dangerous chemical contaminants. The use of hydrogels as water pollutants removal adsorbents presents a powerful opportunity for environmental cleanup efforts. The three-dimensional complex polymeric network of hydrogels allows them to absorb substantial amounts of water, which in turn supports the adsorption of numerous pollutants.

The adsorption properties of hydrogels benefit from their superior absorption capabilities along with designability and compatibility with biological systems, yet suffer from several limitations, which include snapping broken structures and desorption challenges, as well as environmental responsiveness and cost considerations, and sustainability challenges. The selective adsorption of specific molecules through hydrogels occurs through compositional modification of polymers and changes in cross-linking density, and the addition of functional groups. Hydrogels offer both compatibility with living tissue and zero harmful effects, so they work perfectly in biomedical implementations such as tissue engineering systems and medication

Table 7.1 Comparison table of removal of heavy water using different hydrogels

Metal ion	Adsorbent	q1max (mg/g)	Adsorbent 2	q2max (mg/g)	References
Pb^{2+}	Fe_3O_4-chitosan/EDTA adsorbent	220.00	Carboxylated chitosan/carboxylated nanocellulose hydrogel	334.90	Zhu et al. (2019)
	Magnetic chitosan/cellulose microspheres	45.86	Chitosan/cellulose phosphonate composite hydrogel	211.42	Li et al. (2019), Wang et al. (2022)
	Chitosan/alginate/Fe_3O_4@SiO_2 composites	234.77	Aminopropyltriethoxysilane-Fe_3O_4 attapulgite@chitosan	625.34	Wang et al. (2022), Dai et al. (2010)
	Alginate/melamine aerogel	765.30	Alginate/melamine/chitosan aerogel	1331.60	Wang et al. (2023)
	Sodiumalginate/chitosan hydrogel	100.00	Sodium alginate/chitosan/PVA hydrogel	147.93	Kongarapu et al. (2017)
	Chitosan/alginate	217.24	Chitosan/alginate/Fe_3O_4@SiO_2 composites	245.28	Tian et al. (2020)
	Magnetite chitosan/PVA	117.60	Magnetite chitosan/PVA/carboxylated cellulose nanofibrils	171.00	Bouyahmed et al. (2022a, b)
	Boehmite	17.48	Chitosan–boehmite	33.00	
Cd^{2+}	PVA/chitosan/GO	170.36	PVA/chitosan/polydopamine-GO	204.13	Crini et al. (2019)
	Silicate-titanate nanotubes	172.00	Silicate-titanate nanotubes-chitosan hydrogel	656.00	Ali et al. (2012)
	Saccharomyces cerevisiae/ alginate composites	30.70	Chitin nanofiber aerogel	127.50	Crini and Lichtfouse (2019), Foo and Hamee (2009)
	Carboxymethyl cellulose/ chitosan/dopamine	441.00	Carboxymethyl cellulose/chitosan/dopamine @polyethyleneimine	470.00	Wang et al. (2010)
	Maize tassel-magnetite	52.05	Highly cross-linked thiosemicarbazide chitosan	81.26	Zheng and Wang (2010)
	Chitosan	94.30	Thiosemicarbazide modified chitosan	257.20	Sun et al. (2024)
	Chitosan	67.50	Chitosan/nano-hydroxyapatite composite	128.65	Facchi et al. (2018)
	Chitin hydrogel	105.70	$CaCO_3$/chitin hydrogel	191.58	Liang et al. (2019)

(continued)

Table 7.1 (continued)

Metal ion	Adsorbent	q1max (mg/g)	Adsorbent 2	q2max (mg/g)	References
Cu^{2+}	Chitosan/calcium alginate/Fe_3O_4 hydrogel	145.39	NaOH-treated chitosan/calcium alginate/Fe_3O_4 hydrogel	261.31	Gao et al. (2021)
	Magnetic polyamide 6 microparticles	60.30	Magnetic nanochitin-contained hydrogels	109.80	Zhang et al. (2023)
	Fe_3O_4-chitosan	129.60	Polystyrene/Fe_3O_4/chitosan composite	204.60	Zhou et al. (2014), Chen et al. (2020)
	Magnetic cellulose microspheres	89.67	DTPA-modified magnetic cellulose hydrogel	298.62	Li et al. (2019)
	PVA/chitosan hydrogel	102.32	PVA/chitosan/GO hydrogel	162.11	Quiroga-Flores et al. (2020)
	Chitosan aerogel	35.08	Amino-modified carboxymethyl chitosan aerogel	175.56	De Rossi et al. (2020)
	Sodium alginate/sodium humate	32.14	Sodium alginate/sodium humate@polyacrylamide	40.79	Shahzadi et al. (2023)
	Chitosan-glutaraldehyde	66.55	Chitosan/poly(acrylic acid) glutaraldehyde hydrogel	121.55	Li et al. (2017)
	Magnetic chitosan microspheres immobilized Sydowii	86.70	Magnetic chitosan microspheres immobilized Aspergillus sydowii	118.20	Li et al. (2020)
Ni^{2+}	Chitosan magnetic nanocomposite	52.00	Polylactic acid nonwoven fibers hydrogel composite	243.10	Yi et al. (2018)
	Chitosan immobilized on bentonite	39.04	Magnetic nanochitin-contained chitosan hydrogel	99.70	Wu et al. (2019a, b)
	Pre-micellar condition beads	10.00	Post-micellar conditionbeads	37.82	Xiao et al. (2017)
	Carboxymethyl chitosan microspheres	121.01	Polyacrylic acid functionalized carboxymethyl chitosan Microspheres	200.59	Jiang et al. (2014)
Zn^{2+}	Chitosan hydrogel	40.00	Chitosan hydrogels and Multilayer GO	236.00	Gao et al. (2022)

transport methods. Different processes enable hydrogels to undergo numerous regeneration cycles for adsorption and desorption functions, leading to financial benefits. Their high brittleness during swelling prevents their utilization as durable materials in specific applications. The removal of adsorbed substances from hydrogels poses occasional difficulties because standard operating methods often lack practicality. Environmental factors such as pH, temperature, and ionic strength can influence the stability and adsorption properties of hydrogels. Some hydrogels may be costly and potentially detrimental to the environment, especially those composed of synthetic polymers or containing expensive sorbents sustainable. The aim of recent research has been to enhance hydrogel performance through various modifications. It has been shown that including nanoparticles or creating composite hydrogels improves their mechanical and adsorption properties. Investigations on hydrogel regeneration and reutilization aim to offer cost-effective and eco-friendly water treatment solutions. In conclusion, the significant capacity, versatility, and adaptability of hydrogels render them a viable option for adsorption applications.

References

Alam MN, Islam MS, Christopher LP (2019) Sustainable production of cellulose-based hydrogels with superb absorbing potential in physiological saline. ACS Omega 4(5):9419–9426. https://doi.org/10.1021/ACSOMEGA.9B00651

Ali M, Asim, Khan TA (2012) Low-cost adsorbents for the removal of organic pollutants from wastewater. J Environ Manag 113:170–183. https://doi.org/10.1016/j.jenvman.2012.08.028

Bahram M, Mohseni N, Moghtader M, Bahram M, Mohseni N, Moghtader M (2016) An introduction to hydrogels and some recent applications. Emerg Concepts Anal Appl Hydrogels. https://doi.org/10.5772/64301

Bajpai SK (2001) Swelling studies on hydrogel networks—a review. JSIR 60(06):1–462. http://nopr.niscpr.res.in/handle/123456789/26500

Bouyahmed F, Muller F, Richard A, Mostefaoui TA, Belabbas I, Warmont F, Roulet M, Reinert L, Duclaux L, Delpeux-Ouldriane S (2022a) Chitosan-multilayered graphene oxide hybrid beads for Zn2+ and metoprolol adsorption. C R Chim 25(G1):205–223. https://doi.org/10.5802/CRCHIM.183/

Bouyahmed F, Muller F, Richard A, Mostefaoui TA, Belabbas I, Warmont F, Roulet M, Reinert L, Delpeux-Ouldriane S (2022b) Chitosan-multilayered graphene oxide hybrid beads for Zn^{2+} and metoprolol adsorption. C R Chim 25(3):205–223. https://doi.org/10.5802/crchim.183

Chang C, Zhang L (2011) Cellulose-based hydrogels: Present status and application prospects. Carbohyd Polym 84(1):40–53. https://doi.org/10.1016/J.CARBPOL.2010.12.023

Chen Y, Cai W, Dang C, Fan J, Zhou J, Liu Z (2020) A facile sol-gel synthesis of chitosan-boehmite film with excellent acid resistance and adsorption performance for Pb(II). Chem Eng Res des 161:332–339. https://doi.org/10.1016/j.cherd.2020.07.018

Crini G, Lichtfouse E (2019) Advantages and disadvantages of techniques used for wastewater treatment. Environ Chem Lett 17(1):145–155. https://doi.org/10.1007/s10311-018-0785-9

Crini G, Lichtfouse E, Wilson LD, Morin-Crini N (2019) Conventional and non-conventional adsorbents for wastewater treatment. Environ Chem Lett 17(1):195–213. https://doi.org/10.1007/s10311-018-0786-8

Dai J, Yan H, Yang H, Cheng R (2010) Simple method for preparation of chitosan/poly(acrylic acid) blending hydrogel beads and adsorption of copper(II) from aqueous solutions. Chem Eng J 165(1):240–249. https://doi.org/10.1016/j.cej.2010.09.024

De Rossi CVT, Rigueto A, Dettmer A, Colla LM, Piccin JS (2020) Synthesis, characterization, and application of Saccharomyces cerevisiae/alginate composite beads for adsorption of heavy metals. J Environ Chem Eng 8(104009). https://doi.org/10.1016/j.jece.2020.104009

Facchi DP, Cazetta AL, Canesin EA, Almeida VC, Bonafe EG, Kipper MJ, Martins AF (2018) New magnetic chitosan/alginate/Fe_3O_4@SiO_2 hydrogel composites applied for removal of Pb(II) ions from aqueous systems. Chem Eng J 337:595–608. https://doi.org/10.1016/j.cej.2017.12.142

Foo KY, Hameed BH (2009) Value-added utilization of oil palm ash: a superior recycling of the industrial agricultural waste. J Hazard Mater 172(2–3):523–531. https://doi.org/10.1016/j.jhazmat.2009.07.091

Funami T, Hiroe M, Noda S, Asai I, Ikeda S, Nishinari K (2007) Influence of molecular structure imaged with atomic force microscopy on the rheological behavior of carrageenan aqueous systems in the presence or absence of cations. Food Hydrocolloids 21(4):617–629. https://doi.org/10.1016/J.FOODHYD.2006.07.013

Gao C, Wang X-L, An Q-D, Xiao Z-Y, Zhai S-R (2021) Synergistic preparation of modified alginate aerogel with melamine/chitosan for efficiently selective adsorption of lead ions. Carbohydr Polym 256(117564). https://doi.org/10.1016/j.carbpol.2020.117564

Gao J, Zhang L, Liu S, Liu X (2022) Enhanced adsorption of copper ions from aqueous solution by two-step DTPA-modified magnetic cellulose hydrogel beads. Int J Biol Macromol 211:689–699. https://doi.org/10.1016/j.ijbiomac.2022.05.073

Gulrez SKH, Al-Assaf S, Phillips GO, Gulrez SKH, Al-Assaf S, Phillips GO (2011) Hydrogels: methods of preparation, characterisation and applications. Progress in molecular and environmental bioengineering—from analysis and modeling to technology applications. https://doi.org/10.5772/24553

Gupta S, Webster TJ, Sinha A (2011) Evolution of PVA gels prepared without crosslinking agents as a cell adhesive surface. J Mater Sci Mater Med 22(7):1763–1772. https://doi.org/10.1007/S10856-011-4343-2

Hoffman AS (2002) Hydrogels for biomedical applications. Adv Drug Deliv Rev 54(1):3–12. https://doi.org/10.1016/S0169-409X(01)00239-3

Hussein MA (2017) Role of cross-linking process on the performance of PMMA. Int J of Biosens Bioelectron 3(3). https://doi.org/10.15406/IJBSBE.2017.03.00065

Jaipan P, Nguyen A, Narayan RJ (2017) Gelatin-based hydrogels for biomedical applications. MRS Commun 7(3):416–426. https://doi.org/10.1557/MRC.2017.92

Jiang W, Chen X, Pan B, Zhang Q, Teng L, Liu L (2014) Spherical polystyrene-supported chitosan thin film of fast kinetics and high capacity for copper removal. J Hazard Mater 276:295–301. https://doi.org/10.1016/j.jhazmat.2014.05.032

Jones A, Vaughan D (2005) Hydrogel dressings in the management of a variety of wound types: a review. J Orthop Nurs 9(SUPPL. 1). https://doi.org/10.1016/S1361-3111(05)80001-9

Khan F, Atif M, Haseen M, Kamal S, Khan MS, Shahid S, Nami SAA (2022) Synthesis, classification and properties of hydrogels: their applications in drug delivery and agriculture. J Mater Chem B 10(2):170–203. https://doi.org/10.1039/D1TB01345A

Khan Z, Shahid D, Khan AM, Khan MK (2025) Nanoscale solutions for water treatment and purity enhancement. Breaking Boundaries Pioneer Sustain Solut Mater Technol 193–211. https://doi.org/10.1007/978-981-97-9827-8_9

Kim S, Shin BH, Yang C, Jeong S, Shim JH, Park MH, Bin CY, Heo CY, Lee K (2018) Development of Poly(HEMA-Am) polymer hydrogel filler for soft tissue reconstruction by facile polymerization. Polymers 10(7):772. https://doi.org/10.3390/POLYM10070772

Kongarapu RJ, Mahamallik P, Pal A (2017) Surfactant modification of chitosan hydrogel beads for Ni@NiO core-shell nanoparticles formation and its catalysis to 4-nitrophenol reduction. J Environ Chem Eng 5(2):1321–1329. https://doi.org/10.1016/j.jece.2017.02.017

Kuo CK, Ma PX (2001) Ionically crosslinked alginate hydrogels as scaffolds for tissue engineering: part 1. Structure, gelation rate and mechanical properties. Biomaterials 22(6):511–521. https://doi.org/10.1016/S0142-9612(00)00201-5

Li R, Liang W, Li M, Jiang S, Huang H, Zhang Z, Wang JJ, Awasthi MK (2017) Removal of Cd(II) and Cr(VI) ions by highly cross-linked thiocarbohydrazide-chitosan gel. Int J Biol Macromol 104:1072–1081. https://doi.org/10.1016/j.ijbiomac.2017.07.005

Li S-S, Wang X-L, An Q-D, Xiao Z-Y, Zhai S-R, Cui L, Li Z-C (2020) Designing carboxyl methylcellulose and chitosan-derived nanostructured sorbents for efficient removal of Cd(II) and Cr(VI) from water. Int J Biol Macromol 143:640–650. https://doi.org/10.1016/j.ijbiomac.2019.12.053

Li T, Liu X, Li L, Wang Y, Ma P, Chen M, Dong W (2019) Polydopamine-functionalized graphene oxide compounded with polyvinyl alcohol/chitosan hydrogels on the recyclable adsorption of Cu(II), Pb(II), and Cd(II) from aqueous solution. J Poly Res 26(1971). https://doi.org/10.1007/s10965-019-1971-6

Liang XX, Ouyang X-K, Wang S, Yang L-Y, Huang F, Ji C, Chen X (2019) Efficient adsorption of Pb(II) from aqueous solutions using aminopropyltriethoxysilane-modified magnetic attapulgite@chitosan (APTS-Fe_3O_4/APT@CS) composite hydrogel beads. Int J Biol Macromol 137:741–750. https://doi.org/10.1016/j.ijbiomac.2019.06.244

Maitra J, Shukla VK (2014) Cross-linking in hydrogels—a review. Am J Poly Sci 4(2):25–31. https://doi.org/10.5923/J.AJPS.20140402.01

Malik NS, Ahmad M, Minhas MU, Tulain R, Barkat K, Khalid I, Khalid Q (2020) Chitosan/Xanthan gum based hydrogels as potential carrier for an antiviral drug: fabrication, characterization, and safety evaluation. Front Chem 8:50. https://doi.org/10.3389/FCHEM.2020.00050

Pan S, Xia M, Li H, Jiang X, He P, Sun Z, Zhang Y (2020) Transparent, high-strength, stretchable, sensitive and anti-freezing poly(vinyl alcohol) ionic hydrogel strain sensors for human motion monitoring. J Mater Chem C 8(8):2827–2837. https://doi.org/10.1039/C9TC06338B

Peppas NA, Hilt JZ, Khademhosseini A, Langer R (2006) Hydrogels in biology and medicine: from molecular principles to bionanotechnology. Adv Mater 18(11):1345–1360. https://doi.org/10.1002/ADMA.200501612

Quiroga-Flores R, Noshad A, Wallenberg R, Onnby L (2020) Adsorption of cadmium by a high-capacity adsorbent composed of silicate-titanate nanotubes embedded in hydrogel chitosan beads. Environ Technol 41:3043–3054. https://doi.org/10.1080/09593330.2019.1596167

Shah MP (ed) (2024) Emerging innovative trends in the application of biological processes for industrial wastewater treatment. Elsevier

Shahzadi Y, Wu Y, Lin H, Huang J, Zhao Z, Chen C, Shi X, Deng H (2023) Yeast biomass ornamented macro-hierarchical chitin nanofiber aerogel for enhanced adsorption of cadmium(II) ions. J Hazard Mater 453(131312). https://doi.org/10.1016/j.jhazmat.2023.131312

Shoukat H, Noreen S, Asghar S (2020) Hydrogels as controlled drug delivery system: a brief review of properties classification and synthesis. Global Pharma Sci Rev V(I):40–49. https://doi.org/10.31703/GPSR.2020(V-I).05

Sun J, Hu R, Zhao X, Liu T, Bai Z (2024) A novel chitosan/cellulose phosphonate composite hydrogel for ultrafast and efficient removal of Pb(II) and Cu(II) from wastewater. Carbohydr Polym 336(122104). https://doi.org/10.1016/j.carbpol.2024.122104

Tian T, Bai Z, Wang B, Zhao S, Zhang Y (2020) Facile fabrication of polyacrylic acid functionalized carboxymethyl chitosan microspheres for selective and efficient removal of Ni(II) from multicomponent wastewater. Colloids Surf A PhysChemical Eng Asp 597(124676). https://doi.org/10.1016/j.colsurfa.2020.124676

Varghese SA, Rangappa SM, Siengchin S, Parameswaranpillai J (2019) Natural polymers and the hydrogels prepared from them. Hydrogels Based Nat Polym 17–47. https://doi.org/10.1016/B978-0-12-816421-1.00002-1

Wang Q, Li L, Kong L, Cai G, Wang P, Zhang J, Zuo W, Tian Y (2022) Compressible amino-modified carboxymethyl chitosan aerogel for efficient Cu(II) adsorption from wastewater. Sep Purif Technol 293(121146). https://doi.org/10.1016/j.seppur.2022.121146

Wang Q, Xie X, Zhang X, Zhang J, Wang A (2010) Preparation and swelling properties of pH-sensitive composite hydrogel beads based on chitosan-g-poly (acrylic acid)/vermiculite and sodium alginate for diclofenac-controlled release. Int J Biol Macromol 46:356–362

Wang Z-K, Li T-T, Ren H-T, Peng H-K, Shiu B-C, Lin J-H, Lou C-W (2023) Construction of a novel hydrogel composite with polylactic acid nonwoven fibers: rapid and effective removal of Pb(II) and Ni(II). Text Res J 93(2):266–279. https://doi.org/10.1177/00405175221114637

Wu J, Cheng X, Yang G (2019a) Preparation of nanochitin-contained magnetic chitosan microfibers via continuous injection gelation method for removal of Ni(II) ion from aqueous solution. Int J Biol Macromol 125:404–413. https://doi.org/10.1016/j.ijbiomac.2018.11.212

Wu J, Cheng X, Li Y, Yang G (2019b) Constructing biodegradable nanochitin-contained chitosan hydrogel beads for fast and efficient removal of Cu(II) from aqueous solution. Carbohyd Polym 211:152–160. https://doi.org/10.1016/j.carbpol.2019.01.004

Xiao C, Liu X, Mao S, Zhang L, Lu J (2017) Sub-micron-sized polyethylenimine-modified polystyrene/Fe_3O_4/chitosan magnetic composites for the efficient and recyclable adsorption of Cu(II) ions. Appl Surf Sci 394:378–385. https://doi.org/10.1016/j.apsusc.2016.10.116

Xu J, Liu X, Ren X, Gao G (2018) The role of chemical and physical crosslinking in different deformation stages of hybrid hydrogels. Eur Polymer J 100:86–95. https://doi.org/10.1016/J.EURPOLYMJ.2018.01.020

Yi X, Yang M, Mo L, Xu W, Wang S, He J, Gu J, Ou M, Xu X (2018) Modification of chitosan/calcium alginate/Fe_3O_4 hydrogel microsphere for enhancement of Cu(II) adsorption. Environ Sci Pollut Res 25:3922–3932. https://doi.org/10.1007/s11356-017-0802-8

Zhang H, Ye Z, Liang J, Hu Z, Liu H (2023) Preparation and adsorption study of polyvinyl alcohol-functionalized sodium alginate composite hydrogel for Pb(II) in aqueous solutions. Desalin Water Treat 313:162–172. https://doi.org/10.5004/dwt.2023.30072

Zheng Y, Wang A (2010) Removal of heavy metals using polyvinylalcohol semi-IPN poly (acrylic acid)/tourmaline composite optimized with response surface methodology. Chem Eng J 162:186–193

Zhou Y, Fu S, Zhang L, Zhan H, Levit MV (2014) Use of carboxylated cellulose nanofibrils-filled magnetic chitosan hydrogel beads as adsorbents for Pb(II). Carbohyd Polym 101:75–82. https://doi.org/10.1016/j.carbpol.2013.08.055

Zhu T, Mao J, Cheng Y, Liu H, Lv L, Ge M, Li S, Huang J, Chen Z, Li H, Yang L, Lai Y (2019) Recent progress of polysaccharide-based hydrogel interfaces for wound healing and tissue engineering. Adv Mater Interfaces 6(17). https://doi.org/10.1002/ADMI.201900761

Chapter 8
Advances in Hydrogel-Based Materials for Efficient Dye Sorption and Remediation

Manas Mishra and Prabhat Kumar Patel

Abstract Because of their widespread industrial consumption, synthetic dyes have been a significant class of persistent organic chemicals that are frequently discovered in natural environments. It is thought that there are over one million dyes that exist for industrial usage. Numerous these compounds are recognized to be hazardous or malignant, and they are often found to be pollutants. The general population views colored industrial effluents as a sign of harmful contamination. As an aesthetic pollutant, colors are very noticeable even at very low levels. As a chemical pollutant, they alter marine life and the nutritional supply. Water contamination due to dyes has become a global issue, and treating wastewater before releasing it into the environment has gained priority. Dyes are challenging to cure because of their intricate origin and composition. It is common practice to employ a variety of chemical, biological, and physical methods in conjunction with one another to achieve the intended level of water quality. Nevertheless, more efficient, environmentally friendly, and industrially feasible elimination tactics and decolorization processes need to be developed. The deployment of biological substances as efficient dye-removal sorbents is an emerging area of research at present. The most frequently utilized biosorbents among all those identified are cross-linked chitosan-based hydrogels. Several preliminary studies have focused on these polymeric matrix structures. With a focus on hydrogels' adsorption ability for dyes, this chapter showcases current breakthroughs in hydrogel alterations and their varieties. We additionally addressed the modified hydrogels' associated removal mechanisms and regeneration methods using the contrasting approach. It provides a full summary of the research conducted by several authors on synthesis, composition, functioning, and pollutant removal mechanisms. Additionally, it offers a comprehensive analysis of the benefits and drawbacks associated with the utilization of hydrogels in the context of effluent treatment.

M. Mishra · P. K. Patel (✉)
Department of Civil Engineering, Institute of Engineering and Technology Lucknow, Uttar Pradesh, Lucknow, India
e-mail: pprabhat1991@gmail.com

P. K. Patel
School of Sciences, Harbin Institute of Technology Shenzhen, Shenzhen, China

S. Bardhan and M. P. Shah (eds.), *Applications of Hydrogels in Modern Wastewater Treatment*, Advances in Wastewater Research,
https://doi.org/10.1007/978-981-95-1598-1_8

Keywords Hydrogel · Wastewater treatment · Adsorption · Elimination mechanisms · Dyes

Abbreviations

4VBS	Sodium 4-vinylbenzenesulfonate
AA	Acrylic Acid
AAM	Acryl Amide
AB-10B	Amido black 10B
Alg	Sodium Alginate
AMPS	2-Acrylamido-2-methyl-1-propanesulfonic acid
AMPS	2-Acrylamido-2-methyl-1-propan-sulfonic Acid
APTMACl	(3-Acrylamidopropyl) trimethylammonium chloride
AR	Acid Red
AsA	Aspartic Acid
BmImBr	1-Butyl-3-methylimidazolium bromide
CBH	Chitosan-Based Hydrogel
CCGA	Carboxymethyl Cellulose/Chitosan/Glutamic Acid
CGC	Coffee Grounds Cellulose
CLP	Clinoptilolite
CMC	Carboxymethyl Chitosan
CMCNFs	Chitin Nanofibrillar
CMH	Carboxymethyl Hemicellulose
CNTs	Carbon Nanotubes
CQDs	Carbon Quantum Dots
CR	Congo Red
CSt	Cassava Starch
CV	Crystal Violet
DCH	Dual Cross-Linked Hydrogel
DMAEM	2-Dimethylamino ethyl Methacrylate
Dy	Dysprosium
EBT	Eriochrome Black T
GG	Guar Gum
GO	Graphene Oxide
HGs	Hydrogels
IC	Initial Concentration
IPAM	N-Isopropylacrylamide
LN	Lignin
MB	Methylene Blue
MG	Malachite green
MMT	Montmorillonite
MO	Methyl Orange
MOEAMCl	[2-(Methacryloyloxyloxy)ethyl] trimethylammonium chloride

MP	Methoxy Pectin
NF	Nanofiltration
NPs	Nanoparticles
NWs	Nanowires
OD	Organic Dyes
PAA	Poly Acrylic Acid
PAAc	Poly Acrylacetate
PAM	Polyacrylamide
PEGMEM	Polyethylene Glycol Methyl Ether Methacrylate
PEI	Polyethylene Imine
pHPZC	PH at the point of zero charge
PNaSS	Poly(sodium p-styrenesulfonate)
PP	Polyvinyl Pyrrolidone
PrAm	Propenamide
PVA	Polyvinyl Alcohol
RB19	Reactive Blue 19
Rh-B	Rhodamine B
RO	Reverse Osmosis
RR	Reactive Red
SCH	Single cross-linked Hydrogel
SCH-CN	Scaffold–chitosan Hydrogel with Carbon Nanotubes
TA	Tannic Acid
TART	Tartrazine
VPHEM	4-Vinyl pyridine, 2-hydroxyethylmetacrylate

8.1 Introduction

Having access to potable water has become one of the foremost important problems facing humankind in the twenty-first century. Globally, there has been a major issue regarding accessibility to clean, potable water. Global awareness of the scarcity of fresh, potable water is growing. Global population growth, shifts in consumer behavior, growing standards of living, and the expansion of agricultural irrigation have been the main causes of the heightened need for water globally (Mekonnen et al. 2016). This rendered an already unstable situation involving water resources even more exacerbated. Specifically in developing nations, water supplies are scarce or insufficient in inhabited areas worldwide with minimal or no precipitation (Scott 2000).

The water issue has been deteriorating as a result of contamination, making safe water to drink even more scarce. The UN estimates that every day, over two tonnes of trash created by humans are discarded into our waterways. Seventy percent of this is human-generated industrial trash that is negligently thrown into oceans, rivers, and lakes, polluting or destroying water supplies, according to UN data. Because of human activities, a number of contaminants have been released into resources containing freshwater, including water bodies, groundwater, and surface water. Conventional toxins produced by human activity include inorganic contaminants, solvents, emerging micropollutants, pharmaceutical remnants, and other resilient organic substances/pollutants. Certain contaminants and chemicals are ineffectively removed from water by conventional methods of sewage and water purification. More affordable and effective technologies are being used to disinfect and clean water for point-of-use operations. For pastoral area residents, to have the availability of clean water, a water purification technique such as this one is required.

According to Schwarzenbach et al. (2010) and Pandey et al. (2020), the term "water pollution" refers to the poisoning of water with pollutants or additional foreign components that are harmful to plant and animal creatures. Pest control products, fertilizers from crop residue, phenolic compounds from pharmaceutical or manufacturing byproducts, and liquid or solid residues are a few examples of organic pollutants (Bhatnagar et al. 2009). Cr, Pb, Cd, As, Hg, and other metals are examples of inorganic contaminants (Malaviya et al. 2011). According to Mekonnen et al. (2016), over four billion people lack a source of healthy water to drink, putting their lives in jeopardy.

The destruction of forests, growing urbanization, construction of dams, loss of wetland areas, mining due to drainage of acid mines, farming, energy use, and accidental water contamination from spills of oil are some of the factors that contribute to water shortage (Shah 2024). The difficulty of obtaining clean water has become more and more pressing for developing nations as the global population rises. According to Pandey et al. (2020), all of this has led to a rise in contamination and the degradation of river basins. Since there is a glaring lack of potable water, it is crucial to supervise the affordable, simple, and effective purification of water and effluent.

To lower the level of dyes in contaminated commercial releases, a variety of techniques may be used, including flocculation, chemical degradation, membrane filtration, photochemical purification, and biological degradation (Patel et al. 2024b; Patel and Uppaluri 2025). These abatement strategies work, but they've also taken heat for making toxic sludge, being expensive to run, and lacking variety. However, the sorption procedure was praised as a promising purification technique because of its exceptional effectiveness, cost, and convenience for implementation (Patel et al. 2025). Despite producing leftovers, the procedure can handle large flows and offer excellent performance, making it applicable to a variety of applications (Patel et al. 2023b, d). The use of polymers, zeolites, metal oxides, clay minerals, waste byproducts, active carbon, and graphene oxide as adsorbents to remove toxic chemicals from wastewater is described in the literature (Patel et al. 2023a, 2024a). Activated charcoal is an especially commonly utilized sorbent for removing dyes in water-based systems due to its large surface area (Mezohegyi et al. 2012; Patel et al. 2023c). Activating carbon dioxide sources requires expensive substances, tools, and effort. Dedicated reactive carbon's inability to renew is an additional disadvantage.

Most of the described adsorbents have substantial downsides as well, such as ineffectiveness and non-biodegradability (Nagireddi et al. 2025; Patel et al. 2022). Because "green" hydrogel sorbents are accessible, affordable, and biodegradable, they might be employed to get around these restrictions. Recently, there has been an increase in interest in the use of affordable wastewater treatment methods. Similarly, biopolymers, which are inexpensive and environmentally benign, have been employed in cleanup procedures. Many researchers and scholars have chosen to employ biopolymer-based hydrogel to treat wastewater, and their results demonstrate that these substances are efficient in eliminating dyes from water sources (Pandey et al. 2020).

Nowadays, hydrogels derived from biopolymers have primarily been utilized for the mitigation of dyes (Wang et al. 2008). Nonetheless, the unique framework of each dye prevents the generalization of hydrogel development. Hydrogels exhibit significant sorption properties independently; however, intricate desorption procedures become necessary once saturated for effective reuse. As a result, investigations are underway focused on the creation of biopolymer-based hydrogels infused with nanoparticles, aiming to tailor and adapt every single hydrogel to be distinct and suitable for the mitigation of dyes. A number of scholarly articles have documented the sorption of dyes using hydrogel-based nano adsorbents (Mittal et al. 2014). Conversely, there is a limited number of reviews available in this field; most have concentrated on broad summaries regarding "specific dye," "nanoparticles," and "polysaccharides." Based on our understanding, a thorough review is necessary, focusing on the synthesis method in detail, including insights into particular benefits and drawbacks, dye elimination effectiveness, regeneration ability, and the restrictions of different hydrogels across various treatment methods. The application of hydrogel-based treatment in essential fields has generated significant interest. The primary purpose of this investigation is to provide a synopsis of the most important aspects of the field of hydrogel technology for the removal of dyes from water. In a similar vein, these ideas will be expanded upon by emphasizing the advantages of hydrogels and the latest developments in this area. This study did not limit the topic to any particular dye, filler, biopolymer, or hydrogel matrix. We have introduced a conceptual and critical examination of the most frequently utilized reinforcing material in the formulation of biopolymer-based hydrogels for various dyes.

8.2 Dyes—General Aspects

As one of the leading categories of chemicals, synthetic dyes can be utilized in many fields, including textiles, suede, presses, pharmaceuticals, beauty products, food, and biological and chemical investigations, as well as in photo-electrochemical cells and light-harvesting arrays. The worldwide market for dyes has been projected to exceed \$15 billion by the year 2026, according to an analysis by Allied Market Research (2019). Over 10,000 distinct dyes are readily accessible in markets, and they all absorb electromagnetic energy in the visible spectrum. Also, Azo variants make up over 70% of this group. There are discrete categorization schemes for dyes, predicated on: (i) their molecular configuration (auxochrome/chromophore); and (ii) their commercial application technique (metal complex, acid, dispersion, direct, reactive, basic, vat, and mordant) (Benkhaya et al. 2020). The molecular variety of synthetic dyes allows them to demonstrate a wide range of biological, chemical, and physical characteristics. The substantial degradation rate of dyes during the production procedure is the primary concern with their commercial utilization. For instance, around 10–15% of synthetic dyes were wasted by the textile sector (Pereira et al. 2021). Some of it ends up in industrial effluents, which, if not properly treated, may end up in lakes, rivers, and other waterways. A number of these colors are not biodegradable, which means they might be very harmful to people and our surroundings. The existence of very small amounts of manufactured dyes in water (less than 1 ppm) may affect the gas solubility and transparency of water bodies. Also, the development of aquatic organisms gets hampered, and photosynthesis suffers because these colored chemicals reflect and absorb the sunlight that reaches the water. Several studies show that manufactured colors can have both short-term and long-term effects on living things, based on how much they are exposed to them and how long they are exposed. Table 8.1 provides a brief overview of some key facts of various dyes frequently employed as substitutes for laboratory-scale adsorption investigations.

It is vital to notify individuals that a significant portion of the global textile dye industry is comprised of basic or cationic dyes, which are employed as models in sorption investigations. Additionally, its antiquity restricts its applicability to contemporary textile production. Reactive, distributed, and direct classes will be the most crucial ones in the next years.

There is a widespread consensus in recent studies that colored effluents need to be thoroughly treated in order to eliminate all dyestuffs before being disposed of. Figure 8.1 summarizes the basic techniques for treating dye-contaminated streams.

High operating expenses, inadequate effectiveness, restricted adaptability, cumbersome handling, and additional contamination are usually cited as the primary disadvantages of several of these procedures. However, sorption has been suggested by some researchers as a more secure and affordable substitute for traditional methods of color extraction from wastewater and water (Katheresan et al. 2018). Because of its cheap cost, simple construction, and convenience of use, sorption, a phenomenon, in which a solute clings to a solid sorbent, has garnered a lot of interest. In terms of removing dyes from water, sorption offers several benefits, such as: (i) employing dense systems; (ii) being insensitive to daily fluctuations; (iii) having a minimal impact on toxic substances; (iv) offering more design and operational adaptability; (v) effectively removing organic pollutants; and (vi) not producing any adverse secondary substances (Katheresan et al. 2018). Aside from the solutes being adsorbed, there are also important considerations about the sorbent (the solid) and its possible recycling, as well as the extraction of the dye. Adsorption, despite the fact that it has a multitude of fascinating characteristics, as was said earlier, is also accompanied by a number of challenges. These challenges comprise a restricted variety of adsorbents that can be reused, as well as the requirement for chemical and energy response in order to renew the sorbent. In several textile sectors, sorption procedures have been adopted as a subsequent treatment approach after applying preparatory treatments (Yaseen et al. 2019). In addition, there have been reports of composite sorption techniques linked to electrically discharged plasma (Shang et al. 2019).

Table 8.1 Various dyes utilized in adsorption studies (lab-scale)

Sl. no.	Dye	Class	Molecular weight (g/mol)	Chemical formula	Contamination sources
1	Brilliant green	Triphenylmethane	482.64	$C_{27}H_{33}N_2.HO_4S$	Pharmaceutical, pigments, chemical, dyestuff industries
2	Methyl orange	Azo	327.33	$C_{14}H_{14}N_3O_3S$	Paper, foodstuffs, leather, textile industries, lab and laundry activity
3	Rhodamine B	Xanthene	472.09	$C_{28}H_{31}ClN_2O_3$	Paper, food, dyestuffs, plastic, leather, textile industries, laundry activity, natural sources
4	Methyl red	Azo	269.30	$C_{15}H_{15}N_3O_2$	Dyestuffs, chemical, leather, textile industries, laundry activity
5	Methylene blue	Thiazine	319.85	$C_{16}H_{18}ClN_3S$	Paper, food, dyestuffs, plastic, leather, textile, pharmaceutical industries, laundry activity
6	Congo red	Azo	696.66	$C_{32}H_{22}N_6Na_2O_6S_2$	Dyestuffs, leather, textile industries
7	Malachite green	Triphenylmethane	364.91	$C_{23}H_{25}N_2Cl$	Leather, textile, dyestuff industries, lab and laundry activity
8	Crystal violet	Triphenylmethane	407.99	$C_{25}N_3H_{30}Cl$	Leather, textile, dyestuff industries, laundry activity

Various sorbent materials have been developed and employed, either directly or through indirect methods, due to the appealing properties of sorption. There are a plethora of recent studies detailing the mitigation of synthetic colors from water using various sorbents. Additionally, the most significant developments in this scientific field have been documented in prior art (Buruga et al. 2019). Of course, it is inconceivable to ignore the fact that the experimental environment plays a significant role in the physicochemical process's efficacy. It seems that hydrophilic and functional materials promote superior sorption outcomes when considering the fluids tainted with dyes. Research into the development of hybrid hydrogels for sorption applications has increased over the last several decades in response to this trend, since these materials often exhibit better characteristics than traditional hydrogels or other hydrophilic substances.

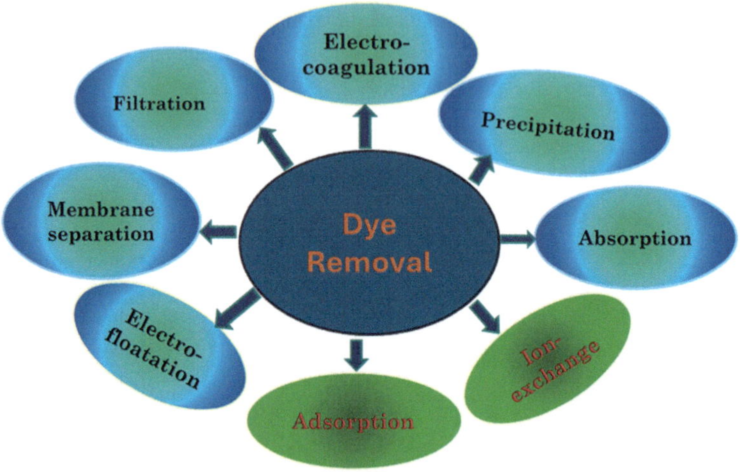

Fig. 8.1 Various wastewater treatment methods utilized for dye elimination

8.3 Dye Ecotoxicity

It is common practice to dispose of dyes, which are composed of non-aromatic and aromatic substances, into the surroundings without undergoing the prior appropriate treatment. In the natural world, oxidation dyes, which have been utilized for coloring hair, as well as other colors that are employed by other enterprises, are poisonous. Many other kinds of chemicals, including acids, soapy products, nitrates, and metallic substances, including heavy metals, are included in the sewage that is released by different types of organizations. The severity of deleterious effects is amplified when all of these compounds are combined with color derivatives. About 40% of colourants include organically bonded chlorine, which makes them potentially hazardous. When dye-containing effluents are not properly treated and dumped in a non-ethical manner, they cause significant modifications in the marine environment. A wide range of variables, including surface color and temperature, pH, photosynthesis, and biological oxygen demand (BOD) levels, may contribute to these changes, which can be either chemically or biologically. Toxic and carcinogenic substances such as aromatic substances, benzamine, and naphthalene are byproducts of dye degradation in various ecological and artificial environments (Saini 2017). When these substances come into contact with water, they have the potential to decrease the amount of dissolved oxygen and sunlight that can reach the water. All of these activities eventually affect the development of other organisms that live in or near water, which upsets the entire food web in our ecology (Bhardwaj et al. 2021). According to reports, plants may experience damage from oxidation when exposed to chromium, which is frequently employed in textile dyes. Dye recalcitrance significantly hinders plant development and photosynthetic activity. As a result, both terrestrial and marine ecosystems are impacted by CO_2 sinks (Bhardwaj et al. 2021). Dyes have a tendency to transform into refractory compounds in aerobic environments or in traditional treatment facilities, which leads to their biological accumulation in the soil and other nearby assets (Vikrant et al. 2018). Metallic dyes, often employed in the textile industry, are composed of heavy metals such as Cu, Ni, Co, and Cr, and they have an exceptional endurance for adverse environmental circumstances. There have been reports that the infiltration of certain metallic ions into waters may lead to dyes being resistant to the environment and, in some cases, even accumulate in fish gills (Vargas et al. 2009). As a result, dyes and their byproducts may negatively impact human wellness by entering the food chain. There have been reports of skin irritation, dermatitis, central nervous system disorders (Khan et al. 2018), and carcinoma of the bladder in people after coming into proximity to dyes. There are cases of acute poisoning via oral consumption and breathing, along with skin and eye irritation (Bide 2007). Many disorders, including occupational asthma, nasal congestion, conjunctive allergy, and skin eruptions, may be caused by inadequate care of reactive dyes and other dye byproducts in work environments (Ribeiro et al. 2014). Different colors' mutagenicity, teratogenicity, and gene toxicity will affect both humans and animals for a very long time. Regulatory agencies are vigilant in their oversight of the correct management and disposal of effluents due to the detrimental effects of synthetic dyes.

8.4 Utilization of Hydrogel in Dye Removal

The elimination of pollutants from wastewater has been mostly accomplished via the use of biological, chemical, and physical methods. Sometimes, to render water treatment even better, a mix of the two approaches is implemented (Donkadokula et al. 2020). The impact of this is the widespread usage of hydrogels based on a physical technique based on sorption. Hydrogel sorption effectiveness is highly dependent on surface functionality and 3D topography of networks. The specifics of each procedure are described below.

8.4.1 Adsorption

One way to remove toxic substances from water is by sorption, which is the attraction of sorbates to the outer layer of a sorbent through physical, chemical, or electrostatic forces. Adsorbents with large surface area, high selectivity, strong regenerative capacity, and low cost are promising (Nagireddi et al. 2025; Patel et al. 2023a, 2024a). Sorption is a very popular and efficient way to treat wastewater because it's easy to use, cheap, and doesn't produce any hazardous waste (FBA et al. 2016). Because of its excellent sorption capacity, activated carbon is widely used in wastewater treatment to remove contaminants. Activated carbon is a good material for wastewater treatment, but its high price and complex regeneration process hinder its wide use

(Liheng Liu et al. 2013). Using these sorbents for large-scale wastewater treatment is still challenging, even though many have been found to remove dyes and heavy metals from industrial wastewater in recent studies (Patel, Pandey, et al. 2025; Patel and Uppaluri 2025). 3D polymer structures or HGs have gained much attention recently for their ability to remove pollutants from wastewater at a high rate (B. Peng et al. 2020). The hydrophilic functional groups present in HGs, such as –OH, COOH, NH_2, SO_3H, and $CONH_2$, and the porous nature of the material can absorb and retain a lot of water during treatment and eventually remove and extract aqueous dyes.

8.4.2 *Ion Exchange*

One of the most common ways to soften water is through ion exchange. This water treatment method involves the reversible transfer of ions in an aqueous medium between hydrophobic ion exchange resin and ions of the same polarity (Patel et al. 2022, 2023a). Cationic and anionic categories of ion exchangers are determined by their electron transfer properties. Toxic multivalent cations (Pb and Cd) can be replaced with monovalent cations (Na and K) by using a cation exchanger. Toxic dyes and toxic metals can be removed from wastewater using easily available cation exchange resins like SP Sepharose, Capto S, CM Sepharose, and TSK gel CM-5PW (Akter et al. 2021). Also, developing environmentally friendly ion exchangers that are affordable and biodegradable has been a major focus recently. Anion exchangers with hydroxyl and chloride groups can quickly remove multivalent anions like cyanide, sulfate, and chromate for their monovalent equivalents. Toxic transitional metals can be removed from water using Amberlite PWA5, which was found to have excellent anion exchange capacity. This method is very effective, has fast kinetics, and what's more, the resin structure, charge density, and ionic radius of the metal ions can be adjusted to change the ion exchanger's selectivity. But because of its high price, it is not prioritized for large-scale industrial water treatment (Guida et al. 2021).

8.4.3 *Membrane Filtration*

Among the many potential methods for purifying water, membrane separation stands noteworthy. A membrane-based filter is used in this pressure-driven technique to remove bacteria and other contaminants from water. A portion of molecules have been permitted to pass through the cell membrane depending on their size, while the remainder are prohibited. Heavy metals, germs, microbes, tiny colloids, divalent ions, and monovalent ions are some of the most typical contaminants that it may eliminate. The fact that it is simple to use, quite efficient, and uses very little energy has made it very popular. Micro, ultra, nanofiltration, and reverse osmosis are the four categories that have been established for membrane filtration procedures (Hube et al. 2020). These classifications are based on the size of the openings. NF and RO membranes are two of the four kinds of membranes that may be manufactured or transformed using hydrogels. This is because hydrogels have the capacity to give the specific pore size and sorption capability that is necessary for the elimination of ions.

8.5 Synthesis of Hydrogel

In response to the numerous problems caused by the overuse of non-biodegradable materials and petroleum-based products, researchers are looking into regenerative and sustainable materials. Polymers are being used massively in many industries like food packaging, sewage treatment, pharmaceuticals, and agriculture (Patel et al. 2023c, 2024a). Polymeric hydrogels are the most exciting sorbent materials for the sorption of toxic dyes from wastewater because of their larger surface area, good solubility in organic solvents, enhanced functionality, affordability, biodegradability, recyclable properties, higher sorption capacity, and ease of preparation. Moreover, these hydrogels are suitable for wastewater treatment because of their hydrophilic nature (Thakur et al. 2018). But the physicochemical properties of the sorbent material are important (Malik et al. 2017). The sorption process starts with preparing a hydrogel-based sorbent material that can absorb dyes from wastewater. A monomer, an initiator, and a cross-linker are the three main components required to prepare a hydrogel. There are many types of monomers used to prepare hydrogels. These include PEGMEM, APTMACl, PP, PVA, AA, DMAEM, IPAM, AMPS, AAM, and VPHEM. Different molecules have different strengths, abilities to bind substances, and other qualities. Scientists found a way to get beyond the limitations of certain monomers in the fabrication of hydrogels. Liu et al. (2012) synthesized chitosan-derived hydrogel for dye and metal sorption using cellulose as the merger polymer to minimize the physical instability of chitosan.

To synthesize polymeric hydrogels, cross-linking agents or cross-linkers are essential. These substances strengthen the necessary positions between the purposeful monomer and the aimed molecule, which helps to build up the 3D network of polymers. Polymers' shape, strength, specificity, and hydrophilic/hydrophobic characteristics are therefore affected by cross-linkers (Yan et al. 2006). During synthesis, a cross-linker of either inorganic or organic composition may be employed. In addition, hydrogel sorbents are often made using inorganic cross-linkers rather than organic ones since the latter have a lower swelling capacity and less durability, making them unsuitable for usage in stressed environments (Haraguchi et al. 2005). In addition, hydrogels may have different defining characteristics whether the cross-linking between their links is non-covalent or covalent. Hydrogels cross-linked via covalent bonds have connections that are persistently cross-linked. However, temporary connections are present in hydrogels that are cross-linked by non-covalent interactions, such as ionic contact, hydrophobic interaction, or hydrogen bonding (Darban et al. 2022). In order to facilitate the polymerization process and the formation of stable 3D structures from linear polymer chains, a cross-linker has to include many active functional groups. The durability of polymeric materials is negatively impacted by a low degree of cross-linking, which is caused by an insignificant quantity of cross-linker. The outcome is a large number of non-specific openings caused by the distortion of the hydrogel's 3D structure and the disruption of sorption sites. The generation of a densely packed 3D mesh structure with exceptional durability and an unusually high mass transfer number occurs when the grade of cross-linking is elevated. Dyes are unable to fully permeate the hydrogel surface due to the decrease in sorption sites and the degree of swelling. Consequently, it is best to keep the cross-linker amount within the permitted range. It is common practice to employ polymers with a cross-link ratio of over 80% (Yan et al. 2006). Hydrogel synthesis relies on the use of initiators, which are chemicals that kick off the polymerization process.

8.6 Mechanisms Involved in Dye Adsorption Utilizing Hydrogels

Sorption is often a surface-related phenomenon in which scattered molecules in a liquid aggregate on the surface of a sorbent. Both physisorption and chemisorption are ways to categorize the sorption process. The process of physisorption is characterized by weak forces such as electrostatic interactions, hydrogen bonding, hydrophobic contacts, van der Waals forces, and π-π interactions; it is not a covalent relationship in which the sorbent and sorbate share electrons (Sivakumar et al. 2022). The adsorbed molecules in a liquid form a strong covalent connection with the sorbent surface via coordination interactions in chemisorption. A detailed discussion is given of how these connections interfere with the elimination of organic dyes (ODs). Figure 8.2 summarizes the reaction mechanisms governing dye adsorption.

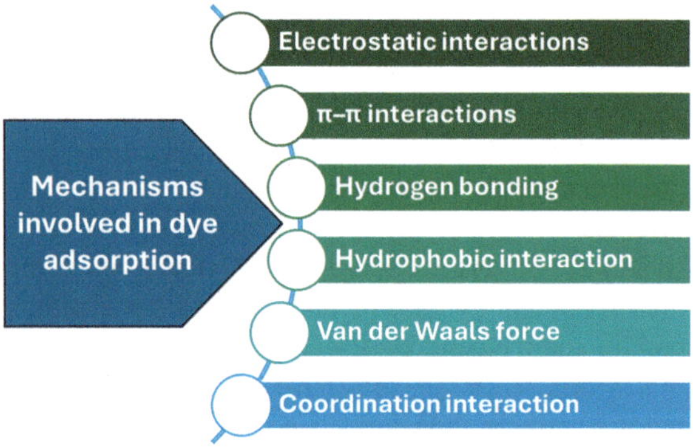

Fig. 8.2 Various mechanisms involved in dye adsorption

8.6.1 Electrostatic Interactions

Compounds with charged substances collaborate with one another in this phenomenon. The cation–anion connection describes the natural tendency for molecules with opposing charges to bind with one another. Therefore, for ionic contaminants to be electrostatically removed, the ions present in the sorbent material must possess charges that are opposite to those of the contaminants. Importantly, this process is very pH-dependent; at pH values greater than 8, functional groups like hydroxyl (–OH) and carboxyl (–COOH) in the adsorbent deprotonate. This allows cationic dyes, such as MB, to be easily sorbed onto the surface (Pei et al. 2017). The protonation of functional groups, especially amine (–NH) and thiol (–SH), causes anionic dyes like crystal violet to effectively concentrate on the sorbent surface at lower pH values (<4).

8.6.2 π–π Interactions

The presence of certain functional groups in the sorbent and sorbate determines the strategy's preference. In most cases, the π-electron connection between the molecules acting as donor and acceptor results in a noncovalent engagement known as π-π interaction. Additionally, these connections are unique to graphene-based sorbents due to the fact that each carbon atom within graphene's honeycomb lattice arrangement contains π-electrons. The π-π interaction causes the aromatic ring of ODs to be efficiently adsorbed on the sorbents' surfaces (Zhuang et al. 2016).

8.6.3 Hydrogen Bonding

Hydrogen bonding refers to the connections of dipoles between sulphur (S), nitrogen (N), oxygen (O), and fluorine (F) atoms, which are negatively charged. A "D-HA bond" is formed when the negatively charged acceptor molecule (A) interacts with the covalently bonded hydrogen atom (D-H). The covalent link is shown by the solid line, while the hydrogen bond is denoted by the dotted line. Also, unlike the van der Waals force and the covalent bond, the strength of a hydrogen bond depends on the characteristics of the provider and recipient atoms. Importantly, hydrogen bonding may occur either between molecules or inside them. Nevertheless, hydrogen bonding within molecules is favored in OD sorption. To illustrate, according to (Wang et al. 2020), the sorbent may accommodate a variety of functional groups, including –COOH, –OH, and –NH, and the negatively charged N-atom in MB can establish hydrogen bonds with these groups.

8.6.4 Hydrophobic Interaction

The hydrophobic effect causes nonpolar compounds to clump heavily in water-based solutions, and this phenomenon is efficiently used in sewage treatment to eliminate organic dyes (ODs). Because of the weak van der Waals contact, nonpolar compounds known as hydrophobes have more alkyl chains and they cannot react with water significantly. Moreover, there is a significant hydrophobic contact between the methyl group in MB and the alkyl chains on the sorbent surface. Hence, MB can easily be isolated from the wastewater (N. Peng et al. 2016).

8.6.5 Van der Waals Force

A contact among molecules or atoms that depends on proximity is known as the van der Waals force. This connection differs from covalent or ionic bonding in that it does not arise from a chemical electronic link. This force is noticeably weakened and vanishes at any sufficiently large distance between atoms. Because the surface area and permeability of sorbents generate the van der Waals force, sorbates can proceed through the sorption process. According to Yue et al. (2016), hydrogels made of polysaccharides have certain characteristics that make them ideal for the specific sorption of ODs.

8.6.6 Coordination Interaction

When two functional groups with distinct signs make a strong covalent link, this is known as a coordination engagement. For example, the cationic dye in wastewater may preferentially combine with a molecule's lone pair of electrons. In addition, cationic dyes (sorbates) bind strongly to hydrogels (sorbents) functionalized with –NH, –OH, or –COOH, since the lone pairs in the hydrogels' outer shells induce the ODs to amass on their surfaces (Godiya et al. 2020).

8.7 Amendment in HGs for Enhanced Dye Elimination

Hydrogels have been frequently employed as excellent sorbents to eradicate diverse toxic synthetic dyes from polluted streams. In order to improve the dye sorption efficacy of distinct HGs, several natural polymer-based matrices such as cellulose, lignin, salecan, and chitosan have been altered in different manners to incorporate desirable properties. These properties included mechanical strength, hydrophilicity, stability across a wide pH range, surface area, permeability, resistance to heat, etc. (Sinha et al. 2019).

To remove harmful dyes, researchers are modifying HGs in a variety of ways, including adding certain short peptide sequences (di-, tri-) (Basak et al. 2017), creating metal-laden organic gel (Jiao et al. 2015), and using nanocomposites (Varaprasad et al. 2018; Zhijiang et al. 2018). Nandi et al. (2017) synthesized H-Leu-Phe-OMe modified HG, which demonstrated quick and effective sorption affinities for several hazardous chemical dyes (Brilliant Blue, Congo Red, Malachite Green, Rhodamine B). In 2018, Bello et al. developed a super sorbent HG from banana pseudo-stems. This material was very good at absorbing methylene blue (MB) and methyl orange (MO), with the most prominent absorption rates of 333.3 mg/g for MB and 124 mg/g for MO (Bello et al. 2018). A comprehensive summary of various hydrogels for the implementation of dye eradication has been provided in Table 8.2.

8.8 Variables Influencing the Adsorption Capability of Hydrogel

8.8.1 *Crosslink Density*

Instead of talking about the number of crosslinks, "crosslink density" means the number of links or sections that connect multiple nodes in a polymer matrix. Crosslink density of polymeric pieces has a big impact on how well CBHs stick to things (Fei et al. 2000; Kono et al. 2012; Wach et al. 2001). In general, CBHs with low crosslink densities have the best sorption capabilities, and vice versa (Wach et al. 2001). But crosslink density must be at least a certain level so the absorbent materials won't break down mechanically or dissolve completely (Guilherme et al. 2005). To make sure most of the polymeric links are attached to the material by at least one covalent bond, it is recommended that the HGs have a high molecular weight in the area of minimal crosslinking density. Researchers found out that CMC HGs with higher molecular weight have more space inside their polymeric chains, which makes them stick better to things (Kono et al. 2012). Since water content decreases during crosslinking, DCH has a higher crosslink density than SCH. A significant reduction in topography and cross-sectional permeable structure of DCH was seen as a result of the enhancement in crosslink density in the polymer system (Ye et al. 2019). This proximity of the neighboring chains of HG materials was because of the increase in crosslink density.

8.8.2 *Initial Concentration (IC)*

IC has been a key factor in determining the amount of contaminants adsorbed on the surface of the sorbent. With a fixed amount of sorbent and sorbate volume,

Table 8.2 Summary of literature on dye eradication utilizing various hydrogels

Dye	Hydrogel	Optimal uptake (mg/g or %)	References
MB	PAA-sorbent NC hydrogel	2100	Hu et al. (2018)
MB	TiO_2-2D MMT-chitosan-PAA hydrogel	100%	Wang et al. (2019)
Acid blue 93	Acrolein/chitosan-based hydrogel	93–89%	Liu et al. (2022)
MB Acid orange II	Cellulose/CMC	120 100	Wang et al. (2021)
AB 9 AR	SCH-CN	357.3 319	Pauletto et al. (2020)
Basic Red 46	Chitosan-g-PAAM/MMT	1813	Ferfera-Harrar et al. (2022)
MB	CS-BmImBr	86%	Karimi-Maleh et al. (2021)
MB EBT	CS-Fe_3O_4/GO	95.3 96.3	Jamali et al. (2021)
RR 195	CS/Oxalic acid	110.7	Pérez-Calderón et al. (2020)
MB	CS-AsA-HG_2	99.88%	Mustafa et al. (2023)
Rh-B	Dy-Alg-Ch (1:2)	273.4	Malik et al. (2023)
MB	CS-MAlg	61%	Aksu Demirezen et al. (2023)
MB	CMC/GG/MTC	598	Yang et al. (2023)
MB Acid orange	CMCNFs/Ch NFs	140.5 372	Jung et al. (2023)
MB	Pectin-PAAc/CNC	576	Al-Gorair et al. (2022)
CV	Pectin@Fe_3O_4-Bentonite	909	Beigi et al. (2023)
CV	Pectin/PEI	192	Shiekh et al. (2023)
MB CR	MP/AA/PVA-2	222 316	Huang et al. (2023)
MB	Alg/CMC	1005	Zhang et al. (2023)
MG	Alg-GG-TA-GO	203	Rather et al. (2023)
MB	LN-NH-SA@3	388.8	Wang et al. (2023)
MB	AlgMA/PNaSS	2300	Yin et al. (2023)
RB19	GG-g-PAM-cl-Borax	1073	Mahto et al. (2022)
Basic violet 1	P(PrAm/AMPS/CLP/GG)	92.10–96.25%	Karadağ et al. (2021)
MB	CSt-g-PAA	1044	Srikaew et al. (2023)
MB Eosin Y	Starch/poly(α-L-lysine)	90.04 105	Hu et al. (2023)

(continued)

Table 8.2 (continued)

Dye	Hydrogel	Optimal uptake (mg/g or %)	References
CR MB	CS/CMC-PEG	1053.88 331.72	Zhu et al. (2023)
MB CR	CGC/Alg	400.504 11.45	Kasbaji et al. (2023)
AB-10B	CCGA-Zn	84.77	Zia et al. (2024)
MG	Pn-g-poly(AAm)	120.77	Nair et al. (2024)
MB TART	4VBS-MOEAMCl	137 16	Bariş et al. (2025)

the number of molecules that adsorb increases as the IC of the effluent increases (Akpomie et al. 2015). As a result, more sorbate particles attach to the functional locations on the sorbent. This, in turn, increases the inclusion of toxicants onto the sorbent surface. This is due to the governing factor of the quantity gradient increases and ultimately sorption capacity increases (Wong et al. 2020). However, several subsequent studies found that sorption decreased as pollutant concentration increased (Pourjavadi et al. 2015). Sorption usually starts slowly and then quickly increases until it reaches a plateau. Once sorption is optimized, increasing IC does not enhance it, and sorption yield decreases (Wang et al. 2017). According to the most recent studies, this is because CBH has a higher adsorption capacity at low contaminant concentration since there is a big ratio of the initial number of moles of contaminant ions to the available sites of CBH. Conversely, available sorbent sites decrease as pollutant quantity increases and thus pollutant removal efficiency decreases (Abdelwahab et al. 2016). Shi et al. (2013) reported how initial MB levels exaggerated uptake utilizing 1.8 to 2.7 mg/mL MB. A linearity was reached at 2.5 mg/mL, and then the adsorption capacity varied linearly with initial dye concentration. As concentration increased above 2.5 mg/mL, adsorption capacity decreased. A similar pattern of sorption capacity versus IC has been described in current investigations on several CBHs for elimination of heavy metals and dyes.

8.8.3 Surface Charge

A very important factor in the sorption process is the pH at the point of zero charge (pHPZC). This can change how well sorbents chelate by changing how much they grow and how they interact with ions (El-Hag Ali 2012). When the pH drops below pHPZC, the absorbent surface becomes positively charged because the percentage of H+ in the water rises, which is called protonation. When anions come into contact with the positively charged surface, essential electrostatic relationships ensue. On the contrary, the surfaces of aqueous solutions are negatively charged when pH is greater than pHPZC, allowing them to interact with cations (Eskandari et al. 2018).

The majority of the deprotonation and protonation processes take place at various functional groups, such as carboxylic or amine, as well as the deposition of atoms in HGs. To a greater extent, CBHs are able to adsorb dyes and heavy metal ions when the solution pH is more basic; in fact, there is a pH range for every dye within which the optimum sorption happens. El-Hag Ali (2012) studied the adsorptive properties of CMC in relation to solution pH to mitigate heavy metals such as Co^{2+}, Cu^{2+}, Fe^{3+}, and Mn^{2+} from wastewater. At increasing pH levels, the sorption capability of CMC HGs rises, according to the results. At very low pH levels, an excess of H^+ ions competes with metal ions for the cohesion sites of the CMC, resulting in reduced metal ion absorption. To remove colors from wastewater using CBHs, one may use the same sorption procedure. As an example, a new CBH had its greatest sorption capacity for removing the anionic dye Congo Red (CR) at pH 4.0, and this capacity decreased as the pH increased. The sorption of CR dye molecules is limited at increased pH levels because the additional OH- covers the functional locations of the sorbent. On the other hand, sorbent surfaces that contain cationic dyes, and metal ions are attracted to higher pH levels (Akter et al. 2021).

8.8.4 Temperature

Because increasing the temperature drastically changes the kinetics of the adsorbate molecules in the water solution, temperature has a profound effect on the adsorption capacity of CBHs. While exothermic adsorption causes a decrease in CBH adsorption capacity with increasing temperature, endothermic adsorption causes the exact reverse to occur (Mahmoodi et al. 2010). As a general rule, CBHs have a temperature-dependent adsorption capability for dye eradication from water specimens (P. Liu et al. 2005). Higher temperatures significantly improve the mobility of dye molecules, which in turn increases the ability to strengthen the connections between the dye and the sorbent (Lin Liu et al. 2015). Consequently, the adsorbent surface experiences less dye desorption, resulting in an increase in uptake capability. In addition, when the temperature rises, the CBH structure may undergo a swelling effect, which allows the dyes to penetrate the adsorbent surface even deeper (Mahmoodi et al. 2010). The desorption properties of CBH molecules take centre stage when adsorption occurs at temperatures greater than the equilibrium temperature. This is because the adsorption capacity is somewhat reduced due to the excessive molecular motion (Liu et al. 2015). In addition, CBHs behave similarly to adsorptive dye removal from wastewater in that they have a larger endothermic uptake for metal ion elimination (Dai et al. 2020), but a much lower exothermic adsorption capacity (Cao et al. 2017).

8.8.5 Ionic Strength

The ionic strength of an aqueous solution is the concentration of ions in it. Salts in water dissociate into ions in the solution. Simply put, the concentration of salts in a solution is directly proportional to its ionic strength (Solomon 2001). Industrial wastewater contains various organic and inorganic contaminants besides salts, including NaCl, KCl, NH_4Cl, $CaCl_2$, $MgSO_4$, $AlCl_3$, etc., and these salts have a significant impact on the adsorption capacity of CBH when the wastewater is treated (Muthukumar et al. 2004). Dye removal from water-based solutions is the area of research on ionic strength and adsorption capacity. By using varying concentrations of NaCl salt, Liu et al. (2015) found that the dye eradication competence of CBH grafted with acrylamide and acrylic acid was altered by the ionic strength of the specimen. Because of the competing consequences between the salt ions (Cl^- and Na^+) and the prevailing dyes on the CBH surface, which includes functional groups ($-COO^-$, $-NH_3^+$ and OH^-), they found that the dye removal efficiency decreased as the concentration of NaCl increased. Adsorbents' adsorption capacity decreases as the concentration of NaCl increases because the defensive outcome of Cl^- and Na^+ ions for the ionized dye particles is enhanced (Mahmoodi et al. 2011). A higher amount of other salts like $CaCl_2$, $AlCl_3$, NH_4Cl, and KCl reduced the absorbency of CBH, similar to NaCl. Among these salts, in the presence of monovalent cations, the adsorption capacity of CBH decreases in the order of $NH^{4+} > K^+ > Na^+$. When looking at the swelling capacity of CBH depending on ion valence, the presence of trivalent cations resulted in the highest decrease, followed by divalent cations and then monovalent cations (Alam et al. 2019).

8.8.6 Coexistence of Ions

The adsorption effectiveness of HG for each ion species is affected in different ways by the presence of different ions in the water solution. The adsorption of some ions in a solution may be reduced by others as a result of competition, or it can be increased by all of the ions working together as a result of cosorption (Antić et al. 2016). Researchers Antić et al. (2016) looked at Pb^{2+} sorption with and without other ions, such as Ni^{2+}, Cd^{2+}, Cu^{2+}, Zn^{2+}, and Co^{2+}.

Adsorption of Pb^{2+} was reduced by 5.27 percent in the binary system and by an additional 11.1 percent in the tertiary system as a result of competition. In addition, heavy metal ions like Cd^{2+}, Cu^{2+}, and Pb^{2+} contend for the same surface area as the adsorbent molecules when interference ions like K^+, Na^+, Mg^{2+}, and Ca^{2+} are present in the solution, which hinders the adsorption of these metals (Ma et al. 2017). Another possible explanation for the poor heavy metal adsorption might be that existing ions in the solution share binding sites with HGs (Hwang et al. 1997). Additionally, some ions with certain ionic characteristics, such as ionic radius, electronegativity, and ionization potential, have been selectively adsorbed by functional groups (e.g., –COOH, $-NH_2$, etc.) trapped in HG polymeric chains. In a nanocomposite HG, for example, the –COOH group has a greater affinity for Pb^{2+} adsorption than for Ni^{2+}, Cd^{2+}, Cu^{2+}, or Zn^{2+} (Jiang et al. 2014). This is because these ions have different characteristics. When compared to other ions, metal ions have a stronger tendency to bind to HGs as their radius increases. Accordingly, the binding capacity is proportional to the ionic radius (Hwang et al. 1997). Both the uptake and the competence of metal ion removal are critically exaggerated by the ionic hydration radius. Research has shown that ions with a lower hydration radius have a negligible effect on adsorption capacity, regardless of the presence of additional ions in the specimens, including Na^{+}, Ca^{2+}, and Ba^{2+}, which might cause interference (Dwivedi et al. 2015). Furthermore, the sorption discernment of the sorbent for certain metals is meaningfully exaggerated by the variation of reactive groups in the COBH structure. Researchers Amr El-Hag Ali and colleagues studied the copolymerization and crosslinking of CMC/AMPS-HG using radiation to study the simultaneous adsorption of Mn^{2+}, Co^{2+}, Cu^{2+}, and Fe^{3+}. Changing the AMPS concentration allowed him to functionalize the adsorbent structure. The adsorption capacity of the CMC/AMPS adsorbent was found to improve substantially with increasing AMPS concentration, as shown by the adsorption of metal ions for different AMPS concentrations. The pH of the solution is one factor that contributes to this behavior. When the reactive groups in the assembly of CBH are changed, the solution pH often changes as well (Fritz 2005). An increase in electrostatic repulsion causes the number of dissociated groups to grow, which in turn causes the concentration of AMPS in the HG to climb, leading to a larger network.

8.9 Limitations of Commercializing Hydrogel Technologies

Everyone agrees that commercializing HGs, especially for wastewater treatment plants, would be good. But before HG engineering can fully be implemented, there are some areas that need further research. For example, there must be thorough testing for robustness, pH sensibility, expansion ratio, heat security, and versatility (Sinha et al. 2019). Since modified HGs can bypass the problems of traditional HGs, which weaken and lose mechanical strength when inflated many times (De France et al. 2017), it is imperative to enhance HGs' mechanical durability while boosting their self-healing property after swelling (Gjipalaj et al. 2017). To overcome the limitations of traditional HGs, researchers are exploring various chemical and physical networking methods to improve the gel's properties, such as adding GO, NPs, CNTs or CQDs (Thakur et al. 2017). Mechanical toughness, dye adsorption, and reusability are all improved when GO is added to PAM nanocomposite HGs because a strong hydrogen bond forms between the two groups (GO sheets and polymer chains) (Cui et al. 2015). The composite HG was shown to have good elasticity and toughness after cross-linking with Al (III) when GO is mixed with PAM/CMH-HG in another study (Sinha et al. 2019). To develop more affordable, thermostable, and HGs with larger swelling capacity, researchers are investigating various NPs (such as silicate and metal-oxide NPs) (Yu et al. 2018). To enhance the characteristics of chitin-based HGs, many studies have used nanocomposites that include carbon quantum dots (CDs) (Konwar et al. 2015), cellulosic nanofibers (Abdul Khalil et al. 2016), and zinc oxide nanowires (ZnO NWs) (Chi et al. 2019). Adding ZnO NWs to chitosan and acrylic acid enhanced their interaction, which in turn increased the gel's strength and thermostability. Therefore, hybrid composites made from NPs mixed with various HGs are durable and reusable; nevertheless, in order to overcome the present limits and drawbacks of HG engineering, further research into the interaction between the polymer matrix and NPs is required (Sinha et al. 2019).

8.10 Recovery and Regeneration of Hydrogel

Hydrogels have the ability to regenerate by desorbing the sorbed dyes and toxic metals, which is an added advantage to their good sorption efficiency. One of the key aspects to consider when evaluating the practicality of a sorbent material is its regeneration and reuse. Multiple dye desorption from the spent hydrogel 3D mesh framework after each removal cycle has been the focus of many studies. Various environmental factors, such as changes in electric and magnetic fields, temperature, pH, and others can trigger dye desorption (Mohammadi et al. 2017). In their study, Liu et al. (2012) showed how magnetized cellulose-chitosan hydrogel desorb heavy metals was affected by pH. Desorption was 83–86% efficient at very low pH between 1.0–2.0. This is an advantage of using hydrogels that are pH dependent for heavy metals like chromium and arsenic, and for their recovery after desorption. Other studies on hydrogel sorbent recovery have used various eluents such as strong and weak acids (HCl, HNO_3, CH_3COOH, H_2SO_4, etc.) (Darban et al. 2022). Both the hydrogel stability and desorption ability are affected by the type of acid used in the desorption process (Weerasundara et al. 2021). Researchers have developed a highly efficient and reusable multi-functionalized hydrogel mimicking a chelator. The heavy metals adsorbed by the hydrogel structure were removed by using a small amount of hydrochloric acid, and the hydrogel can be reused. Five sorption/desorption cycles resulted in a clearance rate of more than 60% (Mohammadi et al. 2017). Pourjavadi et al. (2015) synthesized a new hydrogel utilizing chitosan, AA, and nano-silica that has been functionalized with an amine group. The hydrogel that was laden with Pb was recovered using a 1 M hydrochloric acid solution in this study. Before the sorption cycle, the hydrogel has been redeveloped by filtration and cleaning with Millipore water. Regenerated hydrogel maintained 685–715 mg/g after 3 cycles. Magnetic HGs are one of the most effective sorbent constituents for removing heavy metals from streams. Since the magnetic adsorbent is sensitive to acidity, the eluent acidity must be controlled during recovery. After several sorption cycles, the hydrogel's adsorption capacity decreases due to damage to the binding sites by highly concentrated eluents (Zhang et al. 2021). A magnetic hydrogel was developed by Tang et al. that adsorbed Cr^{6+} at 200 mg/g. Since it can be redeveloped in NaCl solution, the HG is cost-effective to retrieve (Tang et al. 2014). For any magnetic hydrogel sorbent to work in polluted water, two things must happen: the concentration of pollutants in the solution must increase, and the amount of recovery solution must decrease. In summary, a magnetic separation unit is used to capture and isolate the treated polluted water from the hydrogel. NaCl in varying concentrations was added to the hydrogel to regenerate it, and the remaining solution was collected by isolating the hydrogel from the magnetic field. NaCl was added step by step in the regeneration experiments. Then the extracted solution was obtained and treated further by adding a NaCl solution to it. The investigations showed that the rate of recovery was upheld high for 20 cycles, which gave 97–98% removal of Cr^{6+}. Based on the results, the recovery solution contained Cr of 500–600 mg/L, which corresponds to an effluent-to-recovery volume ratio of 40:1 (Tang et al. 2014). The recyclability of hydrogel is one of the most important factors for many applications, but it's a challenge for traditional hydrogel sorbents since they have limited strength after swelling in an aqueous environment. To keep the dye-laden hydrogel's sorption capacity at its target level, strengthening its mechanical durability is needed. In summary, the hydrogel is the best sorbent for removing dyes and is toxic and cheap.

8.11 Conclusions and Future Outlook

Wastewater includes a variety of contaminants, the most prevalent of which are dyes generated by numerous organizations. These pollutants threaten aquatic life, human well-being, and the natural environment. Consequently, to create a healthier atmosphere, it is crucial to remove contaminants from wastewater. One highly efficient and successful way to treat wastewater is via the sorption process, which takes advantage of various sorbents. Researchers are motivated to discover and utilize sorbents that originate from natural resources because the majority of current sorbents employed for treatment purposes are synthetic and do not biodegrade. Yet another concern regarding the environment is the handling of the spent sorbent. The synthetic production of hydrogels, the sorption process, and variables that enhance sorption capabilities are some of the essential but under-discussed areas of hydrogel-based wastewater management that this study fills in. Furthermore, taking into account the constraints of the literature reviewed, it is suggested that future studies investigate the following areas. Firstly, Remedies of single-pollutant laboratory effluent, rather than actual industrial effluent, have dominated the literature. The sorption of a single ion may be affected by the presence of other ions in the effluent, according to several studies (Antić et al. 2016). The sorption effectiveness of some effluents should be investigated further in addition to the targeted contaminants; these wastewaters often include additives or auxiliary substances. Hence, hydrogels' adsorption efficacy in a real-world system requires substantial further investigation. One possible approach is to process commercial wastewater on a smaller scale, or simulate effluent with numerous contaminants to make it more like genuine industrial wastewater. Secondly, wastewater treatment relies on hydrogels because of their robust structure and efficient swelling, which is particularly important when dealing with industrial waste streams that are heated. Repetitive swelling may reduce the durability of certain hydrogels. Hydrogels are often more stable and effective adsorbents when they have a structurally combined system. It would be beneficial to explore the possibility of enhancing the hydrogel sorption capability and swelling characteristics by including magnetic particles, nanoparticles, and other chemical catalysts. The potential for hydrogels to enhance their mechanical reliability via increased self-healing capabilities after swelling should be further investigated. Thirdly, the effectiveness of hydrogels in treating wastewater from industries is mostly determined by their entire shape and physicochemical content. To overcome obstacles and weaknesses, including inadequate turnover number, reduced resistance, and mechanical durability, a well-designed composition and optimized synthesis circumstances are key elements in the creation of hydrogels. Furthermore, before it can be used on a large scale, problems with its thermal resistance, swelling ratio, and susceptibility to pH must be resolved.

References

Abdelwahab HE, Hassan SY, Mostafa MA, El Sadek MM (2016) Synthesis and characterization of glutamic-chitosan hydrogel for copper and nickel removal from wastewater. Molecules 21(6). https://doi.org/10.3390/molecules21060684

Abdul Khalil HPS, Saurabh CK, Adnan AS, Nurul Fazita MR, Syakir MI, Davoudpour Y, Rafatullah M, Abdullah CK, Haafiz MKM, Dungani R (2016) A review on chitosan-cellulose blends and nanocellulose reinforced chitosan biocomposites: properties and their applications. Carbohydr Polym 150:216–226. Elsevier Ltd. https://doi.org/10.1016/j.carbpol.2016.05.028

Akpomie KG, Dawodu FA, Adebowale KO (2015) Mechanism on the sorption of heavy metals from binary-solution by a low cost montmorillonite and its desorption potential. Alex Eng J 54(3):757–767. https://doi.org/10.1016/j.aej.2015.03.025

Aksu Demirezen D, Demirezen Yılmaz D, Yıldız YŞ (2023) Magnetic chitosan/calcium alginate double-network hydrogel beads: Preparation, adsorption of anionic and cationic surfactants, and reuse in the removal of methylene blue. Int J Biol Macromol 239. https://doi.org/10.1016/j.ijbiomac.2023.124311

Akter M, Bhattacharjee M, Dhar AK, Rahman FBA, Haque S, Ur Rashid TU, Kabir SMF (2021) Cellulose-based hydrogels for wastewater treatment: a concise review. Gels 7(1). https://doi.org/10.3390/gels7010030

Alam MN, Islam MS, Christopher LP (2019) Sustainable production of cellulose-based hydrogels with superb absorbing potential in physiological saline. ACS Omega 4(5):9419–9426. https://doi.org/10.1021/acsomega.9b00651

Al-Gorair AS, Sayed A, Mahmoud GA (2022) Engineered superabsorbent nanocomposite reinforced with cellulose nanocrystals for remediation of basic dyes: isotherm, kinetic, and thermodynamic studies. Polymers 14(3). https://doi.org/10.3390/polym14030567

Antić KM, Babić MM, Vuković JS, Onjia AE, Filipović JM, Tomić SL (2016) Uklanjanje pb2+ IZ vodenih rastvora pomoću P(HEA/IA) hidrogela. Hemijska Industrija 70(6):695–705. https://doi.org/10.2298/HEMIND151225006A

Bariş SY, Asan G, Delibaş A (2025) Synthesis, characterization and use of amphoteric hydrogels in cationic and anionic dye adsorption. J Mol Struct 141754. https://doi.org/10.1016/j.molstruc.2025.141754

Basak S, Nandi N, Paul S, Hamley IW, Banerjee A (2017) A tripeptide-based self-shrinking hydrogel for waste-water treatment: removal of toxic organic dyes and lead (Pb2+) ions. Chem Commun 53(43):5910–5913. https://doi.org/10.1039/c7cc01774j

Beigi P, Ganjali F, Hassanzadeh-Afruzi F, Salehi MM, Maleki A (2023) Enhancement of adsorption efficiency of crystal violet and chlorpyrifos onto pectin hydrogel@Fe_3O_4-bentonite as a versatile nanoadsorbent. Sci Rep 13(1). https://doi.org/10.1038/s41598-023-38005-z

Bello K, Sarojini BK, Narayana B, Rao A, Byrappa K (2018) A study on adsorption behavior of newly synthesized banana pseudo-stem derived superabsorbent hydrogels for cationic and anionic dye removal from effluents. Carbohyd Polym 181:605–615. https://doi.org/10.1016/j.carbpol.2017.11.106

Benkhaya S, M'rabet S, El Harfi A (2020) Classifications, properties, recent synthesis and applications of azo dyes. Heliyon 6(1). Elsevier Ltd. https://doi.org/10.1016/j.heliyon.2020.e03271

Bhardwaj D, Bharadvaja N (2021) Phycoremediation of effluents containing dyes and its prospects for value-added products: a review of opportunities. J Water Process Eng 41. Elsevier Ltd. https://doi.org/10.1016/j.jwpe.2021.102080

Bhatnagar A, Sillanpää M (2009) Applications of chitin- and chitosan-derivatives for the detoxification of water and wastewater—a short review. Adv Colloid Interface Sci 152(1–2):26–38. https://doi.org/10.1016/j.cis.2009.09.003

Bide M (2007) Environmentally responsible dye application. Environ Asp Text Dye 74–92. Elsevier Inc. https://doi.org/10.1533/9781845693091.74

Buruga K, Song H, Shang J, Bolan N, Jagannathan TK, Kim KH (2019) A review on functional polymer-clay based nanocomposite membranes for treatment of water. J Haz Mater 379. https://doi.org/10.1016/j.jhazmat.2019.04.067

Cao J, Fei D, Tian X, Zhu Y, Wang S, Zhang Y, Mao Q, Sun M (2017) Novel modified microcrystalline cellulose-based porous material for fast and effective heavy-metal removal from aqueous solution. Cellulose 24(12):5565–5577. https://doi.org/10.1007/s10570-017-1504-6

Chi H, Qiao Y, Wang B, Hou Y, Li Q, Li K, Liu Z (2019) Swelling, thermal stability, antibacterial properties enhancement on composite hydrogel synthesized by chitosan-acrylic acid and ZnO nanowires. Polym Plast Technol Mater 58(15):1649–1661. https://doi.org/10.1080/25740881.2018.1563138

Cui W, Ji J, Cai YF, Li H, Ran R (2015) Robust, anti-fatigue, and self-healing graphene oxide/hydrophobically associated composite hydrogels and their use as recyclable adsorbents for dye wastewater treatment. J Mater Chem A 3(33):17445–17458. https://doi.org/10.1039/c5ta04470g

Dai H, Huang Y, Zhang H, Ma L, Huang H, Wu J, Zhang Y (2020) Direct fabrication of hierarchically processed pineapple peel hydrogels for efficient Congo red adsorption. Carbohydr Polym 230. https://doi.org/10.1016/j.carbpol.2019.115599

Darban Z, Shahabuddin S, Gaur R, Ahmad I, Sridewi N (2022) Hydrogel-based adsorbent material for the effective removal of heavy metals from wastewater: a comprehensive review. Gels 8(5). MDPI. https://doi.org/10.3390/gels8050263

De France KJ, Hoare T, Cranston ED (2017) Review of hydrogels and aerogels containing nanocellulose. Chem Mater 29(11):4609–4631. https://doi.org/10.1021/acs.chemmater.7b00531

Donkadokula NY, Kola AK, Naz I, Saroj D (2020) A review on advanced physico-chemical and biological textile dye wastewater treatment techniques. Rev Environ Sci Biotechnol 19(3):543–560. Springer. https://doi.org/10.1007/s11157-020-09543-z

Dwivedi C, Pathak SK, Kumar M, Tripathi SC, Bajaj PN (2015) Preparation and characterization of potassium nickel hexacyanoferrate-loaded hydrogel beads for the removal of cesium ions. Environ Sci Water Res Technol 1(2):153–160. https://doi.org/10.1039/c4ew00021h

El-Hag Ali A (2012) Removal of heavy metals from model wastewater by using carboxymehyl cellulose/2-acrylamido-2-methyl propane sulfonic acid hydrogels. J Appl Polym Sci 123(2):763–769. https://doi.org/10.1002/app.34470

Eskandari S, Tate G, Leaphart NR, Regalbuto JR (2018) Nanoparticle synthesis via electrostatic adsorption using incipient wetness impregnation. ACS Catal 8(11):10383–10391. https://doi.org/10.1021/acscatal.8b03435
Fei B, Wach RA, Mitomo H, Yoshii F, Kume T (2000) Hydrogel of biodegradable cellulose derivatives. I. Radiation-induced crosslinking of CMC. In J Appl Polym Sci 78
Ferfera-Harrar H, Benhalima T, Sadi A (2022) Development of functional chitosan-based superabsorbent hydrogel nanocomposites for adsorptive removal of basic red 46 textile dye. Polym Bull 79(8):6141–6172. https://doi.org/10.1007/s00289-021-03795-7
Fritz JS (2005) Factors affecting selectivity in ion chromatography. J Chromatogr A 1085(1):8–17. https://doi.org/10.1016/j.chroma.2004.12.087
Gjipalaj J, Alessandri I (2017) Easy recovery, mechanical stability, enhanced adsorption capacity and recyclability of alginate-based TiO_2 macrobead photocatalysts for water treatment. J Environ Chem Eng 5(2):1763–1770. https://doi.org/10.1016/j.jece.2017.03.017
Godiya CB, Xiao Y, Lu X (2020) Amine functionalized sodium alginate hydrogel for efficient and rapid removal of methyl blue in water. Int J Biol Macromol 144:671–681. https://doi.org/10.1016/j.ijbiomac.2019.12.139
Guida S, Rubertelli G, Jefferson B, Soares A (2021) Demonstration of ion exchange technology for phosphorus removal and recovery from municipal wastewater. Chem Eng J 420. https://doi.org/10.1016/j.cej.2021.129913
Guilherme MR, Campese GM, Radovanovic E, Rubira AF, Feitosa JPA, Muniz EC (2005) Morphology and water affinity of superabsorbent hydrogels composed of methacrylated cashew gum and acrylamide with good mechanical properties. Polymer 46(19 SPEC. ISS.):7867–7873. https://doi.org/10.1016/j.polymer.2005.06.068
Haraguchi K, Li HJ, Matsuda K, Takehisa T, Elliott E (2005) Mechanism of forming organic/inorganic network structures during in-situ free-radical polymerization in PNIPA-clay nanocomposite hydrogels. Macromolecules 38(8):3482–3490. https://doi.org/10.1021/ma047431c
Hu XS, Liang R, Sun G (2018) Super-adsorbent hydrogel for removal of methylene blue dye from aqueous solution. J Mater Chem A 6(36):17612–17624. https://doi.org/10.1039/c8ta04722g
Hu C, Wei H, Hua B, Zhang Y, Wang G, Guo T (2023) Facile fabrication of a broad-spectrum starch/poly(α-L-lysine) hydrogel adsorbent with thermal/pH-sensitive IPN structure through simultaneous dual-click strategy. Carbohydr Polym 309. https://doi.org/10.1016/j.carbpol.2023.120672
Huang Y, Shen B, Zheng C, Huang B, Zhang G, Fei P (2023) Preparation of amphoteric double network hydrogels based on low methoxy pectin: adsorption kinetics and removal of anionic and cationic dyes. Int J Biol Macromol 252. https://doi.org/10.1016/j.ijbiomac.2023.126488
Hube S, Eskafi M, Hrafnkelsdóttir KF, Bjarnadóttir B, Bjarnadóttir MÁ, Axelsdóttir S, Wu B (2020) Direct membrane filtration for wastewater treatment and resource recovery: a review. Sci Total Environ 710. Elsevier B.V. https://doi.org/10.1016/j.scitotenv.2019.136375
Hwang D-C, Damodaran S (1997) Metal-chelating properties and biodegradability of an ethylenediaminetetraacetic acid dianhydride modified soy protein hydrogel. J Appl Polym Sci 64. John Wiley and Sons, Inc
Jamali M, Akbari A (2021) Facile fabrication of magnetic chitosan hydrogel beads and modified by interfacial polymerization method and study of adsorption of cationic/anionic dyes from aqueous solution. J Environ Chem Eng 9(3). https://doi.org/10.1016/j.jece.2021.105175
Jiang L, Liu P (2014) Design of magnetic attapulgite/fly ash/poly(acrylic acid) ternary nanocomposite hydrogels and performance evaluation as selective adsorbent for Pb^{2+} Ion. ACS Sustain Chem Eng 2(7):1785–1794. https://doi.org/10.1021/sc500031z
Jiao T, Guo H, Zhang Q, Peng Q, Tang Y, Yan X, Li B (2015) Reduced graphene oxide-based silver nanoparticle-containing composite hydrogel as highly efficient dye catalysts for wastewater treatment. Sci Rep 5. https://doi.org/10.1038/srep11873
Jung S, Kim J, Bang J, Jung M, Park S, Yun H, Kwak HW (2023) pH-sensitive cellulose/chitin nanofibrillar hydrogel for dye pollutant removal. Carbohydr Polym 317. https://doi.org/10.1016/j.carbpol.2023.121090

Karadağ E, Ercan D, Üzüm ÖB, Kundakcı S (2021) Swelling equilibria of novel propenamide/2-acrylamido-2-methyl-1-propanesulfonic acid/guar gum/clinoptilolite biohybrid hydrogels and application as a sorbent for BV1 removal. Polym Bull 78(7):3625–3649. https://doi.org/10.1007/s00289-020-03285-2

Karimi-Maleh H, Ranjbari S, Tanhaei B, Ayati A, Orooji Y, Alizadeh M, Karimi F, Salmanpour S, Rouhi J, Sillanpää M, Sen F (2021) Novel 1-butyl-3-methylimidazolium bromide impregnated chitosan hydrogel beads nanostructure as an efficient nanobio-adsorbent for cationic dye removal: kinetic study. Environ Res 195. https://doi.org/10.1016/j.envres.2021.110809

Kasbaji M, Mennani M, Grimi N, Oubenali M, Mbarki M, El Zakhem H, Moubarik A (2023) Adsorption of cationic and anionic dyes onto coffee grounds cellulose/sodium alginate double-network hydrogel beads: isotherm analysis and recyclability performance. Int J Biol Macromol 239. https://doi.org/10.1016/j.ijbiomac.2023.124288

Katheresan V, Kansedo J, Lau SY (2018) Efficiency of various recent wastewater dye removal methods: a review. J Environ Chem Eng 6(4):4676–4697. Elsevier Ltd. https://doi.org/10.1016/j.jece.2018.06.060

Khan S, Malik A (2018) Toxicity evaluation of textile effluents and role of native soil bacterium in biodegradation of a textile dye. Environ Sci Pollut Res 25(5):4446–4458. https://doi.org/10.1007/s11356-017-0783-7

Kono H, Fujita S (2012) Biodegradable superabsorbent hydrogels derived from cellulose by esterification crosslinking with 1,2,3,4-butanetetracarboxylic dianhydride. Carbohyd Polym 87(4):2582–2588. https://doi.org/10.1016/j.carbpol.2011.11.045

Konwar A, Gogoi N, Majumdar G, Chowdhury D (2015) Green chitosan-carbon dots nanocomposite hydrogel film with superior properties. Carbohyd Polym 115:238–245. https://doi.org/10.1016/j.carbpol.2014.08.021

Liu P, Peng J, Li J, Wu J (2005) Radiation crosslinking of CMC-Na at low dose and its application as substitute for hydrogel. Radiat Phys Chem 72(5):635–638. https://doi.org/10.1016/j.radphyschem.2004.03.090

Liu Z, Wang H, Liu C, Jiang Y, Yu G, Mu X, Wang X (2012) Magnetic cellulose-chitosan hydrogels prepared from ionic liquids as reusable adsorbent for removal of heavy metal ions. Chem Commun 48(59):7350–7352. https://doi.org/10.1039/c2cc17795a

Liu L, Lin Y, Liu Y, Zhu H, He Q (2013) Removal of methylene blue from aqueous solutions by sewage sludge based granular activated carbon: adsorption equilibrium, kinetics, and thermodynamics. J Chem Eng Data 58(8):2248–2253. https://doi.org/10.1021/je4003543

Liu L, Gao ZY, Su XP, Chen X, Jiang L, Yao JM (2015) Adsorption removal of dyes from single and binary solutions using a cellulose-based bioadsorbent. ACS Sustain Chem Eng 3(3):432–442. https://doi.org/10.1021/sc500848m

Liu Y, Zhang YF, Tang J, Tan Y, Lei H, Li Y, Huang P, Li Y (2022) Efficient removal and recycle of acid blue 93 dye from aqueous solution by acrolein crosslinked chitosan hydrogel. Colloids Surf A PhysChemical Eng Asp 632. https://doi.org/10.1016/j.colsurfa.2021.127825

Ma J, Zhou G, Chu L, Liu Y, Liu C, Luo S, Wei Y (2017) Efficient removal of heavy metal ions with an EDTA functionalized chitosan/polyacrylamide double network hydrogel. ACS Sustain Chem Eng 5(1):843–851. https://doi.org/10.1021/acssuschemeng.6b02181

Mahmoodi NM, Hayati B, Arami M (2010) Textile dye removal from single and ternary systems using date stones: kinetic, isotherm, and thermodynamic studies. J Chem Eng Data 55(11):4638–4649. https://doi.org/10.1021/je1002384

Mahmoodi NM, Salehi R, Arami M, Bahrami H (2011) Dye removal from colored textile wastewater using chitosan in binary systems. Desalination 267(1):64–72. https://doi.org/10.1016/j.desal.2010.09.007

Mahto A, Mishra S (2022) The removal of textile industrial Dye-RB-19 using Guar gum-based adsorbent with thermodynamic and kinetic evaluation parameters. Polym Bull 79(5):3353–3378. https://doi.org/10.1007/s00289-021-03663-4

Malaviya P, Singh A (2011) Physicochemical technologies for remediation of chromium-containing waters and wastewaters. Crit Rev Environ Sci Technol 41(12):1111–1172. https://doi.org/10.1080/10643380903392817

Malik DS, Jain CK, Yadav AK (2017) Removal of heavy metals from emerging cellulosic low-cost adsorbents: a review. Appl Water Sci 7(5):2113–2136. Springer Verlag. https://doi.org/10.1007/s13201-016-0401-8

Malik SA, Dar AA, Banday JA (2023) Rheological, morphological and swelling properties of dysprosium-based composite hydrogel beads of alginate and chitosan: a promising material for the effective cationic and anionic dye removal. Colloids Surf A PhysChemical Eng Asp 663. https://doi.org/10.1016/j.colsurfa.2023.131046

Mekonnen MM, Hoekstra AY (2016) Sustainability: four billion people facing severe water scarcity. Sci Adv 2(2). https://doi.org/10.1126/sciadv.1500323

Mezohegyi G, van der Zee FP, Font J, Fortuny A, Fabregat A (2012) Towards advanced aqueous dye removal processes: a short review on the versatile role of activated carbon. J Environ Manag 102:148–164. Academic Press. https://doi.org/10.1016/j.jenvman.2012.02.021

Mittal H, Parashar V, Mishra SB, Mishra AK (2014) Fe_3O_4 MNPs and gum xanthan based hydrogels nanocomposites for the efficient capture of malachite green from aqueous solution. Chem Eng J 255:471–482. https://doi.org/10.1016/j.cej.2014.04.098

Mohammadi Z, Shangbin S, Berkland C, Liang JT (2017) Chelator-mimetic multi-functionalized hydrogel: highly efficient and reusable sorbent for Cd, Pb, and As removal from waste water. Chem Eng J 307:496–502. https://doi.org/10.1016/j.cej.2016.08.121

Mustafa FHA, Gad ElRab EKM, Kamel RM, Elshaarawy RFM (2023) Cost-effective removal of toxic methylene blue dye from textile effluents by new integrated crosslinked chitosan/aspartic acid hydrogels. Int J Bio Macromol 248. https://doi.org/10.1016/j.ijbiomac.2023.125986

Muthukumar M, Sargunamani D, Selvakumar N, Nedumaran D (2004) Effect of salt additives on decolouration of acid black 1 dye effluent by ozonation. Ind J Chem Technol 11

Nagireddi S, Patel PK, Golder AK, Uppaluri RVS (2025) Evaluation of a triethylenetetramine-chitosan derivative for the removal of palladium ion from an intricate adsorbate system. J Mol Struct 1323. https://doi.org/10.1016/j.molstruc.2024.140782

Nair A, Kumawat YK, Choudhary S, Nath J, Sharma K, Rasool T, Sharma V, Kumar V (2024) Malachite green dye adsorption from wastewater using pine gum-based hydrogel: kinetic and thermodynamic studies. J Mol Struct 1295. https://doi.org/10.1016/j.molstruc.2023.136671

Nandi N, Baral A, Basu K, Roy S, Banerjee A (2017) A dipeptide-based superhydrogel: removal of toxic dyes and heavy metal ions from waste water. Biopolymers 108(1). https://doi.org/10.1002/bip.22915

Pandey S, Do JY, Kim J, Kang M (2020) Fast and highly efficient removal of dye from aqueous solution using natural locust bean gum based hydrogels as adsorbent. Int J Biol Macromol 143:60–75. https://doi.org/10.1016/j.ijbiomac.2019.12.002

Patel PK, Nagireddi S, Uppaluri RVS, Pandey LM (2022) Batch adsorption characteristics of Dowex Marathon MSA commercial resin for Au(III) removal from synthetic electroless plating solutions. Mater Today Proc 68:824–829. https://doi.org/10.1016/j.matpr.2022.06.258

Patel PK, Pandey LM, Uppaluri RVS (2024a) Cyclic adsorption and desorption characteristics of citric acid-chitosan variant resins for Pb, Fe, and Zn removal from simulated mining and agricultural wastewater system. J Polym Environ. https://doi.org/10.1007/s10924-024-03343-0

Patel PK, Uppaluri RVS (2025) Adsorption of emerging pollutants utilizing chitosan derivatives: recent advances and future perspective. Int J Biol Macromol 299. Elsevier B.V. https://doi.org/10.1016/j.ijbiomac.2025.140203

Patel PK, Pandey LM, Uppaluri R (2023a) Multi-metal adsorption and cyclic desorption characteristics of Zn^{+2} and Cu^{+2} constituting multi-component synthetic wastewater system using commercial resins. Sustain Environ 3–27. Springer Nature Singapore. https://doi.org/10.1007/978-981-19-8464-8_1

Patel PK, Pandey LM, Uppaluri RVS (2023b) Adsorptive removal of Zn, Fe, and Pb from Zn dominant simulated industrial wastewater solution using polyvinyl alcohol grafted chitosan variant resins. Chem Eng J 459. https://doi.org/10.1016/j.cej.2023.141563
Patel PK, Pandey LM, Uppaluri RVS (2023c) Cyclic desorption based efficacy of polyvinyl alcohol-chitosan variant resins for multi heavy-metal removal. Int J Biol Macromol 242. https://doi.org/10.1016/j.ijbiomac.2023.124812
Patel PK, Pandey LM, Uppaluri RVS (2023d) Synthesized carboxymethyl-chitosan variant composites for cyclic adsorption-desorption based removal of Fe, Pb, and Cu. Chemosphere 340. https://doi.org/10.1016/j.chemosphere.2023.139780
Patel PK, Pandey LM, Uppaluri RVS (2024b) Highly effective removal of multi-heavy metals from simulated industrial effluent through an adsorption process employing carboxymethyl-chitosan composites. Environ Res 240. https://doi.org/10.1016/j.envres.2023.117502
Patel PK, Pandey LM, Uppaluri RVS (2025) Conceptual techno-economic analysis of chitosan derivative resin and commercial sorbents for multi-heavy metal removal. Sep Purif Technol 362. https://doi.org/10.1016/j.seppur.2025.131732
Pauletto PS, Gonçalves JO, Pinto LAA, Dotto GL, Salau NPG (2020) Single and competitive dye adsorption onto chitosan–based hybrid hydrogels using artificial neural network modeling. J Colloid Interface Sci 560:722–729. https://doi.org/10.1016/j.jcis.2019.10.106
Pei Y, Chu S, Chen Y, Li Z, Zhao J, Liu S, Wu X, Liu J, Zheng X, Tang K (2017) Tannin-immobilized cellulose hydrogel fabricated by a homogeneous reaction as a potential adsorbent for removing cationic organic dye from aqueous solution. Int J Biol Macromol 103:254–260. https://doi.org/10.1016/j.ijbiomac.2017.05.072
Peng N, Hu D, Zeng J, Li Y, Liang L, Chang C (2016) Superabsorbent cellulose-clay nanocomposite hydrogels for highly efficient removal of dye in water. ACS Sustain Chem Eng 4(12):7217–7224. https://doi.org/10.1021/acssuschemeng.6b02178
Peng B, Yao Z, Wang X, Crombeen M, Sweeney DG, Tam KC (2020) Cellulose-based materials in wastewater treatment of petroleum industry. Green Energy Environ 5(1):37–49. KeAi Publishing Communications Ltd. https://doi.org/10.1016/j.gee.2019.09.003
Pereira AGB, Rodrigues FHA, Paulino AT, Martins AF, Fajardo AR (2021) Recent advances on composite hydrogels designed for the remediation of dye-contaminated water and wastewater: a review. J Clean Prod 284. Elsevier Ltd. https://doi.org/10.1016/j.jclepro.2020.124703
Pérez-Calderón J, Santos MV, Zaritzky N (2020) Synthesis, characterization and application of cross-linked chitosan/oxalic acid hydrogels to improve azo dye (Reactive Red 195) adsorption. React Funct Polym 155. https://doi.org/10.1016/j.reactfunctpolym.2020.104699
Pourjavadi A, Tehrani ZM, Salimi H, Banazadeh A, Abedini N (2015) Hydrogel nanocomposite based on chitosan-g-acrylic acid and modified nanosilica with high adsorption capacity for heavy metal ion removal. Iran Polym J (Eng Ed) 24(9):725–734. https://doi.org/10.1007/s13726-015-0360-1
Rahman FBA, Akter M (2016) Removal of dyes form textile wastewater by adsorption using shrimp shell. Int J Waste Res 6(3). https://doi.org/10.4172/2252-5211.1000244
Rather RA, Bhat MA, Shalla AH (2023) Development of sodium alginate-guar-gum-tannic acid based nanocomposite hydrogel beads for removal of malachite green dye. Mater Today Commun 36. https://doi.org/10.1016/j.mtcomm.2023.106358
Ribeiro AR, De G, Umbuzeiro A (2014) Effects of a textile azo dye on mortality, regeneration, and reproductive performance of the planarian, Girardia tigrina. http://www.enveurope.com/content/26/1/22
Saini RD (2017) Textile organic dyes: polluting effects and elimination methods from textile waste water. Int J Chem Eng Res 9(1). http://www.ripublication.com
Schwarzenbach RP, Egli T, Hofstetter TB, Von Gunten U, Wehrli B (2010) Global water pollution and human health. Annu Rev Environ Resour 35:109–136. https://doi.org/10.1146/annurev-environ-100809-125342
Scott DV (2000) Advanced materials for water handling : composites and thermoplastics. Elsevier Advanced Technology

Shah MP (ed) (2024) Emerging innovative trends in the application of biological processes for industrial wastewater treatment. Elsevier

Shang K, Li J, Morent R (2019) Hybrid electric discharge plasma technologies for water decontamination: a short review. Plasma Sci Technol 21(4). Institute of Physics Publishing. https://doi.org/10.1088/2058-6272/aafbc6

Shi Y, Xue Z, Wang X, Wang L, Wang A (2013) Removal of methylene blue from aqueous solution by sorption on lignocellulose-g-poly(acrylic acid)/montmorillonite three-dimensional cross-linked polymeric network hydrogels. Polym Bull 70(4):1163–1179. https://doi.org/10.1007/s00289-012-0898-4

Shiekh SS, Waqas Ahmad M, Wasim M, Sabir A, Shafiq M, Ullah Khan R (2023) Designing pectin and hyperbranched polyethylene imine based hydrogel for the removal and adsorption of ionic dyes. J Mol Liq 390. https://doi.org/10.1016/j.molliq.2023.122964

Sinha V, Chakma S (2019) Advances in the preparation of hydrogel for wastewater treatment: a concise review. J Environ Chem Eng 7(5). Elsevier Ltd. https://doi.org/10.1016/j.jece.2019.103295

Sivakumar R, Lee NY (2022) Adsorptive removal of organic pollutant methylene blue using polysaccharide-based composite hydrogels. Chemosphere 286. https://doi.org/10.1016/j.chemosphere.2021.131890

Solomon T (2001) The definition and unit of ionic strength. Res Sci Educ J Chem Ed. chem.wisc.edu 78(12)

Srikaew M, Jumpapaeng P, Suwanakood P, Kaiyasuan C, Promarak V, Saengsuwan S (2023) Rapid synthesis and optimization of UV-photopolymerized cassava starch-based superabsorbent hydrogels as a biodegradable, low-cost, and effective adsorbent for MB removal. J Ind Eng Chem 118:53–69. https://doi.org/10.1016/j.jiec.2022.10.045

Tang SCN, Yan DYS, Lo IMC (2014) Sustainable wastewater treatment using microsized magnetic hydrogel with magnetic separation technology. Ind Eng Chem Res 53(40):15718–15724. https://doi.org/10.1021/ie502512h

Thakur S, Govender PP, Mamo MA, Tamulevicius S, Thakur VK (2017) Recent progress in gelatin hydrogel nanocomposites for water purification and beyond. Vacuum 146:396–408. https://doi.org/10.1016/j.vacuum.2017.05.032

Thakur S, Sharma B, Verma A, Chaudhary J, Tamulevicius S, Thakur VK (2018) Recent approaches in guar gum hydrogel synthesis for water purification. Int J Polym Anal Charact 23(7):621–632. https://doi.org/10.1080/1023666X.2018.1488661

Varaprasad K, Nunez D, Yallapu MM, Jayaramudu T, Elgueta E, Oyarzun P (2018) Nano-hydroxyapatite polymeric hydrogels for dye removal. RSC Adv 8(32):18118–18127. https://doi.org/10.1039/c8ra01887a

Vargas AMM, Paulino AT, Nozaki J (2009) Effects of daily nickel intake on the bio-accumulation, body weight and length in tilapia (Oreochromis niloticus). Toxicol Environ Chem 91(4):751–759. https://doi.org/10.1080/02772240802541353

Vikrant K, Giri BS, Raza N, Roy K, Kim KH, Rai BN, Singh RS (2018) Recent advancements in bioremediation of dye: current status and challenges. Bioresour Technol 253:355–367. Elsevier Ltd. https://doi.org/10.1016/j.biortech.2018.01.029

Wach RA, Mitomo H, Yoshii F, Kume T (2001) Hydrogel of biodegradable cellulose derivatives. II. Effect of some factors on radiation-induced crosslinking of CMC. In J Appl Polym Sci 81

Wang L, Zhang J, Wang A (2008) Removal of methylene blue from aqueous solution using chitosan-g-poly(acrylic acid)/montmorillonite superadsorbent nanocomposite. Colloids Surf A 322(1–3):47–53. https://doi.org/10.1016/j.colsurfa.2008.02.019

Wang Y, Zhang C, Zhao L, Meng G, Wu J, Liu Z (2017) Cellulose-based porous adsorbents with high capacity for methylene blue adsorption from aqueous solutions. Fibers Polym 18(5):891–899. https://doi.org/10.1007/s12221-017-6956-7

Wang W, Bai H, Zhao Y, Kang S, Yi H, Zhang T, Song S (2019) Synthesis of chitosan cross-linked 3D network-structured hydrogel for methylene blue removal. Int J Biol Macromol 141:98–107. https://doi.org/10.1016/j.ijbiomac.2019.08.225

Wang W, Hu J, Zhang R, Yan C, Cui L, Zhu J (2021) A pH-responsive carboxymethyl cellulose/chitosan hydrogel for adsorption and desorption of anionic and cationic dyes. Cellulose 28(2):897–909. https://doi.org/10.1007/s10570-020-03561-4
Wang H, Li J, Ding N, Zeng X, Tang X, Sun Y, Lei T, Lin L (2020). Eco-friendly polymer nanocomposite hydrogel enhanced by cellulose nanocrystal and graphitic-like carbon nitride nanosheet. Chem Eng J 386. https://doi.org/10.1016/j.cej.2020.124021
Wang C, Feng X, Shang S, Liu H, Song Z, Zhang H (2023) Lignin/sodium alginate hydrogel for efficient removal of methylene blue. Int J Biol Macromol 237. https://doi.org/10.1016/j.ijbiomac.2023.124200
Weerasundara L, Gabriele B, Figoli A, Ok YS, Bundschuh J (2021) Hydrogels: novel materials for contaminant removal in water—a review. Crit Rev Environ Sci Technol 51(17):1970–2014. https://doi.org/10.1080/10643389.2020.1776055
Wong S, Ghafar NA, Ngadi N, Razmi FA, Inuwa IM, Mat R, Amin NAS (2020) Effective removal of anionic textile dyes using adsorbent synthesized from coffee waste. Sci Rep 10(1). https://doi.org/10.1038/s41598-020-60021-6
Yan H, Ho Row K (2006) Characteristic and synthetic approach of molecularly imprinted polymer. Int J Mol Sci 7:155–178. www.mdpi.org/ijms/
Yang L, Bao L, Dong T, Xie H, Wang X, Wang H, Wu J, Hao C (2023) Adsorption properties of cellulose/guar gum/biochar composite hydrogel for Cu^{2+}, Co^{2+} and methylene blue. Int J Biol Macromol 242. https://doi.org/10.1016/j.ijbiomac.2023.125021
Yaseen DA, Scholz M (2019) Textile dye wastewater characteristics and constituents of synthetic effluents: a critical review. Int J Environ Sci Technol 16(2):1193–1226. Center for Environmental and Energy Research and Studies. https://doi.org/10.1007/s13762-018-2130-z
Ye D, Chang C, Zhang L (2019) High-strength and tough cellulose hydrogels chemically dual cross-linked by using low- and high-molecular-weight cross-linkers. Biomacromol 20(5):1989–1995. https://doi.org/10.1021/acs.biomac.9b00204
Yin T, Zhang X, Shao S, Xiang T, Zhou S (2023) Covalently crosslinked sodium alginate/poly(sodium p-styrenesulfonate) cryogels for selective removal of methylene blue. Carbohydrate Polymers 301. https://doi.org/10.1016/j.carbpol.2022.120356
Yu K, Wang D, Wang Q (2018). Tough and self-healable nanocomposite hydrogels for repeatable water treatment. Polymers 10(8). https://doi.org/10.3390/polym10080880
Yue Y, Han J, Han G, French AD, Qi Y, Wu Q (2016) Cellulose nanofibers reinforced sodium alginate-polyvinyl alcohol hydrogels: core-shell structure formation and property characterization. Carbohyd Polym 147:155–164. https://doi.org/10.1016/j.carbpol.2016.04.005
Zhang Z, Abidi N, Lucia L (2023) Smart superabsorbent alginate/carboxymethyl chitosan composite hydrogel beads as efficient biosorbents for methylene blue dye removal. J Mater Sci Technol 159:81–90. https://doi.org/10.1016/j.jmst.2023.02.045
Zhang K, Luo X, Yang L, Chang Z, Luo S (2021) Progress toward hydrogels in removing heavy metals from water: problems and solutions a review. ACS ES T 1(5):1098–1116. American Chemical Society. https://doi.org/10.1021/acsestwater.1c00001
Zhijiang C, Ping X, Cong Z, Tingting Z, Jie G, Kongyin Z (2018) Preparation and characterization of a bi-layered nano-filtration membrane from a chitosan hydrogel and bacterial cellulose nanofiber for dye removal. Cellulose 25(9):5123–5137. https://doi.org/10.1007/s10570-018-1914-0
Zhu H, Chen S, Duan H, He J, Luo Y (2023) Removal of anionic and cationic dyes using porous chitosan/carboxymethyl cellulose-PEG hydrogels: optimization, adsorption kinetics, isotherm and thermodynamics studies. Int J Biol Macromol 231. https://doi.org/10.1016/j.ijbiomac.2023.123213
Zhuang Y, Yu F, Chen J, Ma J (2016) Batch and column adsorption of methylene blue by graphene/alginate nanocomposite: comparison of single-network and double-network hydrogels. J Environ Chem Eng 4(1):147–156. https://doi.org/10.1016/j.jece.2015.11.014
Zia H, Kumar A, Ishaque A, Parveen M, Nami SAA (2024) Synthesis, morphology and dye removal studies of ternary hydrogels bearing carboxymethyl cellulose, chitosan and glutamic acid. J Mol Struct 1298. https://doi.org/10.1016/j.molstruc.2023.137003

Chapter 9
Applications of Hydrogel in Modern Wastewater Treatment for Eradication of Heavy Metals in Municipal Sludge

Sumanta Bhattacharya

Abstract Hydrogels are being increasingly recognized as an effective solution for removing heavy metals from municipal sludge in wastewater treatment. These cross-linked hydrophilic polymer systems have a unique structure that allows them to absorb large volumes of water and selectively bind to toxic heavy metals. The functionalization of hydrogels enables the incorporation of functional groups that have a high affinity for heavy metal ions. Recent advancements in hydrogel technology have focused on enhancing their hydrophilicity, adsorption capability, and mechanical properties. The benefits of using hydrogels in wastewater treatment include their potential to be made from renewable materials, biodegrade, and recover heavy metals. However, challenges remain, including scaling up production, regeneration, and reuse, as well as addressing environmental concerns. To address the above-mentioned learning objectives, more consideration has to be accorded to questions concerning the economic possibility of hydrogel production, regeneration, and reuse, as well as potential environmentally sensitive aspects. Further research work is required in the area of life-cycle evaluations would go a long way in popularizing hydrogel technology. The formation of standards and guidelines for hydrogel utilization, with additional promotion of research and development, may lead to improved inventions in this field and ensure the security and efficiency of therapeutic processes.

Keywords Hydrogels · Regulatory framework · Wastewater · Heavy metal pollution · Graphene oxide · Municipal sludge

Abbreviations

HAP	Hospital-acquired pneumonia
COD	Chemical oxygen demand

S. Bhattacharya (✉)
Asian International University, Manipur, India
e-mail: sumanta.21394@gmail.com

S. Bardhan and M. P. Shah (eds.), *Applications of Hydrogels in Modern Wastewater Treatment*, Advances in Wastewater Research,
https://doi.org/10.1007/978-981-95-1598-1_9

BET	Brunauer-Emmett-Teller
SEM	Scanning electron microscopy
TEM	Transmission electron microscopy
FTIR	Fourier-transform infrared spectroscopy
NMR	Nuclear magnetic resonance
XPS	X-ray photoelectron spectroscopy
DSC	Differential scanning calorimeter
TGA	Thermogravimetric analyzer
AFM	Atomic force microscope
CNTs	Carbon nanotubes
VOCs	Volatile organic compounds
MWTPs	Municipal wastewater treatment plant
WwT	Wastewater treatment

9.1 Introduction

The onset of the industrial revolution and modernization over the past few decades has led to the generation of huge volumes of wastewater, mainly from urban centres. Of all the pollutants within the municipal sludge, the heavy metals of concern include lead, cadmium, mercury, and arsenic in view of their toxicity, non-biodegradability and potential to accumulate in the ecosystem. The conventional treatment methods of wastewater that include the physical chemical process of removing pollutants by organic and nutrient elimination fail to offer much in removing heavy metals that cause pollution and may lead to health complications. Wastewater management is therefore very important in order to minimize the risks to the general public and the environment (Hubeny et al. 2021). The heavy metals is another huge problem in wastewater treatment, especially in the case of municipal sludge. The European Union defines heavy metals, for instance lead, mercury and arsenic as hazardous substances that can be found in sludge, and if not well handled, they cause a lot of harm to the environment and human beings. Such GHG can pollute the soil and water and get into the food chain to affect humans and livestock. The problems associated with heavy metals in municipal sludge are complex as explained next. First of all, it is worthy to note that the metals are characterized by high density, immobility and non-biodegradable nature, which causes a lot of problems in sludge. Second, they are known to interfere with certain biological processes that are needed in the treatment of wastewater, hence slowing down treatment plants. Thirdly, the management of sludge with such compounds as heavy metals is a major concern since the disposal causes environmental pollution and legal implications. Last but not least, the regulation on the handling of heavy metals in sludge is normally elaborate and dynamic and in this way, creating a challenge for the treatment plants to conform to new standards or policies. Solving these issues entails the use of creativity and some of the solutions include aspects like treatment technologies, sludge management, and stakeholders in the different fields. In this light, it is possible to protect the environment and human health from adverse effects by adequately organizing heavy metals' management in municipal sludge. The effects of heavy metal pollution on the wastewater, obviously heavy metal pollutants, are hazardous chemicals that are potentially dangerous to human life, animals, and the environment (Kesari et al. 2021). Conventional treatment techniques such as chemical coagulation and sedimentation have been found to be ineffective in the removal of heavy metals from wastewaters. These methods usually possess low removal efficiency, and the generated sludge can contain high concentrations of heavy metals, making it to poses environmental risks and liabilities. Besides, conventional techniques are often energy-consuming, expensive and secondarily polluting, which makes them irrational and inefficient to combat the increasing problem of heavy metal pollution. Furthermore, the nature and content of wastewaters are diverse and rather intricate, and new and potent pollutants, so-called 'emerging pollutants', have appeared, and it became clear that the problem requires more flexible and efficient solutions. Prolonged application of conventional treatment techniques has called for innovations, biological treatment, sophisticated oxidation processes, and nanotechnology applications that show a potential of eradicating heavy metals from wastewater (Shah 2024). However, any more advanced research and improvements to the technology are still needed to improve upon the scalability and utility of such technologies in real-life applications. Control of heavy metal pollution in wastewaters and its impacts on the environment can only be addressed by changing from a traditional approach to a modern technique that is more efficient and flexible to adapt to the challenge and produce safer water and environment (Razzak et al. 2022).

New materials, which are from the field of materials science as hydrogels, have been identified as potential solutions for wastewater treatment. These are the hydrogels that are three-dimensional polymeric structures comprised of hydrophilic chains and can swell in water through rearrangement of their aqueous networks, while they can also selectively interact with dissolved metal ions. Because of this property, they are suitable for the removal of metal ions from wastewater, especially those that include heavy metals. Hydrogels also offer a significant advantage over other conventional treatment methods because they enhance metal ion removal with high effectiveness and efficiency. These surfaces can take up metal ions by possibly either of the following ways: electric attraction, coordination and exchange. Also, it is possible to design hydrogels in a way that enables the selective adsorption of metal ions, thus appropriately addressing the problem of contamination. Besides, they are reusable and/easily regenerated, thus have little to no impact on the environment. Hydrogels are a good candidate for wastewater treatment since they possess a high water absorption rate and have selective interaction with metal ions. With the current trends of research and development, more hydrogels will be utilized to treat the increasing cases of heavy metal polluted wastewater (Sinha and Chakma 2019).

9.2 Effects of Heavy Metals in Water and the Role of Hydrogel in Eradication of Heavy Metal Contamination

Another major problem of wastewater treatment is the heavy metals which have high toxicity and resistance to biodegradation. The sources of heavy metals in wastewater are diverse and can be categorized into two main groups: and sources of emissions of these gases to the atmosphere of anthropogenic and natural origin. They include the mining, smelting, and refining processes, domestic uses, and commercial battery disposal and pesticide usage. Natural sources include platform sources such as weathering and erosion that discharge heavy metals from the earth's crust. It is, however, vital to note that the prevalent heavy metals in wastewater include lead, mercury, cadmium and arsenic. Lead enters the environment mostly through industrial activities and disposal of lead-acid batteries, while mercury is connected to mining and metallurgical operations. Cadmium and arsenic are generally released in water through industrial effluents and pesticides, and arsenic comes as a product of natural geological activities and from mines and metal industries. The following are amongst the heaviest of metals known to cause severe health conditions and pollution of the environment, such as neurologically impaired disorder, failure in vital organs, and disruption of ecosystems. This paper presents information regarding sources and types of heavy metals in wastewater, and to understand it and comply with the issues associated with it, treatment and management of heavy metals are also discussed (Jaishankar et al. 2014).

Conventional techniques of heavy metals elimination in wastewaters involve physical, chemical, biological and combined processes. Sedimentation, filtration and electrochemical precipitation are some of the physical techniques that are used to remove body through the mechanisms of gravity separation, filtration and electrical current-induced precipitation. Chemical processes such as chemical precipitation, ion exchange, adsorption and oxidation/reduction employ the use of chemicals and materials for precipitation, exchange and reduction of soluble heavy metals to less toxic forms. Biosorption bioaccumulation and bioremediation introduce the concepts of using microorganisms and biomass to adsorb and accumulation and transform heavy metals (Vidu et al. 2020). Integrated systems, on the other hand, use several techniques in one system to achieve maximum removal of heavy metals, while AOPs, on the other hand, are complex systems, which utilize different processes in order to generate more powerful oxidation methods and finally, nanotechnology-based solutions, which use smaller particles than normal technology to accomplish the process. Recommended treatment methods include chemical, physical and biological for certain characteristics of wastewater, types and concentrations of heavy metals, as well as the existing laws in the area of interest (Cuerda-Correa et al. 2020). Hence, the present conventional methods range from moderate efficiency to high energy demands and cost, but constant research and development attempt to enhance and discover better approaches to heavy metal removal in the wastewater. Among the different methods used in the treatment of heavy Metals in wastewater, chemical precipitation, Ion exchange and adsorption stand out. Sedimentation in wastewater treatment involves the addition of chemicals to the wastewater so as to coagulate heavy metals, including lead, mercury and cadmium, among others. This process can be very useful, yet it usually requires extensive amounts of chemicals and may produce secondary pollution. Ion exchange, on the other hand, employs the use of resins to replace the bad heavy metal ions in the wastewater with less hazardous ions such as sodium or potassium. This method is good for the removal of heavy metals such as mercury and arsenic, but the resins used will get used up and need to be regenerated. Adsorption applies the use of adsorbents such as activated carbon, zeolites or biomass by having surface properties that will attract heavy metals. This has proved efficient in the elimination of a variety of large groups of heavy metals, such as lead, copper, as well as zinc, and may be cheaper than other techniques. Thus, there is an obvious disadvantage that the adsorbent material can get exhausted and needs to be regenerated or replenished. In general, the above-mentioned three methods could be efficient for the treatment of heavy metals in wastewater; however, their application depends on a number of factors such as the nature of wastewater, concentrations of heavy metals, as well as conditions of the process (Turkmen et al. 2022).

Hydrogels are defined as the soft, wet and flexible polymer products that have the characteristics of swelling in water and are able to absorb large amounts of water and are very useful in different applications in the environmental engineering field. This makes them good in water treatment and conservancy situations since hydrogels can swell to a hundred times their mass in water. Hydrogels also exhibit stimuli-responsive properties; they can swell in water and deswell in response to differences in temperature or pH or any other change in the environment for their application in sensors and actuators. Hydrogels can be synthesized to possess different degrees of mechanical stiffness, and thus can be useful in diverse applications (Thang et al. 2023). Depending on the formulation, the hydrogels can be designed to exhibit electrical conductivity, thus having potential use in sensors and energy storage devices. Depending on the composition and structure of hydrogels can be developed to be biocompatible, that is, to be used in biomedical as well as environmental applications (Hong et al. 2022). These properties make hydrogels suitable for use in environmental engineering, for instance, in the treatment of wastewaters and removal of pollutants and heavy metals from the water. Further, it is used in the reclamation processes of soil, where they are used to absorb and degrade the pollutants and in water management, where it is employed to absorb and release water in required amounts. They also offer future employment in detecting and measuring various pollutants in the environment, as well as in a mimicking style of cleaning of pollutants from the environment. In addition, hydrogels can be made biodegradable and non-toxic, thus, can be introduced in environmentally sensitive engineering applications as a green material. In conclusion, hydrogels show a great potential for countering a vast range of issues in environmental engineering; additionally, their attractive characteristics make the hydrogels a rather interesting object for research and development (Bahram et al. 2016) (Fig. 9.1).

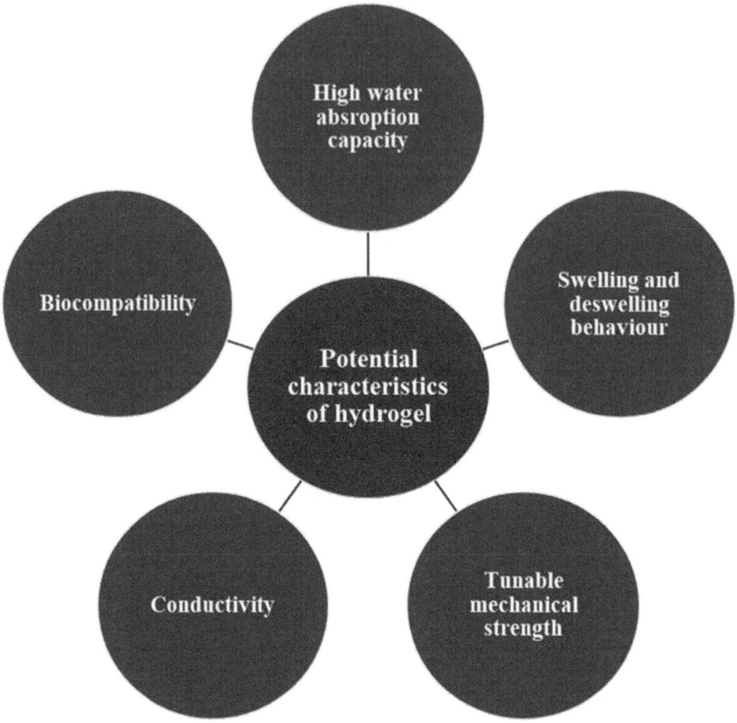

Fig. 9.1 Potential characteristics of hydrogel necessary for wastewater treatment

A lot of research has been done on the application of hydrogels for the removal of pollutants in water, especially in wastewater treatment. Wastewater treatment through hydrogel involves a process or a sequence of processes that enhances the overall purification of water. Firstly, the hydrogels capture the pollutants such as heavy metals, organic compounds and nutrients by adsorption due to weak forces of attraction and hold them within a network. Also, hydrogels swell and are capable of absorbing water, and thus, the pollutants get trapped by the large surface area produced. Here, chemical reactions also happen between hydrogels and pollutants, in the process converting them to forms that are less hazardous or that are easier to wash off. In addition, hydrogels also work as a filter that prevents the passage of particles and bulk phase impurities through the hydrogels' matrix (Chang et al. 2023). There is also biological degradation, in which hydrogels encourage the growth of microorganisms that break down organic pollutants. Furthermore, it is noted that hydrogels may translocate with ions of pollutants and remove them from the water; in addition, they form complexes with pollutants, which makes it easier to remove them. Last of all, certain hydrogels are photoreceptive and produce reactive oxygen species that destroy pollutants by photocatalysis. As such, with all these combined mechanisms, it can be said that hydrogels have the potential for treating wastewater since they can absorb almost any pollutant or contaminant known to man. Some of the possible heavy metals include lead, mercury and arsenic that have been found to be effective in water purification through hydrogels based on adsorption and absorption processes. Adsorption, absorption and degradation are the main end-point processes that have been used by hydrogels in the elimination of organic pollutants like dyes, pharmaceuticals and pesticides within wastewaters. A common use of hydrogels has been in the elimination of excess nutrients such as phosphorus and nitrogen from the wastewater, reducing the eutrophication effect. The removal of bacteria and viruses has been demonstrated by hydrogels and hence can be used in wastewater disinfection. One study has demonstrated that hydrogels have the ability to lower chemical oxygen demand (COD), which suggests the hydrogels' ability to eliminate organic content in wastewater. Hydrogels, in numerous cases, are reusable with the capacity to regenerate, therefore making the concept environmentally sustainable for wastewater treatment. When hydrogels are incorporated with other treatment technologies like membranes or biological systems, there are observed improved removal processes are observed along with mutual interaction. Advanced research works have found the use of hydrogel to be viable at an industrial scale in its application to the treatment of wastewaters. These developments show the resilience of hydrogels as materials enabling the handling of diverse issues related to wastewater treatment. Further HAP research is oriented toward improving the characteristics of hydrogels, expanding the range of disease interventions utilizing hydrogels, and incorporating them into current therapeutic practices (Ahmaruzzaman et al. 2023).

Chemical synthesis of hydrogels is made through polymerization of monomers, forming a polymeric network that creates hydrogels. The examples of biopolymers employed in the formation of hydrogels are cellulosic materials, chitosan, alginates, while the examples of synthetic polymers are polyacrylamide, polyacrylic acid, polyethylene glycol among others. Various cross-linking techniques, including chemical cross-linking, physical cross linking and radiation cross-linking, are used to form a stable and porous hydrogel network. The two major types of cross-linking are chemical one where cross-linking agents such as glutaraldehyde, formaldehyde or epichlorohydrin are used to connect the polymer chains. For the instance of physical cross-linking, it relies on a change in temperature or pH to form a physical interconnection of the polymer chains. Radiation cross-linking attempts to cross-link the polymers by making use of ionizing radiation in order to form the covalent bonds among the polymer chains. Modification techniques are employed with the aim of improving the selectivity and extent of metal adsorption through the addition of functional groups that can form a coordination complex with metal ions. Some of these methods include grafting, copolymerization and surface modification. Grafting means that functional groups are anchored onto the hydrogel main chain, while copolymerization means that MP-ligand monomers are integrated into the hydrogel structure. Another process is the process known as surface activation, which entails the use of chemicals on the hydrogel surface that react with the metal ions. Consequently, techniques such as dopamine coating, polydopamine coating, and bio-inspired modifications are intended for the fabrication of hydrogels with great metal adsorption capabilities. These modified hydrogels have higher adsorption capacities of heavy metals such as lead, mercury and arsenic, making them prospective materials for water treatment (Bashir et al. 2020) (Fig. 9.2).

It will be remarkable to establish and quantify properties such as mechanical, thermal, electrical, and optical properties, which are determinant to the performance of hydrogels in different fields. In order to assess the swelling ability of the synthesized hydrogels, the samples are placed in water with subsequent measurement of their weight change or volumetric increase. The general BET method, like nitrogen adsorption, is considered useful in determining the surface area of the hydrogels. Porosity is measured by a mercury porosimeter or electron microscopy, which can be either scanning electron microscopy (SEM) or transmission electron microscopy (TEM). Other measurable characteristics are tested through rheology or tensile and/or

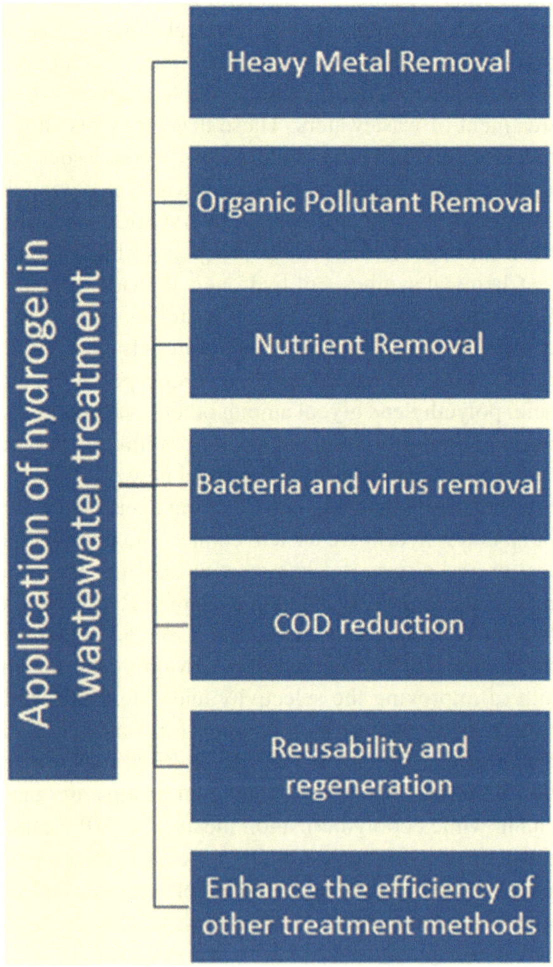

Fig. 9.2 Application of hydrogel in wastewater treatment

compression tests. They are determined by Fourier-transform infrared spectroscopy (FTIR), Nuclear Magnetic Resonance (NMR) spectroscopy or X-Ray Photoelectron Spectroscopy (XPS). Thermal properties are analyzed by Differential Scanning Calorimeter (DSC) or Thermogravimetric Analyzer (TGA). The surface charge of hydrogels is analysed through the zeta potential of the hydrogel samples. The moisture content is found by using the thermogravimetric analysis (TGA) or by drying techniques. The microstructure is investigated with a Scanning Electron Microscope (SEM), Transmission Electron Microscope (TEM) or Atomic Force Microscope (AFM). Lastly, degradation behaviour is studied through enzymatic or hydrolytic degradation assays. Thus, the use of the described methods of characterization allows one to develop an accurate picture of what hydrogel characteristics are needed for certain applications and how these characteristics can be achieved through the modification of gel precursors and fabrication approaches (Azeera et al. 2019).

9.3 Roles of Hydrogel in Improvement of the Wastewater Treatment Technique

Hydrogels are gradually being incorporated into the current wastewater treatment system with emphasis on the removal of heavy metals in municipal sludge. Most of these materials have the ability to absorb large amounts of water and can be tailor-made to capture only certain metals, making them useful in depolluting water (Alsaka et al. 2025).

Functionalization of hydrogels can also be done in a way that they can selectively bind to specific chemical groups that prefer specific heavy metals, including lead, cadmium, mercury and chromium. Such groups include carboxyl, hydroxyl and amine groups, which have a tendency to form chelates and sequester metal ions in wastewater. Hydrogels are characterized by their large surface area to volume ratio and thus possess a large water and heavy metal uptake capability. Due to this property, they earn a name for being efficient in the treatment of large volumes of wastewater. For instance, carboxyl, hydroxyl and amine groups can be incorporated into the structure of hydrogel and these functional groups chelate metal ions such as lead, cadmium, mercury and chromium. This chelation process includes the formation of steady complexes between the metal ions and the functional groups, which neutralize the heavy metals, thus cleaning the water. In addition, due to the developed porous structure and a highly developed surface, hydrogels can absorb large amounts of water and heavy metals, so their use can be justified for the treatment of large flows of wastewater. Therefore, hydrogels have proven to be very effective in the removal of heavy metals from wastewaters, thus providing a remedy in minimizing the effects of such toxic substances on the environment and human health. Thus, using such characteristics of hydrogels as absorbency, biocompatibility, and self-healing ability, it is possible to effectively contribute to the creation of wastewater treatment processes and technologies (Zhu et al. 2023).

These hydrogels are able to be regenerated with the help of changing pH or with the help of some chemical reagents, so the heavier metals will be released and can be used for the regeneration of the hydrogel. This leads to the saving of a lot of resources, which makes the process more economical. These hydrogels provide a method for pollution control as they capture the heavy metals and an additional bonus is that the metals can be reclaimed and recovered for further use. This also means that the heavy metals could be retrieved, as well as the hydrogel, for further use since it is recyclable in nature. Additionally, the heavy metals extracted can be recycled for use in several sectors such as manufacturing or construction, therefore are an opportunity to turn waste into worth. Besides offsetting the environmental cost of heavy metal pollution, it also offers the advantages of resource recycling and recovery. For this reason, hydrogels are beneficial not only for pollution control but also for recycling of the resources to a great extent which gives a big advantage for sustainable wastewater treatment. Through the incorporation of hydrogels with self-healing properties, various industries will be in a position to reduce cases of wastage while, at the same time, enhancing the usage of available resources.

There are also opportunities to incorporate hydrogels with other materials, such as activated carbon or membranes, to increase higher functional capacity of their application in filtration systems. Such integrated systems can offer improvements in the mechanical characteristics of the filter as well as in the removal performance of the heavy metals. They can also be utilized as one or more of the components of filter beds; hydrogels act as media that trap heavy metals as water goes through a filter bed. For instance, hydrogels can be combined with activated carbon, which is widely known for its superior adsorption properties, to produce a highly effective filtration medium for a higher capture rate of such metals. Further, hydrogels can be incorporated into the membrane, which can serve as a strong and selective barrier that will not let heavy metals go through and at the same time, permit water to pass through. These kinds of hybrid arrangements can afford enhanced mechanical strength, durability and removal effectiveness of the heavy metals, therefore qualify for use in industrial wastewater treatment systems. In addition, hydrogels can be used in the filter bed since they are a material that, when placed in a matrix, enables one to filter heavy metals for unbroken periods of time as water passes through the filter. When incorporated with other materials, hydrogels present endless possibilities of creating new and efficient filtration systems to deal with the complexity of heavy metal removal in wastewater (Hossain et al. 2023) (Fig. 9.3).

Hydrogels can be utilized to condition municipal sludge, which results in enhanced dewaterability and hence smaller volumes of sludge. This process not only removes heavy metals but also enhances the physical and chemical characteristics of the sludge to make it more manageable as well as disposability. In the treated sludge, the hydrogels are able to capture the heavy metals and thus restrict their seepage in the environment within a period of time when the sludge is either dumped or used for manure. Besides achieving the goal of desorption of heavy metals, the steps involved in the process also make it possible to change the nature of the sludge to a desirable form that is easy to handle. Moreover, hydrogels have the ability to encapsulate heavy metal ions within the treated sludge, and thus, restrain the metals

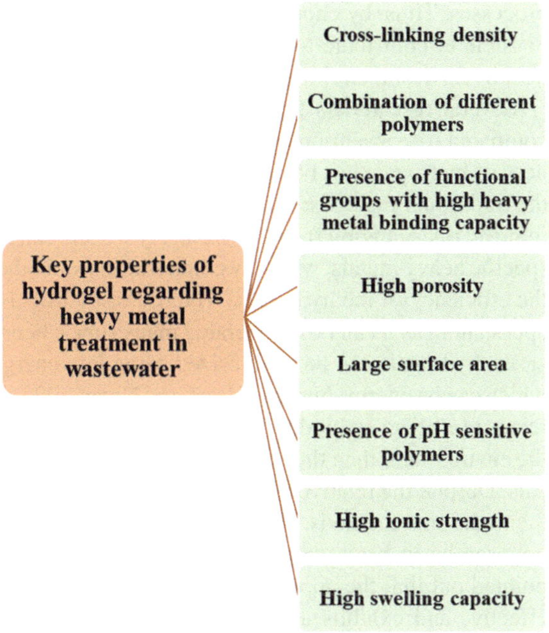

Fig. 9.3 Key properties of hydrogel regarding effective heavy metal treatment in wastewater

from releasing into the ecological system on application for sludge disposal or as a manure. This is important as a number of such metals are highly toxic to the environment and human health when present in water sources or as soil pollutants. In effect, hydrogels prevent the movement of such toxic substances to other areas, thus making the treated sludge safer for disposal and reuse, hence minimizing the social cost of sludge management. However, this is a new approach that can save economic costs when it comes to sludge treatment and disposal hence making it revolutionary in the wastewater industry (Darban et al. 2022).

Some hydrogel applications can be biodegradable which makes it easy for them to decompose after use, thus reducing the mess left behind. They hold the idea that hydrogels, when used in wastewater treatment, are eco-friendly because they can minimize the dependency on strong chemicals and energy costs. Unlike other artificial materials, most of the hydrogels have been developed to disintegrate after their use, and this makes their impact on the environment low. This biodegradability also means that hydrogels do not further complicate the issue of the ever-rising microplastic pollution in water systems. Similarly, the application of hydrogels in wastewater treatment is environmentally friendly as it is a civil and sustainable method compared to severe and energy-consuming methods. It is evident from the above discussion that through the application of hydrogels, the industries can minimize the use of toxic chemicals and minimize the energy demand in treating the wastewater, hence making it a sustainable process for the environment. Therefore, application of hydrogel in wastewater treatment offers a plausible way of changing the face of this industry by providing newer, efficient and environmentally friendly processes. Thus, by choosing hydrogel technology as one of the major focuses, it is possible to contribute to the salvation of the planet's water resources and create a healthy environment for the next generations (Garcia-Garcia et al. 2024).

Several researchers have established that hydrogels provide high efficiency as compared to conventional treatment methods in the following ways. The effectiveness of hydrogels has been reported to eliminate up to 99% of heavy metals, while the traditional methods, such as chemical precipitation and adsorption, have shown removal rates ranging from 50–80%. Hydrogels can be tailored to selectively remove specific heavy metals, whereby competition from other ions is eliminated, enhancing the efficiency of the hydrogels. The use of hydrogels doesn't have to be a one-time application, as it can be regenerated many times, hence cutting on expenses. In many cases, hydrogels can be synthesized using less energy than conventional processes, which can be energy-hungry (Visan and Negut 2025). The use of hydrogels as a slow-release fertilizer is biodegradable as well as non-toxic; thus, it has less impact on the environment than the conventional methods that require the use of strong chemicals. Despite the relatively high cost, hydrogels yield significantly better results than the traditional methods, especially in large-scale use. The treatment's use of hydrogels can be in batch, continuous, as well as column systems. All in all, it can be pointed out that the application of hydrogels for heavy metal elimination is highly effective and exhibits a number of advantages, including enhanced performance, selectivity, and eco-friendliness as compared to conventional treatment techniques. Nevertheless, a number of the issues associated with hydrogel limitations require actual efforts and additional investigation in order to enhance the efficiency of the material (Flores-Valenzuela et al. 2023).

The properties of the hydrogels have a direct effect on the efficiency of the adsorption of the heavy metals. An analysis of the results infers that samples with higher cross-linking density will have low adsorption capacity because less reactive surface area is available for interaction. It is worth noting that the affinity of different polymers toward heavy metals is different. For instance, polyacrylamide and polyacrylic acid are efficient in the removal of metals such as lead and copper. The heavy metal binding can be enhanced by functional groupings such as carboxyl, hydroxyl and or amine functional groups. To some extent, adsorption capacity in CNTs can also be enhanced by increasing porosity because of an increased number of attachment sites. Increased surface areas are beneficial in the case of the adsorption of heavy metals. This type of hydrogel may involve the use of pH-sensitive polymers with a tendency to influence the adsorption of heavy metals in defined pH levels. Thus, it has been established that hydrogels with high ionic strength enhance HE adsorption through the increase in electrostatic interactions. Hydrogels that have high swelling ability have the capacity to adsorb more of the heavy metals. The proposed study will help identify the specific properties of each type of hydrogel and how they affect the adsorption of heavy metals, and then, the respective hydrogels could be further improved for use in their intended purpose; in this case, removal of heavy metals from wastewater (Jiang et al. 2019).

Hydrogel treatment also changes the sludge parameters, where the water content is reduced and the solid content increased thus making easily be manage and dispose of. The process also enhances the dewatering characteristics of the sludge through the hydrogel absorption process, which in turn has reduced the treatment cost and has less adverse effects on the environment. In addition, the hydrogel treatment can filter out the heavy metals, which constitute a menace to lives and the environment and change the pH levels that may impact the microorganisms and the nutrient availability. It also affects the Microbial community, whose implication is on the biodegradation processes and increases biodegradability by degrading large carbon chains of the organic matter. Further, hydrogel sludge treatment decreases the emission of volatile organic compounds (VOCs) that have negative effects on the environment and public health, as well as enhances the stability of the sludge, thus preventing its re-suspension and pollution of the environment. In a nutshell, hydrogel treatment can contribute to sustainable sludge management since it allows for resource recycling from sludge, such as nutrients or energy. This information helps researchers and practitioners to perfect the hydrogel treatment process, thereby enhancing sludge management as well as resource recovery (Sava et al. 2024) (Fig. 9.4).

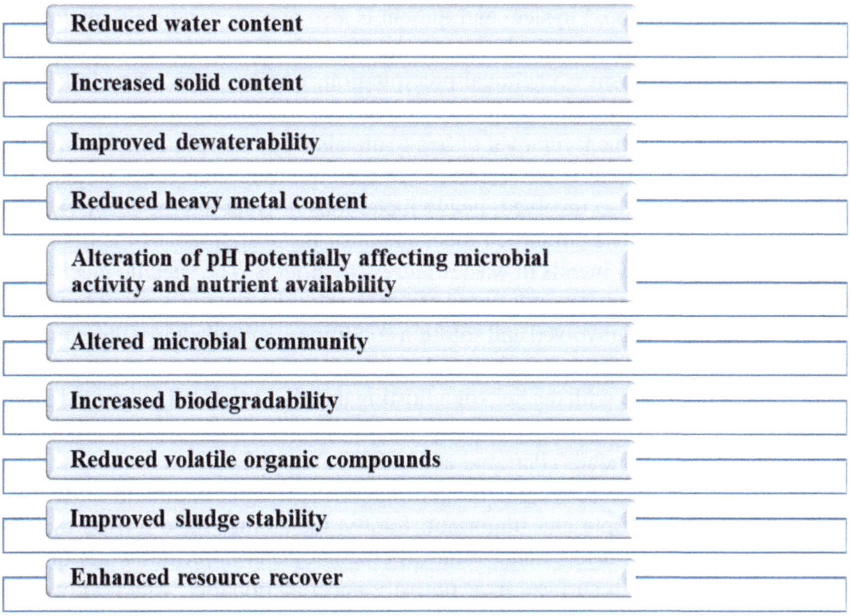

Fig. 9.4 Altered characteristics of sludge after treatment with hydrogel

9.4 Effectiveness of Hydrogel-Based Wastewater Treatment Process Over the Conventional Wastewater Treatment Processes

Specifically, hydrogels are proficient in the removal of heavy metals based on a synergistic interaction of physical and chemical processes. Of particular importance is the force of electrostatic attraction since metal ions, especially those of relatively large atomic numbers, are attracted to charged functional groups like carboxyl or amine functionality in hydrogel. These functional groups play the role of coordination sites to enable the formation of strong coordination bonds with the heavy metal ions in the hydrogel. Moreover, practical applications of hydrogels include selectivity through the ability of the metal ions to interact with the ligands through chelation with polyacrylic acid. Another mode of uptake of heavy metals by the hydrogels is through Ion exchange. This is achieved through ion exchange, in which heavy metal ions unseat other ions, such as sodium or calcium, as members of the hydrogel matrix (Yang et al. 2019). This makes it possible for the hydrogel to be used for adsorbing heavy metals from the solution. Physical entrapment is also observed, whereby metal ions are locked within the pores of the hydrogel and cannot be released back into the solution. Another mechanism is chemical precipitation, which occurs when functional groups of the hydrogel interact with heavy metal ions to form insoluble precipitates. This results in a permanent removal of heavy metals from the solution, thus making water less toxic. Van der Waals forces, hydrogen bonding, and other weak intermolecular forces are also involved; heavy metal ions interact intermolecularly with the hydrogel matrix. The combined mechanism of these approaches facilitates the adsorption of the heavy metals in wastewater by hydrogels. The specific mechanism offered by a given system depends on several factors, including the types of hydrogel, the type of heavy metal involved and solution conditions. Knowledge of these mechanisms is therefore imperative in tailoring the hydrogel in its efficiency on heavy metal adsorption. Altogether, with the help of adjusting the properties and functional group of the hydrogel, the researchers can synthesize excellent and selective adsorbents for the removal of heavy metals (Perumal et al. 2021) (Fig. 9.5).

The advantages of employing hydrogels ahead of the other treatments to remove heavy metals are multiple and important. Firstly, the studies have proved that the synthesis of hydrogels offers better removal efficiency and improved water quality. This is because of high surface area, porosity and functionality, which allows the MOFs to selectively interact with the heavy metal ions. Also, in terms of cost-effectiveness, hydrogels are more economical than conventional treatment processes. They usually cost less to make and run than their disposable counterparts and are reusable or reclaimable in most instances, giving rise to decreased expenses and waste products. This makes them ideal for use by industries as well as municipalities that are searching for ways of minimizing their impact on the environment as well as cutting costs. The other advantage of hydrogels is their safety to the environment. In contrast to those currently employed, which may include the use of toxic chemicals and produce toxic wastes, hydrogels are non-toxic, biodegradable

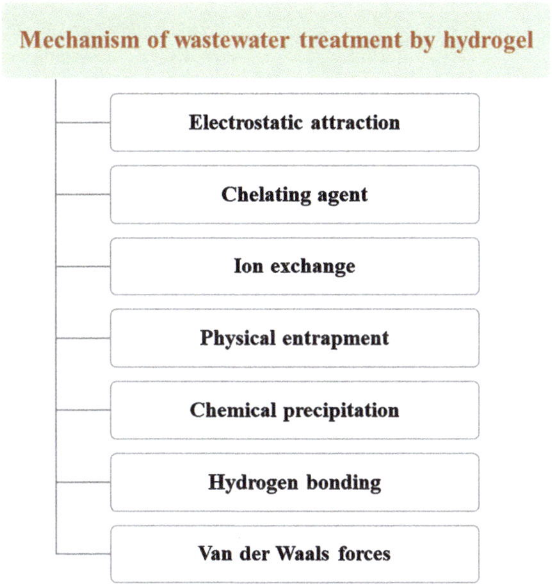

Fig. 9.5 Mechanism of wastewater treatment by hydrogel

and non-hazardous (Sharma et al. 2025). For this reason, they are more environmentally friendly and here refer to the well-being of the human being. Moreover, hydrogels can be selective to those heavy metals without being affected by other ions and, therefore, exhibit increased efficacy. Such selectivity is especially beneficial in industrial processes, where several heavy metals may be present. Other advantages of hydrogels include better sludge management, where the amount of sludge produced is less and the quality of sludge is also better. This makes sludge management easier and less expensive and recovery of valuable resources like the heavy metals can also be achieved. Therefore, hydrogels are a more effective, economical and eco-friendly system for the removal of heavy metals than the traditional processes. Their special features, selectivity and recyclability make them a promising method for usage in various sectors such as industrial wastewater treatment and in environmental cleanup (Mu et al. 2020) (Fig. 9.6).

In capturing heavy metal ions, hydrogels present several benefits; however, there are some barriers and limitations of hydrogel application to be considered. One of the concerns is regeneration—although in the case of using hydrogels, it might be needed to regenerate the material after it has been saturated with heavy metals, which may be a lengthy and expensive process. Furthermore, hydrogels may not be biodegradable, hence they may not have a good environmental influence and disposal will be a problem. Another limitation is the scalability because it might be challenging to scale up hydrogels for industrial-level uses. This can restrict its application, particularly where large volumes of wastewater have to be treated within industries (Song et al. 2022). Additionally, the hydrogels may not bear high stress conditions, for example, a change in pH value or temperature of the environment, and

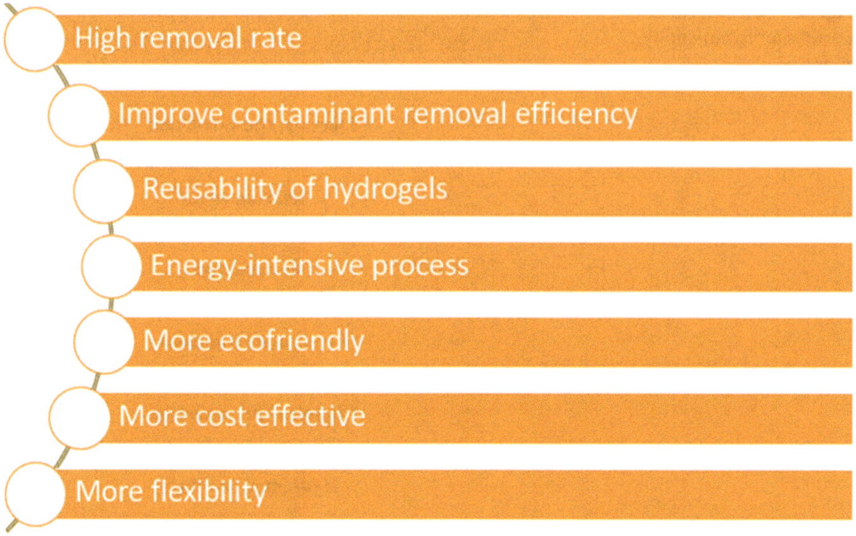

Fig. 9.6 Benefits of hydrogels over traditional wastewater treatment processes

thus reduce the efficiency and durability of the hydrogels. Another disadvantage is the aspect of selectivity: the group of hydrogels may not be very selective and capture heavy metals specifically in wastewater. It can result in the loss of non-target metals, and this will decrease the rate of treatment efficiency. There are some disadvantages, which are associated with fouling; hydrogels can be easily fouled and can lose their properties and efficiency and thus may need replacement after regular intervals of time. Last but not least, cost emerges as the final difficulty, as hydrogels can cost more than regular treatment technologies, so companies and municipalities would not be eager to employ them. Yet, several challenges exist to the use of hydrogels in research, which include the design of reutilizable hydrogels, use of biodegradable hydrogels, huge-scale application, poor stability, low selectivity, fouling issues and cost factor. Hydrogels can be made more effective and more realistically fit for wide use, especially in the removal of heavy metals in different industries, by overcoming these challenges and limitations (Khoo et al. 2024).

9.5 Challenges of the Application of Hydrogel in Wastewater Treatment

The major issue is how to retain the uniformity and high quality of the hydrogel material, particularly around the manufacturing of a large number of products. In laboratory and pilot-scale experiments, hydrogels are prepared in small quantities so that the composition of hydrogels and their properties are easily controllable.

However, often, during the scaling up of production, controls may become less stringent, and this is important when it comes to the use of hydrogels in treating wastewater, as their effectiveness mainly depends on the extent to which pollutants have been removed beforehand. One of them is the requirement of actual equipment and facilities, which are necessary for the industrial handling of large volumes of hydrogels. In a large municipal wastewater treatment plant, use of hydrogels would imply pumping, mixing and separation of hydrogels from the product water, which would call for major investment in new equipment and facilities (Alonso et al. 2021). Also, some characteristics of the hydrogels could be influenced by the storage conditions and handling, which may affect the performance ability of the hydrogels. Additionally, if the application of hydrogels is increased, then the demand for some of the raw materials required for the preparation of hydrogels will have to be ramped up, which may eventually show inefficiency in the supply chain and may lead to increased cost. Furthermore, large-scale manufacturing of hydrogels would pose an environmental issue to control and regulate before embarking on large-scale production. Last of all, the incorporation of hydrogels into conventional processes for wastewater treatment would call for some radical alterations to plant functioning and management. Operators of plants where hydrogels are to be used would require sensitization as to how to handle the hydrogels in order to avoid contamination and members of the maintenance crew would need to be trained on how to solve any problems that may arise with the hydrogel systems. Accumulatively, hydrogels have proven to have the potential for wastewater treatment with certain degrees of limitations, which require extensive research and development for MWTPs (Persano et al. 2024) (Fig. 9.7).

One of the most important challenges and a disadvantage of using hydrogels is the cost involved in the production and distribution of large amounts of hydrogels. At the moment, hydrogel manufacturing can be considered quite costly, primarily because of the expensive reagents and challenges associated with synthesizing the product. Third, the large-scale production of hydrogels from lab to industrial scale is not without issues, such as quality and property control, as well as manufacturing costs. In order to overcome these difficulties, researchers as well as manufacturers should work on the production of meteorites at a lower cost and should find ways and means of reducing the costs of synthesis, using cheaper and eco-friendly starting materials, and enhancing the manufacturing process. Moreover, decreasing the cost of larger-scale PHA in wastewater treatment plants remains a challenge; to achieve this, proper and efficient techniques of hydrogel application and methods of its dosage and placement must be found and the costs associated with its maintenance and replacement also have to be reduced. In addition, the reuse and recyclability of hydrogels could tremendously decrease the overall costs and therefore make them more complied to traditional treatment processes. Also, the revenue from recovering valuable commodities, which include heavy metals and nutrients, could help in paying for the cost of developing and releasing hydrogel. In general, therefore, minimizing the cost of manufacturing and delivering the hydrogels on a commercial scale remains a vital agenda for the same to be adopted as a supplement or replacement to other conventional methods of treatment. If such difficulties are solved and

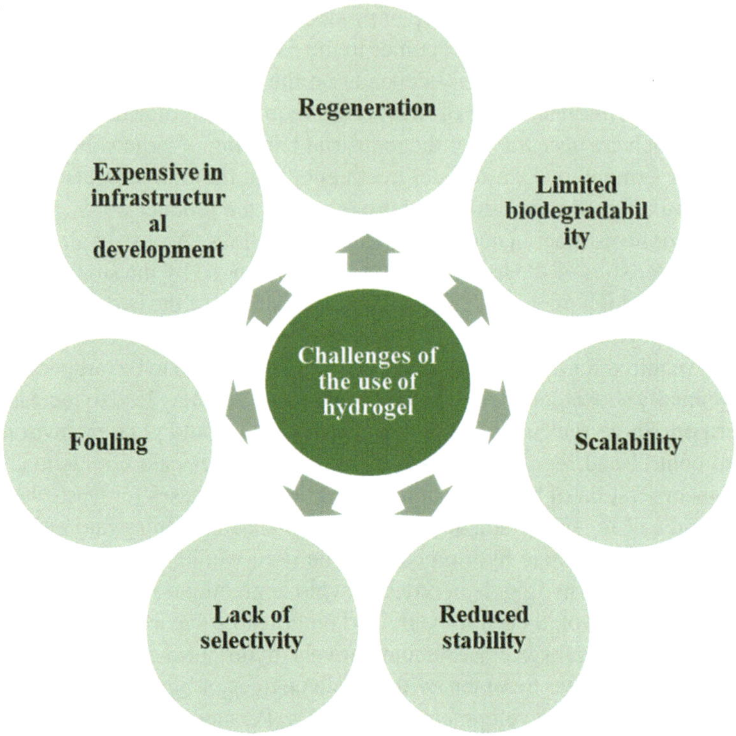

Fig. 9.7 Challenges of the use of hydrogel in wastewater treatment

efficient and cheap methods of producing hydrogels are found, they can be more effective in removing contaminants from wastewater (Trombino et al. 2023).

Scientists are still developing stimuli-responsive hydrogels that are capable of responding directly to the pH or temperature of the environment in order to effectively remove the heavy metals in wastewater. New generations of these hydrogels are being developed to show viscoelasticity changes when exposed to predefined stimuli in order to adjust properties and functionality as needed. For example, an overall structure of a hydrogel could respond to the change in pH by swelling or shrinking and thus be able to store or release the heavy metal ions. Smart hydrogels come in a variety of types, depending on the approach used to prepare them, one of which is by using stimuli-responsive polymers that undergo conformation changes upon exposure to certain environmental stimuli (Koetting et al. 2015). For instance, researchers might employ a change of pH-responsive polymers where the hydrogel alters its absorption rate of the heavy metals or the release of the same. The third approach involves incorporating the sensors or actuators into the hydrogel matrix for it to function as a sensing and responding system. This could entail responsive features, such as when the level of heavy metal is sensed by a device, then the hydrogel form responds by changing its level of absorption. Smart hydrogels will be the most helpful in improving the productivity and efficiency of heavy metal removal from the wastewater. These advanced materials are capable of adapting to their physical environment, which results in enhancement of efficiency in terms of removal rates, energy utilization, as well as operating expenses. Also, smart hydrogels could be used in display applications and directly monitoring water quality in real time and feedback control of WwTs (Zhang et al. 2021).

9.6 Future Perspectives

One possible idea for further research is the creation of new hydrogel composites containing additional substances to improve the metal adsorption capability. This could involve the new development of materials for incorporation into the synthesis of hydrogel with better selectivity, stability and regeneration properties. The last development area is the coupling of hydrogels with another treatment system, the membrane filtration process or the biological treatment process, so that the best of both worlds can be enjoyed.

Another important research area is the scaling up of hydrogel treatment systems for large-scale pilot studies and specific field applications. This would include evaluating the steady state and self-healing capacity of hydrogels in practical application settings, as well as enhancing system development and configuration for the best operating environment. Another interesting direction of research is the improvement of the selectivity and specificity of hydrogels towards target heavy metals in complex water matrices. This could also include the synthesis of new functionalised hydrogels or possibly enhancing functionalized hydrogels that are in existence to improve their efficiency when it comes to capturing specific heavy metals.

Biodegradable hydrogels can be viewed as another interesting field of study and development since they are more ecological as a treatment method. Modelling and simulation can also be used significantly in prioritizing hydrogel treatment systems and in modelling their probable performance. Real-world data from field trials of hydrogel treatment systems and implementation of such systems in typical municipal sludge treatment plants is, therefore, crucial. Last but not least, the cost–benefit analyses will provide an overall assessment of the economic commercial feasibility of hydrogel treatment systems. Such research directions, proposed in this review, can contribute to the further development of hydrogels' usage in the context of contemporary wastewater treatment and, in broader terms, help to progress more in the elimination of heavy metals in municipal sludge.

Dependable modelling and simulation form one of the most significant research avenues in enhancing the hydrogel treatment systems and calculating their feasibility. This entails the establishment of mathematical models of the hydrogel treatment process, system characteristics, wastewater content and conditions of operation. These models can then be employed to model different situations, assess the results of treatment and design the best system and operational parameters. Through constructing sound models and simulations, researchers are able to examine and assess the influence of those various parameters, which include but are not limited to hydrogel dosage, contact time and pH. It can be used for the enhancement of the system design, diminishing the costs of experiments, and for the facilitation of the process of developing proper treatments. Also, models can effectively translate the hydrogel treatment systems from laboratory scale to pilot scale through models, thus making it easy to implement the technology. In addition, the models and simulations help determine whether the hydrogels used in treatment systems would be stable and would regenerate in the long run. This could also assist in defining potential requirements of maintenance and in tuning up the schedule for maintenance of these systems. All in all, by building the models and simulations with adequate validation, the researchers can construct a very useful tool to analyse the hydrogel treatment systems design and performance, as well as to enhance the usage of hydrogels in the modern wastewater treatment technology.

Field trials and actual implementation of hydrogel treatment systems in operational municipal sludge treatment plants will, therefore, be essential to establish the effectiveness and realities of this technology. Field tests include making the hydrogel treatment system in an actual environment and with the Municipal sludge and operating conditions. This enables the researchers to determine the capabilities of a system and the areas where there could be probable problems in determining effective system designs and system implementations. In the case of municipal sludge treatment plants, the operation of the hydrogel treatment system entails having it incorporated into the pre-existing treatment processes. A successful implementation of this needs a close working relationship with plant operators, engineers and other stakeholders. Through field trials and practical application of hydrogel treatment systems, they are also able to provide an avenue for proving not only the effectiveness of the concept in the removal of heavy metals in municipal sludge but also reducing its impact on the environment and enhancing the operational efficiency of treatment systems. Furthermore, both field trials and implementation also offer chances to train plant operators, establish maintenance procedures, and optimize the design and operation of the system generated from authentic performance experience. The positive results obtained by the field trials and implementation of the hydrogel treatment systems can go a long way in inspiring many more industries to embrace the systems as part of wastewater treatment methods, henceforth working towards the enhancement of the environment as well as people's health. Thus, the decision of taking this step allows researchers to take the technology from the laboratory setting into a real-world context where positive impacts are possible.

Cost and cost-effectiveness assessments are crucial in the assessment of the viability of hydrogel-based treatment systems. This involves a planned approach of evaluating the costs that are incurred in developing the hydrogel treatment systems and the cost of operating as well as maintaining the system, besides the cost of disposal. Similarly, the value of hydrogel treatment systems in the form of averted costs and/or tangible benefits such as in the form of the heavy metal removed, the reduction in environmental impacts, efficiency improvement, and the possibility of generating income through recovered resources. This way, the researchers can ascertain a comparative benefit–cost analysis of hydrogel treatment systems and learn of the factors that could be enhancement. Consequently, this analysis is useful in decision-making for stakeholders, particularly the wastewater treatment plant operators, policymakers, and industries. The overall costs and benefits could present an evaluation of the effective and properly proportional scale of hydrogel treatment systems to the environment and proper utilization of this type of technology without possible negative repercussions. Moreover, carrying out the evaluation of costs and benefits, it is possible to compare the effectiveness of hydrogel treatment systems with traditional treatment technologies. Thanks to this, it will be possible to promote further development of more sustainable and efficient methods of wastewater treatment; this will have a positive impact on the environment and the population. Great emphasis has been placed on conducting a feasibility study of the proposed hydrogel treatment systems, whereby the insights gained will help advance research and application of this technology in wastewater treatment.

The inclusion of hydrogel treatment systems with renewable energy options is an innovative way of going about sustainable wastewater treatment. This amalgamation aims at integrating the environmental impacts of hydrogel technology and renewable energy technologies so as to develop a system with fewer impacts on the environment. This aspect will minimize the use of fossil fuels in the treatment process and our preferred hydrogel and energy sources, such as solar, wind or biogas energy, will enhance the reduction of greenhouse gas emissions. In addition, incorporating the hydrogel treatment systems with other renewable power sources can help in cutting the costs of power and make the system more efficient. For example, in solar hydrogel systems, thermal energy can be employed for the removal of heavy metals, while in wind systems, electrical energy can be used in the treatment processes. Furthermore, the biogas produced in the wastewater treatment process can be utilized for operating the hydrogel systems, thus making it an energy self-sustaining unit. It also presents a possibility of decentralized wastewater management since the communities can effectively manage and treat their wastewater on their own. The integration of renewable energy sources in treatment systems has the advantage of being off-grid, making them reliable since they do not rely on large water installations. In addition, integrating hydrogel technology with renewable energy creates a new economy that enables the development and innovation of new business opportunities. Such opportunities can include a wastewater treatment plant that can generate income from selling excess electricity or recovered material and appendix as an eco-friendly wastewater treatment firm. Integrating the hydrogel treatment systems with renewable energy sources is a novel approach that will revolutionize the treatment of wastewaters as a green, efficient and clean process. By adopting this paradigm shift, we are opening up the possibility to design for future, sustainable, and restorable water management systems.

9.7 Conclusion

These applications of hydrogels in the removal of heavy metals and water purification offer an enormous potential in the field is not limited to a single area of application. This has the potential to change the future concept of handling waste and especially with the wastewater in a way that is still unimaginable. In this way, the hydrogels are also effective in reducing the discharge of water pollution and can make wastewater reusable and Recyclable. In addition, there are other general social impacts of developing and applying hydrogel technology, which are: Since hydrogels can enhance the access to clean water and minimize pollution, the health impacts of wastewater disposal are improved health resulting from easy accessibility of clean water is equally enhanced among the population. This is particularly because of the significance that political factors have in the purification of wastewater through hydrogels. Hydrogels, in addition to being an appropriate response to the existing and future world's water problems, can be an effective tool in uniting the nations around an efficient approach to the world water issues. In general, there is an indication that hydrogels application in the process of wastewater management will be able to revamp the current approaches of water utilization, thus improving prospects that are related to sustainable water use. Therefore, as more research and development activities are conducted in this field, it is equally important to go a step further and look at the social impact of hydrogel technology and make sure that the technology is shared fairly. The use of hydrogel-based treatment methods in the municipal wastewater plants implies certain strategies which can be adopted following certain key considerations. First, a feasibility study should be conducted in order to establish the possibility and effectiveness of using hydrogel technology in the given wastewater treatment application. This should be succeeded by the choice of the right kind of hydrogel material and a proper treatment system that would be compatible with the existing installations. After that, an integrated training program should be designed to work with plant operators and maintenance personnel to avoid conflicts with treatment processes currently in place. Another option is to roll out the system from one region to another, as this makes the implementation more controlled and effective, in that the performance of the system may be monitored and enhanced along the process. Further, there are issues related to the disposal of spent hydrogel material or its regeneration, which will require further research and possible emphasis on cooperation with other scientists and experts from the industry to address this and other issues in the development of hydrogel technology. The last integration model is the cost–benefit analysis, which should take into consideration other long-term economic relations of hydrogel technology.

References

Ahmaruzzaman M, Roy P, Bonilla-Petriciolet A, Badawi M, Ganachari SV, Shetti NP, Aminabhavi TM (2023) Polymeric hydrogels-based materials for wastewater treatment. Chemosphere 331:138743, https://doi.org/10.1016/j.chemosphere.2023.138743

Alonso JM, Andrade Del Olmo J, Perez Gonzalez R, Saez-Martinez V (2021) Injectable hydrogels: from laboratory to industrialization. Polymers 13(4):650. https://doi.org/10.3390/polym13040650

Alsaka L, Alsaka L, Altaee A, Zaidi SJ, Zhou J, Kazwini T (2025) A review of hydrogel application in wastewater purification. Separations 12(2):51. https://doi.org/10.3390/separations12020051

Azeera M, Vaidevi S, Ruckmani K (2019) Characterization techniques of hydrogel and its applications. Cellulose-based superabsorbent hydrogels, pp 731–761

Bahram M, Mohseni N, Moghtader M (2016) An introduction to hydrogels and some recent applications. In: Majee SB (ed) Emerging concepts in analysis and applications of hydrogels. Intechopen. https://doi.org/10.5772/64301

Bashir S, Hina M, Iqbal J, Rajpar AH, Mujtaba MA, Alghamdi NA, Wageh S, Ramesh K, Ramesh S (2020) Fundamental concepts of hydrogels: synthesis, properties, and their applications, MDPI. MDPI Polym 12(11):2702. https://doi.org/10.3390/polym12112702

Chang H, Zhao H, Qu F, Yan Z, Liu N, Lu M, Liang Y, Lai B, Liang H (2023) State-of-the-art insights on applications of hydrogel membranes in water and wastewater treatment. Sep Purif Technol 308:122948, https://doi.org/10.1016/j.seppur.2022.122948

Cuerda-Correa EM, Alexandre-Franco MF, Fernández-González C (2020) Advanced oxidation processes for the removal of antibiotics from water. Overview Water 12(1):102. https://doi.org/10.3390/w12010102

Darban Z, Shahabuddin S, Gaur R, Ahmad I, Sridewi N (2022) Hydrogel-based adsorbent material for the effective removal of heavy metals from wastewater: a comprehensive review. MDPI Gels 8(5):263. https://doi.org/10.3390/gels8050263

Flores-Valenzuela LE, González-Fernández JV, Carranza-Oropeza MV (2023) Hydrogel applicability for the industrial effluent treatment: a systematic review and bibliometric analysis. MDPI Polym 15(11):2417. https://doi.org/10.3390/polym15112417

Garcia-Garcia A, Muñana-González S, Lanceros-Mendez S, Ruiz-Rubio L, Alvarez LP, Vilas-Vilela JL (2024) Biodegradable natural hydrogels for tissue engineering, controlled release, and soil remediation. Polymers 16(18):2599. https://doi.org/10.3390/polym16182599

Hong Y, Lin Z, Yang Y, Jiang T, Shang J, Luo Z (2022) Biocompatible conductive hydrogels: applications in the field of biomedicine. Int J Mol Sci 23(9):4578. https://doi.org/10.3390/ijms23094578

Hossain MS, Hossain MM, Khatun MK, Hossain KR (2023) Hydrogel-based superadsorbents for efficient removal of heavy metals in industrial wastewater treatment and environmental conservation. Environ Funct Mater 2(2):142–158. https://doi.org/10.1016/j.efmat.2024.01.001

Hubeny J, Harnisz M, Korzeniewska E, Buta M, Zieliński W, Rolbiecki D, Giebułtowicz J, Nałęcz-Jawecki G, Płaza G (2021) Industrialization as a source of heavy metals and antibiotics which can enhance the antibiotic resistance in wastewater, sewage sludge and river water. PLoS ONE 16(6):e0252691. https://doi.org/10.1371/journal.pone.0252691

Jaishankar M, Tseten T, Anbalagan N, Mathew BB, Beeregowda KN (2014) Toxicity, mechanism and health effects of some heavy metals. Interdiscip Toxicol 7(2):60–72. https://doi.org/10.2478/intox-2014-0009

Jiang C, Wang X, Wang G, Hao C, Li X, Li T (2019) Adsorption performance of a polysaccharide composite hydrogel based on crosslinked glucan/chitosan for heavy metal ions. Compos B Eng 169:45–54. https://doi.org/10.1016/j.compositesb.2019.03.082

Kesari KK, Soni P, Jamal QMS, Tripathi P, Lal JA, Jha NK, Siddiqui MH, Kumar P, Tripathi V, Ruokolainen J (2021) Wastewater treatment and reuse: a review of its applications and health implications. Water Air Soil Pollut 232(208)

Khoo PS, Ilyas RA, Aiman A, Wei JS, Yousef A, Anis N, Zuhri MYM, Abral H, Sari NH, Syafri E, Mahardika M (2024) Revolutionizing wastewater treatment: a review on the role of advanced functional bio-based hydrogels. Int J Biol Macromol 135088. https://doi.org/10.1016/j.ijbiomac.2024.135088

Koetting MC, Peters JT, Steichen SD, Peppas NA (2015) Stimulus-responsive hydrogels: theory, modern advances, and applications. Materials science and engineering. R Rep Rev J 93:1–49. https://doi.org/10.1016/j.mser.2015.04.001

Mu R, Liu B, Chen X, Wang N, Yang J (2020) Hydrogel adsorbent in industrial wastewater treatment and ecological environment protection. Environ Technol Innov 20:101107. https://doi.org/10.1016/j.eti.2020.101107

Persano F, Malitesta C, Mazzotta E (2024) Cellulose-based hydrogels for wastewater treatment: a focus on metal ions removal. MDPI Polymers 16(9):1292. https://doi.org/10.3390/polym16091292

Perumal S, Atchudan R, Edison TNJI, Babu RS, Karpagavinayagam P, Vedhi C (2021) A short review on recent advances of hydrogel-based adsorbents for heavy metal ions. Metals 11(6):864. https://doi.org/10.3390/met11060864

Razzak SA, Faruque MO, Alsheikh Z, Alsheikhmohamad L, Alkuroud D, Alfayez A, Hossain SMZ, Hossain MM (2022) A comprehensive review on conventional and biological-driven heavy metals removal from industrial wastewater. Environ Adv 7:100168, https://doi.org/10.1016/j.envadv.2022.100168

Sava C, Iluţiu-Varvara DA, Mare R, Roman MD, Rada S, Pică EM, Jäntschi J (2024) Physico-chemical characterization and possible uses of sludge processed from an urban sewage treatment plant. Heliyon 10(8):e29576. https://doi.org/10.1016/j.heliyon.2024.e29576

Shah MP (ed) (2024) Emerging innovative trends in the application of biological processes for industrial wastewater treatment. Elsevier

Sharma K, Choudhary P, Majeed A, Guleria S, Kumar M, Rana AK, Rajauria G (2025) Cellulose based membranes, hydrogels and aerogels for water treatment application. Ind Crop Prod 225:120474. https://doi.org/10.1016/j.indcrop.2025.120474

Sinha V, Chakma S (2019) Advances in the preparation of hydrogel for wastewater treatment: a concise review. J Environ Chem Eng 7(5):103295. https://doi.org/10.1016/j.jece.2019.103295

Song M, Wang J, He J, Kan D, Chen K, Lu J (2022) Synthesis of hydrogels and their progress in environmental remediation and antimicrobial application. Gels (Basel, Switzerland) 9(1):16. https://doi.org/10.3390/gels9010016

Thang NH, Chien TB, Cuong DX (2023) Polymer-based hydrogels applied in drug delivery: an overview. Gels 9(7):523. https://doi.org/10.3390/gels9070523

Trombino S, Sole R, Di Gioia ML, Procopio D, Curcio F, Cassano R (2023) Green chemistry principles for nano- and micro-sized hydrogel synthesis. MDPI Mol 28(5):2107. https://doi.org/10.3390/molecules28052107

Turkmen D, Bakhshpour M, Akgönüllü S, Asir S, Denizli A (2022) Heavy metal ions removal from wastewater using cryogels: a review. Front Sustain. https://doi.org/10.3389/frsus.2022.765592

Vidu R, Matei E, Predescu AM, Alhalaili B, Pantilimon C, Tarcea C, Predescu C (2020) Removal of heavy metals from wastewaters: a challenge from current treatment methods to nanotechnology applications. Toxics 8(4):101. https://doi.org/10.3390/toxics8040101

Visan AI, Negut I (2025) Environmental and wastewater treatment applications of stimulus-responsive hydrogels. Gels 11(1):72. https://doi.org/10.3390/gels11010072

Yang X, Wan Y, Zheng Y, He F, Yu Z, Huang J, Wang H, Ok YS, Jiang Y, Gao B (2019) Surface functional groups of carbon-based adsorbents and their roles in the removal of heavy metals from aqueous solutions: a critical review. Chem Eng J Lausanne, Switzerland: 1996) 366:608–621. https://doi.org/10.1016/j.cej.2019.02.119

Zhang K, Luo X, Yang L, Chang Z, Luo S (2021) Progress toward hydrogels in removing heavy metals from water: problems and solutions—a review. ACS ESandT Water 1(5):1098–1116. https://doi.org/10.1021/acsestwater.1c00001

Zhu H, Chen S, Luo Y (2023) Adsorption mechanisms of hydrogels for heavy metal and organic dyes removal: a short review. J Agri Food Res 12:100552. https://doi.org/10.1016/j.jafr.2023.100552

Chapter 10
Metal Nanoparticles Incorporated Antimicrobial Hydrogels for Pathogen Control

Chandrima Kundu and Sougata Ghosh

Abstract Hydrogels are three-dimensional (3D) network chains prepared from various natural and synthetic polymers with diverse biomedical applications. Hydrogels can be used as ideal dressing material because of their excellent biocompatibility, water absorption, gas exchange, adhesion without causing skin irritation, antimicrobial nature, biodegradability, self-healing and controlled drug release properties. Antimicrobial hydrogels can play a critical role in the prevention of water-borne infections. This chapter highlights the recent advances on antimicrobial hydrogels impregnated with silver nanoparticles, titanium dioxide nanoparticles, zinc oxide nanoparticles, carbon dots, graphene oxide and cerium oxide quantum dots. The polymeric counterparts of the hydrogels are mainly comprised of €-polylysine, carboxymethyl cellulose, chitosan, dextran, hyaluronic acid, pectin, pluronic-127, polycaprolactone, polyethylene glycol, polypropylene glycol, polyvinyl alcohol, and ulvan. The hydrogels with loaded nanoparticles (NPs) can help in trapping the pathogens and sustained release of antimicrobial agents and their longer retention for enhanced microbicidal activity. Local administration of the antimicrobial hydrogels can efficiently control the spread of the pathogens. Their promises to control multidrug-resistant pathogenic microbes and the underlying mechanism are also discussed. Hence, nanoparticles incorporated antimicrobial hydrogels can be a powerful wastewater treatment strategy for controlling water-borne infections.

Keywords Hydrogel · Polymer · Nanoparticles · Pathogens · Antimicrobial · Antibiofilm

Abbreviations

3D	Three-dimensional

C. Kundu · S. Ghosh (✉)
Department of Microbiology, School of Science, RK. University, Rajkot, Gujarat, India
e-mail: ghoshsibb@gmail.com

S. Bardhan and M. P. Shah (eds.), *Applications of Hydrogels in Modern Wastewater Treatment*, Advances in Wastewater Research,
https://doi.org/10.1007/978-981-95-1598-1_10

AA	Acrylic acid
AEEA	*N*-(2-hydroxyethyl)ethylenediamine
AFM	Atomic force microscope
APS	Ammonium persulfate
ATR-FTIR	Attenuated total reflectance infrared spectroscopy
BDDE	1,4-Butanediol diglycidyl ether
CA	Citric acid
CA	Citric acid monohydrate
CDs	Carbon dots
CEA	Cellulose acetate
CeO_2-QDs	Cerium dioxide quantum dots
CHAgNPs	Chitosan-PEG hydrogel
CMC	Carboxy methyl cellulose
CS	Chitosan
CTAB	Cetrimonium bromide
DDW	Double distilled water
DMF	*N*, *N* dimethyl formamide
DMSO	Dimethyl sulfoxide
DNA	Deoxyribonucleic acid
EDC	1-Ethyl-3-(3- dimethylaminopropyl)carbodiimide hydrochloride
EDX	Energy dispersive X-ray
EGDMA	Ethylene glycol dimethacrylate
FA	Folic acid
FTIR	Fourier transform infrared spectroscopy
G'	Storage modulus
G''	Loss modulus
GM	Gentamicin
GQDs	Graphene QDs
GSC	GOQDs@silk fibroin/chondroitin sulfate
GTMAC	Glycidyl trimethylammonium chloride
H&E	Hematoxylin and eosin
H_2O_2	Hydrogen peroxide
HA	Hyaluronic acid
HCl	Hydrochloric acid
HDF	Human dermal fibroblast
HGFs	Human gingival fibroblasts
KA	Keratin
KM	Kunming male mice
L-MA	L-malic acid
MA-PDMS-MA	Macromonomer bis-alpha, omega-(methacryloxypropyl)polydimethylsiloxane
DMA	*N,N*-Dimethyl acrylamide
MEFs	Mouse embryonic fibroblasts
MIC	Minimum inhibitory concentration
MPTMS	3-Mercaptopropyltrimethoxysilane

MRSA	Methicillin-resistant *S. aureus*
MRST	Multidrug-resistant *Staphylococcus typhimurium*
MS	Mesoporous silica shell
MSA	Mercapto succinic acid
$NaBH_4$	Sodium borohydride
NF	Nanofibrous
NHS	*N*-Hydroxy succinimide
NMR	Nuclear magnetic resonance
Nor-Dex	Norbornene-functionalised dextran
NPs	Nanoparticles
NVP	*N*-Vinyl pyrrolidone
P	Pectin
PAA	Polyacrylic acid
PAAm	Polyacrylamide
PADM	Porcine dermal extracellular matrix
PBS	Phosphate buffer saline
PCL	Polycaprolactone
PEA	Polyethyleneimine/polyacrylic acid
PEG	Polyethylene glycol
PEG	Polyethylene glycol
PEGDA	PEG dimethacrylate
PEO	Poly-ethylene oxide
PG	Propylene glycol
PHEMA	Poly (2-hydroxyethyl methacrylate)
PLA	Poly-lactic acid
Plys	€-Polylysine
PMMA	Polymethyl methacrylate
PVA	Poly-vinyl alcohol
PVP	Polyvinyl pyrrolidone
QCS	Quanternized chitosan
RNA	Ribonucleic acid
ROS	Reactive oxygen species
SDS	Sodium dodecyl sulfate
SiHCFs	Silicone hydrogel composite films
SOSG	Singlet oxygen sensor green
SSTIs	Skin and soft tissue infections
TA	Tannic acid
TEOT	Tetraethyl orthosilicate
TEOT	Titanium ethoxide
TGA	Thermogravimetric analysis
TRIS	3-Methacryloxypropyl tris (trimethyl soloxy silane)
UHF	Ulvan hydrogel film
VAN	Vancomycin
VMA-CDs	Vancomycin-malic acid-CDS
WFA	Wound fluid absorption

WHO	World Health Organization
WVTR	Water vapor transmission rate
XPS	X-ray photoelectron spectroscopy
ZnO-MCM-41	Zinc oxide-impregnated mesoporous silica
ZnONC	ZnO/zeolite nanocomposite
ZOI	Zone of inhibition

10.1 Introduction

Hydrogels were first reported in the 1960s, for the synthesis of poly (2-hydroxyethyl methacrylate) (PHEMA) for contact lenses. They are cross-linked three-dimensional meshwork structures, made of hydrophilic polymers containing $–SO_3H$, –OH, $–NH_2$, –COOH, and $–CONH_2$ groups. Due to high water absorption and retention capability (almost 20% of its total weight), hydrogels exhibit high swelling properties in water as well as biological fluids. (Li et al. 2012; Ghorpade et al. 2019; Dimatteo et al. 2018; Nie et al. 2019). Hydrophilic polymers can be either natural or synthetic. Natural polymers can be neutral, cationic, and anionic. They are easily accessible, well-defined, abundant, cheap, non-toxic, and biodegradable (Iresha and Kobayashi 2020; Hu et al. 2018; Kwiecien and Kwiecien 2018). Natural polymers such as alginate, carrageenan, cellulose, chitin, chitosan, gums, and starch are used in developing biocompatible hydrogels. Nucleic acid (DNA), polyamides (collagen), polyphenols, polyanhydrides, and polyesters are other chemically distinct natural polymers (Hasnain and Nayak 2018). Natural hydrogels or hydrogels containing natural polymers have plenty of biomedical applications. They can be used for targeted drug delivery and triggered release, tissue engineering, and wound healing.

Synthetic polymers have controllable physical and chemical properties, which makes them flexible and mechanically robust. Hydrogels can be designed using several synthetic polymers, among which the most commonly used are polycaprolactone, polyvinyl alcohol (PVA), polylactic acid (PLA), polyvinyl pyrrolidone (PVP), polyethylene glycol (PEG), polyethylene oxide (PEO), polyacrylamide (PAAm), PHEMA, and polyacrylic acid (PAA) (Bashir et al. 2020). Hydrogels have significant mechanical properties like tensile strength, elongation percent to break, toughness, Young's modulus, swelling, softness, fluffiness, and viscoelasticity. These properties are dependent on pH, temperature, and ionic strength of the surrounding environment (Guo et al. 2018; Ahmad et al. 2022).

Due to the versatile properties of hydrogels, it has biomedical, electrochemical, food packaging, and agricultural applications (Jose et al. 2023). Among various biomedical applications, hydrogels can be applied in three-dimensional (3D) cell culture by growing cells in vitro in all directions. This can mimic in vivo conditions for controlled drug delivery by releasing the drug at predetermined rates. Hydrogels can also be applied in wound dressings, for rapid healing by stopping the bleeding, infection control, and repair of tissue damage. It can be applied in tissue engineering for promoting the regeneration of bone and neural tissues. Hydrogel-based biosensors can investigate and report the biophysical and biochemical alterations in the tissues for rapid diagnosis and monitoring of diseases (Ho et al. 2022; Bai et al. 2018; Ehrenhofer et al. 2020).

Waterborne infections pose a potential threat to society. According to the World Health Organization (WHO), diseases linked to polluted water increase by 5 million every year (Cabral 2010). Among all the different types of waterborne infections, skin and soft tissue infections (SSTIs) are very common. SSTIs are recognized as an ecological disaster by the WHO after the 2022 devastating flood in KwaZulu-Natal in South America (Naidoo et al. 2023). The organisms responsible for such infections are *Aeromonas hydrophilia, Edwardsiella tarda, Vibrio vulnificus, Mycobacterium marinum, Eryspelothrix rhusiopathiae, Clostidirum tetani, Aspergillus* sp., and *Rhizopus* sp. (Naidoo et al. 2023). According to the WHO report of 2019, a minimum of 700,000 deaths occur worldwide every year due to drug-resistant bacteria (Mattar et al. 2020). These conventional therapies cannot provide a complete cure, which causes an increase in bacterial resistance, death, and increases the cost of treatment.

10.2 Problem of Antibiotic Resistance

Antibiotic resistance is one of the greatest worldwide challenges associated with high morbidity and mortality (Church and McKillip 2021; Frieri et al. 2017; Baquero 2021). Since the discovery of penicillin in 1928, the overuse and misuse of antibiotics have contributed severely to the problem of multidrug-resistant bacterial (MDR) nosocomial infections. ESKAPE pathogens, namely, *Mycobacterium tuberculosis, Enterococcus faecium, Staphylococcus aureus, Klebsiella pneumoniae, Acinetobacter baumannii, Pseudomonas aeruginosa*, and *Enterobacter* spp. Along with other MDR bacteria, they cause nearly 70,000 annual deaths worldwide by surviving against the extremely potent antibiotics. It is evident from previous studies that antibiotic resistance may have existed for 2 billion years, while the modern era of antibiotic resistance began in the late 30 s after the large-scale introduction of sulfonamide in 1937. These resulted in the rise of the resistant strains associated with fatal infections by surviving the therapeutic dosages of the antibiotics. The antibiotic resistance gene transfer via bacterial conjugation helps other bacteria to become resistant to the particular antibiotics. This genetic transfer rapidly and efficiently spreads the antimicrobial resistance throughout the bacterial population. Bacteria develop antibiotic resistance in various ways. It can be developed by them by altering the drugs chemically using enzymes that can fundamentally change the antibiotic through acetylation, adenylation, or phosphorylation. The bacteria like *S. aureus* can also develop antibiotic resistance using the extended spectrum β-lactamases (ESBL) enzymes to destroy and hydrolyze the β-lactam rings of antibiotics like penicillin, including its derivatives, third-generation cephalosporins, monobactams, while showing no effectiveness against cephamycin and carbapenems. Further, the complex, multilayer membrane in Gram-negative bacteria makes it difficult for antibiotics to enter. However, hydrophilic antibiotics like β-lactam can enter through the specific membrane channel porins, while less expression or alteration of these channels can prevent the entry of these antibiotics. There are some substrate-specific specialized proteins or efflux pumps in the bacterial membrane that can actively pump specific antibiotics, like macrolides or β-lactams, out of the cell. Bacteria can also develop antibiotic resistance by protecting the antibiotic target site from being affected via non-genetic modification, genetic mutation, replacement or bypass, and destruction by producing enzymes. *S. aureus* and *P. aeruginosa* develop biofilms by closely associating with each other by embedding themselves in a slimy, extracellular matrix composed of polysaccharides, proteins, and nucleic acids. This helps to protect themselves from several conventional antibiotics, making the infections life-threatening, complicated, and untreatable (Church and McKillip 2021). Antibiotic resistance contributes to direct and indirect mortality of patients. Nearly 25,000–33,000 annual deaths in Europe and 23,000 annual deaths in the USA are due to the direct mortality caused by antibiotic resistance. The Centers for Disease Control (CDC) estimated that the 7 billion annual deaths worldwide due to MDR bacterial infections can reach approximately 10 million deaths by 2050 if no alternatives to antibiotics are made available (Baquero 2021; Church and McKillip 2021).

10.3 Development of Antimicrobial Hydrogels and Modification Techniques

Conventional antimicrobial agents like antibiotics have significantly reduced the deaths caused by infections. However, over- and inappropriate use of antibiotics has developed antimicrobial resistance in the pathogens, leading to the formation of MDR bacteria like methicillin-resistant *S. aureus* (MRSA), carbapenemase-producing *K. pneumoniae*, metallo-beta-lactamase-producing *A. baumannii*, *P. aeruginosa* (MBL-PA), and ESBL-producing *Enterobacter* spp., *Clostridium difficile, Escherichia coli*, and *Klebsiella oxytoca.* To solve this issue, antimicrobial hydrogels play a crucial role in treating infectious diseases due to their compatibility with host tissues and controlled release of antimicrobial agents at the site of infection (Carpa et al. 2022; Ahovan et al. 2022). They are also called smart antimicrobial hydrogels due to their specific responses to physical and chemical environmental stimuli. Hydrogels can deliver antimicrobial agents like antibiotics, metal nanoparticles (NPs), antimicrobial peptides (AMPs), biological extracts, and coatings to prevent infections. The properties of the antimicrobial hydrogels depend upon the type of polymeric precursor, pH, temperature, and the degree of cross-linking (Tallet et al. 2021). The inherently antimicrobial hydrogels are made up of natural polymers derived from microbes (dextran, xanthan gum, and gellan gum), algae (alginate, agarose, and carrageenan), animals (chitosan, hyaluronic acid, and chondroitin sulfate), and plants (cellulose, guar gum, and locust bean gum). These polycationic materials disrupt the membrane, making it difficult for bacteria to develop antibiotic resistance. This property of the antimicrobial hydrogels makes them ideal to treat MDR bacterial infections. The antimicrobial efficacy can be increased by different modification techniques. The chitosan hydrogel can be modified by adding tertiary amino groups along the backbone to increase its efficacy. The chitosan or alginate-based hydrogels in containing honey can provide wound healing, pain reduction, and surgical wound prevention. Other inherently antimicrobial hydrogels are made up of synthetic polymers like polyacrylamide, polyvinyl acetate, and polyethylene glycol. The antimicrobial efficacy of the hydrogels can be enhanced by the addition of quaternary ammonium groups, and AMPs like poly-lysine that can disrupt the bacterial membrane electrostatically and hinder the growth of Gram-positive and Gram-negative bacteria. The amphoteric gels, like polynorbornene, similar to AMPs, containing both acidic and basic groups, can interact electrostatically with the bacterial membrane, eventually inhibiting it. The hydrophilicity, porosity, and antimicrobial efficiency of the hydrogels can be modulated by rational selection and incorporation of specific antibiotics, AMPs, bioactive principles, polysaccharides, and NPs. Alginate-based hydrogels in the chitosan-hydrochloride solution, a combination of chitosan and polyacrylic acid-co-hydroxyethyl methacrylate, and dextrin grafted poly(2-hydroxyethyl methacrylate) were reported as modified composite hydrogels with enhanced antimicrobial activity (Carpa et al. 2022).

As a solution, the nanoparticle-impregnated hydrogels are the most potent enchanting system for delivering the antimicrobial agents in a controlled manner and to the particular target site. It can disrupt the physical membrane of the pathogens by interfering with the microbial cell membrane, leading to cell lysis and death (Ruiz et al. 2023). NPs can generate reactive oxygen species (ROS), causing damage to cellular components like DNA, RNA, and proteins (Dong et al. 2022). It can prevent biofilm formation by the pathogens. Further, hydrogels can act as a potential carrier of the NPs that can be released in a controlled and sustainable manner, resulting in enhanced antimicrobial activity (Ruiz et al. 2023). This chapter gives an elaborate account of the nanoparticles impregnated hydrogels with enhanced antimicrobial activity as listed in Table 10.1.

10.4 Metal Nanoparticles Incorporated Hydrogels

Nanomaterials, mainly the metal NPs, are 3D crystalline materials having 1–100 nm size. They have a high surface area to volume ratio, small sizes, shapes, stability, low toxicity, are economically attractive, and have physicochemical properties (Sheikh-Oleslami et al. 2023). Metal NPs hybrid hydrogels (MNHHs) enhance the stability and the circulation time of the metal NPs, while reducing their toxicity (Wang et al. 2024). The MNHHs also exhibit a response to a wide range of physicochemical stimuli, including sound, light and electric field. The physical and mechanical properties, like stretchability and rheology, of the MNHHs increase due to the thorough distribution of the metal NPs as a prominent structural component, which improves

Table 10.1 Composition and applications of antimicrobial hydrogels

Hydrogel type	Composition	NPs	Shape of NPs	Size of NPs (nm)	Antibacterial activity against pathogens	Wound healing in animal model	Ref
Silicone hydrogel composite films (SiHCFs)	*N,N*-dimethyl acrylamide (DMA), 3-methacryloxypropyl tris (trimethyl soloxy silane) (TRIS) (98%), bis-alpha, omega-(methacryloxypropyl)polydimethylsiloxane (MA-PDMS-MA), *N*-vinyl pyrrolidone (NVP), ethylene glycol dimethacrylate (EGDMA)	AgNPs	Spherical	–	*E. coli*, *P. aeruginosa* and *S. aureus*	–	Mourad et al. (2019)
Chitosan polyethylene glycol (PEG) hydrogel	Chitosan (CS), polyethylene glycol (PEG)	AgNPs	–	99.1 ± 2.3	*E. coli*, and *P. aeruginosa*, *B. subtilis* and *S. aureus*	Wound in alloxan-induced diabetic rabbits	Masood et al. (2019)
Ulvan hydrogel film (UHF)	Ulvan (U)	AgNPs	Spherical to flower-shaped	13.87–21.55 nm (UHF-AgNP0.5) 22.44–34.77 (UHF-AgNP1)	*S. epidermidis*, *E. coli*, *S. aureus* and *P. aeruginosa*	Second-degree thermal burn in male Wistar rats	Sulastri et al. (2023)
Porcine dermal extracellular matrix (PADM) hydrogel	Porcine skin tissue	AgNPs	–	5	*S. aureus*, *E. faecalis* and *E. coli*	Skin wounds in six-month-old rats infected with *S. aureus*	Dong et al. (2022)
Pluronic-127 (PF-127) hydrogel	Pluronic-127, propylene glycol	AgNPs	–	2.98 ± 0.85	Antibiofilm activity against *P. aeruginosa*, *S. epidermidis* and *S. aureus*	–	Haidari et al. (2020)
Triple-layered nanofibrous (TNF) bandage	Polyvinyl alcohol (PVA), cellulose acetate (CEA), polycaprolactone (PCL)	AgNPs	–	–	*E. coli* and *S. aureus*	Partial thickness wound model in Sprague–Dawley rats	Dugam et al. (2024)

(continued)

Table 10.1 (continued)

Hydrogel type	Composition	NPs	Shape of NPs	Size of NPs (nm)	Antibacterial activity against pathogens	Wound healing in animal model	Ref
Chitosan (CH)/ polypropylene glycol (PPG)/ titanium dioxide (TiO_2) composite hydrogel	Chitosan (CH), polypropylene glycol (PPG)	TiO_2NPs	Spherical	40	*E. coli*, *S. aureus* and *C. lipolytica*	–	Ulu et al. (2020)
(Nor-Ald-Dex)/ quanternised chitosan (QCS)/silica coated TiO_2 (NorAld-Dex/ QCS/ TiO_2 @ MS-SH) hydrogel	Norbornene (Nor) functionalized dextran (Dex), quaternized chitosan (QCS)	TiO_2NPs	–	29.4 (TiO_2), 34.3 (MS-SH TiO_2 NPs)	*E. coli*, and *S. aureus*	–	Yan et al. (2024)
Zinc oxide-impregnated carboxy methyl cellulose (CMC/ ZnO-MCM-41) hydrogel	Carboxymethyl cellulose (CMC)	ZnONPs	Spherical	–	*E. coli* and *S. aureus*	–	Rakhshaei and Namazi (2017)
Zinc oxide/zeolite nanocomposite chitosan (ZnONC-CS) hydrogel	Chitosan (CS)	ZnONPs	Spherical	30	Antibiofilm activity against *S. mutans*	–	Afrasiabi et al. (2021)
Polyethyleneimine/ polyacrylic acid-cerium dioxide quantum dots (PEA@CeO_2-QDs) hydrogels	Polyethyleneimine, polyacrylic acid	CeO_2QDs	–	1.5–2.5	*S. aureus* and *E. coli*	–	Hou et al. (2024)

(continued)

Table 10.1 (continued)

Hydrogel type	Composition	NPs	Shape of NPs	Size of NPs (nm)	Antibacterial activity against pathogens	Wound healing in animal model	Ref
Spermidine- carbon dots- acrylic acid-pectin (SCDs-AP) hydrogel	Acrylic acid (AA), pectin (P)	CDs	–	–	*A. hydrophila*, *E. coli* O157:H7, *S. aureus* ATCC 29,213, *Y. enterocolitica*, *P. fluorescens*, MRSA and MRST	Subcutaneous wound healing in male rabbits	Cui et al. (2021a, b
Carbon dot embedding soft hyaluronic acid hydrogel (CDs gel)	Hyaluronic acid (HA)	CDs	–	2–10	*S. aureus* ATCC 25,923 and *E. coli* KCTC 2571	Female C57BL6 mice skin wound infection	Lee et al. (2022)
Carbon dots-polylysine hydrogel (CDs-Plys gel)	€- polylysine (Plys)	CDs	Spherical	6.96 ± 0.3	*E. coli* and *S. aureus*	Full-thickness wound defect in female mice	Mou et al. (2022)
Garlic carbon dots-pectin-acrylic acid hydrogel (GCDs-PAA)	Pectin (P), acrylic acid (AA)	GCDs	–	3.1 ± 0.23	MRSA and MRST	Epidermal wound infection in female BALB/c mice	Sheng et al. (2024)
Polyvinyl alcohol-sodium alginate-vancomycin-malic acid- carbon dot (PVA-SA-VMA-CDs) hydrogel	Polyvinyl alcohol (PVA), sodium alginate (SA)	CDs	Spherical	3.72	*S. aureus*, MRSA, *B. subtilis* and *E. coli*	–	Zhao et al. (2024)
Chitosan-folic acid-carbon quantum dots (CS-FA-CQDs) hydrogel	Chitosan (CS)	CQDs	–	–	*E. faecalis*, *P. aeruginosa*, *S. mutans*, *S. aureus*, *K. pneumoniae* and *E. coli*	–	Kazeminava et al. (2022)

(continued)

Table 10.1 (continued)

Hydrogel type	Composition	NPs	Shape of NPs	Size of NPs (nm)	Antibacterial activity against pathogens	Wound healing in animal model	Ref
Graphene quantum dots-bacterial cellulose (GQDs-BC) composite hydrogel	Bacterial cellulose (BC)	GQDs	Disc like	10 ± 2	*S. aureus*, MRSA, *E. coli*, *P. aeruginosa* and *S. agalactiae*,	–	Zmejkoski et al. (2022)
Graphene oxide-silk fibroin, chondroitin sulfate (GSC) hydrogel	Silk fibroin (SF), chondroitin sulfate (CS)	GOQDs	–	30	*S. aureus* and *E. coli*	Full-thickness skin defect in Kunming mice	Deng et al. (2024)
Tannic acid/keratin-graphene oxide quantum dots (TA/KA-GOQDs) hydrogel	Keratin (KA)	GOQDs	Spherical	10	*S. aureus* and *E. coli*	Excision wound in rats	Ren et al. (2020)

the crosslinking. The viscoelastic modulus increases while the frequency dependency decreases due to the formation of the microstructures and their permeation alters the overall gel behaviour, making it stiffer or solid-like, rendering it resistant to deformation. Composite hydrogel with AgNPs, gelatin and chitosan exhibited superior mechanical properties with higher water absorption, moisture retention, and biocompatibility. Muscle-like movement and flexibility of poly (hydroxyethyl methacrylate) (pHEMA) hydrogel were observed after being impregnated by cobalt NPs (CoNPs), which increased the loading efficiency by reducing the leaching effects. The self-healing ability of the MNHHs is due to the non-covalent interactions between the polymer chains and the metal ions. The self-assembly and the optical activity of the metal NPs are also utilized by the MNHHs to gain their self-repairing ability. The interfacial interactions of MNHHs are determined by the surface ligands, morphology, and the size of the metal NPs. Composite hydrogel (Gel-DA@AgNPs) synthesized using dopamine-modified gelatin exhibited superior injectability and self-healing capabilities. Incorporation of the metal NPs can increase the environmental responsiveness, biomarker monitoring, and imaging capabilities of the MNHHs (Wang et al. 2024).

An antimicrobial hydrogel was fabricated by impregnating silver nanoparticles (AgNPs) in the silicone hydrogel to prevent microbial keratitis (Mourad et al. 2019). Initially, a 3 mL mixture of hydrophobic monomer 3-methacryloxypropyl tris (trimethylsiloxy silane) (TRIS) (98%), macromonomer bis-alpha, omega-(methacryloxypropyl)polydimethylsiloxane (MA-PDMS-MA), and hydrophilic monomer *N,N*-dimethyl acrylamide (DMA), in a ratio of 4:2:1, was added to 0.18 mL of *N*-vinyl pyrrolidone (NVP) to increase water content and 15 μL of ethylene glycol dimethacrylate (EGDMA) for controlled cross-linking. Dry nitrogen (N_2) was circulated through the mixture for 15 min. Silver nitrate ($AgNO_3$) was suspended in ethanol at various concentrations (0–80 ppm) that resulted in a transparent brown colour, indicating the formation of AgNPs. This mixture was combined with the previous mixture, followed by the addition of ethanol (0.3 mL) and the photoinitiator Darocur 1173 (8 mg), to facilitate free radical bulk polymerization of the monomers. The mixture was stirred for 5 min and was injected into a mold of two 5 mm-thick glass plates, with a silicone composite spacer in between. It was then exposed to UV light for 50 min at 365 nm, resulting in the formation of silicone hydrogel composite films (SiHCFs). Excess $AgNO_3$ and unreacted monomers were removed by washing the mixture in 70% ethanol, followed by air-drying. Under transmission electron microscope (TEM), the uniformly dispersed AgNPs within the polymer matrices of SiHCFs exhibited a spherical shape with a core–shell-like structure. UV–visible spectroscope analysis showed characteristic peaks at 400 nm for the copolymerizing monomer mixture solutions and the prepared SiHCFs, indicating an increase in bandwidth, intensity, and peak position with an increase in AgNPs content. The ophthalmic transmittance of SiHCFs exceeded 90% between 750 and 550 nm. Under the scanning electron microscope (SEM), slight aggregation was observed as the AgNPs concentration increased.

Energy dispersive X-ray spectroscope (EDX) analysis confirmed the presence of carbon (C), oxygen (O), silicon (Si), and nitrogen (N) elements in SiHCFs. Fourier transform infrared (FTIR) spectra exhibited C=O group associated vibrations at 1740, 1665, 1277, and 1146 cm^{-1}. Stretching vibrations of C-H and absorption band of C–N were observed at 2957 cm^{-1}, and 1453 cm^{-1}, respectively. Thermal and structural stability of SiHCFs with a higher amount of AgNPs was confirmed by thermogravimetric analysis (TGA), where the Young's modulus increased from 1.06 to 1.38 MPa with an increase in AgNPs concentration. The SiHCFs showed inhibition of Gram-negative bacteria *Escherichia coli* ATCC-25922 and *Pseudomonas aeruginosa* strain OS4, as well as Gram-positive bacteria *Staphylococcus aureus* ATCC-47077. The relative colony numbers for bacterial growth in phosphate-buffered saline (PBS) surrounding 80 ppm $AgNO_3$-containing SiHCFs discs, were 0% for *E. coli*, *P. aeruginosa* and *S. aureus*. The bacterial adhesion assay demonstrated a dose-dependent response of AgNPs containing SiHCFs. Interestingly, 10, 20, and 30 ppm $AgNO_3$ containing SiHCFs killed only 20–55% of microbes, whereas 40, 60, and 80 ppm $AgNO_3$ containing SiHCFs killed a higher number of microbes. It also effectively inhibited bacterial biofilm formation on the surface of SiHCFs. Maximum inhibition was noted against *P. aeruginosa* followed by *E. coli* and *S. aureus*. The viability of normal human epithelial amnion cells (WISH) was 96.3%, 86.7%, 84% and 81% when treated with SiHCFs prepared from 0, 40, 60 and 80 ppm concentrations of $AgNO_3$, respectively. This confirmed the biocompatibility of the composites. Viability of fibroblast-origin cells (BJ1) was 92%, 91%, 89% and 83.3% when treated with SiHCFs prepared from 0, 40, 60, and 80 ppm concentrations of $AgNO_3$, respectively. These findings suggested that after 24 h of incubation, all WISH and BJ1 cells remained viable with relative cell viability exceeding 80%. Therefore, these SiHCFs did not exhibit significant toxicity to the tested cells, indicating their potential biocompatibility and safety for medical applications.

Another antimicrobial hydrogel was developed by impregnating AgNPs in the chitosan (CS) polyethylene glycol (PEG) hydrogel to enhance the wound healing process in diabetic patients (Masood et al. 2019). Initially, CS was immersed in acetic acid with various concentrations (1, 2, and 5%), followed by stirring on a hot plate for 2 h to facilitate dissolution. $AgNO_3$ (0.1 g) was added to the CS solution, followed by the addition of PEG solution, at 1:1, 2:1 and 5:1 ratios. Then, 2% glutaraldehyde was mixed for crosslinking the polymers. UV–visible spectroscopy of AgNPs-impregnated chitosan-PEG hydrogel (CHAgNPs) with a CS:PEG ratio of 5:1 showed the strongest molar absorbance at 432 nm. The FTIR analysis for the plain chitosan hydrogel (CH) showed OH/NH stretching, indicating peaks at 3000 to 3550 cm^{-1}, while CHAgNPs showed C–H vibrational stretching at 2880 cm^{-1} because of PEG. Stretching vibration of CH_2 asymmetric stretches and C–H stretches for both CS and PEG chains were observed at 1093 and 1023 cm^{-1}. Similarly, for C=O and N–H/C–N, the bands were noted at 1642 cm^{-1} and 1341 cm^{-1}, respectively, in plain CH. C=O stretch associated bands were found at 1093 cm^{-1} and 1023 cm^{-1}, while C–H bond displayed bands at 1340 and 1463 cm^{-1}. Absorption bands of C=O, N–H/C–N stretches and CH_2–OH were found at 1642 cm^{-1}, 1341 cm^{-1} and 1568 cm^{-1}, respectively. In the case of CHAgNPs, an additional peak was observed at 955 cm^{-1} for the C–O–C stretch. The binding of Ag with CS was confirmed by the appearance of a peak at 1724 cm^{-1}. Under SEM, a clear, uniform, and flat surface appeared for the plain CH, while an agglomerated white dotted surface was seen for CHAgNPs. The AgNPs present in the hydrogel had a diameter of 99.1 ± 2.3 nm. The porosity of plain CH and CHAgNPs was almost the same, having a high porosity of 70.8 ± 0.4% and 72.2 ± 0.2%, respectively. The decrease in the swelling capacity was obtained in the CHAgNPs when compared with plain CH.

The water carrying capacity was higher in CHAgNPs than in plain CH, signifying higher suitability of using CHAgNPs, as evident from the water vapor transmission rate (WVTR) measurements. The zone of inhibition (ZOI) against *E. coli*, *P. aeruginosa*, *Bacillus subtilis*, and *S. aureus* was 20.2 ± 1.0, 21.8 ± 1.5, 15.5 ± 0.8 and 21.5 ± 0.5, respectively, which was superior in comparison to plain CH and AgNPs. Viable bacterial cell counts for *E. coli* and *S. aureus* were reduced by 99.7 ± 3.1%, and 99.2 ± 1.9% by CHAgNPs, 73.4 ± 3.4%, and 64.1 ± 2.6% by plain CH, and 83.6 ± 3%, and 75.5 ± 4.1% by AgNPs, respectively. Further, wound healing activity was checked against alloxan monohydrate-induced chronic diabetic wounds in rabbits. Five groups of rabbits were treated with none, nitrofurazone, AgNPs, plain CH, and CHAgNPs, respectively. CHAgNPs directed complete re-epithelisation within 12.3 ± 0.8 days, compared to other treatment groups. Hence, promising antibacterial and wound healing activities were exhibited by CHAgNPs due to the synergistic effects of AgNPs and CS.

AgNPs were impregnated in the ulvan hydrogel film (UHF) for burn wound dressing (Sulastri et al. 2023). Initially, ulvan (5, 7.5, and 10%) was immersed in deionized water, followed by the addition of boric acid (0.1%) as a crosslinker. Glycerol (2%) was added as a plasticizer in the mixture and homogenized at 50 °C and room temperature for 15 min and 45 min, respectively. Homogenized hydrogel (10 mL) was poured into petri dishes and dried at 45 °C for 48 h. $AgNO_3$ solutions (0.5 and 1 mM) were then added to the previously prepared homogenous UHF at room temperature. The two different hydrogels were stirred at 300 rpm for 40 min and sonicated for 20 min. The formation of AgNPs was confirmed by the appearance of the brownish-orange colour. The prepared hydrogels, UHF-AgNP0.5 and UHF-AgNP1, were dried at 45 °C for 48 h. The weight (g), thickness (mm), pH, viscosity (Cps), for UHF-AgNP0.5 were 1.15 ± 0.05, 0.73 ± 0.03, 7.09 ± 0.14, 367 ± 14.52, respectively, while the same for UHF-AgNP1 were 1.27 ± 0.08, 0.78 ± 0.07, 7.43 ± 0.2, 356 ± 34.8, respectively. Swelling degrees (%) for UHF-AgNP0.5 were 91.3 ± 9.03 (1 h), 138 ± 9.85 (4 h), 149 ± 5.57 (6 h), and for UHF-AgNP1 were 89.4 ± 9.93 (1 h), 127 ± 6.66 (4 h), 138 ± 4.22 (6 h). WVTR (g/m^2/day) was 1732 ± 23 and 1809 ± 28, for UHF-AgNP0.5 and UHF-AgNP1, respectively. UV–visible spectra showed the absorbance peaks specific to AgNPs at 400.4 nm and 411.86 nm, for the UHF-AgNP0.5 and UHF-AgNP1, respectively. TEM revealed particle sizes of 13.87–21.55 nm for UHF-AgNP0.5 and 22.44–34.77 nm for UHF-AgNp1. Zeta potential was −24.5 ± 1.47 for UHF-AgNP0.5 and −22.43 ± 2.5 for UHF-AgNP1. FTIR exhibited symmetrical and asymmetrical stretching vibrations of the C=O group, at 1631 cm^{-1}, 1431 cm^{-1} for UHF-AgNP0.5, and at 1631 cm^{-1}, 1433 cm^{-1} for UHF-AgNP1. The characteristic glycosidic bond (C–O–S) of ulvan showed a peak at 848 cm^{-1}. Stretching vibration of the sulfate group (S=O) appeared at 1253 cm^{-1} for UHF-AgNP0.5 and at 1255 cm^{-1} for UHF-AgNP1. SEM revealed the presence of mostly spherical to flower-shaped AgNPs in the hydrogels. Energy dispersive X-ray analysis (EDX) confirmed the presence of Ag in both the UHFs, mass and atomic percentages being 5.78% and 3.49%, respectively, for UHF-AgNP1. This indicated a comparatively higher Ag content in UHF-AgNP1. The antibacterial activity was observed against *Staphylococcus epidermidis*, *E. coli*, *S. aureus*, and *P. aeruginosa*, with ZOI of 17.39 mm, 17.35 mm, 16.75 mm, and 13.58 mm, respectively, for UHF-AgNP0.5, and 13.07 mm, 15.36 mm, 14.89 mm and 12.55 mm, respectively, for UHF-AgNP1. Further, the UHF-AgNP0.5 was used to check the wound healing activity in second-degree thermal burn in the male Wistar rats. Wounds closed completely on day 21 for all the treatments, except the first group, which was left untreated (75.36%). Therefore, a strong antibacterial and wound-healing activity was shown by UHF-AgNP0.5 rather than UHF-AgNP1.

An antimicrobial hydrogel was prepared by impregnating two types of AgNPs (ø5-nm and 50 nm) in porcine dermal extracellular matrix (PADM) hydrogel (AgNPs-PADM) to treat infected wounds (Dong et al. 2022). Initially, pig skin was reaped from an adult pig and rinsed with sterile water for 3 h. Followed by freezing and thawing 3 times, between −80 and 37 °C. The tissue was cut into 1×1 cm^2, from which the subcutaneous tissue was obtained and treated with 1% Triton X-100 solution and 0.1% sodium dodecyl sulfate (SDS) for 12 h and 6 h, respectively. The treated tissue was rinsed and lyophilized, leading to the formation of powder. Next, 20 mg mL^{-1} of the powder was digested by pepsin, in dilute hydrochloric acid (HCl), at pH 2–3 for 10 min. The ø5-nm AgNPs solution was added dropwise and stirred rapidly, followed by digestion for 2 h until the gel became translucent and viscous. The gel was refrigerated at 4 °C and PBS was added to maintain the osmotic pressure. Precooled 10 M NaOH was used to maintain the pH at 7–8 and further kept at 37 °C for 20 min, resulting in the formation of AgNPs-PADM hydrogel. Under UV–visible spectroscopy, 5 nm AgNPs, 5 nm AgNPs-PADM hydrogel, and 50 nm AgNPs showed a peak at 400 nm, except 50 nm AgNPs-PADM hydrogel. Indicating uniform distribution of 5 nm AgNPs and agglomeration of 50 nm AgNPs in the hydrogel. TEM analysis revealed a negligible difference in size and physical characteristics of 5 nm AgNPs before addition and after release from the hydrogel, while 50 nm AgNPs significantly increased in size after release from the hydrogel. SEM images showed a mesh-like fiber structure of AgNPs-PADM hydrogel in comparison to PADM hydrogel. Porosities of the PADM and AgNPs-PADM were $61.67 \pm 3.535\%$, $63.57 \pm 3.106\%$ (20 μg mL^{-1} AgNPs solution), 61.30 ± 2.975 (50 μg mL^{-1} AgNPs solution), $56.73 \pm 3.316\%$ (80 μg mL^{-1} AgNPs solution), respectively. This confirmed the negligible effect of AgNPs concentration on the porosity of the hydrogel. Energy dispersive spectroscopy (EDS) mapping detected Ag concentration in an 80 μg mL^{-1} AgNPs solution containing AgNPs-PADM hydrogel, showing the uniform distribution of AgNPs in AgNPs-PADM hydrogel, making it a good dispersion medium for AgNPs. Rheological measurements exhibited similar rheological behaviours for PADM and AgNPs-PADM hydrogels at pH 7, with storage (G′) moduli 9.737 ± 5.057, 6.868 ± 4.274 and loss (G″) moduli 6.119 ± 3.984, 5.989 ± 3.731, for 5 min and 10 min, respectively. Prolonged storage led G′ to increase significantly faster than G″, causing the gradual formation of an elastic solid gel. Gelatinization of AgNPs-PADM was slightly slower than PADM hydrogel and transformed completely to an elastic solid gel within 15–20 min. Both the hydrogels PADM and AgNPs-PADM lost all the internal water at 37 °C after 30 h. Hydrogel water storage was unaffected by AgNPs, leaving the wound wet for a long time, resulting in better healing. The degradation rate of both the hydrogels increased with an increase in temperature. PADM (0.46 g) and AgNPs-PADM (0.54 g) were degraded at 25 °C and 37 °C, respectively, within 2 days. *S. aureus, Enterococcus faecalis,* and *E. coli* were inhibited with a ZOI equivalent to 49.26 ± 6.852 mm, 50.62 ± 10.65 mm, and 152.8 ± 6.183 mm, respectively, when treated with 80 μg mL^{-1} AgNPs solution containing AgNPs-PADM hydrogel. The dilution coating method depicted a lower bacterial count for *S. aureus* on treatment with AgNPs-PADM (205.3 ± 14.53) in comparison to PADM (852.7 ± 6.498) after 12 h. Cytotoxicity of both the hydrogels was determined by calcein staining on coculturing the gels with cervical epithelioid carcinoma cells (HeLa) cells for 48 h. Higher viable cell density on both the hydrogel surfaces confirmed non-cytotoxicity. Infected wound healing activity was checked against the scalpel, and tweezers mediated skin wound infected with *S. aureus* (30 μL) in six-month-old rats. Wound closure in rats treated with PADM and AgNPs-PADM, 83.37 ± 0.6489%, and 94.53 ± 0.8090%, respectively, after 7 days, which was significantly higher than the untreated wound (70.33 ± 0.6064%). Interestingly, AgNPs-PADM hydrogel-treated group showed complete closure of the wound with hair follicle regeneration, and the remaining two groups depicted similar closure of wounds, with a trace of granular tissue, after day 14. Inflammation and fibroblast development were checked with hematoxylin and eosin (H&E) staining. The AgNPs-PADM hydrogel-treated group of rats showed incessant generation of epithelial tissue and less infiltration of inflammatory cells. However, less epithelial tissue generation and significant inflammatory cell infiltration were noted in others.

Another antibacterial biocompatible thermosensitive hydrogel was designed by impregnating ultrasmall AgNPs in pluronic F-127 (PF-127) hydrogel to check biofilm eradication capacity (Haidari et al. 2020). Initially, mercapto succinic acid (MSA) coated AgNPs were prepared by mixing 1 mL of 20 mM $AgNO_3$ with 1 mL of 20 mM MSA and 7.7 mL MilliQ water, followed by vigorous stirring in ice-cold conditions to form Ag(I)-thiolate complex. The mixture was stirred again for 5 min, followed by the addition of 0.3 mL of 0.05 M sodium borohydride ($NaBH_4$), which was already prepared in 20% 1 M NaOH. It was stirred again at 500 rpm for 24 h which resulted in a colour change from pale to bright yellow, confirming the formation of MSA-coated AgNPs. The particles were purified by dialysis against MilliQ water and lyophilized. The pluronic hydrogel was prepared by mixing a pre-weighed quantity of PF-127 and propylene glycol (PG), slowly into MilliQ water, with or without AgNPs. The mixture was kept in a cold bath followed by refrigeration at 4 °C overnight, to secure the complete dissolution of PF-127. Controlled loading of MSA-coated ultrasmall AgNPs at concentrations 25, 50, 100, 150, and 200 μg/g, to PF-127 hydrogel, caused the formation of AgNPs impregnated PF-127 hydrogel. Under UV–visible spectroscopy, AgNPs loaded PF-127 hydrogel depicted similar absorption as AgNPs solution at first and even after 2 months at 4 and 25 °C temperatures, indicating its high stability. TEM images revealed the core diameter of AgNPs as 2.98 ± 0.85 nm. Rheological measurement depicted an increase in sweeps in the oscillatory shear rate of AgNPs-loaded PF-127 hydrogel, with an increase in AgNPs concentration. A decrease in the hydrogel's viscosity for application of stress, and rapid retention of its mechanical properties, on reduction of stress was confirmed. Thermoresponsive property was evaluated by studying the G' and G'', where at a lower temperature the solid-like component was dominant, causing the gelation point to increase. The PF-127 and AgNPs-loaded PF-127 gels showed equal solution/gel transition temperature of the G' and G'', at 28 °C and 34 °C, respectively. The presence of AgNPs elevated the gelation temperature, contributed to H-bonding interactions, and increased the cross-linking density of the hydrogel, as indicated by the temperature sweep measurements. DLS study revealed that the size of NPs remained unchanged over time at 4 °C and 25 °C, from day 0 to day 60. A slight increase in size was observed with the increase in temperature to 37 °C after 2 months. The physical properties of the AgNPs were maintained even after impregnation into the hydrogel, indicating the stability of MSA-coated AgNPs in the hydrogel. This can make the hydrogel a compatible carrier for Ag delivery in vivo. TGA exhibited an initial weight loss of PF-127 between 50 and 130 °C, due to the evaporation of water, while the major weight loss was between 250 and 350 °C, due to the degradation of molecular chains. A higher decomposition temperature (430 °C) with a slower decomposition rate was achieved on the incorporation of AgNPs into the hydrogel, leading to an improved thermal stability. The internal morphology of hydrogels was three-dimensional porous structures with pore diameters for PF-127 and AgNPs loaded PF-127 being 2.01 ± 0.49 μm and 1.40 ± 0.4 μm, respectively as evident from the SEM images in Fig. 10.1. EDS analysis revealed the presence of Ag and its homogenous distribution in the hydrogel framework. The AgNPs-loaded PF-127 hydrogel showed significant ZOI and bacterial cell viability reduction equivalent to >97%, >96%, and 95%, against *P. aeruginosa*, *S. epidermidis*, and *S. aureus*, respectively, as seen in Fig. 10.2. Significant antibiofilm disruption was noted for both single-species and multispecies biofilm using 50 μg/g AgNPs-loaded PF-127 hydrogel in comparison to the same concentration of ciprofloxacin.

An antimicrobial hydrogel was reported by impregnating AgNPs in a triple-layered cellulose acetate (CEA) based hydrogel nanofibrous sheet (TNF) for wound healing (Dugam et al. 2024). Initially, the top polyvinyl alcohol (PVA) layer was formed by mixing 9% w/v PVA and distilled water, followed by constant stirring for 4 h until a homogenous mixture was formed. A 5 mL syringe with an 18 g needle was loaded with the homogenous solution for electrospinning, while the nanofibrous (NF) material was deposited on a collector drum kept 10 cm apart. The electrospinning process was conducted at 30 ± 2 °C, and relative humidity of 40 ± 2% and 60 ± 2%. The NF sheet was peeled off and cross-linked by citric acid (CA, 1–5% w/v) to enhance the mechanical properties. The cross-linked NF mats were treated in a vacuum oven by CA vapours at 90 °C for 15 min. The middle AgNPs-loaded CEA layer was prepared by mixing 10 mM $AgNO_3$ and 3 mM $NaBH_4$, which led to the formation of AgNPs. The formed AgNPs were dispersed in water at a 200 ppm concentration and added to the solution containing

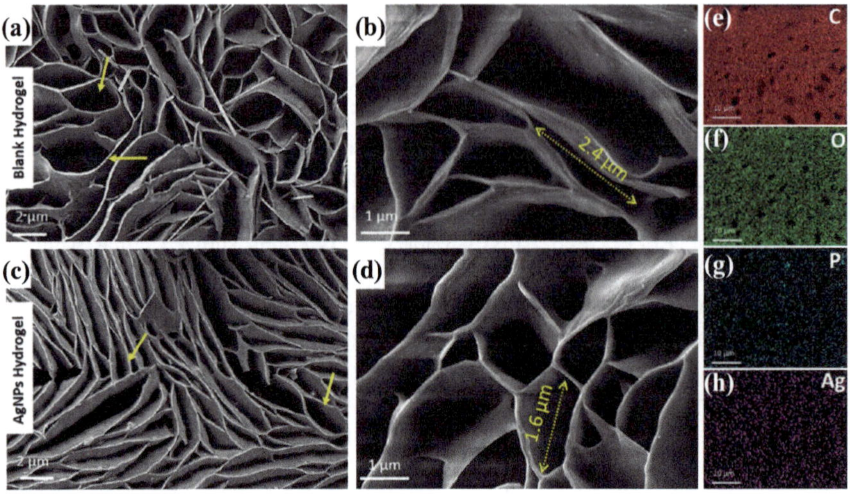

Fig. 10.1 SEM micrograph of the blank hydrogel (**a**, **b**) and the hydrogel loaded with AgNPs (**c**, **d**) at two different magnifications. **e**–**h** EDS mapping of major elements present on the surface of the AgNP-loaded hydrogel. Reprinted with permission from Haidari et al. (2020). Copyright © 2020 American Chemical Society

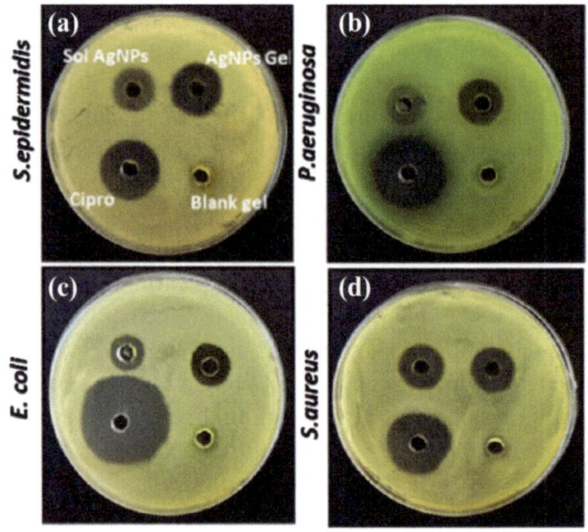

Fig. 10.2 Antibacterial properties assessed via the disk diffusion assay. **a**–**d** Zone of inhibition for **a** *S. epidermidis*, **b** *P. aeruginosa*, **c** *E. coli* and **d** *S. aureus*. Reprinted with permission from Haidari et al. (2020). Copyright © 2020 American Chemical Society

CEA (16% w/v), *N*, *N*-dimethyl formamide (DMF), and acetone mixed at a ratio of 1:3. The mixture was stirred at 500 rpm until a homogenous dispersion of AgNPs in CEA was formed. The mixture was then loaded in a 5 mL syringe and subjected to electrospinning over the preformed PVA layer. The last layer was formed by electrospinning polycaprolactone (PCL) that was prepared by adding 10%w/v PCL in a solution of chloroform and dimethyl sulfoxide (DMSO) mixed in a ratio of

3:1 at 25 °C. SEM revealed that the sizes of PVA NF, CEA NF, and PCL NF were 132.46 ± 15.32 nm, 161.23 ± 12.23 nm, and 212.56 ± 27.23 nm, respectively. Attenuated total reflectance infrared spectroscopy (ATR-FTIR) exhibited OH group, C–H alkyl stretch, alcohol stretch, C–O stretch for PVA and PVA NFs at 3297.60 cm^{-1} and 3288.60 cm^{-1}, 2975.62 cm^{-1} and 2975.62 cm^{-1}, 1284.078 cm^{-1} and 1288.15 cm^{-1}, 1033.45 cm^{-1} and 1048.32 cm^{-1}, respectively. Pure CA showed O–H stretch, C–H stretch, and C=O stretch vibrations at 3494.30 cm^{-1}, 3019.00 cm^{-1} and 2558.00 cm^{-1}, 1753.19 cm^{-1} and 1721.83 cm^{-1}, respectively. The crosslinked CA showed C=O stretch at 1733.00 cm^{-1} and 1715.09 cm^{-1}. CEA and CEA NF gave stretching vibrations of C–H, C=O, C–O at 2945 cm^{-1}, 1730–1737 cm^{-1}, and 1050 cm^{-1}, respectively, and bending of C–O and C–H at 1228 cm^{-1} and 1368 cm^{-1}, respectively. PCL and PCL NF showed the presence of CH_2 group stretch, C=O group stretch, and asymmetric with symmetric stretch of C–O–C at 2943–2940 cm^{-1} and 2863–2855 cm^{-1}, 1730 cm^{-1}, 1250–1300 cm^{-1} and 1150–1200 cm^{-1}, respectively. TGA analysis revealed that PVA and PVA NF, CEA degraded in three steps, while CA degraded in one step, and PCL, PCL NFs in two stages. Slower thermal deposition and greater thermal stability were noted in the composites. XRD analysis revealed distinct crystalline peaks for pure PVA at 2Θ degrees of 11.211°, 19.450°, 20.654°, and 23.520°. However, the PVA NF did not show any peaks, indicating the conversion to an amorphous nature from a crystalline nature, due to electrospinning. Pure CA exhibited peaks at 14°, 19°, and 26°, but no peak was observed for cross-linked mats. CEA showed peaks at 14°, 20°, and 25°, though no peak was observed for CEA NF. PCL and PCL NF both displayed peaks at 21.3°, 27°, and 29°, indicating the crystal nature of PCL NF leading to a controlled release of AgNPs. The TNF sheet had sufficient water uptake capacity. WVTR and porosity of PVA CA NF, CEA NF, CEA AgNPs NFs, PCL and TNF, were 192.70 ± 1.020 $g/m^2/24$ h, and 90.52 ± 2.42%, 192.30 ± 4.55 $g/m^2/24$ h, and 87.01 ± 1.36%, 191.2 ± 1.523 $g/m^2/24$ h, and 88.86 ± 2.72%, 200.8 ± 3.800 $g/m^2/24$ h, and 93.6 ± 1.70%, 192.50 ± 3.189 $g/m^2/24$ h, and 85.75 ± 1.80%, respectively. The degradation of TNF was faster initially, but it slowed down due to the presence of the hydrophobic polymer PCL. Mechanical strength of PVA, PVA CA NF, CEA AgNPs NF, PCL NF, and TNF were 2.88 ± 0.25 MPa, 6.82 ± 0.37 MPa, 1.22 ± 0.035 MPa, 1.55 ± 0.45 MPa, 11.90 ± 0.77 MPa, and 12.70 ± 0.606 MPa, respectively. The viability of human dermal fibroblast (HDF) cells on PVA NFs, PVA CA NFs, CEA NFs, AgNPs loaded CEA NFs, PCL NFs and TNF was 97.75112 ± 2.33%, 97.77661 ± 2.8%, 94.4028 ± 1.48%, 98.90055 ± 1.07%, 93.45327 ± 2.6% and 99.17 ± 1.6%, respectively. The viability of keratinocyte (HaCat) cells was 80.47337 ± 1.01%, 72.51 ± 12.16%, 78.6 ± 4.2%, 91.6 ± 1.95%, 79.0 ± 1.52%, and 76.9 ± 5.23%, respectively, on the aforementioned biomaterials. The antibacterial activity was checked against *E. coli* and *S. aureus*, by disc diffusion method, showing ZOI of 2.5 ± 0.23 mm, and 2.06 ± 0.16 mm for AgNPs, 1.98 ± 0.12 mm, and 1.76 ± 0.09 mm for TNF, respectively. Single and double-layered materials exhibited ZOIs equivalent to 1.9 ± 0.4 mm, and 3.18 ± 0.63 mm, for *E. coli*, 5.4 ± 0.3 mm, and 2.36 ± 0.33 mm, for *S. aureus*, respectively. Wound healing activity was checked employing a partial thickness wound model in Sprague–Dawley rats. Wound closure in untreated, cotton gauze, and TNF mats with AgNPs treated groups were 50%, 80%, and 90.28%, respectively, after 10 days.

10.5 Metal Oxide Nanoparticles Incorporated Hydrogels

Titanium dioxide nanoparticles (TiO_2NPs) were impregnated in chitosan/ polypropylene glycol (CH/PPG) hydrogel for enhanced antimicrobial activity (Ulu et al. 2020). Controlled hydrolysis of tetraethyl ortho titanate, also known as titanium ethoxide (TEOT), was carried out under an N_2 gas atmosphere. The hydrolyzed TEOT was mixed with absolute ethanol (50 mL), followed by stirring in an atmosphere of N_2 at room temperature for 3 h. SDS (1.3 g/50 mL ethanol) was added to the mixture and stirred for 3 h at room temperature. The temperature was then elevated to 50 °C and refluxed for 3 h, followed by cooling the mixture. The resulting white precipitate was collected by centrifugation and heated at 120 °C for 3 h. The powder was calcinated at 350 °C to remove the organic compounds. CH was dissolved in 1% v/v acetic acid and stirred overnight until a clear solution was formed at room temperature. PPG was added to the CH solution and mixed by magnetic stirring until a clear solution was formed. CH and PPG were mixed in a 1:1 ratio. TiO_2NPs powder was suspended in water by ultrasonication for 10 min. The nanosuspension was added dropwise into the blended solution at a relative ratio of 0% and 10% to the weight of the blended solution. The mixture was homogenized by vigorous stirring for 24 h and poured into a Teflon mould, followed by drying in a vacuum oven for 48 h. This led to the formation of CH/PPG/TiO_2 composite hydrogel. DLS analysis showed the hydrodynamic diameter of the TiO_2NPs was 40 nm. UV–visible spectroscopy showed an absorbance maximum at 255 nm. SEM images revealed mostly spherical shapes of the TiO_2NPs with a uniform distribution. A homogenous smooth surface was noted for the CH/PPG film. However, a slightly rough surface was observed for CH/PPG/TiO_2 films. Peaks observed in the EDS spectrum indicated the presence of elemental Ti and O in the surface of TiO_2NPs. FTIR spectra exhibited the stretching vibrations at 3354 cm^{-1}, 3284 cm^{-1}, 2914 cm^{-1} and 1070 cm^{-1}, for the O–H curve, N–H curve, CH_2 group and C–O–C bond, respectively. Peaks at 1651 cm^{-1} and 1566 cm^{-1} were attributed to the amide I and amide II of CH, respectively. Stretching vibrations were observed at 2970–2866 cm^{-1}, and 1092 cm^{-1} for the C–H and C–O–C bond of PPG, respectively. Ti–O bending and Ti–OH stretching bands of TiO_2NPs were observed at 483 cm^{-1} and 1623 cm^{-1}, respectively. XRD analysis data of TiO_2NPs appeared at 2Θ degrees of 25.8°, 31.0°, 38.1°, 47.8°, 55.1° and 63.4°. CH/PPG exhibited a broad peak at 20° due to its amorphous nature. Peak intensity of CH/PPG/TiO_2NPs decreased in comparison to CH/PPG. TiO_2NPs' presence gave characteristic diffraction peaks at 25.8°, indicating the crystal planes of TiO_2NPs. The presence of TiO_2NPs with a homogenous distribution in the CH/PPG matrix resulted in a minor alteration in the peak intensity. TGA curves demonstrated that the weight loss in the PPG structure and groups were between 170 °C–276 °C and 190 °C–330 °C, respectively. Pure TiO_2NPs had weight loss at 100 °C (3%)

and 230 °C–460 °C (6.7%), while degradation of the CH groups was noted at 330 °C–450 °C. The decomposition of CH/PPG and CH/PPG/TiO_2NPs was up to 3% at 80–150 °C which was attributed to the water removal. Total residual amounts for CH/PPG film and CH/PPG/TiO_2NPs composite film structures were 67.24% and 72.45% at 700 °C, respectively. Thermal decomposition of PPG and CH groups was noted between 220–330 °C and 335–465 °C, respectively. Atomic force microscope (AFM) analysis confirmed the uniform distribution of TiO_2NPs over the CH/PPG film surface. EDS analysis confirmed the presence of carbon (C), nitrogen (N), oxygen (O), and Ti in the CH/PPG/TiO_2NPs composite film. The swelling ratio for CH/PPG film was >110%, while it was 90% for CH/PPG/TiO_2NPs film, indicating the decrease in the ratio on incorporation of TiO_2NPs. The initial high water retention ability in CH/PPG and CH/PPG/TiO_2NPs films decreased with time. The retention percentage was 77.3 and 68.5% after 60 min which became almost similar after 240 min. Water uptake by the CH/PPG/TiO_2NPs film (50%) was lower than the pure CH/PPG films (59%) as noted after 60 min. CH/PPG and CH/PPG/TiO_2NPs showed 36 and 32% weight loss after 4 weeks of immersion in PBS. The lower weight loss in CH/PPG/TiO_2NPs films was due to the slow degradation property of TiO_2NPs. CH/PPG and CH/PPG/TiO_2NPs composite films had a thickness equivalent to 0.372 ± 0.008 nm and 0.367 ± 0.073 nm, respectively, while the contact angles were $44.7° \pm 1.1°$ and $38.7° \pm 2.1°$, respectively. The composites showed a porosity equivalent to 95.4% and 94.6%, respectively, while their WVTR values were 365 g m^{-2} day^{-1} and 369.3 g m^{-2} day^{-1}, respectively. Mechanical properties were 1.88 MPa (elongation 27%), and 3.0 MPa (elongation 31%), respectively. CH, TiO_2NPs, CH/PPG and CH/PPG/TiO_2NPs composite films exhibited ZOIs of 14.67 ± 0.577 mm, 12.00 ± 1 mm, 16.67 ± 0.577, and 17.67 ± 0.577 mm, respectively, against *E. coli* ATCC 25,922. The ZOIs against *S. aureus* ATCC 25,923 were 16.33 ± 0.577 mm, 15.00 ± 0.00 mm, 20.00 ± 1.00 mm, and 22.33 ± 0.577 mm, respectively. *Candida lipolytica* was inhibited as well with ZOIs equivalent to 16.00 ± 1.00 mm, 15.33 ± 0.577 mm, 19.33 ± 0.577 mm, and 21.00 ± 0.000 mm, respectively. This confirmed the superior antimicrobial activity of the CH/PPG/TiO_2 biocomposite film.

Both exogenous and endogenous antibacterial nanocomposite hydrogel films were developed by impregnating TiO_2NPs in norbornene-functionalized aldehyde containing dextran (Nor-Ald-Dex)/quaternized chitosan (QCS)/silica coated TiO_2 hydrogel (NorAld-Dex/QCS/ TiO_2 @ MS-SH) (Yan et al. 2024). Initially, norbornene-functionalized dextran (Nor-Dex) was prepared by conjugating carbic anhydride at pH 9.8–10.5 in water. Then sodium periodate ($NaIO_4$) was used for further oxidation for 24 h which resulted in the addition of the aldehyde groups. This led to the formation of NorAld-Dex. QCS was prepared by slow dripping of glycidyl trimethylammonium chloride (GTMAC) into an aqueous CS solution under acidic conditions. Mesoporous silica shell (MS) coated TiO_2NPs were synthesized by sol–gel reaction using tetraethyl orthosilicate (TEOT) and cetrimonium bromide (CTAB). Thiol groups modified MS-coated TiO_2NPs on reaction with 3-mercaptopropyltrimethoxysilane (MPTMS) led to the formation of MS-SH TiO_2NPs. Three types of hydrogels were prepared by mixing NorAld-Dex and QCS, in the absence or presence of TiO_2NPs, which were NorAld-Dex/QCS, NorAld-Dex/

QCS/TiO_2NPs, and NorAld-Dex/QCS/TiO_2NPs@MS-SH hydrogels. ^{1}H Nuclear magnetic resonance (^{1}H NMR) analysis revealed peaks for NorAld-Dex at 3.4–3.9 ppm and 4.9 ppm for the polymeric backbone of dextran. The peaks at 1.4 ppm and 3.0–3.3 ppm were for bridge protons and carboxyl adjacent protons of carbic groups, respectively. The peaks at 6.1–6.3 ppm were for alkene protons of norbornene structure on carbic anhydride. These different peaks for NorAld-Dex indicated the successful implantation of carbic anhydride on dextran. Hemiacetalic protons of the aldehyde group were recognized by the peak at 5.1–5.6 ppm, which indicated the oxidation of hydroxyl groups of Nor-Dex to aldehyde groups. A peak at 3.2 ppm was due to the protons from the quaternary ammonium groups of GTMAC. FTIR spectra of NorAld-Dex exhibited stretching vibrations at 1730 and 1565 cm^{-1} for the C=O group and at 1645 cm^{-1} for the C=C group of the norbornene groups. The peak at 1475 cm^{-1} was due to the asymmetric angular bending of methyl groups in quaternary hydrogen. Peaks for MS-SH coated TiO_2NPs were noted at 2560, 1078, and 960 cm^{-1} which were due to the S–H bonds, Si–O–Si bonds, and Ti–O–Ti bonds, respectively. This indicated that the thiolated silica shell was successfully coated on the TiO_2NPs. The presence of imine crosslinks was confirmed by the peak at 1640 cm^{-1} showing the bond C=N in all three hydrogels. The oxidative degree of the NorAld-Dex was 24% while the remaining amine groups on QCS were almost 8.5%. This indicated that the synthesized QCS possessed 76.5% substitution of GTMAC. The compressive modulus of the hydrogel prepared were 8.69 ± 1.68, 8.21 ± 0.61, 8.90 ± 1.43, 19.47 ± 2.78, 11.01 ± 2.71, 10.46 ± 1.14, and 9.13 ± 2.96 kPa with the sonication time of 3, 5, 10, 30, 60, 120, and 180 min, respectively. SEM images showed that the aggregation behaviours of TiO_2NPs were significantly reduced due to the coating by MS-SH shells. TEM revealed the particle size of TiO_2NPs as 29.4 nm, while the hydrodynamic size was 252 nm. The particle size and hydrodynamic size of MS-SH TiO_2NPs were 34.3 nm, and 306 nm, respectively. X-ray photoelectron spectroscopy (XPS) analysis revealed that MS-SH-coated TiO_2NPs were composed of 27.15% C, 54.05% O, 2.81% Ti, 15.46% Si, and 0.54% S. Ti–O–Si bond was observed at 534.6, 460.1, and 102.1 eV, under O 1 s, Ti 2p, and Si 2p spectra, respectively. This confirmed the core–shell structure of MS-SH-coated TiO_2NPs. NorAld-Dex/QCS/TiO_2NPs@MS-SH hydrogel sonicated for 0, 10, and 30 min (NCH0, NCH10, NCH30 hydrogels) showed pore sizes equivalent to 112.04 ± 26.20, 117.49 ± 28.61, and 62.6 ± 14.53 μm, respectively. Pore diameters for NCH0, NCH10, and NCH30 were 94.98, 72.47, and 43.37 μm, respectively. The swelling ratios were 11.33 ± 0.38%, 10.48 ± 0.23%, and 7.06 ± 1.12%, respectively. The water content was 91.01 ± 0.31%, 89.19 ± 1.82%, and 86.16 ± 1.84%, respectively. Water content above 85% indicated that the hydrogels had good moisturizing properties. The stability was checked by immersing them in PBS solution with pH 7.4 at 37 °C. The result showed that the remaining mass was 60.66 ± 3.61%, 63.81 ± 1.13%, and 69.34 ± 5.76%, respectively. G′ of NCH10, at pH 5.5, 6.5, and 7.4, were 1169, 1282, and 1634 Pa, respectively. This indicated lower G′ in the acidic environment, which was due to the hydrolyzation of imine bonds present in the network, causing the softening of the hydrogel structure. The antibacterial efficacy was checked against *E. coli* and *S. aureus* by the spotting assay and the spread plate method. The NorAld-Dex/QCS/

TiO_2NPs@MS-SH hydrogel showed antibacterial activity against them at pH 7 in the absence of ultrasound irradiation. Higher antibacterial efficacy was seen at pH 5.5 than at pH 6.5, and pH 7.4. The spread plate method quantified the number of viable bacteria, showing a significant reduction in the survival of *S. aureus* by 1.04, 0.99, and 0.66 log units at pH 5.5, 6.5, and 7.4, respectively. Reduction in the viable number of *E. coli* was by 7.86, 2.86, and 1.97 log units, at the respective pH values. The results signified that the NorAld-Dex/QCS/TiO_2NPs@MS-SH hydrogel showed enhanced antibacterial efficacy in an acidic environment. Significant bactericidal activity was seen against *E. coli* in comparison to *S. aureus* due to the thickness of the cell walls and the ability to penetrate the bacterial cell walls by positively charged QCS. Cytocompatibility testing of the three different hydrogels was done by mouse embryonic fibroblasts (MEFs) isolated from C57BL/6N mice embryos. The Alamar Blue assay demonstrated that the MEFs treated with all of the samples had increased metabolic activity with time. The live and dead fluorescent staining showed the proliferation of MEFs up to 7 days, confirming the cytocompatibility of the hydrogels.

An antibacterial hydrogel film was developed with zinc oxide-impregnated mesoporous silica (ZnO-MCM-41) in carboxy methyl cellulose (CMC) (Rakhshaei and Namazi 2017). Initially, 2.74 mmol CTAB and 480 mL NaOH (150 mM) were mixed, followed by the dropwise addition of 22.4 mmol TEOS. The mixture was stirred vigorously for 2 h at 80 °C. The product was isolated by hot filtration, washed with distilled water and methanol, and dried subsequently. The resulting powder was calcinated at 600 °C for 6 h. Then 0.33 g of the calcinated powder was introduced into the aqueous solution (0.20 g/12 mL) of zinc nitrate hexahydrate (Zn $(NO_3)_2 \cdot 6\,H_2O$). The resulting mixture was dried after 1 h at 80 °C, followed by the collection of the precipitate and heating at 400 °C for 3 h to oxidize the impregnated Zn. Then CMC (1.5 g in 250 mL distilled water) and 7.44 g (0.025 mol) of Zn $(NO_3)_2 \cdot 6\,H_2O$ were mixed followed by dropwise addition of 250 mL NaOH solution (0.2 mol/l) under continuous stirring. Further, the mixture was centrifuged at 800 rpm for 20 min, and the settled precipitate was washed several times, dried and calcinated at 540 °C for 7 h. SEM image showed aggregated particles of different sizes for ZnONPs. Similarity for native and ZnO-MCM-41 NPs with no specific difference confirmed no aggregation of ZnONPs in ZnO-MCM-41 NPs. XRD analysis exhibited an intense diffraction peak at 2Θ of 2.2° and low-intensity peaks at 2Θ of 3.8° and 4.4°, respectively for ZnO-modified MCM-41 and MCM-41. This indicated the formation of hexagonal MCM-41 mesoporous material. The pore diameters of native MCM-41 and ZnO-impregnated MCM-41 were 4 nm and 3.5 nm, respectively. Diffraction peaks at 2Θ of 2.55°, 4.29°, and 4.91°, appeared for ZnO-MCM-41 NPs. Zeta potential analysis revealed that the ZnONPs changed the negative superficial charge of the native MCM-41(-20 ± 3.1 mV) to a positive one ($+9.3 \pm 1.2$ mV). A clear and uniform surface morphology was noted for pure CMC hydrogel, unlike MCM-41 and ZnO-MCM-41 NPs, which showed a rough surface with visible spherical particles. The tensile strength of CMC was 50.5 MPa. However, MCM-41 and ZnO-MCM-41 showed an enhancement in the tensile strength by 24% and 12%, respectively. Citric acid cross-linked CMC was stable during 48 h, but CMC/ZnO got disintegrated early. CMC/ZnO-MCM-41 had a higher swelling ratio than CMC and CMC/MCM-41. The addition of 10% ZnO-MCM-41 increased the swelling capacities of CMC hydrogel by 2.2-fold. The addition of MCM-41 to the CMC hydrogel film increased its water vapour and O_2 permeability. The antibacterial activity was checked against *E. coli* and *S. aureus* by the disc diffusion method. The maximum ZOIs against *E. coli* were 10 ± 1 mm, 11 ± 1 mm, and 15 ± 2 mm for pure CMC hydrogel, CMC/MCM-41, and CMC/ZnO-MCM-41, respectively. Maximum ZOIs against *S. aureus* were 9 ± 1 mm, 11 ± 1 mm, 16 ± 2 mm for pure CMC, CMC/MCM-41, and CMC/ZnO-MCM-41, respectively.

An antibacterial hydrogel was fabricated by impregnating ZnO/zeolite nanocomposite (ZnONC) in CS hydrogel (Afrasiabi et al. 2021). Initially, zeolite powder was suspended in distilled water and shaken for 1 h. The solution was then filtered, and the residue was dried at 80 °C, causing an exchange of zeolite by Zn^{2+}. Then, 10 g of zeolite powder was mixed with 7 g of zinc acetate in an aqueous solution with continuous stirring at 60 °C for 1 h. ZnONPs were fabricated on a zeolite bed with the addition of 1 M NaOH until pH 12 was obtained. The nanocomposite was collected by filtration and washed thoroughly with distilled water to remove excess Zn. The filtrate was dried in an oven at 80 °C and calcinated at 400 °C for 2 h. The CS solution in 1% v/v acetic acid was autoclaved at 121 °C for 15 min. Further, ZnONC-CS was prepared by mixing 5 mL of the resulting CS solution and 5 mg of ZnONC in the presence of stirring at 37 °C for 60 min. White precipitate appeared with the slow addition of NaOH (1 M), until the pH reached 8–8.5. The final product was kept at 4 °C for 24 h for settling. The antibacterial activity was checked against *Streptococcus mutans*, where the minimum inhibitory concentration (MIC) of ZnONC, ZnONC-CS, and CS were 78.1 μg/mL, 78.1 μg/mL, and 156.2 μg/mL, respectively. The toxicity of CS gel increased with an increase in concentration (>156.2 μg/mL), while the incubation period of 48 h did not show any significant effect on cell viability of human gingival fibroblasts (HGFs). ZnONC showed no cytotoxicity up to 78.1 μg/mL. However, it showed cytotoxicity between 156.2 and 625 μg/mL, with an incubation period of 48 h and 72 h. ZnONC-CS inhibited biofilm formation in *S. mutans* up to 33%. Treatment with ZnONC, ZnONC-CS, and CS resulted in the reduction of the metabolic activity by 19.75%, 30%, and 45%. FE-SEM analysis of the biofilm revealed that the number of bacteria was reduced on treatment with CS gel and ZnONC. However, treatment with ZnONC-CS resulted in a reduction in the bacterial cell number, single cells, and short chains. Quantitative real-time PCR (RT-PCR) showed downregulation of the genes *gtfB*, *gtfC* and *ftf*, by 1.73-, 1.84- and 1.49-fold, respectively, on treatment with ZnONC. Likewise, the expression of the aforementioned genes was reduced by 4.16, 4.40, and 2.88-fold by ZnONC-CS.

Another antimicrobial hydrogel with polyethyleneimine/polyacrylic acid (PEA) and cerium dioxide quantum dots (CeO_2-QDs) (Hou et al. 2024). Initially, polymethyl methacrylate (PMMA) was synthesized using 700 mL of water, 0.18 g of potassium persulfate powder ($K_2S_2O_8$), and methyl methacrylate at 70 °C for 60 min. The suspension was centrifuged, washed, and dried to obtain PMMA spheres. Then, 1.2 g of PMMA spheres were dispersed in a mixture of 50 mL ethanol and 50 mL water, followed by the addition of 10 mL ceric ammonium nitrate solution (3 g/L). Then, 0.6 g dopamine hydrochloride in 20 mL DI water was added and the pH was maintained at 8.5. After 24 h, the resulting dark grey suspension was filtered, washed and dried overnight at 60 °C, followed by calcination at 650 °C in the presence of Ar gas for 3 h. The final product was CeO_2-QDs. Further, PEA hydrogel was prepared by mixing 10 wt% of each polyethyleneimine and polyacrylic acid in a 1:1 ratio to ensure maximum mechanical and adhesive strength. The mixture was immersed in liquid N_2 for 15 min to freeze, which was then pulverized into a powder. The PEA@ CeO_2-QDs hydrogel was prepared by mixing polyethyleneimine and polyacrylic acid with different doses of CeO_2-QDs suspension. SEM and TEM images revealed a 3D-honeycomb-like- relatively smooth texture with interconnected, uniformly dispersed pores and a higher exposed surface. The CeO_2-QDs were 1.5 to 2.5 nm in size. XPS spectra revealed the presence of elemental C, N, O, and Ce^{3+} in CeO_2-QDs. FTIR spectra of PEA@CeO_2-QDs hydrogel showed a peak shift from 1563 cm^{-1} to 1560 cm^{-1}, and from 1635 cm^{-1} to 1630 cm^{-1} due to the amine group, and carboxylic acid, respectively. The highest tensile strength of 55 kPa was observed in case of 75 mg/mL CeO_2-QDs containing hydrogel, which was 31.25% less than the pure PEA hydrogel. PEA and PEA@CeO_2-QDs were viscoelastic, signifying good wound adaptability, which is crucial for fabricating wound dressing. Superior adhesion on the porcine skin was observed with an increase in the concentration of CeO_2-QDs in the hydrogel. The lap shear stress of the hydrogel decreased from 75 to 73 kPa, and 47 kPa. Likewise, the strain of the hydrogel decreased gradually from 4.6 to 2.3, 2.2, and 0.8 mm/mm, after complete dislocation from porcine skin. G'' varied from 8,000 to 2,500 Pa, and G' varied from 17,500 to 2,500 Pa, for PEA hydrogel. The value of G' was reduced to 8500–600 Pa, 5700–500 Pa, and 5000–400 Pa, whereas G'' was reduced to 4700–600 Pa, 2500–500 Pa, and 3200–700 Pa for PEA@25CeO_2-QDs, PEA@50CeO_2-QDs, PEA@75CeO_2-QDs, respectively. The swelling ratio for PEA hydrogel was 117.1 ± 0.8% after 24 h, while the same for PEA@25CeO_2-QDs, PEA@50CeO_2-QDs, PEA@75CeO_2-QDs were 119.4 ± 1.6%, 117.6 ± 2.9%, and 117.6 ± 1.5%, respectively. This confirmed that the addition of CeO_2-QDs had no negative effect on the swelling rates of the PEA hydrogel. Cytotoxicity was checked against CCK-8 cells that showed considerable proliferation even after 72 h which was dependent on the concentrations of CeO_2-QDs. *S. aureus* and *E. coli* were significantly inhibited even after 2 h in co-culture when treated with PEA@CeO_2-QDs hydrogel. PEA@50CeO_2-QDs and PEA@75CeO_2-QDs hydrogels containing 50 mg/mL and 75 mg/mL CeO_2-QDs, respectively, showed superior antibacterial activity that was attributed to the rupturing of the bacteria by the generated reactive oxygen species (ROS).

10.6 Carbon-Based Nanoparticles Incorporated Hydrogels

Highly cationic carbon dots (CDs) were impregnated in acrylic acid (AA), and pectin (P) triggered hydrogel (Cui et al. 2021a, b). Initially, CDs were prepared by mixing *O*-phenylenediamines/spermidine and O-phenylenediamines/α-lipoic acid in a 1:1 ratio and dissolved in water at different concentrations. Thereafter, 30 mL of the solution was transferred to a polytetrafluoroethylene-lined autoclave to synthesize the CDs, spermidine CDs (SCDs), and lipoic acid CDs (LCDs) at 220 °C for 3 h. The hydrogel was prepared by mixing P, AA, ammonium persulfate (APS), PEG dimethacrylate (PEGDA), and CDs (1 mg/mL) which was transferred to a mold and heated at 40 °C, resulting in the formation of AP hydrogel, SCDs-AP hydrogel, CDs-AP hydrogel, and LCDs-AP hydrogel. Under UV–visible spectroscopy, La-Y-CDs, Spe-Y-CDs, and Y-CDs, showed emission peaks at 600, 320, and 400 nm, while the excitation peaks were at 400, 320, 420 nm, and the absorption peaks were at 240, 290 and 400 nm, respectively. Fluorescence quantum yields were 0.36%, 3.82% and 2.15%, respectively. TEM images revealed respective size distributions of 5–10 nm, with average diameters of 9.68 ± 0.21, 5.02 ± 0.43, and 7.80 ± 0.68 nm. Under a 365 nm laser, they showed red, blue, and yellow fluorescent emission, respectively. XPS indicated the presence of S (169 eV), C (285 eV), and O (531 eV), on their surfaces. FTIR exhibited absorption peaks at 1560 cm^{-1}, and 1385 cm^{-1}, for La-Y-CDs that were much higher than others, indicating the presence of carboxylate. XRD analysis showed a broad peak at 2θ values of 21.0°, and 26.6°, due to the layered spacing of the graphite structure of 0.41 and 0.34 nm. Zeta potentials were +17.06, +51.20, and +12.28 mV for the three CDs, respectively. The fluorescent properties of LCDs-AP, CDs-AP, and SCDs-AP, hydrogels were orange-red, yellow, and blue, respectively. The stability of the swelling behaviour of SCDs-AP hydrogels, with Spe-Y-CDs, and AP hydrogels, without CDs, was much higher than the other two CDs, with a swelling ratio of 4–5. The swelling behaviour was stable due to the static electricity when the CD content was >75%. Under SEM, different CD hydrogels showed identical microstructure after swelling and freeze-drying, but differed after the addition of various CDs. A 25 mg/mL concentration was slightly bacteriostatic for lipoic acid and the La-Y-CDs and completely bacteriostatic for spermidine and the Spe-Y-CDs. MIC against *E. coli*, *Pseudomonas fluorescens*, *Yersinia enterocolitica*, methicillin-resistant *S. aureus* (MRSA) and multidrug-resistant *Staphylococcus typhimurium* (MRST), were 64 μg/mL, while it was 32 μg/mL for other bacterial strains. In case of Spe-Y-CDs. MIC of spermidine (25 mg/mL) was much higher than the MIC of Spe-Y-CDs, while Spe-Y-CDs had the same or even better MIC than the standard antibiotic, kanamycin. The antibacterial activity was checked against MRSA and MRST for four hydrogels, giving clear respective ZOI of 21.7 and 17.7 mm. The antibacterial activity was attributed to the damaging and wrinkling of the cell surface or cell membrane after 3 h of incubation with SCDs-AP hydrogel, which showed 108.5 to 88.5-fold higher ZOI compared to AP hydrogel, as evident from Fig. 10.3. Biocompatibility and cell affinity of the hydrogels were checked against fibroblast cells L929. Best cell adhesion was noted for SCDs-AP hydrogel compared to others. Although the viability of cells was higher in LCDs-AP, CDs-AP, and SCDs-AP hydrogels than in AP hydrogels, SCDs-AP hydrogels showed the best cell viability, due to the presence of CDs and P. Further, the in vivo antibacterial activity of SCDs-AP hydrogel was checked using a male rabbit subcutaneous infection model. SCDs-AP hydrogel was implanted subcutaneously on the backs of the rabbits, followed by injection with 1 mL *S. aureus* suspension at the site. After 4 days, macrophages, fibroblasts, and new blood vessels appeared at the wound site, providing growth factors, new extracellular

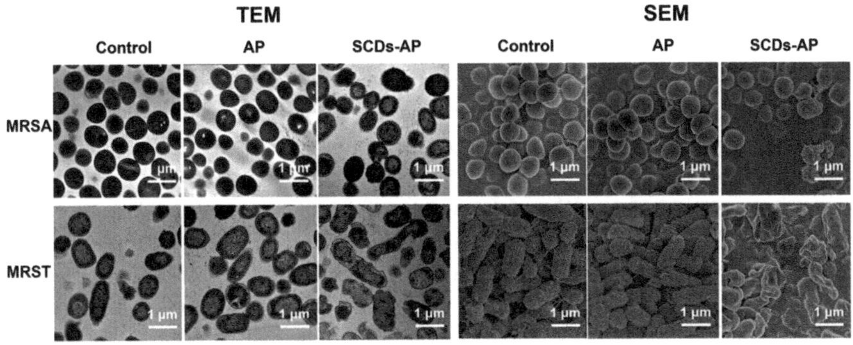

Fig. 10.3 TEM (left) and SEM (right) of MRSA and MRST before and after treatment with hydrogels (AP and SCDs-AP). Reprinted with permission from Cui et al. (2021a). Copyright © 2020 Elsevier B.V

matrix, oxygen, and nutrients. However, the wounds treated with AP hydrogel were prominent with scars, pustules, edematous tissue, and multinucleated giant cells, indicating a severe inflammatory response caused by the applied organism. Therefore, the SCDs-AP hydrogel showed superior in vitro and in vivo antibacterial activity.

An antibacterial hydrogel film using hyaluronic acid (HA) was reported to involved the impregnation with CDs (Lee et al. 2022). Initially, CDs were prepared by mixing 1.5 g ammonium citrate and 714 μL polyethyleneimine, followed by dissolving in 15 mL water. The mixture was placed in a hydrothermal reactor at 180 °C for 2 h. The unreacted monomers were removed by dialysis for 1 day, after which it was lyophilized. The CDs gel was prepared by mixing 30 g of CDs, 60 mg HA, and 120 mg of gelatin in 3 mL of 0.25 M NaOH solution. The mixture was stirred for 18 h under mild agitation, followed by the addition of 200 μL of 1,4-butanediol diglycidyl ether (BDDE). The solution was again stirred for 1 h at 4 °C for further gelation. After incubation in water and PBS with pH of 7.4, the gel was swollen to around 250% of its original weight. CDs showed an excitation wavelength at 450 nm. TEM images revealed that the size of the CDs ranged from 2 to 10 nm, and the size of the freshly prepared hydrogels was 50–100 μm. The storage modulus (G′) was significantly higher than the loss modulus (G″) for the CDs hydrogel. SEM images revealed that after overnight swelling in PBS with pH 7.4 in the absence of HAase, the CDs gel pore size increased from 50–100 to 200–300 μm without affecting the hydrogel network. The photodynamic capability of CDs was analyzed by singlet oxygen sensor green (SOSG) measurements, which showed an emission at 488 nm. The cytotoxicity was evaluated against L929 cells, a murine fibroblast cell line. CDs gel showed negligible cytotoxicity up to concentrations of 50 mg mL^{-1} after 24 h incubation. A hemolysis assay confirmed the biocompatibility of the hydrogel. Hydrogel-treated blood samples exhibited a hemolysis degree of 1%, which was much lower than the hemolysis degree limit. Under the light, CD gel was highly effective, causing 99% and 97% mortality of *S. aureus* ATCC 25,923 and *E. coli* KCTC 2571, respectively. This was attributed to the disruption of the cell membrane by the photodynamically generated O_2, released by CDs. Only the HA gel showed negligible bacterial colony mortality, which was about 0–10%. Further, in vivo wound healing activity was checked against the six-week-old female C57BL6 mice with an infected skin wound. Wounds (8 mm) were created in the dorsal region and 50 μL bacterial suspension was introduced at the wound site. After day 3, the infected wounds became noticeably smaller in the groups treated with the CDs gel than in the control group without the CDs gel, regardless of the presence of light. Wound closure was about 90% after 6 days of treatment with the CDs gel, while complete closure was noted after day 11. However, in the case of untreated wounds, <50% of wounds were closed on day 6 but showed no complete wound closure even after day 13. Photodynamic therapy showed a significant effect against Gram-positive bacteria, rather than Gram-negative bacteria. Histological analysis revealed that the treatment with the CDs gel under light resulted in an intact epidermis with a smooth interface at the wound site instead of an open injury.

An antibacterial hydrogel film was developed that was composed of €-polylysine (Plys) impregnated with anionic CDs, especially CD 31 (Mou et al. 2022). A mixture of glutaraldehyde (300 μL), and PEG (100 μL) was prepared in 5 mL of ethanol that was autoclaved for 140 min at 150 °C. The mixture was then cooled to room temperature, resulting in a brown solid that was dispersed in deionized water, and centrifuged at 10,000 rpm for 20 min to remove carbon aggregates. The product was dialyzed against water for 24 h and brown CDs were obtained by freeze-drying. CDs-Plys gel was prepared by mixing 400 mg/mL Plys, and 200 mg/mL CDs, followed by gelation. TEM images revealed the spherical shape of CD31 with an average diameter of 6.96 ± 0.3 nm. UV–visible spectroscopy of CD31 showed absorption bands at 233 and 292 nm, along with maximum emission at 483 nm, under optical excitation of 400 nm. CD-Plys showed a maximum emission at 530 nm. FTIR exhibited peaks for CD31, at 3000–2800 cm^{-1} for –OH, and $–CH_2$, at 1700 cm^{-1} for –CHO, at 1100 cm^{-1} for –C–O. CD-Plys hydrogel exhibited peaks at 1664 cm^{-1}, 1700 cm^{-1} specific to imine bonds and aldehyde groups, respectively. XRD analysis showed a 2θ peak at 24.2° for CD31, a strong peak at 27.4°, and a broad peak at 40.3°, for the CD-Plys hydrogel. Zeta potential of CD31, Plys, and CD-Plys were −9.0, 9.29, and 9.18 mV, respectively. SEM images showed a porous structure of the hydrogel, with an interwoven network between CD31 and Plys. Rheological analysis revealed the G'' of the hydrogel was significantly lower than G' (3.3 MPa), indicating the hydrogel with higher mechanical strength. Cell viability of L929 cells was 95% even after direct contact with the hydrogel for 48 h. The antibacterial activity was checked against *E. coli* and *S. aureus*, by disc diffusion assay. ZOIs with 200 mg/mL CDs were 2.0 ± 0.02 cm and 3.0 ± 0.06 cm against the aforementioned bacteria, respectively. A higher antibacterial activity against *S. aureus* was attributed to the disruption of the cell membrane with efflux of cytoplasmic contents, while *E. coli* cells exhibited shrinking and cracking of the membrane. In vivo studies in female mice indicated regeneration of the tissue and healing in the epidermis of infected wounds in the two CD-Plys treated groups on day 3. The wounds were nearly closed on day 7 with 92% healing. However, the control groups showed severe inflammatory cell infiltration.

CD-Plys gel showed well-organized lamellar epithelium with orderly granular tissue, along with a large number of new blood vessels and hair follicles after day 7.

An antibacterial hydrogel was reported that was produced by impregnating garlic CDs (GCDs) in P-AA-based gel (Sheng et al. 2024). Initially, 1:4 w/v garlic cloves were mixed with ultra-pure water and blended to obtain the garlic juice that was filtered and heated at 210 °C for 4 h. GCDs were centrifuged at 14,000 rpm for 20 min, and passed through a 0.22 μm filter. The mixture was dialyzed in a 500 Da dialysis bag against ultra-pure water for 6 h, followed by freeze drying under vacuum. Then, 10 mL of GCDs (1.6 mg/mL), 15 mL of NaOH, 5 mL of AA, 150 μL of PEGDA, and 0.4 g of P were mixed till a homogenous solution was formed. It was mixed with 0.4 g of APS and heated at 50 °C for 20 min. The mixture was transferred to a mold and dried at 60 °C for 30 min, which led to the formation of GCD-PAA hydrogel. The control PAA gel was prepared by mixing the GCDs solution and 10 mL of NaOH (2.5 mmol/L). GCDs had varied particle sizes between 1.5 and 4.5 nm, with an average of 3.1 ± 0.23 nm, and a zeta potential of +8.53 ± 0.41 eV. XPS analysis revealed the presence of C (73.62%), O (16.32%), N (9.62%), and S (0.44%) in the GCDs. SEM images showed a continuous porous structure of PAA and GCDs-PAA hydrogel that had an average pore size of 1.08 ± 0.51, and 1.00 ± 0.49 μm, respectively. GCDs-PAA showed a denser pore structure with numerous small pores of size ≤0.5 μm. FTIR exhibited peaks at 2930 cm^{-1} for the stretching vibration of C–H, and at 1698 cm^{-1} for the ester group (C=O), signifying the crosslinking of P and AA with PEGDA. Bands at 1447 and 1232 cm^{-1} were specific to the asymmetric stretching vibrations of $-COOCH_3$. Absorption peaks at 1160 and 1013 cm^{-1} indicated the C–O stretching vibrations of the C–O/C–C, and C–OH of the ring present in pectin. The adhesion properties were tested on human skin and showed that the skin tissue was tightly adhered to the hydrogel, whether it was in an upright or inverted position. The stress of PAA and GCDs-PAA hydrogels increased slowly with an increase in strain (<60%), while the stress increased rapidly with an increase in strain (>60%). The final maximum stress for the aforementioned hydrogels was 91.02%, and 88.68%, respectively, whereas the strain was 271.6 kPa and 282.3 kPa, respectively. Young's modulus for CDs-PAA hydrogel was 5.59 kPa which was double of PAA hydrogel. Swelling equilibrium was reached at 28 h and 32 h, with the equilibrium swelling ratios of 7.12 and 6.21, respectively. Finally, the antibacterial activity was checked against MRSA and MRST, where no significant ZOI was noted for the PAA hydrogel. However, GCDs-PAA hydrogel showed ZOIs of 10.1 mm and 6.2 mm against the aforementioned pathogens, respectively. GCDs-PAA gel showed a bactericidal rate of 99.99% after 24 h of exposure. It was speculated that penetration within the bacteria via electrostatic interaction, cellular structure disruption, and induction of the bacterial stress response were the key mechanisms behind the antibacterial activity. The GCDs-PAA hydrogel reduced biofilm formation by MRSA and MRST up to 62% and 45%, respectively. NIH3T3 treated with PAA and GCDs-PAA hydrogel for 24 h did not show any significant difference from the control group. In spite of an increase in the incubation time, the viability of cells was >90%. Female BALB/c mice with 5 mm diameter incision wounds were infected with 15 μL of MRSA. 12 mice were divided randomly into four groups and treated respectively with 50 μL saline, 50 μL kanamycin sulfate at 2 mg/mL, PAA hydrogel, and GCDs-PAA hydrogel. The maximum wound healing was 93.29% on treatment with GCDs-PAA hydrogel. However, large wounds and abscesses were seen in the control group.

An antimicrobial hydrogel was prepared with PVA and sodium alginate (SA), impregnated with vancomycin-malic acid-CDS (VMA-CDs) (Zhao et al. 2024). Initially, 0.3 g citric acid monohydrate (CA), and 0.4 g L-malic acid (L-MA) were dissolved in 20 mL of water, and 0.3 g *N*-(2-hydroxyethyl)ethylenediamine (AEEA) was added. After stirring for 10 min under a N_2 environment, the mixture was autoclaved at 200 °C for 4 h. The resulting yellow solution was then centrifuged for 30 min, followed by freeze drying to obtain the solid MA-CDs. The VMA-CDs were then prepared by reacting 50 mg/mL of MA-CDs solution, 0.1 g of 1-Ethyl-3-(3-dimethylaminopropyl)carbodiimide (EDC) hydrochloride/N-Hydroxy succinimide (NHS) and 100 mg of vancomycin (VAN) for 24 h. The mixture was ultracentrifuged for 15 min and freeze-dried to get VMA-CDs. SA-PVA (SP) gel was prepared by mixing 10% PVA and 1% SA solution, followed by injection into the mold. The pure SP solid gel was supplemented with different concentrations of VMA-CDs, resulting in VMA-PVA-SA (V-SP) hydrogel after repeated freeze-thawing. Under UV–visible spectroscopy, the absorption peak was observed at 350 nm and 280 nm for MA-CDs and VAN, respectively. The same peaks were observed in the case of VMA-CDs, signifying the successful coupling of VAN and MA-CDs. Photoluminescence properties showed fluorescence emission wavelength redshift for MA-CDs from 440 to 510 nm, with the highest fluorescent intensity at 1 mg/mL concentration. VMA-CDs, at the same concentrations, showed a similar shift from 440 to 490 nm with the highest fluorescent intensity at 3 mg/mL. TEM images revealed that MA-CDs were spherical and well distributed with a uniform size of 3.27 nm, while the size of VMA-CDs was 3.72 nm. HRTEM images revealed that the lattice spacings of MA-CDs and VMA-CDs were 0.34 nm and 0.32 nm, respectively. FTIR exhibited absorption peaks at 1647 cm^{-1}, 3380 cm^{-1}, 1560 cm^{-1}, 2950 cm^{-1}, 1390 cm^{-1} for C=O, –OH/–NH, N–H/C = C, $-CH_3$, $-CH_2$, respectively. The bands at 1779, and 1708 cm^{-1} were specific to C=O, while 1487 cm^{-1}, 1080 cm^{-1}, and 1250 cm^{-1} were for C–N=, C–O, C–N, respectively. The bands at 1016, 816, and 997 cm^{-1} were attributed to the C–O–C stretching vibrations in case of both CDs. Zeta potential for MA-CDs and VMA-CDs were +4.09 mV and +7.78 mV, respectively. XPS analysis showed the presence of elemental C (61.18%), O (29.17%), and N (9.65%) in MA-CDs. Similarly, the percentage of C, O, N, and Cl were 62.35, 23.07, 12.60 and 1.98% in VMA-CDs. XRD analysis exhibited a broad peak at 2Θ values of 22.5° and 24.8° for MA-CDs, and VMA-CDs, respectively. The MIC against *S. aureus*, *B. subtilis* and *E. coli* were 3.9, 7.8, and 2.0 μg/mL, respectively. Maximum ZOI of 15.83 mm against *S. aureus* was obtained at 30 μg/mL of V-SP hydrogel, while no ZOI was observed against *E. coli*. This indicated selective antibacterial activity against Gram-positive bacteria.

An antimicrobial hydrogel was synthesized using CS and folic acid (FA) impregnated with CQDs and gentamicin (GM) (Kazeminava et al. 2022). Initially, the FA solution (10 mg in 15 mL water) was sonicated for 5 min and carbonized by heating for 8 h at 200 °C. Subsequently, the clear light yellow–brown solution was centrifuged for 15 min at 11,000 rpm and the supernatant was collected. Different amounts of CQDs (5%, 10%, and 15%) and GM (0.010 g) were added individually to the CS solution and stirred, followed by the addition of 0.010 g L-Arg and glycerol (0.5% w/v) which was again stirred for 4 h at 50 °C. The homogeneous, and transparent paste (25 mL) was cast into a polystyrene plate with a 10 cm diameter and cured at 60 °C for 24 h. FTIR exhibited peaks at 1604 cm^{-1} and 1590 cm^{-1} for CS that were specific to C=O stretching vibrations of amide moieties, and N–H bending vibrations of amines, respectively. CS, CS-CQDs nanocomposite hydrogel films showed a peak at 1520 cm^{-1} corresponding to the stretching vibrations of the aromatic rings. The peaks between 3200 and 3600 cm^{-1} indicated the stretching vibrations of O–H that became wider for the CS-CQDs due to the presence of H-bonding between CS and CQDs. This confirmed the presence of CQDs in the CS polymeric network. UV–visible spectroscopy of CQDs showed a maximum absorption peak at 450 nm, while a steady blue shift was observed for CS-CQDs nanocomposites. CS-CQDs nanocomposites showed an emission wavelength of 570 nm in PL spectroscopy. SEM images revealed uniform surface morphology for the nanocomposites. A rough texture was noted when the concentration of the CQDs increased in the nanocomposites. Tensile strength of the CS film was 3.43 MPa, while CS-CQDs with 5%, 10%, and 15% CQDs were 4.26 MPa, 5.86 MPa, and 6.60 MPa, respectively. The swelling ratio was directly proportional to the immersion time of the nanocomposites in PBS solution. Blood clotting ability was highest in 15% CQDs containing nanocomposite film, while the maximum haemolytic rate was 34% at the highest concentrations of CQDs in the nanocomposite film. Cytotoxicity was checked for both films against HFF-1 cell lines, which confirmed a cell viability over 80% even after 48 h. The antimicrobial activity was checked against *E. faecalis*, *P. aeruginosa*, *S. mutans*, *S. aureus*, *K. pneumoniae* and *E. coli*. The highest inhibition against *E. coli* was noted on treatment with the CS-FA-CQDs nanocomposite.

Bacterial cellulose (BC) was impregnated with graphene QDs (GQDs) to design an antimicrobial hydrogel (Zmejkoski et al. 2022). Initially, graphite foil strips were immersed in NaOH/ethanol as an electrolyte, followed by evaporation of the volatile ethanol. BC composite hydrogels were dipped into GQDs suspension with concentrations equivalent to 0.2 and 2 mg/mL for 2 days at an ambient temperature, resulting in the formation of GQDs-BC0.2 and GQDs-BC2 hydrogels, respectively. UV–visible spectroscopy showed the peak for GQDs at 330 nm. The loading percentage of GQDs in BC after 48 h was 11.7 wt%. AFM images revealed the disc-like shape of GQDs with an average size of 10 ± 2 nm and a height of 1.8 nm. XPS analysis showed high C content and low N and O contents with an increase in characteristic C–C and O–C=O bonds. Scratch assay for GQDs-BC composite hydrogels depicted a positive effect on cell migration by inducing a significant migratory potential in the HGF cells. Wound fluid absorption (WFA) was 78.2% and 117.7% for GQDs-BC0.2 and GQDs-BC2, respectively, after 48 h, while it was 49.5% and 67.4% after 24 h and 48 h, respectively in BC hydrogel. The BC hydrogel got dehydrated just after 4 h, with a water retention of 25% in the GQDs-BC hydrogel. No cytotoxicity of BC and GQDs-BC hydrogels was noted against cells seeded directly on material discs, even after 24 h. Both the composite hydrogels inhibited *S. aureus*, MRSA, *E. coli*, *P. aeruginosa*, and *Streptococcus agalactiae*, while bactericidal effects were evident against MRSA, *E. coli*, and *P. aeruginosa.* This was speculated to be attributed to the rupture of the bacterial cell membrane.

In another study, an antibacterial hydrogel was fabricated with silk fibroin (SF), and chondroitin sulfate (CS), impregnated by graphene oxide quantum dots (GOQDs), which is schematically illustrated in Fig. 10.4. (Deng et al. 2024). Initially, 5 g of cocoons of *Bombyx mori* were cut and boiled in 1 L of 0.02 M Na_2CO_3 solution for 30 min. This process was repeated 3 times for the successful removal of sericin. Further, it was dried overnight in an oven at 60 °C. The resulting degummed fibers were dissolved in a ternary solution at 100 °C. The SF solution was then dialyzed against DI water for consecutive 3 days using a cellulose dialysis tube, followed by lyophilization. GOQDs were mixed were synthesized by reacting 40 mg of GO suspended in 100 mL of DI water, 40 mL sulphuric acid and 20 mL nitric acid at 70 °C for 3 h. After dialysis, the GOQDs were recovered and lyophilized. GSC hydrogel was prepared by mixing CS powder and lyophilized SF that was dissolved in PBS at pH 7.4 to form a 10 wt% solution. The SF, hydrogen peroxide (H_2O_2), CS, horseradish peroxidase (HRP), and different concentrations of GOQDs were mixed and incubated at 37 °C for gelation. This resulted in the formation of GO-SF-CS hydrogel, denoted as GSC. TEM images of the GOQDs showed a diameter of 30 nm, while HRTEM showed lattice fringes. AFM 2D linear scanning showed a height of 2 nm, and XRD analysis revealed the peak at 2Θ near 25.5° for GOQDs. Most of the GOQDs had a diameter <50 nm and XPS analysis depicted the presence of C (284.96 eV), and O (531.96 eV), in the GSC10 that had 0.1 mg/mL of GOQDs. FTIR exhibited bands at 1640 cm^{-1}, 1520 cm^{-1}, and 1420 cm^{-1} corresponding to amide I, amide II, and amide III, respectively. Additionally, a strong absorbance band at 1245 cm^{-1} was attributed to the O–SO_3 group of CS. Wide and strong absorption bands were observed at 3400 and 1100 cm^{-1} due to the vibrations of O–H and C–O groups in GOQDs and GSC. SEM images revealed a porous internal structure with a pore size of 100 μm. This architecture is favourable for cell attachment, infiltration and growth.

The water contact angle was <50°, indicating superior hydrophilicity due to the presence of hydrophilic carboxyl, hydroxyl, and epoxy groups on the surface of GOQDs, enhancing the overall hydrophilicity of GSC hydrogels. The swelling ratio was 60% for GSC10 and 80% for the SC group while the degradation was 60% after 14 days, which was significantly higher than the pure SF hydrogel. Mechanical properties and stability analysis revealed a tensile strength of 50% prolongation of their original length. GSC10 hydrogel had a Young's modulus of 81.35 kPa and a compression modulus of 136.23 kPa. The temperature of GSC hydrogels increased up to 40 °C within 2 min and was maintained at around 50 °C when irradiated with 808 nm NIR laser. At 2.0 W. cm^{-2} the hydrogel reached a temperature of 51 °C after 5 min of NIR irradiation, which was high enough to kill bacteria with negligible impact on the normal tissue. GSC hydrogel showed no notable cell death in L929 cells and HUVECs, even after treatment for 5 days. The antibacterial activity against *S. aureus* and *E. coli* under NIR irradiation was accompanied by membrane rupture,

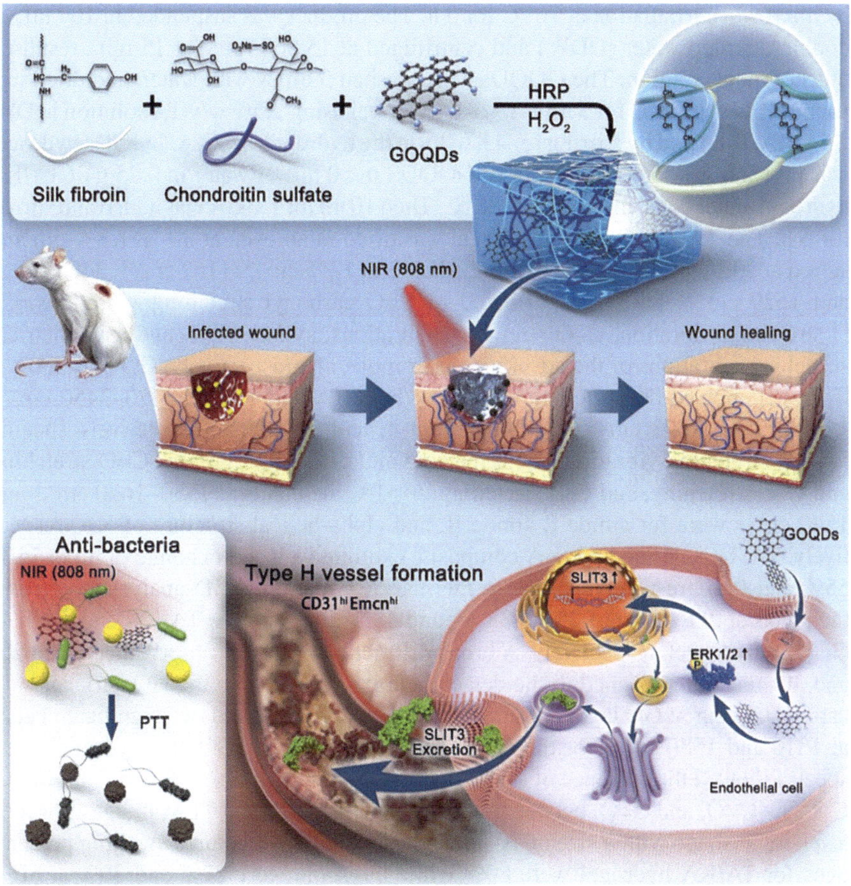

Fig. 10.4 Schematic illustration of the design strategy and application of GOQDs@silk fibroin/chondroitin sulfate (GSC) hydrogel as a multifunctional dressing for bacteria-infected skin wound healing. Reprinted with permission from Deng et al. (2024). Copyright © 2024 Elsevier

bacterial deformation, and cell shrinkage. The bacterial survival rate was 0% in the case of NIR irradiation (2.0 W cm^{-2}). Negligible haemolysis (<5%) confirmed high compatibility of the hydrogel with blood components. Further, wound healing activity was checked using a full-thickness skin defect in Kunming (KM) male mice. The wound (diameter 5 mm) was symmetrically created on each side of the spine and infected by *S. aureus*. The wound closure percentages were 96.9%, and 99.2% after 14 days when treated with GSC10 and GSC10 + NIR, respectively. Bacterial colonization was also checked on day 2 and revealed a bacterial survival ratio of 53.3%, and 2.6% for after treatment with GSC10 and GSC10 + NIR, respectively.

An antimicrobial hydrogel was reported that was composed of tannic acid (TA), keratin (KA), and GOQDs (Ren et al. 2020). GOQDs were prepared by mixing 20 mg GO, 50 mL DI water, 20 mL sulfuric acid, and 10 mL nitric acid, followed by microwave irradiation at 70 °C for 3 h. The product was suspended in 100 mL of double-distilled water (DDW) and centrifuged at 15,000 rpm for 15 min, resulting in a colloidal solution. The GOQDs were washed 5 times with water, and dried in a vacuum oven at 80 °C for 5 min. TA/KE hydrogel using 2.5% w/v TA solution in DW, 2.5 g KA, followed by stirring for 4 h to form the hydrogel. TA/KA/GOQDs hydrogel was prepared by dispersing 0.2 g of GOQDS in 50 mL of water in 0.25 g of TA/KE hydrogel under constant stirring at 80 °C. Then 10 mL of 1% CA was added and stirred for 6 h. The TA/KA-GOQDs hydrogel was dried in an oven at 80 °C for 4–5 h and stored at 20 °C. FTIR spectra exhibited bands at 1745 cm^{-1}, 1162 cm^{-1}, 3435 cm^{-1}, and 1629 cm^{-1} that were attributed to C=O carbonyl stretching vibrations, C–O stretching vibration, C–O stretching vibrations, symmetrical, and asymmetrical stretching vibrations of the adsorbed water molecule, respectively. GOQDs-TA/KA hydrogel showed characteristic peaks for both TA and KA, at 3250–3450 cm^{-1}, <1700 cm^{-1}, 1673 cm^{-1}, 1350–1450 cm^{-1}, and 800 cm^{-1}, respectively specific to -OH stretching vibrations, benzene ring and carboxylic groups, C=O stretching and C–C stretching, and C-H stretching, for TA, the bands at 1550–1670 cm^{-1} and 1540 cm^{-1} were for amide I, amide II, and alpha-helical structure of KA, respectively. TA, GOQDs and TA/KA composite exhibited a weight change of up to 75%, 55% and 63%, respectively, due to thermal degradation. XRD analysis confirmed the peaks at 2Θ values of 25.2°, 38.1°, and 48.2° for GOQDs, TA/KA and TA/KA-GOQDs hydrogel, respectively. A broad diffraction peak was observed between 20° and 30° that was assigned to the detachment of oxygen from the GO surface, indicating stacking of GO layers. Raman spectral analysis showed two significant peaks at 1310 and 1580 cm^{-1} for the G and D bands in TA/KA-GOQDs hydrogel. XPS spectra showed the presence of two different functional groups of C and O like C=C, C–O, O–C=O, and C=O, between 283 and 290 eV. SEM and TEM images revealed a spherical shape with a size of 10 nm and a crystalline structure. Swelling percentages for TA/KA hydrogel were 80% while the same were 27% and 30% at pH 1 and pH 3, respectively. The antibacterial activity was checked against *S. aureus* and *E. coli* by the direct contact method. Bacterial viability was reduced when treated with GOQDs, showing greater bacteriostatic activity against *E. coli* in comparison to *S. aureus*. Further, wound healing activity was checked against the excision wound model created in six healthy rats. The full-thickness wound of 1.5 cm^2 was made using a scalpel blade. The wound closure was 74% for the TA/KA hydrogel-treated group on day 7, and 100% for the TA/KA-GOQDs hydrogel-treated group on day 16.

10.7 Effect of Surface Functionalization

Surface functionalization of a hydrogel can be done using antibiotics, biological extracts, synthetic antibacterial drugs, AMPs, metal NPs, metal oxide NPs, and carbon materials. They show great effectiveness in therapy, wound healing, prevention of colonization on the medical devices, and antibacterial activity against

MDR pathogens such as *S. aureus* and *Staphylococcus epidermidis* (Carpa et al. 2022). Antibiotics like vancomycin, ciprofloxacin, gentamicin, and tobramycin are the most common, effective, low-cost antimicrobial secondary metabolites derived from microbes, plants and animals. They can kill or halt the growth of microorganisms by inhibiting cell wall synthesis, enhancing cell permeability, interfering with protein synthesis and inhibiting replication and transcription of the genetic material. The development of hydrogels functionalized by antibiotics is an alternative approach to mitigate antibiotic resistance via local delivery of the drug in a controlled manner at the infected site by avoiding systemic toxicity (Jia et al. 2023). Ciprofloxacin functionalized hydrogel self-assembled as tripeptide (d-Leu-Phe-Phe) and showed high drug loading efficiency with sustained release and antimicrobial activity against *S. aureus, E. coli, K. pneumoniae.* In another study, the same antibiotic functionalized polyacrylate hydrogel prevented the Ti implant-associated infections with MRSA via prolonged drug release. Gentamicin functionalized gellan gum hydrogel with poly (lactide-co-glycolide) NPs showed inhibited *Staphylococcus saprophyticus* without affecting the bone-forming cells. Vancomycin functionalized hydrogel formed by combining methacrylic anhydride-modified gelatin and metal-based organic backbone-modified oxidized sodium alginate halted the growth of bacterial metabolism by interfering with the cell wall synthesis by disrupting its peptidoglycan. The AMPs are produced by plants and animal cells, showing strong, broad-spectrum antimicrobial activity against Gram-positive and Gram-negative bacteria, fungi, and viruses. They are amphiphilic cations that can destroy the membrane integrity by forming transmembrane ion channels on the bacterial cell membrane and also can inhibit different cellular processes like DNA replication, transcription, translation and enzyme activity. AMPs functionalized Ala5-Tritrp7, ABU-CHRG01, Temporin-A and poly (2-hydroxyethyl methacrylate) hydrogels exhibited enhanced antimicrobial activity. In another study, HHC-36 AMP, cerium oxide NPs (CeONPs), and catechol-functionalized GelMA acted as an efficient antibacterial hydrogel. The most common metal ions like Ag, gold (Au), copper (Cu) and the most used metal oxides, like zinc oxide (ZnO), titanium dioxide (TiO_2) and nickel oxide (NiO) exhibit broad-spectrum antimicrobial properties, summarized in Fig. 10.5 (Kaur et al. 2023). These particles, in combination with enzymes, can attack the DNA and proteins of the bacterial cells, leading to the formation of reactive oxygen species (ROS), which cause bacterial death due to oxidative stress (Ahmad et al. 2020). The antimicrobial activity against *S. aureus, E. coli,* and *Candida tropicalis* was shown by the carbon nanotube (CNTs) incorporated hydrogels. A strong antibacterial activity against Gram-positive and Gram-negative bacteria was shown by the graphene oxide hydrogels. In another study, excellent antibacterial activity against *E. coli* and *S. aureus* was exhibited by Ag-reduced graphene oxide hydrogel (Jia et al. 2023; Carpa et al. 2022).

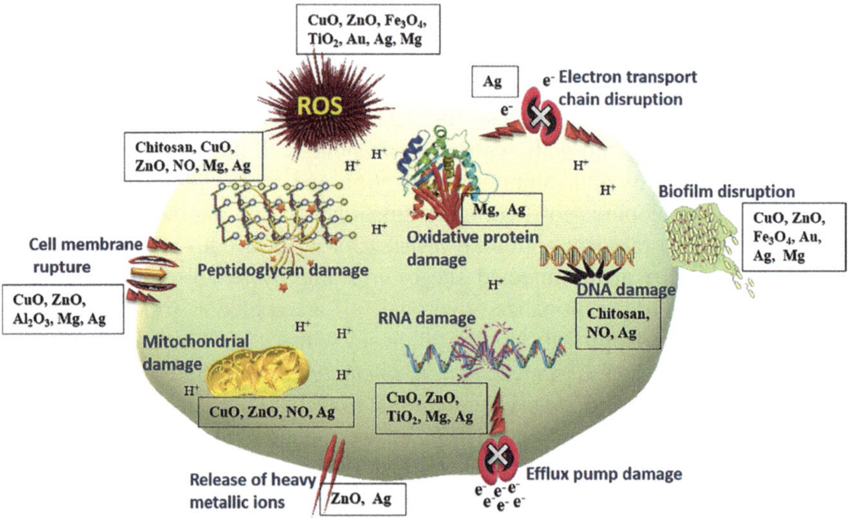

Fig. 10.5 Summary of the different antimicrobial mechanisms of nanomaterials. Reprinted with permission from Kaur et al. (2023). Copyright © 2023 Elsevier Inc.

10.8 Effect of Particle Size

The particle size of the NPs is one of the most significant determinants of the antimicrobial property as it influences the cellular uptake (Menichetti et al. 2023). The small-sized AgNPs can release Ag^{1+} ions that disrupt the cell membrane, deactivate the enzyme, and impair the cellular metabolism of the pathogen. The most common metal oxide-based NPs are ZnONPs, CuONPs, TiO_2NPs, and iron oxide NPs (Fe_2O_3NPs). The ZnONPs are composed of Zn^{2+} and O^{2-} ions that induce oxidative stress in bacteria by generating ROS, making them an excellent candidate as an antimicrobial agent. The TiO_2NPs are also bactericidal in nature. The smaller NPs can generate more ROS that can disrupt the cell membrane, proteins, and DNA of bacterial cells, leading to cell death. Smaller NPs have a broad-spectrum antimicrobial activity against bacteria, fungi, and viruses. The magnesium oxide NPs (MgONPs), based on their small size and high surface area-to-volume ratio, inhibited several bacteria and fungi by disrupting their cell membranes and metabolic processes. Fe_2O_3NPs are composed of various compounds like magnetite (Fe_3O_4), hematite (Fe_2O_3), and maghemite (γ-Fe_2O_3) that exhibited broad-spectrum antimicrobial activity against both Gram-positive and Gram-negative bacteria, along with many fungal species. So, the increased surface area of all the small-sized metal and metal oxide NPs is a significant factor for superior antimicrobial activity by providing more contact points, enhanced drug uptake, efficient interactions with the bacterial cell membrane, and delivering the drug at the target sites in a controlled manner (Mondal et al. 2024).

10.9 Effect of Surface Modification on Stability and Bioavailability

The surface modification of the hydrogel has an important role in determining its stability and effectivity. The surface modification can also affect the hydrophobicity, protein adsorption, mechanical properties, microbial adhesion, and roughness of the hydrogels (Cui et al. 2021a, b). These modifications help to increase bioavailability by increasing cell adhesion. The microgroove containing PEG-based hydrogel prepared via the casting method enhanced the rate of corneal epithelial in vitro cell migration. The cell proliferation can also be regulated by the strength and adhesion of the modified hydrogels by influencing the cell spreading on the substrates. The presence of epoxy resin and poly (dimethylsiloxane) (PDMS) on the surface facilitated the epithelial cell proliferation. Selection of appropriate surface-associated polymers and their crosslinking density modulate the cell differentiation of the self-renewable stem cells. Microgrooved PAM surface of hydrogels exhibited superior neuronal differentiation compared to the human mesenchymal stem cells (hMSCs), and mouse embryonic stem cells (mESCs). Thus, the surface modification of the hydrogels has a significant effect on their stability and bioavailability as well (Bao et al. 2022).

10.10 ROS Generation by NPs in Hydrogels

Metal and metal oxide NPs with various chemical compositions can generate free radicals, which can eventually influence intracellular calcium concentrations, activate the transcription factors, and regulate cytokine production (Canaparo et al. 2020). The electronic properties of the NPs' surface can be altered by their reduced size, leading to the creation of reactive groups. The molecular oxygen at the reactive sites, after interacting with the electron donor and acceptor, generates the ROS additionally by Fenton-type reactions. The surface properties of some metal NPs, like silica (Si) and Zn, due to their identical size and shape, showed cytotoxic responses. The oxide form of Zn, being more chemically active than that of Si, led to the increase of primary ROS formation, which eventually enhanced the oxidative stress in cells. The physicochemical properties like size, shape, chemical structure, surface area to volume ratio, and charge of metal NPs can control the mechanism of generating ROS by triggering sequential pathological events like inflammation, fibrosis, genotoxicity, and carcinogenesis. The toxicity related to NPs can activate inflammatory cells like macrophages and neutrophils by enhancing the expression of pro-inflammatory and fibrotic cytokines, leading to improved generation of ROS. The mechanism of ROS formation varies among the different metal and metal oxide NPs (Dayem et al. 2017).

10.11 Mechanism Behind Biofilm Inhibition by ROS

Biofilms are spatially organized communities of grouped microorganisms that develop on biotic and/or abiotic surfaces. Biofilm-associated bacteria protect themselves from the surrounding hostile environment using extracellular polymeric substances (EPS). They can cause significant infections in plants, animals, and humans. Biofilm-related infections have emerged as a challenging health issue due to their nature of persistence, and resistance. The main biofilm-forming organisms are the ESKAPE group of pathogens, including both Gram-positive and Gram-negative bacteria like *E. faecium*, *S. aureus*, *K. pneumoniae*, *A. baumannii*, *P. aeruginosa* and *Enterobacter* spp. Biofilm formation involves multiple interconnected stages, like surface attachment, monolayer formation, microcolony formation, maturation, and dispersion. The adhesion protein present on the bacterial cell surface first interacts with the carbohydrate molecule of the host cell surface and attaches to initiate the pathogenesis. The complex network of the EPS matrix is made up of polysaccharides and proteins that help in the formation of multiple slime layers. The extracellular DNA (eDNA) of this matrix enhances the structural stability of the biofilms. Further, the whole bacterial population communicates among themselves by quorum sensing to maintain the strong, stable biofilm structure. The metal and metal oxide NPs have shown a significant role in the eradication of biofilm by generating ROS. Among various NPs, AgNPs, ZnONPs, CuONPs, and Fe_3O_4NPs have shown extensive antibiofilm properties. They first penetrate the bacterial cell membrane, interfere with the EPS matrix of the biofilm, and disrupt the quorum-sensing mechanism. They induce the formation of ROS by damaging the bacterial cell membrane, which induces oxidative stress in the bacterial cells. The cells in response to the stress generate hydroxyl radicals, superoxide radicals, and hydrogen peroxide radicals, leading to the damage of the DNA and protein of bacteria. Histidine-functionalized AgNPs inhibited the *K. pneumoniae* biofilm formation (Kumar et al. 2023). The same NPs with antibiotics like vancomycin and ampicillin showed greater antibiofilm activity against Gram-negative bacteria, *P. aeruginosa* and *Shigella flexneri* and Gram-positive bacteria, *S. aureus* and *Streptococcus pneumoniae* (Sahli et al. 2022). ZnONPs synthesized using the flower extract of *Clitoria ternatea* exhibited antibiofilm activity against oral pathogens like *Porphyromonas gingivalis* and *Alcaligenes faecalis*. Likewise, antibiofilm activity of AuNPs functionalized with the enzyme proteinase-K (PK) was reported against *Pseudomonas fluorescens*. PK disrupted and dispersed the matrix, leading to the destruction of bacteria released from the biofilm. The same NPs showed antibiofilm activity against *S. aureus* and *P. aeruginosa* by disrupting the EPS matrix after functionalization with a cationic ligand (Sahli et al. 2022).

10.12 Conclusion and Future Perspectives

Nanoparticles impregnated within hydrogel can enhance the antimicrobial activity significantly. The metal nanoparticles released from the hydrogel can adhere and alter the surface of the pathogen. After entering the cells of the bacteria, they can generate ROS and oxidative stress which can damage the vital biomolecules like DNA and proteins. There is a scope to functionalize the nanoparticles with more positively charged ligands that can efficiently bind to the bacterial cell surface, which is negatively charged. Metal nanoparticles with more affinity to the sulfur proteins associated with the microbial cell wall may be selectively incorporated into the hydrogels to make them more effective. Biologically synthesized nanoparticles with more promising antimicrobial properties may be considered for generating a rapid healing hydrogel with more biocompatibility to the host tissues. Antibiotics can be loaded in the mesoporous silica nanoparticles or carbon nanotubes for targeting the bacterial protein and nucleic acid synthesis. It can ensure sustained release of the antibiotics and from the hydrogel. Likewise, the optically active nanoparticles incorporated in the hydrogel may be promising for photothermal and photodynamic therapy against multidrug-resistant bacterial infections. In conclusion, polymeric hydrogels embedded with antimicrobial nanoparticles can help to develop potential nanomedicine for the healthcare industry.

References

Afrasiabi S, Bahador A, Partoazar A (2021) Combinatorial therapy of chitosan hydrogel-based zinc oxide nanocomposite attenuates the virulence of *Streptococcus mutans*. BMC Microbiol 21(1):62

Ahmad SA, Das SS, Khatoon A, Ansari MT, Afzal M, Hasnain MS, Nayak AK (2020) Bactericidal activity of silver nanoparticles: a mechanistic review. Mater Sci Energy Technol 3(2020):756–769

Ahmad Z, Salman S, Khan SA, Amin A, Rahman ZU, Al-Ghamdi YO, Akhtar K, Bakhsh EM, Khan SB (2022) Versatility of hydrogels: from synthetic strategies, classification, and properties to biomedical applications. Gels 8(3):167

Ahovan ZA, Esmaeili Z, Eftekhari BS, Khosravimelal S, Alehosseini M, Orive G, Dolatshahi-Pirouz A, Chauhan NPS, Janmey PA, Hashemi A, Kundu SC (2022) Antibacterial smart hydrogels: new hope for infectious wound management. Mater Today Bio 17(2022):100499

Bai X, Gao M, Syed S, Zhuang J, Xu X, Zhang XQ (2018) Bioactive hydrogels for bone regeneration. Bioactive Mater 3(4):401–417

Bao Y, He J, Song K, Guo J, Zhou X, Liu S (2022) Functionalization and antibacterial applications of cellulose-based composite hydrogels. Polymers 14(4):769

Baquero F (2021) Threats of antibiotic resistance: an obliged reappraisal. Int Microbiol 24(4):499–506

Bashir S, Hina M, Iqbal J, Rajpar AH, Mujtaba MA, Alghamdi NA, Wageh S, Ramesh K, Ramesh S (2020) Fundamental concepts of hydrogels: synthesis, properties, and their applications. Polymers 12(11):2702

Cabral JP (2010) Water microbiology. Bacterial pathogens and water. Int J Environ Res Public Health 7(10):3657–3703

Canaparo R, Foglietta F, Limongi T, Serpe L (2020) Biomedical applications of reactive oxygen species generation by metal nanoparticles. Materials 14(1):53
Carpa R, Remizovschi A, Culda CA, Butiuc-Keul AL (2022) Inherent and composite hydrogels as promising materials to limit antimicrobial resistance. Gels 8(2):70
Church NA, McKillip JL (2021) Antibiotic resistance crisis: challenges and imperatives. Biologia 76(5):1535–1550
Cui F, Sun J, Ji J, Yang X, Wei K, Xu H, Gu Q, Zhang Y, Sun X (2021a) Carbon dots-releasing hydrogels with antibacterial activity, high biocompatibility, and fluorescence performance as candidate materials for wound healing. J Hazard Mater 406:124330
Cui L, Yao Y, Yim EK (2021b) The effects of surface topography modification on hydrogel properties. APL Bioeng 5(3):031509
Dayem AA, Hossain MK, Lee SB, Kim K, Saha SK, Yang GM, Choi HY, Cho SG (2017) The role of reactive oxygen species (ROS) in the biological activities of metallic nanoparticles. Int J Mol Sci 18(1):120
Deng T, Lu W, Zhao X, Wang H, Zheng Y, Zheng A, Shen Z (2024) Chondroitin sulfate/silk fibroin hydrogel incorporating graphene oxide quantum dots with photothermal-effect promotes type H vessel-related wound healing. Carbohyd Polym 334:121972
Dimatteo R, Darling NJ, Segura T (2018) In situ forming injectable hydrogels for drug delivery and wound repair. Adv Drug Deliv Rev 127(2018):167–184
Dong Q, Zu D, Kong L, Chen S, Yao J, Lin J, Lu L, Wu B, Fang B (2022) Construction of antibacterial nano-silver embedded bioactive hydrogel to repair infectious skin defects. Biomater Res 26(1):36
Dugam S, Jain R, Dandekar P (2024) Silver nanoparticles loaded triple-layered cellulose-acetate based multifunctional dressing for wound healing. Int J Biol Macromol 276(1):133837
Ehrenhofer A, Binder S, Gerlach G, Wallmersperger T (2020) Multisensitive swelling of hydrogels for sensor and actuator design. Adv Eng Mater 22(7):2000004
Frieri M, Kumar K, Boutin A (2017) Antibiotic resistance. J Infect Public Health 10(4):369–378
Ghorpade VS, Dias RJ, Mali KK, Mulla SI (2019) Citric acid crosslinked carboxymethylcellulose-polyvinyl alcohol hydrogel films for extended release of water soluble basic drugs. J Drug Deliv Sci Technol 52(2019):421–430
Guo G, Chen Y, Liu X, Zhu DY, Zhang B, Lin N, Gao L (2018) Tough and durable hydrogels with robust skin layers formed via soaking treatment. J Mater Chem B 6(48):8043–8054
Haidari H, Kopecki Z, Bright R, Cowin AJ, Garg S, Goswami N, Vasilev K (2020) Ultrasmall AgNP-impregnated biocompatible hydrogel with highly effective biofilm elimination properties. ACS Appl Mater Interfaces 12(37):41011–41025
Hasnain MS, Nayak AK (2018) Chitosan as responsive polymer for drug delivery applications. In: Makhlouf ASH, Abu-Thabit NY (eds) Stimuli responsive polymeric nanocarriers for drug delivery applications, volume one: types and triggers discusses. Woodhead publishing, pp 581–605
Ho TC, Chang CC, Chan HP, Chung TW, Shu CW, Chuang KP, Duh TH, Yang MH, Tyan YC (2022) Hydrogels: properties and applications in biomedicine. Molecules 27(9):2902
Hou W, Lin Z, Xia X, Sun S, Niu Z, Liu J, Lu J, Yin D, Li X (2024) Antibacterial and anti-inflammatory polyethyleneimine/polyacrylic acid hydrogel loaded with CeO_2 quantum dots for wet tissue wound dressings. Front Mater 11:1379836
Hu Y, Zhang Z, Li Y, Ding X, Li D, Shen C, Xu FJ (2018) Dual-crosslinked amorphous polysaccharide hydrogels based on chitosan/alginate for wound healing applications. Macromol Rapid Commun 39(20):1800069
Iresha H, Kobayashi T (2020) Smart polysaccharide hydrogels in drug delivery and release. In: Nayak A, Hasnain M (eds) Advanced biopolymeric systems for drug delivery. Advances in material research and technology. Springer, Cham, pp 135–149. https://doi.org/10.1007/978-3-030-46923-8_6
Jia B, Li G, Cao E, Luo J, Zhao X, Huang H (2023) Recent progress of antibacterial hydrogels in wound dressings. Mater Today Bio 19(2023):100582

Jose J, Athira VP, Michel H, Hafeela AR, Bhat SG, Thomas S, Chapter LM (2023) Hydrogels: an overview of the history, classification, principles, applications, and kinetics. In: Thomas S, Sharma B, Jain P, Shekhar S (eds) Sustainable hydrogels: synthesis, properties, and applications. Elsevier, pp 1–22. https://doi.org/10.1016/B978-0-323-91753-7.00005-3

Kaur H, Rauwel P, Rauwel E (2023) Antimicrobial nanoparticles: synthesis, mechanism of actions. In: Guisbiers G (ed) Antimicrobial activity of nanoparticles: applications in wound healing and infection treatment. Advanced topics in biomaterials. Elsevier, pp 155–202. https://doi.org/10.1016/B978-0-12-821637-8.00008-0

Kazeminava F, Javanbakht S, Nouri M, Gholizadeh P, Nezhad-Mokhtari P, Ganbarov K, Tanomand A, Kafil HS (2022) Gentamicin-loaded chitosan/folic acid-based carbon quantum dots nanocomposite hydrogel films as potential antimicrobial wound dressing. J Biol Eng 16(1):36

Kumar L, Bisen M, Harjai K, Chhibber S, Azizov S, Lalhlenmawia H, Kumar D (2023) Advances in nanotechnology for biofilm inhibition. ACS Omega 8(24):21391–21409

Kwiecien I, Kwiecien M (2018) Application of polysaccharide-based hydrogels as probiotic delivery systems. Gels 4(2):47

Lee CH, Song SY, Chung YJ, Choi EK, Jang J, Lee DH, Kim HD, Kim DU, Park CB (2022) Light-stimulated carbon dot hydrogel: targeting and clearing infectious bacteria in vivo. ACS Appl Bio Mater 5(2):761–770

Li X, Kong X, Zhang Z, Nan K, Li L, Wang X, Chen H (2012) Cytotoxicity and biocompatibility evaluation of *N*, *O*-carboxymethyl chitosan/oxidized alginate hydrogel for drug delivery application. Int J Biol Macromol 50(5):1299–1305

Masood N, Ahmed R, Tariq M, Ahmed Z, Masoud MS, Ali I, Asghar R, Andleeb A, Hasan A (2019) Silver nanoparticle impregnated chitosan-PEG hydrogel enhances wound healing in diabetes induced rabbits. Int J Pharm 559:23–36

Mattar C, Edwards S, Baraldi E, Hood J (2020) An overview of the global antimicrobial resistance research and development hub and the current landscape. Curr Opin Microbiol 57:56–61

Menichetti A, Mavridi-Printezi A, Mordini D, Montalti M (2023) Effect of size, shape and surface functionalization on the antibacterial activity of silver nanoparticles. J Funct Biomater 14(5):244

Mondal SK, Chakraborty S, Manna S, Mandal SM (2024) Antimicrobial nanoparticles: current landscape and future challenges. RSC Pharm 1(3):388–402

Moreno Ruiz YP, de Almeida Campos LA, Alves Agreles MA, Galembeck A, Macario Ferro Cavalcanti I (2023) Advanced hydrogels combined with silver and gold nanoparticles against antimicrobial resistance. Antibiotics 12(1):104

Mou C, Wang X, Teng J, Xie Z, Zheng M (2022) Injectable self-healing hydrogel fabricated from antibacterial carbon dots and ε-polylysine for promoting bacteria-infected wound healing. J Nanobiotechnol 20(1):368

Mourad R, Helaly F, Darwesh O, El-Sawy S (2019) Antimicrobial and physicomechanical natures of silver nanoparticles incorporated into silicone-hydrogel films. Contact Lens Anterior Eye J Br Contact Lens Assoc 42(3):325–333

Naidoo S, Zwane AM, Paruk A, Hardcastle TC (2023) Diagnosis and management of severe water-related skin and soft tissue sepsis: a summative review of the literature. Diagnostics 13(13):2150

Nie J, Pei B, Wang Z, Hu Q (2019) Construction of ordered structure in polysaccharide hydrogel: a review. Carbohyd Polym 205:225–235

Rakhshaei R, Namazi H (2017) A potential bioactive wound dressing based on carboxymethyl cellulose/ZnO impregnated MCM-41 nanocomposite hydrogel. Mater Sci Eng, C 73:456–464

Ren Y, Yu X, Li Z, Liu D, Xue X (2020) Fabrication of pH-responsive TA-keratin bio-composited hydrogels encapsulated with photoluminescent GO quantum dots for improved bacterial inhibition and healing efficacy in wound care management: in vivo wound evaluations. J Photochem Photobiol B, Biol 202:111676

Sahli C, Moya SE, Lomas JS, Gravier-Pelletier C, Briandet R, Hemadi M (2022) Recent advances in nanotechnology for eradicating bacterial biofilm. Theranostics 12(5):2383

Sheikh-Oleslami S, Tao B, DSouza J, Butt F, Suntharalingam H, Rempel L, Amiri N (2023) A review of metal nanoparticles embedded in hydrogel scaffolds for wound healing in vivo. Gels 9(7):591

Sheng L, Wang Z, Song L, Yang X, Ye Y, Sun J, Ji J, Geng S, Ning D, Zhang Y, Sun X (2024) Antimicrobial carbon dots/pectin-based hydrogel for promoting healing processes in multidrug-resistant bacteria-infected wounds. Int J Biol Macromol 264(1):130477

Sulastri E, Lesmana R, Zubair MS, Mohammed AFA, Elamin KM, Wathoni N (2023) Ulvan/silver nanoparticle hydrogel films for burn wound dressing. Heliyon 9(7):e18044

Tallet L, Gribova V, Ploux L, Vrana NE, Lavalle P (2021) New smart antimicrobial hydrogels, nanomaterials, and coatings: earlier action, more specific, better dosing? Adv Healthcare Mater 10(1):2001199

Ulu A, Birhanli E, Koytepe S, Ates B (2020) Chitosan/polypropylene glycol hydrogel composite film designed with TiO_2 nanoparticles: a promising scaffold of biomedical applications. Int J Biol Macromol 163:529–540

Wang Y, Zhang M, Yan Z, Ji S, Xiao S, Gao J (2024) Metal nanoparticle hybrid hydrogels: the state-of-the-art of combining hard and soft materials to promote wound healing. Theranostics 14(4):1534

Yan CJ, Yang SR, Yeh YC (2024) Injectable pH-and ultrasound-responsive dual-crosslinked dextran/chitosan/TiO_2 nanocomposite hydrogels for antibacterial applications. Chem Asian J 19(15):e202301151

Zhao D, Liu H, Xu M, Yin C, Xiao X, Dai K (2024) Functional carbon dots-hydrogel complex for selective antibacterial and detection applications. Spectrochimica Acta. Part A Mol Biomol Spectro 314:124195

Zmejkoski DZ, Markovic ZM, Mitic DD, Zdravkovic NM, Kozyrovska NO, Bugarova N, Todorovic Markovic BM (2022) Antibacterial composite hydrogels of graphene quantum dots and bacterial cellulose accelerate wound healing. J Biomed Mater Res Part B, Appl Biomater 110(8):1796–1805

Chapter 11
Biopolymeric Hydrogels for Pathogen Eradication in Wastewater Treatment

Sreehari Suresh, Mridul Umesh, Nilina James, and Adhithya Sankar Santhosh

Abstract Water scarcity and an exponential increase in the demand for clean water have necessitated the development of modern techniques and practices for effective utilization of wastewater wherever possible to highlight a sustainable strategy for water recycling and reuse. The presence of pathogenic microorganisms acts as a major threat to the wastewater recycling strategy, as it may cause harm to the handling person or may result in the entry of pathogenic microorganisms into the environment and food chain when treated water is used for irrigation. The conventional methods of wastewater treatment employing chlorination and other chemical methods sometimes fail with the emergence of more resistant microorganisms. This leads to the research interest in eco-friendly and effective strategies for pathogen eradication from wastewater. Development of membrane processes employing high-end membrane filters or nanomaterial-mediated filtration systems, although highly successful in laboratory studies, fails to achieve the commercial success mainly due to high production and maintenance costs. Biopolymeric hydrogels, derived from agro-waste and food waste, provide eco-friendly, biodegradable solutions for effective pathogen eradication in wastewater treatment, representing a sustainable and scalable method for commercial water purification applications. This chapter describes the production and application of biopolymeric hydrogels for pathogen eradication and wastewater treatment.

Keywords Wastewater treatment · Pathogenic microorganisms · Biopolymeric hydrogels · Eco- friendly solutions · Sustainable water recycling

---

S. Suresh
Centre for Global Health Research, Saveetha Medical College and Hospital, Saveetha Institute of Medical and Technical Sciences (SIMATS), Saveetha University, Chennai, India

M. Umesh (✉) · N. James · A. S. Santhosh
Christ University, Bangalore, Karnataka, India
e-mail: mridul.umesh@christuniversity.in

S. Bardhan and M. P. Shah (eds.), *Applications of Hydrogels in Modern Wastewater Treatment*, Advances in Wastewater Research,
https://doi.org/10.1007/978-981-95-1598-1_11

Abbreviations

CNF	Cellulose Nanofibers
$Cu2^{+}$	Copper Ion
E. coli	*Escherichia coli*
$Fe3^{+}$	Iron Ion
MCL-PHA	Medium-Chain-Length Polyhydroxyalkanoates
Na +	Sodium Ion
Ni(II)	Nickel Ion
P. putida	*Pseudomonas putida*
P. aeruginosa	*Pseudomonas aeruginosa*
$Pb2^{+}$	Lead Ion
PHA	Polyhydroxyalkanoates
PHB	Poly(3-hydroxybutyrate)
ROS	Reactive Oxygen Species
SCL-PHA	Short-Chain-Length Polyhydroxyalkanoates
SDG	Sustainable Development Goals
TiO_2	Titanium Dioxide
UN	United Nations
WHO	World Health Organization
$Zn2^{+}$	Zinc Ion

11.1 Introduction

An increase in the human population has triggered the need for more water and food resources to meet the needs of the accelerating population. This exponential increase in population has often created a scarcity of water in many parts of the world. Recycling and reuse of water and wastewater treatment were focused as sustainable alternatives to cope with the issues associated with water scarcity all over the world. A major hurdle in the reuse of treated wastewater is the presence of pathogens, including bacteria, virus and fungi that pose a threat to human health (Forés et al. 2023). Moreover, as per the recent estimates, about 1.1 billion people consume unsafe water globally, leading to various diseases. As per the statistics given, it was estimated that about 380 trillion litres of wastewater are generated every year globally (Pratap et al. 2023). Although the efforts taken by various countries globally have helped in improving the proportion of the people who have access to safely managed sanitation from 47 to 54% (between 2015 and 2020), there are still over 2 billion people who lack access to clean water sources. The increasing instances of waterborne infections all over the world have stressed the United Nations (UN) to prioritize 'Clean water and sanitation' (Goal 6) as one of its Sustainable Development Goals (SDGs). This goal focuses majorly on reducing the release and contamination of emerging contaminants into the environment. UN's Target 6.3 emphasizes improving water quality by reducing pollution, minimizing waste dumping, reducing the release of harmful chemicals and materials, halving the amount of untreated wastewater and subsequently accelerating the safe reuse and recycling of treated wastewater globally by 2030. The main goal of any wastewater treatment is to reduce the level of pathogens and toxic compounds in wastewater, thereby making it safe for the use of humans. This often requires a combination of physical, chemical and biological treatment methods working in synchrony to remove the pathogens and toxins. Conventional treatment methods like precipitation, coagulation and filtration may not be fully effective in removing the emerging contaminants in wastewater, especially pathogens like viruses and antibiotic-resistant bacterial strains. The environmental issues associated with widely used disinfection methods, like the use of large amounts of chemicals and persistence of toxic compounds in treated water have created an imperative need to develop more eco-friendly and efficient methods for wastewater treatment. The World Health Organization (WHO) has clearly emphasized the treatment of wastewater in properly constructed and properly maintained centralized treatment facilities, as the conventional methods of wastewater disinfection, including coagulation and precipitation were reported to cause accumulation of large volumes of sludge (Rizzo et al. 2020).

Among the various wastewater treatment methods used for pathogen removal from wastewater over the decades, adsorption-based techniques still have potential advantages. This is majorly due to their simplicity, ease in operation and maintenance, cost effectiveness, simple design, diverse application and recycling potential. As there are several factors like the nature of adsorbents and physical conditions that chiefly affect the success of an adsorption process, proper selection and designing of materials used for the elimination of pathogens holds the key to its commercial success and economic viability. Membranes and hydrogels produced from biopolymers are gaining more attention for effective removal of pathogens from wastewater due to their abundance, sustainability, carbon neutrality and affordability. This chapter summarizes the importance of biopolymer-based hydrogels in removing pathogens from wastewater (Ghosh et al. 2023). The chapter focuses mainly on the fabrication and use of hydrogels derived from cellulose, PHA, gelatin and chitosan for effective removal of pathogens from wastewater. A detailed comparison of various methods used for the formulation of hydrogels using these natural polymers was described in detail in this work. Furthermore, the current challenges in the commercialization of biopolymer hydrogels for wastewater treatment and the future directions in this research domain are also discussed in detail in this chapter. Thus, the chapter presents an overall outlook on tuning the sustainable biobased polymers to effectively fabricate hydrogels for their use in pathogen removal from wastewater.

11.2 Emergence of Eco-Friendly Strategies

11.2.1 Need for Eco-Friendly and Effective Pathogen Eradication Methods

In recent years, the main focus has been on sustainability and the problems regarding environmental safety. This has accelerated the knowledge of these issues and an increase in demand for green strategies across all sectors as a countermeasure to environmental predicaments. The sustainable practices will result not only in a decrease in environmental impact but also concomitantly lead to implementing more effective and lower-cost solutions worldwide to decrease hazardous chemicals consumption, conserve natural resources sustainably, as well as promote biodiversity. Traditional water treatments, such as chlorination and ozonation used may generate toxic chemicals in the process that can harm both human health and nature. Safety can, therefore, be increased by switching to less harmful alternatives, as continuing these processes will pose risks. Another key part of sustainability is the conservation of natural resources. Natural coagulants and constructed wetlands are suitable, eco-friendly methods to treat water, resulting in more efficient use of the resources that would require energy-intensive processes. The systems conserve the balance of natural ecosystems and simultaneously reduce production costs. Additionally, by thriving from natural systems and processes in water treatment (wetlands and biotic filters) that inhabit diverse, resilient nature, it becomes imperative to promote biodiversity. Pathogen eradication is required to move towards cleaner measures in response to the limitations and potential perils of traditional strategies. This offers fertile ground for solutions that align with the goals of global sustainability through emerging technologies that are less dependent on hazardous chemicals and increase resource saving. Adopting these new methodologies will provide environmental solutions and innovative ways to effectively and affordably deliver clean and safe water (Ojha 2020).

11.2.2 Limitations of Membrane Processes and Nanomaterial Filtration Systems

In water treatment, membrane processes and nanomaterial filtration systems are effective approaches to remove numerous contaminants, such as pathogens, heavy metals, or organic pollutants. There are certain technologies like reverse osmosis, ultrafiltration, and nanofiltration, which use semi-permeable membranes, making them suitable for desalination purposes or municipal water treatment. Today, nanomaterial-based systems are prepared using various materials (such as carbon nanotubes, graphene oxide, or metal–organic frameworks) that have high surface area per unit mass and adjustable pore sizes; these exhibit strong antimicrobial action. As efficient as these technologies may be, the challenge is that their eco-friendly application has limitations. High energy consumption is a major problem. For example, reverse osmosis requires substantial amounts of energy to generate the pressure necessary to push water through membranes, which results in higher operational costs along with carbon footprints. And, the disposal of concentrated waste streams is another problem that needs to be addressed with these technologies. Nanomaterial filtration systems are produced using chemicals and require energy-intensive methods that also produce toxic waste and climate-change gases. However, the expense associated with these methods and their scalability issues prevent extensive deployment. Within these limitations, there is a necessity for other water treatment options that are efficacious but environmentally friendly. Biopolymeric hydrogels have used biologically derived substitutes for ordinary technologies, representing one potential resolution (Elsaid et al. 2023).

11.2.3 Introduction to Biopolymeric Hydrogels as a Potential Solution

Biopolymeric hydrogels, due to their exceptional functional versatility, represent an invaluable solution in the highly demanding field of water purification. As a new type of biopolymer, hydrogel is usually most stable with less evaluation in solution compared to previous conventional methods such as membrane filtration and nanomaterial-based systems, due to problems like energy consumption, associated environmental impact, cost, etc.). Natural hydrogels are natural polymers that form a three-dimensional network, known as biopolymeric hydrogels, and they include chitosan, alginate, or cellulose used to produce such materials. Biopolymeric hydrogels remove pathogens from wastewater through adsorption, biofilm inhibition, and controlled release of antimicrobial agents (Fig. 11.1). are able to absorb and hold water owing to their hydrophilic properties, making them a high-water material that has gel-like attributes. This characteristic specifically makes them more efficient for capturing and removing contaminants from water. The most promising characteristic of biopolymeric hydrogels is that they may be easily functionalized with a range of active agents, and such agents can include antimicrobial peptides, metal nanoparticles, as well as enzymes. These hydrogels could be functionalized to selectively trap pathogens and contaminants, therefore improving the purification of water. An example is hydrogels containing silver nanoparticles, which kill bacteria and can inactivate viruses very well. Enzyme-encapsulated hydrogels induce the enzymatic degradation of organic pollutants in a more sustainable manner that is suited for varied water treatment applications. In addition, biopolymeric hydrogels work at room temperature and under low pressure, which is associated with reduced energy consumption compared to traditional membrane processes. Being biodegradable, they do not add to environmental pollution, as is common with synthetic materials and nanomaterials that can stay in the environment for a very long time. Therefore,

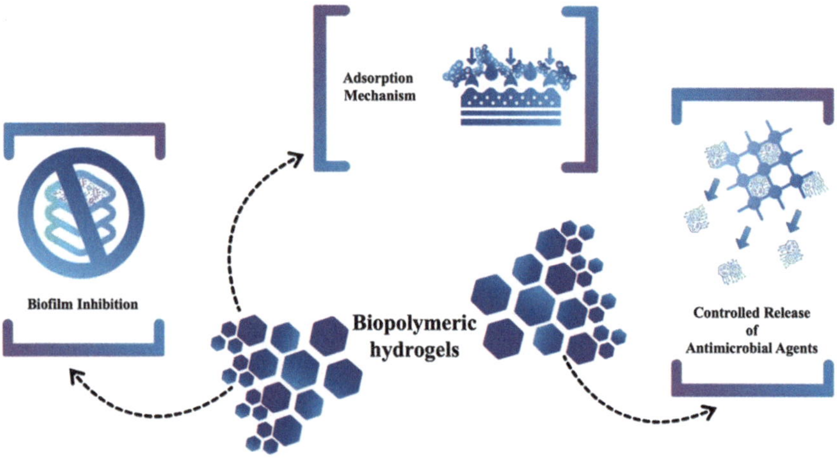

Fig. 11.1 Pathogen removal mechanisms of biopolymeric hydrogels: adsorption, biofilm inhibition and controlled antimicrobial release in wastewater treatment

biopolymeric hydrogels could be a promising candidate to perform sustainable and efficient water purification with negligible environmental hazards (Muir and Burdick 2021).

11.3 Chitosan-Based Hydrogels: Synthesis and Applications

Chitosan is a versatile biopolymer derived from the second most abundant natural polymer, chitin, which is commonly found in the exoskeletons of crustaceans and insects. This extensively studied biopolymer can be used in various applications and shows unique material properties such as biocompatibility, bioresorbability and antimicrobial effects. These properties make chitosan a distinguished biopolymer among researchers and scientists, particularly in the crafting of hydrogels, which focus on improved wastewater treatment and pathogen eradication. This forward-looking attitude in research results in the development of a safe environment where modern environmental challenges can best be addressed through natural solutions. Demineralization, deproteinization and deacetylation are the three major steps involved in the extraction of chitosan. The process of deacetylation converts chitin to chitosan and provides various biological properties. The unique structure of chitosan is important in various applications, especially in the formation of nanobiocomposites and hydrogels because of its cationic nature under acidic and neutral pH levels. This characteristic makes chitosan effectively interact with negatively charged microbial cells and other contaminants, making it an ideal candidate for water purification processes (Sreehari et al. 2022). Its diverse applications in various fields are mainly due to its non-toxic nature. Additionally, chitosan contains large amounts of amino groups and hydroxyl groups in the main chain, which enables it to absorb substantial quantities of water, enhancing the functionality of the hydrogel. This is important in efficiently retaining water content and treating wastewater.

The multiple benefits of chitosan make it one of the major biopolymers in treating wastewater. Chitosan modifications to form various chitosan-based biomaterials can enhance adsorption and overcome its low chemical stability in the unmodified state. These modifications are achieved using chemical or physical processes. Sometimes combinations of both processes are also used to influence the characteristics, surface morphology and adsorption of chitosan. Chitosan-based hydrogels are known among these modifications due to their easy preparation and enhanced loading capacity. These hydrogels show greater porosity and expanded polymer structures, which result in a larger surface area (Qu and Luo 2020). Moreover, they reduce the crystalline nature of chitosan, enhancing the accessibility of internal sorption sites. Chitosan also shows great antimicrobial activity against a number of pathogenic bacteria, fungi and viruses. The positive charge of chitosan is caused due to the surface charge arising from the amino groups on the glucosamine monomer, enabling interactions with negatively charged cell membranes of microbes, which eventually leads to rupture of the cell (Qu and Luo 2020). Incorporating the antimicrobial property of chitosan along with the other features shows its efficiency in water purification and pathogen eradication. The biodegradable property of chitosan also aids in the reduction of secondary pollution, which is one of the major drawbacks of synthetic polymers.

Developments in chitosan-based hydrogel synthesis delve into the use of methods like enzymatic cross-linking and freeze–thaw technique. In the conventional cross-linking methods, chemical cross-linking involves grafting quaternary ammonium salts onto the polymer backbone, directly quaternizing NH_2 groups, initiating graft polymerization from chitosan and forming metal-coordinated complexes to improve its antimicrobial properties and water solubility across a broad range of pH (Ding et al. 2021). On the other hand, physical crosslinking relies on physical interactions such as hydrogen bonding and electrostatic interactions. Physical crosslinking mostly occurs by the physical blending of chitosan with other materials that enhance its properties, such as carbon-based materials and biopolymers, to form hydrogels (Ding et al. 2021). In UV radiation crosslinking, the exposure to UV creates a network structure with higher stability and mechanical properties. Among all the crosslinking methods, enzymatic cross-linking is the most eco-friendly. Enzymatic cross-linking provides high catalytic efficiency, resulting in faster and more accurate cross-link formation. This method has the upper hand over conventional cross-linking methods, as it does not require extreme reaction conditions. The freeze–thaw technique works by alternating cycles of freezing and thawing of the chitosan solution. The alternating cycles promote ice crystallization and melting, pushing polymer chains together and enhancing intermolecular interactions, thereby forming physical cross-links that strengthen the hydrogel's network structure. The list of crosslinking methods and their common crosslinking agents is given in Table 11.1. Recently, studies have shown that blending multiple polymers can improve mechanical strength and other properties of hydrogels (Rodríguez-Rodríguez et al. 2020).

Table 11.1 Chitosan modifications for hydrogel synthesis

Method	Cross-linking agents	References
Physical cross-linking	Molecular entanglements, ionic interaction	Pita-López et al. (2021)
Chemical cross-linking	Glutaraldehyde	Goycoolea et al. (2003)
UV radiation cross-linking	UV light	Wang et al. (2023)
Enzymatic cross-linking	Laccase	Beer et al. (2020)
Freeze–thaw technique	NA	Zhang et al. (2023)

11.4 PHA (Polyhydroxyalkanoates) Hydrogels: Production and Utilization

Polyhydroxyalkanoates (PHAs) are a group of biodegradable polyesters produced by many bacteria as carbon and energy storage compounds. PHAs are bacterial fermentation-derived plastics known as potential biodegradable and more generally environmentally friendly substitutes for petroleum-based polymers. PHA is a type of biodegradable material included in packaging, medical appliances and as environmental remediation materials as it is synthesized from renewable resources. The two main types of PHAs are short-chain-length (SCL)-PHAs like poly(3-hydroxybutyrate) (PHB), and medium-chain-length. This feature can be applied for structural uses like pharmaceuticals, medical implants and packaging. SCL-PHAs are crystalline in nature which gives them rigidity with a higher melting point than MCL-PHA. On the other hand, their characteristic fragility and limited impact resistance restrict usage in more harsh environments. In comparison, MCL-PHAs are amorphous in nature with less crystalline structure and have characteristics like flexibility, toughness and hence impact resistance, rendering them suitable for applications where these properties may be desired (e.g., adhesives or coatings). Metabolic engineering of these microorganisms and optimization of fermentation process conditions allow the tailoring of polymer compositions with different physical properties (Javaid et al. 2020).

11.4.1 Water Treatment Applications of PHA Hydrogels

PHAs modifications, such as the PHA hydrogels, could be a promising candidate for water purification applications. The three-dimensional polymer networks are able to absorb more water and at higher concentrations for use in applications requiring high aqueous capacity or gel-like states. The remarkable properties of PHA for biodegradability and compatibility, combined with the potential to store contaminants by hydrogels, make a bio-hybrid material attractive as an efficient water pollutant capture agent. CH hydrogels that are created via crosslinking with agents such as glutaraldehyde or physical methods such as freeze–thaw cycles can be controlled for their physical properties, like capacity to absorb water and mechanical strength, along with degradation profiles. As such, PHA hydrogels are an interesting material to consider for a range of water treatment processes that involve the capture or adsorption of pollutants within aqueous systems, including wastewater contaminants like heavy metals and pathogens (Sabapathy et al. 2020).

11.4.2 Types of Green Synthesis Method for Preparing PHA Hydrogel

Hydrogels are two-stage products that involve biosynthesis and gelation phases. In the biosynthesis step, microorganisms like *Cupriavidus necator*, *Bacillus cereus* or *Pseudomonas putida* make use of fermentation to convert carbon sources into PHAs. The production preferably would require biological feedstock like agricultural residues (bagasse and corn stover), food waste or plant oils. Waste streams to feedstocks increase productivity and decrease production costs. PHA accumulates as intracellular granules by microorganisms through the fermentation of carbon sources under nutrient-limiting conditions. After fermentation, the PHAs are extracted from microbial biomass by either solvent or mechanical extraction. Although the use of solvent extraction methods such as chloroform and dichloromethane is widely reported, they raise environmental and safety issues. Though mechanical methods, like high-pressure homogenization or ultrasonication, are the more sustainable replacements. The PHAs are purified following extraction to acquire the desired polymer properties.

Purified PHAs are dissolved in a suitable solvent to form a polymer solution and the hydrogel network is constructed by crosslinking during the gelation stage. Crosslinking is a physical process in which the polymer chains bond by forming covalent bonds or they are entangled together through freeze–thaw cycles. The methods for crosslinking and conditions (like temperature) greatly affect their properties, such as water-holding capacity, biodegradability or mechanical strength, depending on the polymer used. Green chemistry techniques are also used to improve the sustainability of PHA hydrogel production. Non-toxic solvents such as ionic liquids or supercritical CO_2 can be used, which decreases the environmental impact and provides increased safety. Production of PHA hydrogels is also being developed with energy-saving processing methods such as microwave-assisted extraction or enzymatic crosslinking to reduce the carbon footprint in producing these bio-based polyesters (Zhang et al. 2019).

11.4.3 Using PHA Hydrogels for Wastewater Treatment and Pathogen Eradication

PHA hydrogels have special features that enable them to be a promising solution for wastewater treatment and pathogen elimination. Both its high-water content and the porous nature can absorb organic contaminants, heavy metals and pathogens from wastewater. The biodegradable properties of PHA hydrogels, free from concerning mineral solubility and not generating solid waste post-use, promise a green alternative to traditional water treatment materials (Fig. 11.2). In addition, PHA hydrogels have potential that can be functionalized with different active agents to improve the efficiency of pollutant removal. For example, the addition of silver nanoparticles to PHA hydrogels lends them powerful antimicrobial properties that can prove beneficial in pathogen elimination in drinking water purification or treating wastewater from hospitals. Moreover, the scaffolds can be designed to enhance bio removal by allowing functionalization with specific contaminants through simple or multiple-layer coatings containing targeted enzymes (or other agents required for their action). Enzyme-modified hydrogels could efficiently and specifically degrade some organic pollutants (such as phenols or dyes) in water treatment. Additionally, the open-mesh polymer structure makes PHA hydrogels well-suited for filtration systems that screen out microorganisms and particulate impurities while allowing only sterilized water to flow (Carvalho et al. 2021).

The biocompatibility nature of PHA hydrogel would avoid integration with treated water to produce undesired products. PHA hydrogels are environmentally friendly as they are derived from renewable feedstocks and suitable packaging materials that undergo biodegradation during a water purification process. Development and research are ongoing in the area of PHAs and hydrogels, which can provide a considerable impact on innovative water treatment technologies. The combination of these results (i.e., promising functionalities and sustainability) indicates that this research could provide powerful approaches to introduce PHA hydrogels in the current systems for water treatment, offering new applications with a reduced impact on public health, water quality and environment (Subash et al. 2023).

11.5 Cellulose-Derived Hydrogels: Properties and Functionalities

The cellulose-derived hydrogels are prepared through different types of physical and chemical cross-linking methods. Cellulose modifications have crucial roles in forming hydrogels with unique properties. The cross-linking in such hydrogels is facilitated by the native hydroxyl groups as well as these modified functional groups. The hydrogel formation is associated with the interconnected network formation between the cellulose matrices. Various physical interactions, including van der

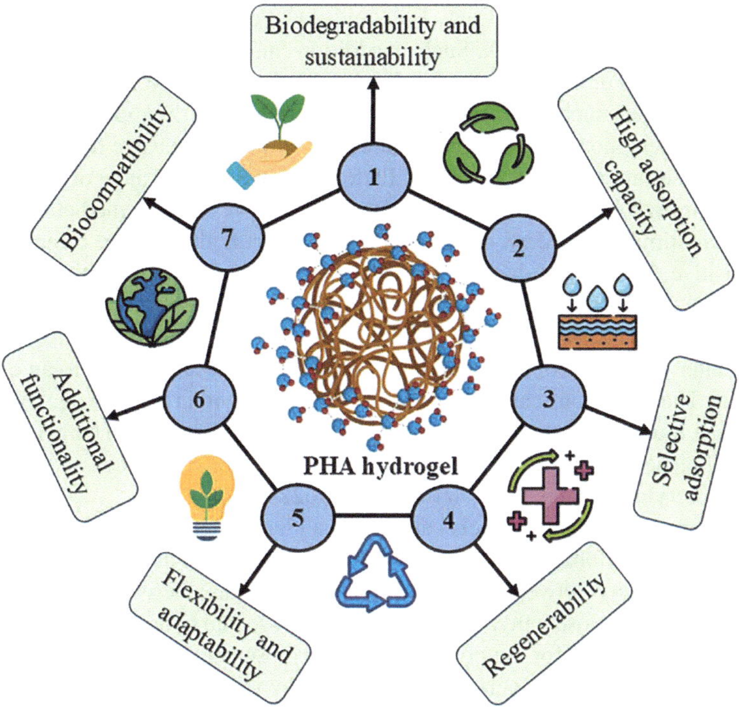

Fig. 11.2 Advantages of PHA hydrogel in water purification

Waals interactions, mechanical chain entanglements, hydrogen bonding, electrostatic associations and hydrophobic aggregations (Ajdary et al. 2021). Interpreting these properties in a cellulose hydrogel is crucial in obtaining hydrogels with unique properties, such as biomedical materials.

11.5.1 Unique Properties of Cellulose-Based Hydrogels

The research on cellulose hydrogel is gaining momentum recently due to its higher biocompatibility combined with the ability to mimic essential in vivo functions. The customizable biochemical, physical and mechanical properties make the cellulose hydrogel an ideal candidate for biomedical and bio remedial applications. Based on the composition, cellulose hydrogels can be synthesized with unique properties.

A hydrogel with a three-dimensional nanofibre network with distinct properties can be obtained when the hydrogel is synthesized using cellulose nanofibres (CNF). These CNF-hydrogels are recognized for their enhanced cytocompatibility and high tolerogenic potential. These hydrogels have potential applications in elastic hydrogels, ionic conduction, and water purification. The addition of divalent or trivalent cations such as Zn^{2+}, Cu^{2+}, Ca^{2+}, Fe^{3+} and Al^{3+}, can lead to hydrogels with an interconnected porous network. Such an addition can enhance the capability of CNF hydrogel for drug delivery systems and encapsulation applications.

The limitations, such as poor mechanical properties, limit the applicability of traditional hydrogels; however, the development of stimuli-responsive hydrogels helps to solve these limitations. These smart hydrogels have a wide range of applications due to their high elasticity, adaptability, and ability to respond immediately to environmental stimuli. Based on the stimuli, these smart hydrogels are categorized as pH sensitive, thermally sensitive, salt sensitive and tiny force-sensitive hydrogels (Deng et al. 2021a, b). The presence of acid and base groups (such as carboxyl and amino groups) is the characteristic of stimulus-based smart cellulose hydrogels. Such groups alter the degree of protonation or deprotonation as a response to an external stimulus. The alterations cause electrostatic variations within the hydrogel, which in turn affect the swelling behaviours exhibited by these hydrogels. The pH-sensitive smart cellulose gel can be obtained by the addition of materials such as polyacrylic acid, polyacrylamide and chitosan (Yin et al. 2022).

The thermosensitive hydrogels can manifest changes in volume as a response to temperature alterations. Temperature changes can influence the hydrogen bondings and hydrophobicity between the macromolecular chains. Derivatives of cellulose, such as methylcellulose, ethyl hydroxyethyl cellulose and hydroxypropyl cellulose, are commonly utilized for the production of thermosensitive hydrogels.

The salt-sensitive hydrogels can be triggered by varying the salt concentrations. The variation in salt concentration in the environment stimulated varying swelling rates and water absorption capacities. When salt concentration in the environment is altered, the covalently bonded positive and negative charged groups tend to shield or disrupt the macromolecular associations. This leads to the molecular chain stretching and alterations in the swelling nature of the hydrogel. The common salt-sensitive hydrogel is the nanocellulose hydrogels, which are modified by the addition of ions such as ClO^- (Hai et al. 2019).

The tiny force-sensitive hydrogels are designed to detect minimal external stimuli and convert them to detectable electrical signals. The applications of such hydrogels are in biomedical fields to detect human motion and skin perception.

11.5.2 Extraction and Modification Techniques

Cellulose can be obtained from all plant sources through basic steps involving alkali treatment and bleaching or ultrasonication (Santhosh and Umesh 2024). Based on the source of extraction, the cellulose can exhibit differences in certain properties. Thus, the hydrogel produced from cellulose extracted from different sources may exhibit different characteristics. The hydrogels, which are formed due to the non-covalent interactions such as hydrogen bonds, van der Waals forces, ionic and hydrophobic interactions, are called physical hydrogels. These gels are also known as thermally reversible hydrogels or pseudo-hydrogels.

The ionic interaction can be commonly seen in hydrogels with polyvinyl alcohol and chitosan. The crystal cross-linking involves strain-induced crystallization. The elongation and contraction of such hydrogels are associated with the crystallization and melting, which are observed in cellulose hydrogel mixed with polyethylene glycol. The hydrogen bond cross-linking is present in cellulose-based hydrogels produced by blending with PVA and tannic acid; such cross-links can be modulated using ethanol (Maiti et al. 2024). The hydrophobic associations which drive the physical crosslinking can be observed in hydrogels modified with groups such as short and long alkyl side chains.

Chemical cross-linking is irreversible and forms three-dimensional networks which are formed through covalent bonds. Such linkages can be seen in hydrogels produced using monomers, polymers or both. The chemical cross-linkages are not spontaneous; they require external stimuli such as ultraviolet light. The common agents utilized for chemical cross-linking of cellulose derivatives include dialdehydes, acetals, polycarboxylic acids, epichlorohydrin, and polyepichlorohydrin. Hydrogel materials formed from such materials have a uniform morphology, lower crystallinity and a highly porous structure with high swelling capacity (Zou et al. 2022).

11.5.3 Role in Pathogen Eradication and Wastewater Treatment Processes

It was well established that a modified cellulose hydrogel can exhibit antimicrobial activity. These hydrogels can be used in wound dressings to prevent pathogen infection in chronic wounds. Wound dressings primarily use silver as an antibacterial agent. In wastewater treatment, hydrogels can be modified with antimicrobial compounds or materials such as nanoparticles to filter out pollutants and eliminate potential pathogens. Wastewater treatment with cellulose hydrogels is accomplished through various mechanisms, which include adsorption, filtration, the addition of antimicrobial agents, catalytic degradation, biofilm inhibition, and controlled release systems. The adsorption and filtration are facilitated by the large surface area and porous surface, which could filter pollutants and absorb contaminants such as heavy metals, dyes, and organic pollutants from wastewater. These properties also tend to adsorb and immobilize the pathogens which are present in contaminated wastewater. As discussed above, the addition of antibacterial agents could kill the pathogens that are present in water. The controlled release systems in cellulose hydrogel can facilitate the slow release of antimicrobial agents and disinfectants, ensuring prolonged pathogen eradication. Incorporating biofilm-disrupting agents into cellulose hydrogels can prevent biofilm formation in wastewater, which facilitates a more effective pathogen elimination (Tavakolian et al. 2020). Additionally, adding catalysts like enzymes and metal oxide nanoparticles can help oxidize pollutants and pathogens, which can be immobilized through the adsorption ability of the

Table 11.2 Different types of cellulose-derived hydrogels and their applications in wastewater treatment

Type of cellulose-derived hydrogel	Insight into wastewater treatment	References
Graphene oxide crosslinked nanocomposites hydrogels of chitosan and carboxymethyl cellulose	The hydrogel (0.4 g/L) was able to remove 99% of methylene blue (50 mg/L)	Mittal et al. (2021)
Cellulose–chitosan hydrogel	A 0.5 g sample of the hydrogel successfully removed 94.75% of crystal violet at a concentration of 60 mg/L, while a 0.13 g sample removed 95.88% of methylene blue through adsorption	Poornachandhra et al. (2023)
Graphene oxide carboxymethylcellulose-alginate hydrogel microspheres	A 5% gel could absorb 64 mg/g of Cu^{2+} ions	Allouss et al. (2020)
Cellulose hydrogel functionalized titanate microspheres	It is a self-cleansing hydrogel which could remove 176.4 mg/g Cu^{2+} ions from oily wastewater	Xiong et al. (2020)
Carboxymethyl cellulose- based cryogel	The prepared cryogel was able to absorb 550, 620 and 760 mg/g of Pb(II), Ni(II) and Co(II), respectively	Li et al. (2020)

hydrogel. Recent studies regarding cellulose hydrogel in wastewater treatment are detailed in Table 11.2.

11.6 Gelatin Hydrogels: Synthesis and Performance

Gelatin is a protein-based natural biopolymer that can be made from animal collagen. It is the most common gelling agent in foods, cosmetics and ointments. The gelatin gels are formed through the cross-linkage between gelatin molecules. The gelatin hydrogel is extremely biocompatible and biodegradable, hence are utilized in biomedical applications such as tissue engineering and drug delivery (Hogan and Mikos 2020). The biodegradability of gelatin hydrogel makes it an ideal candidate for environmental applications, including wastewater treatments.

11.6.1 Gelatin as a Versatile Biopolymer for Hydrogel Production

The market value of gelatin biopolymer was USD 3.07 billion in 2023, and is expected to be USD 5.51 billion by 2032. Gelatin is produced by partial hydrolysis of collagen proteins found in animal tissue such as bones, cartilage, tendons, ligaments, and skin. The non-toxic property and biodegradability of the gelatin biopolymer make it a common utility across various industries, including both food and non-food industries. The common applications of gelatin biopolymer include film formation, emulsification, thickening, and gelation. Gelatin possesses functional groups including terminal $-NH_2$ and –COOH groups, aspartic acid –COOH groups, $-NH_2$ group of lysine, guanidinium group of arginine and imidazolium group of histidine, which add to their functionalities. The presence of additional groups, such as carboxyl and phenolic groups, serves as potential sites for chemical modifications. The cross-linkable and graftable groups in gelatin are highly receptive to alterations, which makes them the ideal candidates for hydrogels that require rapid modifications. These modifications can be achieved physically or chemically, or by adding crosslinkers or copolymers. The versatility of gelatin facilitates adjustments to the structural morphology, which could alter the degradability, functionality and hydrophilicity. Hydrogels produced with gelatin offer consistency and viscosity, and their biodegradable and renewable nature makes them a sustainable option. The current field of application of gelatin biopolymer expands from biomedical and biotechnological fields to various water and waste treatment processes due to their structural and textural properties (Kouhi et al. 2020).

11.6.2 Synthesis Methods and Structural Properties

Collagen loses its native structure when it is transformed into gelatin. Gelatin proteins tend to denature at higher temperatures; hence, hydrogel preparation should not be done at higher temperatures. The normal reaction temperature for gelation of gelatin is 40 °C. For the preparation of hydrogels with materials that have a high dissolution temperature, the compounds should be dissolved before the addition of gelatin. Upon heating, a conformational change happens to the collagen fibres, which leads to the formation of helix-like structures. Simultaneous cooling partially restores the original structure of the gelatin, trapping water within the network and resulting in gel formation. The structure of gelatin changes with the occurrence of gelling and the state of the gel is determined by its spatial arrangements and interactions (Nishinari 2021). The gelatin hydrogel preparation is illustrated in Fig. 11.3.

The gelatin are classified into three forms based on the structure: primary- a single α-chain, secondary- either two single α-chains or a single α-chain with a loop and tertiary- three single α- chains, two single α-chains with one forming a loop, or a single α-chain forming two loops.

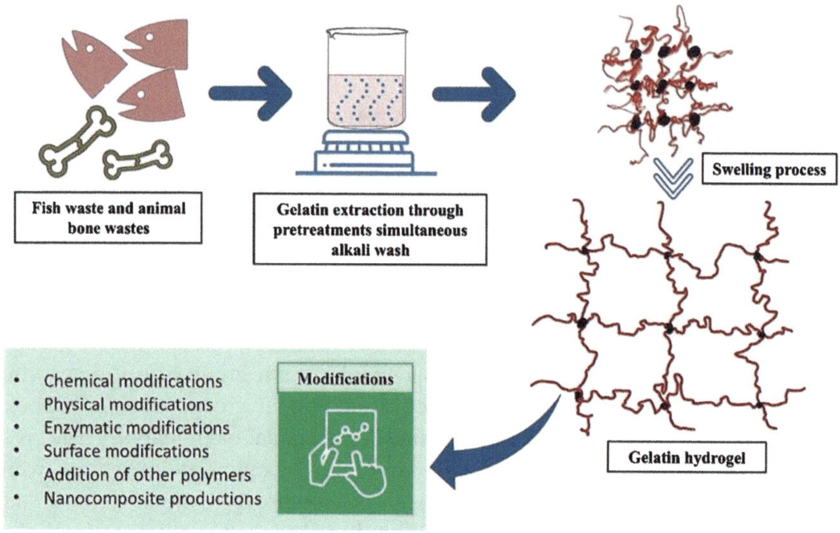

Fig. 11.3 Preparation of gelatin hydrogel

Factors such as the concentration, composition of gelatine, reaction time and temperature affect the formation of these primary, secondary, and tertiary structures in gelatin hydrogel (Milano et al. 2023). Other factors that affect the physicochemical characteristics of gelatin hydrogel are contributed by the composition of amino acids in the native collagen. The presence of amino acids such as proline and hydroxyproline in higher quantities tends to enhance the stability of the triple helices formed during gelation. The melting point of gelatin ranges between 27and 34 °C, is determined by the amino acid compositions. The lower melting point makes gelatin hydrogel highly reversible and mouldable. Additionally, the ability of gelatin to absorb water up to 5 to 10 times its own mass by swelling enhances its applicability.

Gelatin itself serves as the crosslinking against in gelatin hydrogel; however, the addition of materials such as glutaraldehyde, alginate, acrylic acid, N,N'-methylene-bis-acrylamide, chitosan, poly(vinyl alcohol) and acrylamide enhances the crosslinkings (Sapuła et al. 2023). The additions can mould the characteristics of gelatine hydrogel to widen its applicability in various fields.

11.6.3 Performance in Removing Pathogens from Wastewater

The matrix crosslinking and controllable porosity in gelatin hydrogel enable it to trap contaminants by adsorption. Pathogens can be trapped within the gelatin hydrogel, and they can be eliminated by incorporating antimicrobial agents during the hydrogel synthesis, making it suitable for wastewater treatment applications. Hydrogels based on gelatine are currently studied in the removal of azo dyes, toxic metals, phosphates and nitrates, and the oily contents. The removal efficiency of gelatin-based hydrogels can reach up to 99% efficiency.

The adsorption of pathogens on gelatin hydrogels from wastewater can be controlled by altering the functional groups, thereby creating a charge difference for effective adsorption (Elella et al. 2022). The properties of gelatin hydrogels can be precisely customized, which enables them to target specific pathogens or groups of pathogens by not affecting useful microbial communities. Addition of antimicrobial agents such as silver nanoparticles, plant extracts, or other compounds, along with the ability to tailor the controlled material release from gelatin hydrogel, can significantly increase its effectiveness in pathogen eradication (Moreno Ruiz et al. 2023). It can be stated that the gelatin hydrogels are a promising material for the removal of pathogenic microorganisms and pollutants from wastewater when they are engineered to target specific conditions and pathogens.

11.7 Other Biopolymers in Hydrogel Production

The advantages of composite hydrogels and the increasing demand for efficient and sustainable eradication of pathogens from wastewater point out the need for exploring more biopolymers. Even though further advanced research is required for traditional biopolymers, the potential of novel polymers remains unexplored. Alginate, pectin and xanthan gum are some of the promising polymers that need extensive research. These polymers offer environmental benefits and unique properties that can enhance the quality of composite hydrogels or even serve as a standalone biopolymer for hydrogel. Each of these biopolymers has its own properties that widen the application in different sectors.

11.7.1 Alginate

Alginate is a naturally occurring polyanionic heteropolymer derived from brown algae or certain bacteria, consisting of β-D-mannuronate (M) and α-L-guluronate (G) units linked via 1–4-glycosidic bonds. It is a promising biopolymer with high gelation capacity and other properties such as biocompatibility, non-immunogenicity and biodegradability. Alginate is a common emulsifier and thickener used in the food industry. This is due to the high solubility, stability, viscosity and gel-forming properties of alginate. This property provides aid in forming high-quality hydrogels. Alginate hydrogel has high mechanical properties along with the ability to adsorb heavy metals and dyes in a wide range of pH (ALSamman and Sánchez 2021). These hydrogels have also shown multi-levelled pore structure and photochromic ability, which makes it highly applicable in wastewater treatment. Nanoparticles can increase the antimicrobial and antioxidant capacity and they can provide better structural stability. Alginate has shown high potential for incorporating nanoparticles in its matrix by the process of crosslinking. Studies have shown improvement in antibacterial properties and stability by incorporating silver phosphate (Ag_3PO_4) nanoparticles accumulated on the surface of the polymer (Zhang et al. 2020).

The ability of alginate to form composite hydrogels with other polymers is important in fusing the properties of two polymers into a single hydrogel. A Nano-in-micro alginate/chitosan hydrogel system fabricated using electrospray technology, loaded with curcumin, has shown increased antibacterial properties. In another study, an alginate-chitosan hydrogel containing beta-glucan nanoemulsions increased antibacterial activity against *Staphylococcus aureus* and *Escherichia coli*. Nanocellulose also has high interlinking potential with alginate, which shows application in various fields as viscosity modifiers, sheets, nanocellulose and aerogels (Deng et al. 2021a, b).

11.7.2 Pectin

Pectin is a water-soluble anionic polysaccharide typically extracted from fruit and vegetable waste like citrus peels and apple pomace, which is prevalent in the primary cell walls and middle lamella of plants. This carbohydrate polymer is primarily made up of galacturonic acid (Gal-A) units along with three major structural components: rhamnogalacturonan I, rhamnogalacturonan II and homogalacturonan. Commercially, pectin is extracted from sugar beet pulp, mango peel, banana and sunflower seed heads. The gelation property of pectin makes it unique from other polymers. Compared to the gels of other biopolymers, gels made using pectin have faster gel-forming properties along with better thermal and mechanical properties (Said et al. 2023). This makes pectin a suitable candidate for hydrogel formation.

On the basis of the degree of esterification, it is divided into High-Ester Pectin and Low-Ester Pectin, which constitute different forms such as aerogels, hydrogels, cryogels, oleogels and xerogels. Pectin is soluble in nature and in order to change that, pectin is often crosslinked with ionic crosslinking agents, with monovalent, divalent or trivalent cations (Martínez-Sabando et al. 2023). Pectin's ability to form complexes and consist of properties like being inert, biodegradable and biocompatible makes it widely used in wastewater treatment.

11.7.3 Xanthan Gum

Xanthan gum is made from the bacterium *Xanthomonas campestris* and holds very high viscosity with great stability for use in hydrogels at extreme temperatures and pH levels. Xanthan gum has a backbone made up of β-D-glucose linked similarly to cellulose. Hydrogels based on xanthan are extremely effective for application in a dynamic environment of wastewater, where the stability of thermal and mechanical properties is extremely important. Xanthan, being a hydrogel, becomes shear-thinning and for applications in wastewater flow, it can be designed to alter its viscosity for the transportation of pathogens and removal by the hydrogel. The formation of the hydrogels with xanthan gum usually takes place either through physical or chemical cross-linking, which offers the advantage of designing hydrogels that are tough yet degradable in a controlled manner (Hassanzadeh-Afruzi et al. 2023). These hydrogels are designed to release antigens or stabilize heavy metals and thus are relevant to the processes of pathogen destabilization and removal.

11.8 Challenges and Future Directions

Biopolymeric hydrogels come forth as one of the best materials in the treatment of water and other applications, with great potential in eradicating pathogens. These are environmentally beneficial due to their biocompatibility and biodegradability, making them the best choices to be used in sustainable treatment solutions. However, despite this promise, a number of challenges remain before it can achieve the efficiency and reliability needed to work effectively in the field. The main drawback of using natural biopolymers is that their properties are variably inherent and such nature could cause inconsistent performance in hydrogels in wastewater treatment. Their natural origin results in changes of molecular weight, degree of polymerization and compositions of functional groups (Russo et al. 2021). All these differences, in fact, vary the gelation behaviour, swelling properties and mechanical strength of hydrogels, which gives rise to differences in their trapping and inactivation ability of the pathogens in different wastewater environments. The overall inconsistencies could compromise the treatment effectiveness, especially under variable wastewater composition.

11.8.1 Strategies for Overcoming Challenges and Enhancing Performance

In recent times, hybrid hydrogels have been synthesized to enhance the performance of biopolymeric hydrogels in wastewater treatment. The current approach, therefore, consists of designing hydrogels with features that remain unchanged while increasing mechanical properties to resist the harsh conditions in wastewater environments. In addition, new cross-linking techniques and stronger chemical cross-linkers are being developed, while also using nano-enhancers like graphene oxide or biochar to enhance the strength and lifetime of hydrogels in wastewater treatment systems (Verma and Singh 2023). The enhancement would be beneficial for maintaining its functional capability for a long time and under varying environmental conditions.

11.8.2 Future Research Directions and Potential Innovations

The interest for future studies will therefore be the development of 'smart' hydrogels that may become responsive to environmental changes and, in turn, selectively react to particular pathogens. Innovations may range from stimuli-responsive hydrogels, which can swell or shrink due to changes in pH, to hydrogels that are capable of releasing antimicrobial agents upon the detection of specific bacterial strains with increases in temperature, ultimately increasing sensitivity and efficiency in pathogen eradication (Tang et al. 2021). It also holds substantial potential in joining sensor technology with hydrogel systems for the development of smart diagnostic and treatment platforms that could dynamically monitor and respond to the loads of pathogens in real time. Systems of this kind could revolutionize the process of wastewater treatment by providing adaptive, highly targeted, and efficient pathogen eradication.

11.9 Conclusion

Development of cost-effective and efficient methods for the removal of pathogens from wastewater has been a thrust area of research, taking into consideration the amount of wastewater generated and also the scarcity of fresh water for various agricultural and other uses. Conventional wastewater treatment methods often fail to achieve these objectives due to the persistence of bacterial and viral pathogens even after the conventional treatment process. Wastewater treatment using biopolymeric hydrogels has been a successful strategy for pathogen removal due to their cost-effectiveness and ease in preparation and maintenance. Hydrogels seem to have excellent reproducibility and high selectivity, making them ideal candidates for wastewater treatment. Use of biopolymers like chitosan, PHA, cellulose and gelatin that can be produced through waste valorization approach in turn suggests an integrated strategy for waste management and wastewater treatment. Although the studies regarding water treatment using hydrogels are still in infancy and restricted to laboratory studies mainly, extending them to field studies is required to assess their commercial success. Long-term monitoring and analysis of the treated water is required to ensure the safety and efficacy of the hydrogel-based treatment systems. Secondly, designing a proper strategy for the disposal of the hydrogels containing the entrapped pathogens filtered from wastewater should also be focused on ensuring the safety of humans and the environment. Thus, more research focusing on this area can clearly emphasize the importance of developing sustainable biobased materials for wastewater treatment.

References

Ajdary R, Tardy BL, Mattos BD, Bai L, Rojas OJ (2021) Plant Nanomaterials and Inspiration from nature: water interactions and hierarchically structured hydrogels. Adv Mater 33(28):e2001085

Allouss D, Essamlali Y, Chakir A, Khadhar S, Zahouily M (2020) Effective removal of Cu(II) from aqueous solution over graphene oxide encapsulated carboxymethylcellulose-alginate hydrogel microspheres: towards real wastewater treatment plants. Environ Sci Pollut Res Int 27(7):7476–7492

ALSamman MT, Sánchez J (2021) Recent advances on hydrogels based on chitosan and alginate for the adsorption of dyes and metal ions from water. Arab J Chem 14(12):103455

Beer B, Bartolome MJ, Berndorfer L, Bochmann G, Guebitz GM, Nyanhongo GS (2020) Controlled enzymatic hydrolysis and synthesis of lignin cross-linked chitosan functional hydrogels. Int J Biol Macromol 161:1440–1446

Carvalho LT, Vieira TA, Zhao Y, Celli A, Medeiros SF, Lacerda TM (2021) Recent advances in the production of biomedical systems based on polyhydroxyalkanoates and exopolysaccharides. Int J Biol Macromol 183:1514–1539

Deng W, Tang Y, Mao J, Zhou Y, Chen T, Zhu X (2021a) Cellulose nanofibril as a crosslinker to reinforce the sodium alginate/chitosan hydrogels. Int J Biol Macromol 189:890–899

Deng Z, Yu R, Guo B (2021b) Stimuli-responsive conductive hydrogels: design, properties, and applications. Mater Chem Front 5(5):2092–2123

Ding S, Wang Y, Li J, Chen S (2021) Progress and prospects in chitosan derivatives: modification strategies and medical applications. J Mater Sci Technol 89:209–224

Elella MHA, Aamer N, Mohamed YMA, El Nazer HA, Mohamed RR (2022) Innovation of high- performance adsorbent based on modified gelatin for wastewater treatment. Polym Bull 79(12):11217–11233

Elsaid K, Olabi AG, Abdel-Wahab A, Elkamel A, Alami AH, Inayat A, Abdelkareem (2023) Membrane processes for environmental remediation of nanomaterials: potentials and challenges. Sci Total Environ 879:162569

Forés E, Mejías-Molina C, Ramos A, Itarte M, Hundesa A, Rusiñol M, Girones R (2023) Evaluation of pathogen disinfection efficiency of electrochemical advanced oxidation to become a sustainable technology for water reuse. Chemosphere 313:137393

Ghosh S, Falyouna O, Onyeaka H, Malloum A, Bornman C, AlKafaas SS, Suhas (2023) Recent progress on the remediation of metronidazole antibiotic as emerging contaminant from water environments using sustainable adsorbents: a review. J Water Process Eng 51:103405

Goycoolea FM, Heras A, Aranaz I, Galed G, Fernández-Valle ME, Argüelles-Monal W (2003) Effect of chemical crosslinking on the swelling and shrinking properties of thermal and pH-responsive chitosan hydrogels. Macromol Biosci 3(10):612–619

Hai J, Zeng X, Zhu Y, Wang B (2019) Anions reversibly responsive luminescent nanocellulose hydrogels for cancer spheroids culture and release. Biomaterials 194:161–170

Hassanzadeh-Afruzi F, Ranjbar G, Salehi MM, Esmailzadeh F, Maleki A (2023) Thiacalix[4]arene-functionalized magnetic xanthan gum (TC4As-XG@Fe_3O_4) as a hydrogel adsorbent for removal of dye and pesticide from water medium. Sep Purif Technol 306(122700):122700

Hogan KJ, Mikos AG (2020) Biodegradable thermoresponsive polymers: applications in drug delivery and tissue engineering. Polymer 211:123063

Javaid H, Nawaz A, Riaz N, Mukhtar H, Ul-Haq I, Shah KA, Murtaza G (2020) Biosynthesis of polyhydroxyalkanoates (PHAs) by the valorization of biomass and synthetic waste. Molecules 25(23). https://doi.org/10.3390/molecules25235539

Kouhi M, Prabhakaran MP, Ramakrishna S (2020) Edible polymers: an insight into its application in food, biomedicine and cosmetics. Trends Food Sci Technol 103:248–263

Li S-S, Song Y-L, Yang H-R, An Q-D, Xiao Z-Y, Zhai S-R (2020) Carboxymethyl cellulose-based cryogels for efficient heavy metal capture: aluminum-mediated assembly process and sorption mechanism. Int J Biol Macromol 164:3275–3286

Maiti S, Maji B, Yadav H (2024) Progress on green crosslinking of polysaccharide hydrogels for drug delivery and tissue engineering applications. Carbohyd Polym 326:121584

Martínez-Sabando J, Coin F, Melillo JH, Goyanes S, Cerveny S (2023) A review of pectin-based material for applications in water treatment. Materials 16(6). https://doi.org/10.3390/ma16062207

Milano F, Masi A, Madaghiele M, Sannino A, Salvatore L, Gallo N (2023) Current trends in gelatin-based drug delivery systems. Pharmaceutics 15(5). https://doi.org/10.3390/pharmaceutics15051499

Mittal H, Al Alili A, Morajkar PP, Alhassan SM (2021) GO crosslinked hydrogel nanocomposites of chitosan/carboxymethyl cellulose—a versatile adsorbent for the treatment of dyes contaminated wastewater. Int J Biol Macromol 167:1248–1261

Moreno Ruiz YP, de Almeida Campos LA, Alves Agreles MA, Galembeck A, Macário Ferro Cavalcanti I (2023) Advanced hydrogels combined with silver and gold nanoparticles against antimicrobial resistance. Antibiotics (Basel, Switzerland) 12(1). https://doi.org/10.3390/antibiotics12010104

Muir VG, Burdick JA (2021) Chemically modified biopolymers for the formation of biomedical hydrogels. Chem Rev 121(18):10908–10949

Nishinari K (2021) Gelling properties. In: Fang Y, Zhang H, Nishinari K (eds) Food hydrocolloids: functionalities and applications. Springer Singapore, Singapore, pp 119–170

Ojha A (2020) Chapter 19—nanomaterials for removal of waterborne pathogens: opportunities and challenges. In: Vara Prasad MN, Grobelak A (eds) Waterborne pathogens. Butterworth-Heinemann, pp 385–432

Pita-López ML, Fletes-Vargas G, Espinosa-Andrews H, Rodríguez-Rodríguez R (2021) Physically cross- linked chitosan-based hydrogels for tissue engineering applications: a state-of-the-art review. Eur Polymer J 145(110176):110176

Poornachandhra C, Jayabalakrishnan RM, Prasanthrajan M, Balasubramanian G, Lakshmanan A, Selvakumar S, John JE (2023) Cellulose-based hydrogel for adsorptive removal of cationic dyes from aqueous solution: isotherms and kinetics. RSC Adv 13(7):4757–4774

Pratap B, Kumar S, Nand S, Azad I, Bharagava RN, Romanholo Ferreira LF, Dutta V (2023) Wastewater generation and treatment by various eco-friendly technologies: possible health hazards and further reuse for environmental safety. Chemosphere 313:137547

Qu B, Luo Y (2020) Chitosan-based hydrogel beads: preparations, modifications and applications in food and agriculture sectors—a review. Int J Biol Macromol 152:437–448

Rizzo L, Gernjak W, Krzeminski P, Malato S, McArdell CS, Perez JAS, Fatta-Kassinos D (2020) Best available technologies and treatment trains to address current challenges in urban wastewater reuse for irrigation of crops in EU countries. Sci Total Environ 710:136312

Rodríguez-Rodríguez R, Espinosa-Andrews H, Velasquillo-Martínez C, García-Carvajal ZY (2020) Composite hydrogels based on gelatin, chitosan and polyvinyl alcohol to biomedical applications: a review. Int J Polym Mater 69(1):1–20

Russo T, Fucile P, Giacometti R, Sannino F (2021) Sustainable removal of contaminants by biopolymers: a novel approach for wastewater treatment. Current state and future perspectives. Processes (Basel, Switzerland) 9(4):719

Sabapathy PC, Devaraj S, Meixner K, Anburajan P, Kathirvel P, Ravikumar Y, Qi X (2020) Recent developments in Polyhydroxyalkanoates (PHAs) production—a review. Biores Technol 306:123132

Said NS, Olawuyi IF, Lee WY (2023) Pectin hydrogels: gel-forming behaviors, mechanisms, and food applications. Gels (Basel, Switzerland) 9(9). https://doi.org/10.3390/gels9090732

Santhosh AS, Umesh M (2024) Valorization of waste chilli stalks (*Capsicum annuum*) as a sustainable substrate for cellulose extraction: insights into its thermomechanical, film forming and biodegradation properties. Biomass Convers Biorefinery. https://doi.org/10.1007/s13399-024-05370-2

Sapuła P, Bialik-Wąs K, Malarz K (2023) Are natural compounds a promising alternative to synthetic cross-linking agents in the preparation of hydrogels? Pharmaceutics 15(1). https://doi.org/10.3390/pharmaceutics15010253
Sreehari S, Mridul U, Adhithya SS (2022) Wastewater treatment using chitosan and its derivatives: a mini review on latest developments. Notulae Scientia Biologicae 14(4):11369
Subash A, Naebe M, Wang X, Kandasubramanian B (2023) Biopolymer—a sustainable and efficacious material system for effluent removal. J Hazard Mater 443:130168
Tang L, Wang L, Yang X, Feng Y, Li Y, Feng W (2021) Poly(N-isopropylacrylamide)-based smart hydrogels: design, properties and applications. Prog Mater Sci 115(100702):100702
Tavakolian M, Munguia-Lopez JG, Valiei A, Islam MS, Kinsella JM, Tufenkji N, van de Ven TGM (2020) Highly absorbent antibacterial and biofilm-disrupting hydrogels from cellulose for wound dressing applications. ACS Appl Mater Interfaces 12(36):39991–40001
Verma L, Singh J (2023) Production of biochar-based nanocomposites from chemical and biological methods. In: Advances in science, technology and innovation. Springer International Publishing, Cham, pp 47–56
Wang J, Xu W, Zhang W, Da J, Liu L, Huang X, Zhang B (2023) UV cross-linked injectable non-swelling dihydrocaffeic acid grafted chitosan hydrogel for promoting wound healing. Carbohyd Polym 314:120926
Xiong Y, Xu L, Jin C, Sun Q (2020) Cellulose hydrogel functionalized titanate microspheres with self- cleaning for efficient purification of heavy metals in oily wastewater. Cellulose 27(13):7751–7763
Yin H, Song P, Chen X, Xiao M, Tang L, Huang H (2022) Smart pH-sensitive hydrogel based on the pineapple peel-oxidized hydroxyethyl cellulose and the hericium erinaceus residue carboxymethyl chitosan for use in drug delivery. Biomacromol 23(1):253–264
Zhang X, Li Z, Che X, Yu L, Jia W, Shen R, Chen G-Q (2019) Synthesis and characterization of polyhydroxyalkanoate organo/hydrogels. Biomacromol 20(9):3303–3312
Zhang Q, Zhang X, Cheng W, Li Z, Li Q (2020) In situ-synthesis of calcium alginate nano-silver phosphate hybrid material with high flame retardant and antibacterial properties. Int J Biol Macromol 165(Pt B):1615–1625
Zhang K-Y, Li D, Wang Y, Wang L-J (2023) Carboxymethyl chitosan/polyvinyl alcohol double network hydrogels prepared by freeze-thawing and calcium chloride cross-linking for efficient dye adsorption. Int J Biol Macromol 253(Pt 3):126897
Zou P, Yao J, Cui Y-N, Zhao T, Che J, Yang M, Gao C (2022) Advances in cellulose-based hydrogels for biomedical engineering: a review summary. Gels (Basel, Switzerland) 8(6). https://doi.org/10.3390/gels8060364

Chapter 12
Bioflocculants-Based Hydrogels for Wastewater Management

Abhishek Kumar and Deeptimayee Pal

Abstract Sustainable and effective treatment solutions have become more important as industrial effluences, agricultural runoff, and household wastewater contaminate aquatic bodies. Combining the strong adsorption and water retention qualities of hydrogels with the natural flocculating ability of flocculants, bioflocculant-based hydrogels have become increasingly intriguing alternatives for wastewater treatment. Mostly derived from natural biopolymers, including polysaccharides (chitosan, alginate, cellulose), proteins (gelatine, casein), and microbial flocculants (extracellular polymeric substances). Hybrid materials with charge neutralization, bridging, and adsorption capabilities may remove organic pollutants, heavy metals, suspended particles, and new contaminants like microplastics and dyes simultaneously. This chapter delves into the various synthesis procedures for hydrogels based on bioflocculants, including physical and chemical crosslinking strategies that enhance their structural stability and flocculation efficiency. The methods of contaminants removal, effectiveness under different water conditions (pH, ionic strength), and possible synergy with other treatment techniques like adsorption and membrane filtration have been examined. Notwithstanding its benefits, concerns like biodegradability, reusability, mechanical strength, and large-scale manufacturing need further research. Furthermore, the environmental sustainability and economic viability of bioflocculant-based hydrogels as opposed to synthetic hydrogels and conventional flocculants are addressed. Future research should focus on enhancing stability, optimizing hydrogel compositions, and investigating hybrid materials with great efficiency. This study emphasizes bioflocculant-based hydrogels as a potential

A. Kumar · D. Pal
Department of Chemical Engineering, National Institute of Technology Rourkela, Rourkela 769008, Odisha, India
e-mail: deeptimayeenitrkl@outlook.com; deeptimayee081993@gmail.com

A. Kumar (✉)
Information Management & Dissemination Centre (IMDC), Council of Scientific & Industrial Research, National Metallurgical Laboratory (CSIR—NML), Jamshedpur 831007, Jharkhand, India
e-mail: abhisheknitrkl@outlook.com; abhishek.kr.chemical@gmail.com

S. Bardhan and M. P. Shah (eds.), *Applications of Hydrogels in Modern Wastewater Treatment*, Advances in Wastewater Research,
https://doi.org/10.1007/978-981-95-1598-1_12

environmentally friendly wastewater treatment option, supporting sustainable water management and pollution prevention.

Keywords Hydrogels · Bioflocculant-based hydrogels · Cellulose-based hydrogels · Chitosan-based hydrogels · Alginate-based hydrogels · Adsorption

List of Abbreviations

3D	Three-dimensional
APS	Ammonium persulphate
BCH	Bacterial cellulose
BOD	Biochemical oxygen demand
CaAlg	Calcium alginate
CBHs	Cellulose-based hydrogels
CD	Cyclodextrin
CLDH	Calcined layered double hydroxide
CMC	Carboxymethyl cellulose
CMC-AM-GO HGs	Carboxymethyl cellulose-amino-modified graphene oxide hydrogels
CMCNa	Carboxymethyl cellulose sodium
CNTs	Carbon nanotubes
COD	Chemical oxygen demand
$-CONH_2$	Acyl amino
–COOH	Carboxyl
CQDs	Carbon quantum dots
CR	Congo Red
CS	Chitosan
CV	Crystal violet
DFT	Directional freezing–thawing
ECH	Epichlorohydrin
EDA	Electron donor–acceptor
EGDGE	Ethylene glycol diglycidyl ether
GLA	Glutaraldehyde
GL-cl-P(AA-co-AAM)	Gelatin and acrylic acid-co-acrylamide
GNPs	Gold nanoparticles
GO	Graphene oxide
H bonding	Hydrogen bonding
HNTs	Halloysite nanotubes
KPS	Potassium persulphate
MB	Methylene blue
MBA	N,N′-methylenebisacrylamide
Mch-Fe	Magnetic Chitosan-Iron(III)
MDEA	Methyl diethanolamine

MF	Microfibre
MG	Malachite green
MT	Montmorillonite
NaAlg	Sodium alginate
NACs	Nitroaromatic compounds
$-NH_2$	Amino
NIR	Near-infrared
NR	Neutral red
NY	Nitrazine yellow
OH	Hydroxyl
PAA-Fc	Cyclodextrin polymer (P-CD)/ferrocene-modified polyacrylic acid
PAAm	Polyacrylamide
PAHs	Polycyclic aromatic hydrocarbons
PAP	p-aminophenol
PEI	Polyethyleneimine
PGA	Poly-γ-glutamic acid
PGCB	Polyaniline graft chitosan beads
PGO HG	Pre-fabricated hydrogel
PNP	p-nitrophenol
PVA	Polyvinyl alcohol
RB	Rhodamine B
rGOs	Reduced graphene oxides
RSM	Response surface methodology
$-SO_3H$	Sulphonic
ST	Safranine T
TEMED	Tetramethylene-diamine
TGA	Thermogravimetric analysis

12.1 Introduction

Water is the primary natural resource on Earth; nevertheless, just around 3% of the available water reserves constitute freshwater. Moreover, less than one-third of this freshwater is appropriate for agricultural, residential, and industrial use. The significant growth in water consumption leads to the depletion of fresh water owing to heightened pollution, causing water scarcity for modern civilization. The growth of wastewater results from rapid industrialization, uncontrolled urbanization, various human activities, and inappropriate waste disposal (Pal and Sen 2024a). The prevalence of dyes in industrial effluent is widespread and severely impacts the sustainability of ecosystems. Dyes, even in small amounts, hinder sunlight from entering the water, leading to a substantial decrease in dissolved oxygen levels. This poses a severe threat to aquatic organisms and their well-being. Dyes often induce anaerobic digestion, leading to the generation of many carcinogenic compounds. These chemicals may then infiltrate the food chain through marine species. Nonetheless, the volume of color released into many water sources remains considerable. Annually, some 700,000 tons of reactive dyes are synthesized, with an estimated 5–10% of these colors released into industrial effluent. The inherent toxicity of dyes to the environment and human health underscores the need for effective industrial effluent treatment.

Adsorption is widely regarded as the predominant and efficient technique used for wastewater treatment because of its convenience, simplicity, cost-effectiveness, and absence of deleterious by-products (Pal and Sen 2024b). Activated carbon, an efficient adsorbent, is often used in commercial applications to eliminate contaminants from wastewater. Nevertheless, the high expense and intricate regeneration technique restrict the extensive use of activated carbon in wastewater treatment. While there have been recent findings of many substances that can be used to remove dyes from industrial effluent, using these substances for large-scale wastewater treatment remains challenging. Hydrogels, polymer networks with a three-dimensional (3D) structure, are now receiving significant interest for their exceptional ability to remove pollutants from waste streams, due to their high removal efficiency. Hydrogels are polymers with a 3D crosslinked network structure. They are extensively used in many industries, including water treatment, food technology, and biomedical disciplines. Their excellent porosity and abundance of hydrophilic functional groups (such as carboxyl (–COOH), hydroxyl (–OH), amino ($-NH_2$), sulphonic ($-SO_3H$), acyl amino ($-CONH_2$), etc.) allow them to retain and absorb a significant amount of water effectively. Hydrogels have been categorized according to their polymer composition, origin, charge, configuration, responsiveness to stimuli, degradability, and pore size. Hydrogels may be produced from natural, synthetic, or semisynthetic polymers using free radical polymerization, grafting polymerization, radical crosslinking, physical crosslinking, chemical crosslinking, and irradiation crosslinking. Therefore, the preparation techniques may be used to categorize hydrogels. Hydrogels of various forms, such as beads, films, and rings, have been developed for diverse purposes. Geometry may be used to differentiate a specific hydrogel from other types (Deng et al. 2017).

However, most current hydrogels are made from petrochemicals that are nonrenewable and non-degradable (Peng et al. 2020). The primary raw ingredients for hydrogel synthesis are petroleum-derived substances, including N-isopropyl acrylamide, acrylic acid, and hydroxyl ethyl methacrylate. While synthetic polymer hydrogels efficiently eliminate impurities from water, they also pose substantial environmental concerns. Furthermore, synthetic polymers are costly and not easily accessible. Consequently, due to the influence of awareness campaigns and other procedures, there is an increasing inclination to transition to natural substances to manufacture hydrogels (Matei et al. 2022). Cellulose, chitosan, and alginate, which are environmentally friendly carbohydrate polymers, have been extensively studied to produce hydrogels made from polysaccharides. These polymers possess unique characteristics: being cost-effective, readily accessible, biodegradable, renewable, non-toxic, and reusable (Farshidfar et al. 2023).

In wastewater treatment, commercially used polymers have shown satisfactory results. These polymers comprise cheap monomers such as methyl methacrylate, acrylamide, and acrylic acid. However, a main drawback of such polymers is their limited or non-biodegradability, which makes their reuse challenging and raises environmental issues. Bio-based polymers have lately become a possible substitute to reduce environmental damage. However, depending only on biopolymers may not satisfy the particular needs for commercial uses. Hence, chemical changes are necessary to improve their characteristics. Natural availability and nominal production cost have drawn polysaccharide-based polymers such as starch, cellulose, hydroxyethyl starch, guar gum, chitosan, and dextrin have gained a lot of interest. Because of their eco-friendliness and $-NH_2$ groups, chitosan has been investigated closely for composite, nanocomposite, and polymer manufacture. Direct use of chitosan is hampered, therefore, by difficulties with its cost consequences and solubility in polymer solubility and changing reaction conditions. Thus, chemical changes are needed to increase the charge density and water solubility of chitosan while preserving its efficiency and economy in water treatment (Kolya and Kang 2023).

Although many studies claim their methods to be affordable and ecologically friendly, few include the running expenses and biodegradability. Usually, substituting synthetic solutions for real industrial effluent, most research also focuses mainly on removing organic color molecules rather than on metal ions or combinations of them. Supported by thorough research reports to enable their industrial implementation, further studies in this domain should highlight the development of sustainable polymeric materials that are multifunctional, efficient, user-friendly, interference-resistant, and cost-effective. To surpass these limitations and challenges, local researchers must cooperate with the sector to create bio-based polymers for practical use. This partnership should focus on addressing the requirement for economical regeneration approaches, evaluating the regeneration cost of flocculants and adsorbents, and providing a full understanding of the possible performance of the material in actual situations. Focusing on such opportunities enables the development of bio-based products to promote more ecologically friendly and effective wastewater treatment systems. Hence, this chapter focuses on the significance of using bio flocculants-based hydrogels, such as cellulose, chitosan, and alginate hydrogels, for wastewater treatment. The various synthesis routes, such as physical and chemical, have been discussed. The different mechanisms involved in the adsorption of multiple pollutants, such as electrostatic interactions, ion exchange, chelation, π–π interactions, and hydrophobic interactions, have been explained.

12.2 Synthesis of Hydrogels

Hydrogels comprise a wide range of natural and manufactured monomeric and polymeric components, including acrylamide and other polysaccharides such as starch, alginate, cellulose, rice husks, chitosan, and gums. Hydrogels derived from natural polysaccharides have many benefits over synthetic materials, including superior renewability, increased availability, and environmental friendliness. These gels include various hydrophilic groups that get hydrated when they come into contact with moisture, resulting in the formation of a 3D gel structure. Over the last ten years, researchers have investigated various alternative polymer hosts for producing hydrogels (Kumar and Gangawane 2023a). These include glass beads, silica gel, polyamine beads, alginate beads, polyethylene terephthalate, polyurethane, polyvinylalcohol, polyacrylic acid polyester, and others. The various basic materials are interconnected using diverse chemical (optical, thermal, and radiation-induced) or physical methods, resulting in a 3D gel network, as illustrated in Fig. 12.1. Moreover, the behavior and density of links impact the chemical structure and adsorption potential of the produced gel. Crosslinking also enhances tolerance to lower pH solutions and greater temperature ranges than the original polysaccharide. A significant constraint in the synthesis process is the decrease in sorption sites as the crosslinking density in the crystalline areas increases. This subsequently alters the original crystalline structure of the polymeric matrix (Akhtar et al. 2016).

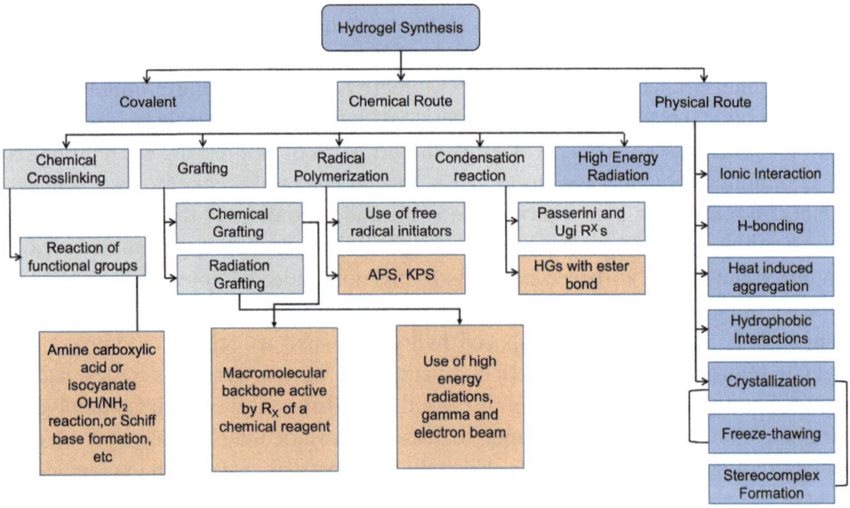

Fig. 12.1 Hydrogel synthesis approaches. Reproduced with permission from (Sinha and Chakma 2019). Elsevier

12.2.1 Chemical Methods for Hydrogel Synthesis

Chemically cross-linked hydrogels include the bonding of polymer chains via covalent bonds. The following sections explain several methods to establish covalent bonds for synthesizing and modifying hydrogels.

Strong chemical bonds between individual and multiple chains are developed using an isocyanate–OH/NH_2 reaction, the amine-carboxylic acid, or Schiff base synthesis procedures. Glutaraldehyde allows hydrophilic units containing eOH groups, such as polyvinyl alcohol, to be cross-linked employing aldehydes. Alternatively, the condensation process in the presence of a cross-linker causes the cross-linking of polymers, including amine and ester groups, to generate Schiff bases. Bis and other cross-linkers with several functional groups undergo extra reactions with the functional groups coupled to polymers to generate various hydrogels (Tran et al. 2017). Many researchers have chemically bonded different hydrophilic functional groups, including eNH_2, eCOOH, $eCONH_2$, and eSO_3H groups, to the basic hydrogel unit. Through better absorbing and separating certain pollutants, these functional groups increase sorption efficiency and selectivity. Attaching different functional components to the surface of a hydrogel could improve the sites accessible for sorption. Furthermore, improving the polarity and hydrophilicity of the surface is this procedure. Introduced into the backbone of polysaccharides, vinyl groups generated new derivatives (Yu et al. 2017). Reticulate polymerization of the new monomers replaced chemical crosslinkers to produce. Moreover, immobilizing with a solid matrix produces a hydrogel with better stability and turnover number. Using polyacrylamide (PAAm), the hydrogel was modified by graft polymerization. This change greatly accelerated the methylene blue (MB) removal from wastewater. With a 92.3 mg/g adsorption capacity, the modified hydrogel showed very remarkable performance (Soleimani et al. 2018).

Free radical polymerization is a well-documented process involving three distinct steps: Initiation, polymeric chain propagation, and termination, which characterize the well-researched process known as free radical polymerization. Compounds like tetramethylene-diamine (TEMED) and potassium persulphate (KPS) cause free radicals in this kinetics, triggering the reaction. These compounds disintegrate under temperature, light, redox reaction, gamma (γ), or UV radiation. The process results in the polymeric chain developing and dispersing. These hydrogels generate a 3D polymeric network through crosslinks between the polymeric chains. Combining or disproportionation reactions eventually bring the step reaction to a standstill. Starting free radical polymerization with ammonium persulphate (APS), a carboxymethyl cellulose-amino-modified graphene oxide hydrogel (CMC-AM-GO HGs) was developed (Varaprasad et al. 2017). With the highest uptake capacity of 185.45 mg/g from wastewater, these hydrogels showed amazing dye (acid blue) removal ability. Several metals (Pb^{2+}, Cu, Cd^{2+}, Cr, Mn, Zn, Fe) were efficiently removed from wastewater using a technology to create a porous double network Jute/polyacrylic acid hydrogel. With the highest values of 542.9 mg/g and 401.7 mg/g for Pb^{2+} and Cd^{2+}, respectively, the hydrogel showed great adsorption capabilities. Several high-grade polymers have been developed using this technique, which has also proved very useful in wastewater treatment (Fang et al. 2018).

Using radiation therapy such as γ radiation, electron beams, and UV light radiation, hydrogel synthesis presents a novel approach to induce crosslinking in polymer chains. This method generates a one-step hydrogel and sterilizes efficiently while generating no waste, offering a benefit over chemically initiated methods. Conducted at sub-ambient or ambient temperatures, the process lacks chemical initiators, crosslinkers, or catalysts. The irradiation dose and time used to start the reaction kinetics control the density of interactions. Highly absorbent hydrogels were synthesized with cellulose derivatives using γ irradiation (a 60Co γ source). The gel formed using radiation exhibited exceptional characteristics when subjected to lower dosages (9000 Gy/h) of γ-radiation. A specialized imprinting hydrogel was developed with high selectivity utilizing crosslinking triggered by irradiation (Mahmoud et al. 2016). These hydrogels were explicitly designed to remove Pb^{2+} and Co^{2+} ions. The copolymerization of acrylamide monomers and *N*-vinyl-2-pyrrolidone was conducted using γ radiation polymerization to produce imprinted hydrogels that strongly attracted the template Pb^{2+} ions. Similarly, a new kind of copolymer hydrogel was made by synthesizing poly(acrylic, polyvinyl alcohol) using γ irradiation at 20 kGy. The gel exhibited a swelling property of 273%, with the highest adsorptive removal of 150 mg/g, 155 mg/g, and 193 mg/g for Ni^{2+}, Co^{2+}, and Cu^{2+} ions. These results were observed at pH 5 and after 24 h. Cong et al. researched using a near-infrared (NIR) laser method to create thermoresponsive poly-γ-glutamic acid (PGA) hydrogel to treat wastewater. The process included the simultaneous development of gold nanoparticles (GNPs) and graphene oxide (GO) reduction into graphene. This was achieved using a pre-fabricated hydrogel (PGO HG) containing auric acid. The gel produced by radiation showed a good turnover number and great efficiency in removing crystal violet (CV) dye. The great catalytic performance may be ascribed to the favorable kinetic conditions of graphene's photothermal energy conversion effect. Compared to other conventional techniques used for wastewater treatment, this one has demonstrated exceptional speed, dependability, and effectiveness in generating gel of higher quality (Cong et al. 2015).

12.2.2 Physical Methods for Hydrogel Synthesis

Non-covalent bonding between molecules or polymeric chains forms physically cross-linked gels (Gong et al. 2016). In most cases, unlike chemical reactions, synthesis takes place under benign conditions, including physiological pH and ambient temperature. Other cohesive forces, including ionic contacts, hydrogen bonding (H bonding), stereo complex development, or hydrophobic interactions, focus first on giving these hydrogels mechanical strength instead of crosslinks. Many polysaccharides, including dextran, chitosan, carboxy methyl curdlan, and pullulan, are employed widely in different publications to create physically crosslinked hydrogels. Including metallic ions in the formation of ionic bonds strengthens hydrogels.

Many studies have used alginate crosslinked with calcium ions. H-bonded hydrogels interact through carboxylic acid groups being protonated. The pH of the fluid determines the extent to which these gels may swell. Crystallization is another method of producing gels in which a durable and highly flexible hydrogel results from a freezing–thawing process. Using this method, polyvinyl alcohol (PVA) hydrogels were physically crosslinked polyvinyl alcohol (PVA) hydrogels by many thawing and freezing cycles. On the other hand, crystallization generates stereo-complexes in extremely glassy materials (Voo et al. 2016).

12.3 Cellulose-Based Hydrogels for Wastewater Treatment

Cellulose-based hydrogels (CBHs) are derived from native/pure cellulose or bacterial cellulose. CBHs can also be derived from cellulose derivatives, including ether derivatives such as ethylcellulose, methylcellulose, hydroxypropyl cellulose, hydroxyethyl methylcellulose, and carboxymethyl cellulose (CMC), as well as ester derivatives like acetate phthalate, acetate trimellitate, hydroxypropyl methyl phthalate, and hydroxypropyl methyl phthalate acetate succinate. These hydrogels possess exceptional superabsorbency, durability, biodegradability, biocompatibility, and non-toxicity. CBHs can also comprise cellulose composites, such as polyelectrolyte complexes, interpenetrating polymer networks, and blends with other polymers. A major obstacle in producing hydrogels using these raw materials is their limited solubility in common solvents. Although some cellulose derivatives may dissolve in water, pure cellulose is not readily soluble in solvents (organic/inorganic) like water. Therefore, to carry out CBH synthesis, it is essential to have an appropriate solvent. Some ideal solvents for this purpose are ionic liquids, LiCl/dimethylacetamide, alkali/urea (or thiourea), and *N*-methyl morpholine-*N*-oxide. The formation of a stable 3D network structure of hydrogel, essential for retaining a large quantity of water in the spaces between its components, is accomplished by crosslinking the polymer chains that constitute the hydrogel (Zainal et al. 2021).

Cellulose has hydrophilic functional groups, namely OH groups, which allow them to create physical crosslinking via chemical crosslinking and electrostatic interactions through covalent interactions with a crosslinker. CBHs exhibit various interactions, depending on the materials and fabrication methods employed. These interactions include (i) electrostatic interaction between cellulose and small cations, (ii) electrostatic interaction between polycations and cellulose, (iii) hydrophobic interaction or H-bond between polymers and cellulose, (iv) covalent crosslinking, (v) self-assembly, and (vi) coordination complex crosslinking. Various interactions occur between cellulose molecules and other polymers or tiny molecules inside a hydrogel matrix. These interactions contribute to developing the crosslinked structure of the hydrogels, which inhibits the destruction and dissolution of the CBHs during swelling. Using different production techniques, cellulose-based hydrogels may be created in many physical forms, such as spheres, cylinders, beads, blocks, microparticles, nanoparticles, and films (Kumar and Gangawane 2023b).

Recent research discovered that the most intense electrostatic contacts occurred between an enhanced-capacity hydrogel containing $-NH_2$ groups and an anionic dye called Acid Black 1 at lower pH settings. These interactions diminished as the pH increased and reached their lowest level at extremely high pH levels. The chemo-mechanical pretreatment of cellulose-based superabsorbent hydrogel demonstrated a comparable electrostatic affinity for removing MB, a cationic dye (Klinpituksa and Kosaiyakanon 2017). A detailed explanation of the H bonding interactions between the sulfur (S) atoms in Acid Blue 93 (anionic dye) and MB (cationic dye) and the hydrogen atoms was provided in a CBH material. Hou et al. showed hydrophobic and electrostatic associations between the functional groups (–SH and –OH) and MB dye in thiol-modified CMC/L-cysteine HG (Hou et al. 2019). The adsorption process of the chitosan/cellulose composite adsorbent for removing Congo Red (CR) dye involves electrostatic adsorption and coordination interaction between the adsorbent molecules and the adsorbate. A high uptake potential of 96.43 mg/g was achieved for Methyl Violet dye using a nano adsorbent with a CMC structure. The high adsorption capacities in this case were attributed to π–π stacking, which involves electrostatic interactions. It is believed that the contaminants transferred pie-electrons to the adsorbent molecules, leading to this phenomenon. Regarding carboxymethyl cellulose sodium (CMCNa)/GO hydrogel microparticles, it was proposed that electrostatic and π–π interactions influence dye adsorption influence the adsorption of dyes. Conversely, the adsorption of heavy metals is a result of the combined impact of ion exchange, surface complexation, and electrostatic interactions (Pal and Sen 2023a).

The impact of the initial concentration of MB dye on adsorption capacity was investigated using MB concentrations ranging from 1800 to 2700 ppm (Shi et al. 2013). The adsorption capacity exhibited a linear rise in correlation with the original dye concentration, reaching a plateau at 2500 ppm. When the concentration exceeded 2500 ppm, the ability to adsorb decreased as the concentration increased. Other recent investigations on other CBHs have also shown a similar pattern in the adsorption capacity with the initial concentration of dyes and heavy metals. The adsorption properties of CMC the pH of the solution were examined. The results revealed that when the pH values increase, the CMC hydrogels' adsorption capability shows an ascending trend. Reduced absorption of metal ions results from a competition between excess H^+ ions at extremely low pH levels with metal ions to bind to the active sites of the CMC. CBHs allow the same adsorption process to eliminate colors from aqueous solution. For example, at a pH of 4.0, the maximum concentration of Congo Red (CR), a new CBH that may adsorb, was discovered. The absorbability of the dye dropped as the pH values rose. The excess of OH^- ions restricts the active sites of the adsorbent molecules when the pH is raised, therefore lowering the adsorbent capacity to adsorb CR dye molecules (Wang et al. 2016). By contrast, cationic dyes such as metal ions are attached to the surface of the adsorbent at raised pH values.

In the endothermic adsorption process, CBHs' adsorption capacity grows with increasing temperature; this is unlike in exothermic adsorption. Usually, the capacity of CBHs to adsorb dyes from a water-based solution is exactly correlated with the temperature, so the adsorption capacity of CBHs likewise increases as the temperature elevates. Increased temperatures greatly increase dye molecule mobility, improving the interactions between the adsorbent surface and dye molecules. Consequently, the release of dye from the surface of the adsorbent is decreased, thereby improving the adsorbability. Furthermore, a temperature rise might induce a swelling effect in the CBH structure, facilitating the extra dye penetration into the adsorbent surface. The effect of ionic strength on the adsorption of acrylamide-grafted CBH and acrylic acid was investigated. Their alteration of NaCl salt concentration changed the ionic strength of the fluid. Dye removal's efficacy dropped as the NaCl level rose. On the CBH surface, the salt ions (Na^+ and Cl^-) compete with the dyes, including functional groups ($-COO^-$, $-NH_3^+$, and OH^-). The decrease in adsorption efficiency of the adsorbents results from the rise in NaCl concentration, enhancing the shielding effect of Na^+ and Cl^- ions on the ionized dye molecules (Liu et al. 2015).

12.4 Chitosan-Based Hydrogels for Wastewater Treatment

Chitosan (CS) is the predominant polymer used to produce hydrogels. The *N*-deacetylation of chitin, a substance in fungi, algae cell walls, insect exoskeletons, mollusks, and crustaceans, forms it. Chitin consists of β-(1 → 4)-α-glucosamine and *N*-acetyl-α-glucosamine units that are scattered randomly. Chitin is the second most prevalent biopolymer in nature, behind cellulose. Chitosan has many reactive NH_2 and OH groups, allowing chemical and physical changes. Chitosan hydrogels have several limitations, such as poor stability in acidic conditions, low mechanical strength, and limited heat stability. Chemical and physical networking techniques have been suggested to address these drawbacks. Chemical networking is a chemical process that develops permanent covalent links between polymer chains and crosslinkers. The primary chemical networking agents often used for synthesizing chitosan hydrogels are glutaraldehyde (GLA), genipin, *N*, *N′*-methylenebisacrylamide (MBA), formaldehyde, ethylene glycol diglycidyl ether (EGDGE), and epichlorohydrin (ECH). Most of these components are hazardous and should be eliminated from the gels, particularly for biomedical uses (Hennink and van Nostrum 2012). Reversible molecular ionic interactions, entanglements, and H bonding sustain chitosan hydrogels in the physical networking approach. Ionic networking is the predominant physical networking approach. This approach utilizes anionic crosslinkers, including sodium tripolyphosphate, sodium citrate, oxalic acid, and sulfosuccinic acid, to create chitosan hydrogels. Chitosan hydrogels crosslinked by ionic bonds exhibit decreased stability in an acidic environment compared to hydrogels crosslinked through covalent connections.

The synthesis of hydrogels is impacted by the degree, type, and condition of crosslinking, which are crucial criteria. Extensive cross-linking develops hydrogels that are prone to brittleness, have a limited ability to expand, and have a poor capacity for adsorption. Therefore, it is necessary to ascertain the optimal level of cross-linking. The cross-linking of chitosan-based hydrogels leads to decreased available NH_2 and OH groups in the chitosan framework, reducing adsorption potential. NH_2 groups, which exhibit higher reactivity than OH groups, participate in certain crosslinking processes. Most crosslinker agents react with the NH_2 groups of chitosan, except ECH, which explicitly targets OH groups (Vakili et al. 2014). Chitosan hydrogels exhibit pH-dependent behavior. The pH of these hydrogels regulates the surface charge. The NH_2 groups of chitosan become positively charged when the pH is below the pKa range of 6.3–7.0. Consequently, the expansion of these hydrogels was amplified because of the electrostatic repulsion between ions that carry the same charge. Currently, chitosan-based hydrogels are highly valued in the industrial wastewater treatment industry because of their rapid kinetics, higher adsorption capacity, and advantageous reusability. Chitosan-based hydrogels adsorb contaminants from aqueous solutions by forming complexes, interacting electrostatically, and bonding via hydrogen. The formation of the metal-chitosan complex is contingent upon the polymer chain length, the deacetylation level, and the crystallinity degree.

Chitosan-grafted polyaniline beads were synthesized to eliminate copper(II) from a liquid solution (Igberase et al. 2014). An investigation was conducted to study the elimination of copper(II) at room temperature and within a pH range of 2.0–8.0. Experimental findings indicated that hydrogen ions chemically interact with adsorption sites on the hydrogel's surface in acidic conditions, reducing adsorption capability. By raising the pH level, the competition between hydrogen ions and copper(II) ions for adsorption diminishes, leading to an increase in both the capacity and rate of adsorption. The pH at which the highest adsorption capacity occurs was 5.0. If the pH exceeds 5.0, copper(II) hydroxide forms solid particles and the adsorption ability declines. To recover adsorbed copper(II) from the hydrogel, HCl/HNO_3 eluents were used at varying concentrations. When comparing two eluents, it was evident that a 0.5 M HCl solution successfully removed copper(II) ions during desorption. Currently, zirconium is used to alter hydrogels. It undergoes hydrolysis readily and binds contaminants via exchanging ligands. Zirconium-chitosan hydrogel beads were fabricated using encapsulation to extract boron(III) from wastewater (Kumar and Gangawane 2022a). The adsorption capability of the modified hydrogel beads rises when the zirconium concentration is raised from 15 to 50%. The researchers discovered that the OH groups of zirconium-chitosan hydrogel beads interact with H_3BO_3, the primary form of boron(III), at a neutral pH, to form complexes. Liu et al. enhanced the properties of chitosan beads by including zirconium and then used these modified beads to adsorb phosphate from agricultural wastewater. The adsorbents showed fast kinetics relative to other adsorbents and attained their maximal adsorption capacity in 50 min. The pseudo-first-order and pseudo-second-order models were found most appropriate for defining the adsorption behavior by a study of the kinetics data using the pseudo-first-order, pseudo-second-order, intraparticle diffusion, and Elowich models. This implies that mechanisms of chemisorption and physisorption were involved. With an estimated maximum adsorption capacity of 60.6 mg/g, the equilibrium produced matches the Langmuir model (Liu and Zhang 2015).

The phase inversion method synthesized chitosan-halloysite nanocomposite hydrogel beads to efficiently remove MB and malachite green (MG) dyes. The effects of nanotubes on the visual properties, dimensions, microscopic structure, and heat resistance of nanocomposite hydrogel beads were investigated by the scientists. Including halloysite nanotubes (HNTs) in hydrogel beads increases the heat stability of adsorbents, according to the results of the thermogravimetric analysis (TGA). The electrostatic interactions and H bonds between the chitosan and HNTs may assist in explaining this development. The SEM results show that the exterior and interior surfaces of the nanocomposite beads are more irregular than those of the chitosan beads. Moreover, some HNTs found their position in the nanocomposite beads. The two dyes' adsorption capacity and rate were raised by increasing the halloysite concentration because the flexible and uneven structures of composite beads helped molecular diffusion. The adsorption rate of MB was higher compared to MG, and the equilibrium adsorption was achieved in around 40 min for MB and around 7 days for MG. The experimental data collected in the batch mode were analyzed using the Langmuir and the Freundlich isotherm models (Peng et al. 2015).

A chitosan/ZnO nanocomposite hydrogel was synthesized using sodium triphosphate pentabasic to eliminate RB5. Sodium triphosphate was used to conduct crosslinking. The nanocomposite beads were analyzed using TEM, SEM, FTIR, XRD, and TGA. The TGA findings of the chitosan and chitosan/nano-ZnO composite bead exhibit two distinct phases, with the first stage commencing at around 210 °C and the second stage at 300 °C. The highest mass is lost at a temperature of 300 °C for chitosan. Therefore, the inclusion of ZnO particles increases the thermal resistance of chitosan. The TEM findings indicate that the diameters of chitosan/nano-ZnO beads fall within the range of 20–40 nm. An investigation of the thermodynamics of RB5 adsorption is conducted at temperatures of 293 and 313 K. The positive value of ΔH° indicates that the adsorption process is endothermic, which absorbs heat from its surroundings. On the other hand, the negative value of ΔG° indicates that the adsorption process is spontaneous, meaning it occurs naturally without the need for external energy input (Çınar et al. 2017).

Using magnetic chitosan iron (III) (MCh-Fe) hydrogels, researchers investigated Cr^{6+} adsorption. In less than 30 min, their results show a rapid and efficient clearance of Cr^{6+} that achieves total removal. According to research regarding how pH affects the removal of Cr^{6+} the pH of 3.0 has the maximum degree of adsorption. Cr^{6+} ions are present in solution as $HCrO_4^-$ within the pH range of 2.0 to 5.0. By a ligand-exchange mechanism, these ions might chelate with Fe^{3+} on MCh-Fe hydrogels. Apart from the ligand-exchange mechanism, another process in charge of the adsorption of negatively charged Cr^{6+} ions is electrostatic contact. This results from the chitosan surface's various functional groups, including NH_2 and OH, through protonation in a highly acidic solution; functional groups of chitosan enable them to bind and adsorb negatively charged Cr^{6+} ions via electrostatic attraction. Increased CrO_4^{2-} concentration, which the hydrogel cannot absorb, causes a reduction in adsorption capabilities at pH values above 5.0. By use of adsorption isotherm and kinetic models, respectively, the Langmuir and pseudo-second-order models fairly depict the experimental results (Yu et al. 2013).

12.5 Alginate-Based Hydrogels for Wastewater Treatment

Comprising linear chains of (1,4)-linked-D-mannuronic acid (M) and their C-5 epimer, -L-guluronic acid (G), alginate is a polysaccharide. These residues connect covalently in several blocks or sequences. Derived from brown algae, the chemical might be generated by bacterial biosynthesis by Azotobacter and Pseudomonas. Its biocompatibility and biodegradability render it significant in the biomedical and pharmaceutical domains as a drug delivery system. Water treatment depends on extremely important alginate-based gels. The researchers found significant capacity for the heavy metal adsorption and elimination of organic compounds. Alginate and polymers with suitable functional groups have recently created many hydrogels (Kamoun et al. 2015). Conversely, poly(4-vinylpyridine), often known as P4VP, is a hydrophobic polycationic polymer. At lower pH levels, this polymer becomes hydrophilic by protonating the pyridine group; it is pH-sensitive. The hydrogels' adsorption for the maxilon blue basic dye and the Congo red acidic dye were examined. The results showed that compared to the hydrogel produced from calcium alginate, the adsorption of Congo red dye was greater with increasing P4VP levels. This may be ascribed to the pyridine rings' increasing intrinsic basicity. On the other hand, the calcium alginate adsorbed the maxilon blue basic dye more effectively than the hydrogels made of alginate/P4VP. The carboxylate anion's acidic qualities help to explain its anionic behavior and so the conclusion is reasonable. Improving the basic character of the hydrogels, especially by increasing the P4VP concentration, would certainly assist in reducing the absorption of alkaline dyes (Soliman et al. 2017).

The impact of NaAlg concentration, dosage of crosslinker, crosslinking temperature, and bead size on the efficacy of carboxylate alginate beads was investigated. Based on their research, the combination of 2 wt% alginate and $CaCl_2$ showed remarkable effectiveness in adsorbing dyes. Neither temperature nor bead size significantly affected the dye adsorption process. The drying period of the beads might impact the MB adsorption by influencing the porosity and the subsequent diffusion process. The dried alginate/ZnO beads exhibited superior adsorption effectiveness compared to the fresh beads, owing to their reduced weight and increased surface area. This finding aligns with prior research that used dehydrated calcium alginate (CaAlg) beads to eliminate heavy metal and organic acid from lean methyl diethanolamine (MDEA) solvent (Eldin et al. 2019).

Because of their unique properties and advantages, carbon nanoparticles, including GO, are often employed to produce high-performance hydrogels based on sodium alginate (NaAlg). For instance, it was shown that employing NaAlg's higher specific surface area and average pore size of 6.983 m^2/g and 24.62 nm, respectively, GO significantly affected the adsorption process. This was shown by contrasting the acrylamide/graphic NaAlg system's adsorption capability with both GO included and absent (Kumar et al. 2022). Observing the change in the composite's distinctive peak, the FTIR study validates the dye-composite material interaction. The C–H stretching of the hydrogel changed from 2922 to 2928 cm^{-1} and then from 2925 to

2845 cm^{-1} after dye adsorption. One might ascribe the change to the interactions between adsorbents and dye molecules. To maximize the adsorbent-dye interaction, one must design a suitable degree of ionic strength. This is so because the ionic strength considerably influences the adsorption capacity of the composite.

When the hydrogel loaded with GO was used to treat textile wastewater, it demonstrated an approximate removal efficiency of 80% for CV and decreased biochemical oxygen demand (BOD), metal concentration, and chemical oxygen demand (COD). Furthermore, the addition of iron oxide-modified GO to a hydrogel composed of NaAlg and CS was discovered to significantly enhance crucial characteristics of an adsorbent system, including electrostatic interaction, magnetism, porosity, and surface area (Ma et al. 2021). The presence of functional groups, primarily carboxyl groups, was shown to cause high electrostatic interaction with cationic dyes such as neutral red (NR), ST, and MB, exemplifying their important features (Pal and Sen 2023b). The hydrogel bead was effectively used to eliminate dyes from water because of its beneficial characteristics. Subsequent adsorption isotherm investigations showed the chemical connection. Additionally, they demonstrated their capacity to eliminate dye.

A nanocomposite hydrogel was developed by crosslinking tetraammine copper (II) sulfate, NaAlg, and acrylic acid, to adsorb CV and MG. Incorporating NPs in the hydrogel enhanced the chemical associations with the MB molecules and increased the surface area for adsorption, facilitating dye adsorption. The MoS_2-doped NaAlg/Fe hydrogel is another remarkable example; it efficiently removed 98% of MB dye molecules in only 15 min (Xie et al. 2022). The remarkable photocatalytic activity of MoS_2 was the reason the hydrogel was so highly removed. The hydrogel is suitable for treating real wastewater as it also shows good recycling properties. The great regenerating capacity of the hydrogel suggests the possibility of improving its applicability in large-scale industrial use. Ultimately, many research groups have documented the presence of hydrogels made of several nanocomposites derived from inorganic nanoparticles.

A dual-functioning hydrogel was developed of diatomite and silver NPs meant for water treatment (Zhao et al. 2021). Using silver nanoparticles considerably improved the adsorption capacity and produced antibacterial properties. Furthermore, the application of diatomite greatly raised the surface area and porosity of CaAlg, improving its capacity for adsorbing MB. Strong sterilizing activity and great adsorption capacity of the hydrogel were its most important characteristics. As an additional instance of multiple nanocomposite hydrogels based on inorganic nanomaterials, the effectiveness of MB adsorption in a hydrogel composed of NaAlg poly (acrylic acid)/zinc oxide (NaAlgpoly (AA)/ZnO) was compared to that of a hydrogel made of NaAlg-poly (AA). This comparison aimed to investigate the impact of zinc oxide on MB adsorption. The main advantage of NaAlg-poly (AA)/ZnO NPs is their uniform dispersion, resulting in a higher adsorption capacity compared to NaAlg-poly (AA), such as 1529.6 mg/g and 1129.0 mg/g, respectively (Makhado et al. 2020) (Table 12.1).

Table 12.1 Adsorption capacities of bioadsorbent hydrogels for pollutant removal from wastewater

Hydrogels	Pollutants	Adsorption capacity (mg/g)	References
Polyacrylamide grafted quaternized cellulose	Congo Red	380.08	Wang et al. (2016)
Polyacrylamide grafted quaternized cellulose	Eriochrome Blue Se	349.28	Wang et al. (2016)
Magnetic Chitosan-Iron(III) Hydrogel (Mch-Fe)	Chromium(VI)	144.9	Yu et al. (2013)
Chitosan (CS)/reduced graphene oxide (RGO)/montmorillonite (MT) porous composite hydrogel	Chromium(VI)	87.03	Yu et al. (2017)
Porous cellulose-based bioadsorbent	Acid Blue 93	1372	Liu et al. (2015)
Porous cellulose-based bioadsorbent	MB	1372	Liu et al. (2015)
Gelatin and acrylic acid-co-acrylamide [GL-cl-P(AA-co-AAM)]	Organic Pollutants	11,368.8	Fosso-Kankeu et al. (2017)
Hybrid hydrogel composite	Organic Pollutants	12,787	Fosso-Kankeu et al. (2017)
Lignocellulose-g-poly(acrylic acid)/montmorillonite (LNC-g-PAA/MMT) hydrogel nanocomposites	MB	1994.38	Shi et al. (2013)
Chitosan-montmorillonite hydrogel	Cu(II)	132.74	Ngwabebhoh et al. (2016)
Chitosan-montmorillonite hydrogel	Nitrazine Yellow	144.41	Ngwabebhoh et al. (2016)
Polyaniline graft chitosan beads (PGCB)	Cu(II)	100	Igberase et al. (2014)
Cyclodextrin polymer (P-CD)/ ferrocene-modified poly acrylic acid (PAA-Fc) hydrogel	Bisphenol A (BPA)	23.61	Hou et al. (2019)
Cyclodextrin polymer (P-CD)/ ferrocene-modified poly acrylic acid (PAA-Fc) hydrogel	MB	24.64	Hou et al. (2019)
Novel polyethylenimine-functionalized ion-imprinted hydrogel	Cu(II)	40	Wang and Li (2015)
GO-based slide ring hydrogel	MB	92.3	Soleimani et al. (2018)
Magnetic graphene oxide-containing chitosan-sodium alginate hydrogel beads	MB	21.325	Ma et al. (2021)

(continued)

Table 12.1 (continued)

Hydrogels	Pollutants	Adsorption capacity (mg/g)	References
Magnetic graphene oxide-containing chitosan-sodium alginate hydrogel beads	Neutral red (NR)	44.654	Ma et al. (2021)
Magnetic graphene oxide-containing chitosan-sodium alginate hydrogel beads	Safranine T (ST)	44.313	Ma et al. (2021)

12.6 Significance of Bioflocculant-Based Hydrogels for Wastewater Treatment

This work aims to synthesize hydrogel-based bioflocculants, specifically gelatin and acrylic acid-co-acrylamide [GL-cl-P(AA-co-AAM)] and a hybrid hydrogel composite with significant adsorption capacity, thus addressing biodiesel wastewater. The two hydrogels were studied using SEM, XRD, and FTIR. It was discovered that the composite included the functional groups of the constituent components during the synthesis process. At a dose of 12.5 mg/L, the hydrogels showed their peak effectiveness, which was around 65% BOD elimination. The kinetic investigations of BOD removal by all the hydrogels followed the pseudo-second-order kinetic model; the hybrid hydrogel composite had an adsorption capacity of 12,787 mg/g, and the GL-cl-P(AA-co-AAM) had an adsorption capacity of 11,368.8 mg/g. It seems that the addition of attapulgite to the composite led to the formation of a superior coagulant since the flocs produced by polyferric chloride and the hybrid hydrogel composite were more stable than those produced by the GL-cl-P(AA-co-AAM) and MgOH. Hence, for large-scale deployment, additional pilot-scale investigation into the hybrid hydrogel composite's potential for treating biodiesel wastewater may be conducted (Fosso-Kankeu et al. 2017).

Inadequate conventional wastewater treatments make microfibers (MFs), the most common microplastics in ecosystems, a major environmental threat. This study sought a sustainable solution to the widespread issue of MFs polluting water bodies, recognizing the necessity to address this environmental dilemma. As a possible bioflocculant for polyacrylonitrile MFs, the leftover BC was crushed into a hydrogel form of bacterial cellulose (BCH). By using the response surface approach computer modeling, we were able to understand and predict how different operational and environmental variables affected the flocculation efficiency. Important variables in flocculation effectiveness were the mixing intensity and the BCH: MFs ratio. BCH was durable across different environmental conditions, with a removal rate of 93.6%. The flocculation tests conducted in chlorinated water produced an efficiency of 86.1% in polluted wastewater, respectively, while the BCH's impressive MF retention only released 8.3% of the MFs after a 24-h wash. Because of its remarkable flocculation capability and versatility, BCH offers a practical, long-term solution to the issue of MF contamination in water. This groundbreaking research establishes an emerging benchmark for sustainable wastewater treatment using BCH as a bioflocculant to remove MFs. It also accelerates studies on fibrous contaminant remediation, crucial for protecting the environment (Rodrigues et al. 2024).

12.7 Mechanism of Hydrogel Interaction with Pollutants

Advancement and modification of the properties of different heterogeneous materials depend on a complete understanding of the adsorption process and the methods by which different contaminants are eliminated from different materials. However, the intricate composition of hydrogels makes the characterization of metal ion sorption processes difficult, requiring a thorough knowledge of process parameters. Interactions such as electrostatic interaction, ion exchange, H bonding, complexation, hydrophobic interaction, coordination/chelation, and acid–base interactions occur depending on the various functional groups of hydrogels, experimental conditions provided (salt concentrations, pH, ligands, temperature), and pollutant chemistry. Explaining adsorption processes in most heterogeneous catalysts involves simultaneously operating many interactions to different extents. Physisorption and chemisorption forces occur concurrently in starch-composed hydrogels during research experiments, facilitating the binding of metal ions through acid–base interactions, H-bonding, chelation, and/or ion exchange (Mittal et al. 2018). Chitin hydrogels include several interactions, individually or in combination, influenced by the chemical composition and operating circumstances. Coordination reactions happen in heterogeneous catalyst hydrogels, including NH_2 and vicinal OH groups. Protonated NH_2 groups substitute metal anions for counter ions, facilitating ion exchange. Electrostatic interactions are known to be the dominant force under acidic circumstances. Within cyclodextrin (CD) molecules, the process of forming inclusion complexes occurs between the specific metal contamination and the CD molecule. This interaction involves host–guest interactions, where hydrophobic interactions play a dominant role in capturing the metal. Moreover, the effectiveness of performance and the ability to selectively choose rely primarily on the chemical network inside hydrogels. The various sorption mechanisms of pollutants by heterogeneous materials have been illustrated in Fig. 12.2.

12.7.1 Electrostatic Interactions

Electrostatic interactions include producing hybrid materials with monomeric units containing certain functional groups that carry oppositely charged ions. These ions are intended to sorb or attract the corresponding metal ions. Electrostatic interactions

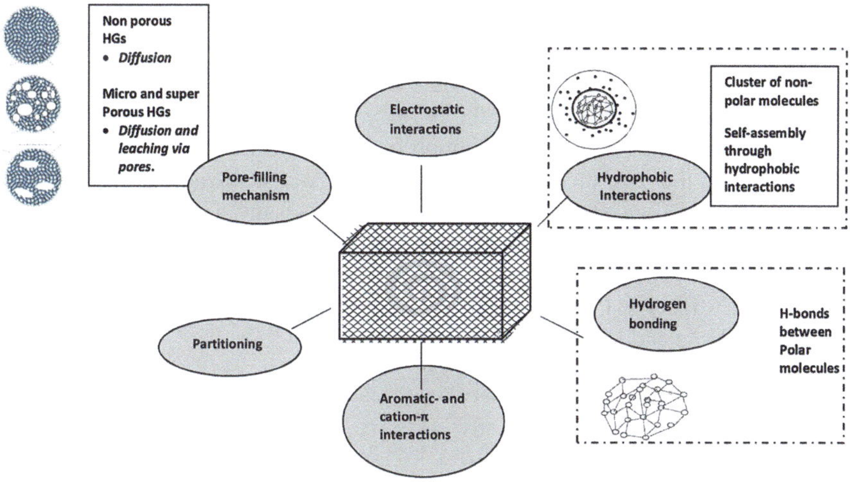

Fig. 12.2 Adsorption mechanism of pollutants using hydrogel. Reproduced with permission from (Sinha and Chakma 2019). Elsevier

may effectively produce hybrid hydrogels consisting of polymers and NPs. In addition to enhancing the strong connection between the ionic polymers, it also allows for easy modification of the mechanical characteristics of the high-gloss materials. In several investigations, nanomaterial hybridization was used to synthesize hydrogels crosslinked via beneficial electrostatic interactions. These highly gifted individuals demonstrated exceptional skills in autonomously rejuvenating themselves, transmitting reusable attributes. Various studies have shown that electrostatic interactions are the dominant adsorption force that removes ionic dyes in various heterogeneous materials. Guo et al. developed GO/polyethyleneimine (PEI) hydrogels to eliminate organic dyes such as MB and rhodamine B (RB). The composition of the hydrogel was adjusted and maintained by electrostatic interactions and H-bonding between the GO films and amine. The hydrogel showed exceptional capacity and characteristics for dye adsorption (Guo et al. 2015).

Furthermore, manipulating electrostatic interactions allows for controlling and modifying the microstructure and particle characteristics of biopolymer hydrogel particles. The research was conducted to produce hydrogel nanocomposites of AA-grafted alginate and TiO_2 NPs to remove the cationic dye MB. The gel material exhibited a maximum sorption capacity of 99.4%, positively correlated with the acrylamide (AA) dose. The formation of the association complex is ascribed to electrostatic interactions between the cationic ions of MB and the eCOOH groups on the hydrogel. Additionally, there have been reports of secondary interactions via H bonding between imine groups (eRCH]NR) of the dye and (eOH) groups of the hydrogel. An investigation used radiation therapy (γ rays) to manufacture Gum ghatti-cl-poly (AA) and Gum ghatti-cl-poly (AA-IPN-aniline) hydrogels to remove methyl orange (MO). The dominant adsorption process in MO sorption is believed to be the electrostatic interactions between the NH_2 and OH groups of polyaniline and polyacrylic and the anionic dye. A study on producing gel beads made from sodium alginate-CMC cellulose to remove Pb^{2+} ions was conducted. The analysis findings indicate that the primary adsorption mechanism of Pb^{2+} ions onto hydrogels involves the complexation between the eOH and COOe functional groups, where the lone electron pair is shared. It was also shown that secondary interactions occur via electrostatic attraction between the negatively charged gel beads and positively charged Pb^{2+} ions. A combination of free radical and cationic hybrid polymerization was used to create an ion-imprinted hydrogel (Cu(II)-IIH) to remove Cu(II). The Cu(II) interacted with the amide and ether groups found in hydrogels (Wang and Li 2015).

12.7.2 Ion Exchange

Studies have indicated that ion exchange is the predominant adsorption mechanism when high-grade materials are coupled to surfaces with acidic or basic functions. Recently, a calcined layered double hydroxide (CLDH)-alginate hydrogel was developed to eliminate wastewater that contains p-nitrophenol (PNP) and p-aminophenol (PAP). The primary factors contributing to the elimination of PNP were anionic exchange and H bonding. The elimination mechanism of PAP was primarily driven by H-bonding. The method of interaction between hydrogel beads for PAP and PNP was analyzed using the results of characterization tests conducted before and after the adsorption trials. The investigation revealed that the FTIR spectra shifted towards the blue end of the spectrum for the OeH bond, from 3460 cm^{-1} to 3455 cm^{-1} and 3459 cm^{-1} after adsorption. This shift proved the existence of a hydrogen connection between the surface OeH of CLDH and the N or O of PAP and PNP during the adsorption process (Shan et al. 2018).

In addition, the adsorption enthalpy values of PNP and PAP are 4504 kJ/mol and 11,349 kJ/mol, respectively, providing more evidence of chemisorption, namely the ion exchange process. This conclusion is supported by the expected influence of pH on the experimental data. In addition, the interaction mechanism of the molecules was elucidated using theoretical calculations employing the directional freezing–thawing (DFT) approach. The PNP and PAP interacted via distinct groups on the surface of CLDL. Within the context of PNP, a bonding contact was seen between the oxygen atom in NO_2 and the OH group of the benzene ring, specifically with the hydrogen atom on the surface of CLDL.

Conversely, in PAP, the nitrogen atom in NH_2 and the oxygen atom in OH formed a bond with the hydrogen atom on the surface of CLDH. Analysis of energies and bond length demonstrated a more stable and substantial link between the CLDH surface and PNP. This finding aligns with the most excellent adsorption capabilities of CLDH-alginate hydrogel beads for PNP (98.23 mg/g) and PAP (36.46 mg/g). A composite material called cellulose-graft polyacrylamide/hydroxyapatite hydrogel was synthesized to remove Cu(II) ions. Hydroxyapatite in hydrogel enhanced the ion-exchange process, wherein Cu(II) was either bound to the surface of hydroxyapatite or penetrated within it, facilitating the release of Ca^{2+} ions. Hua and Li researched to develop a new hydrogel with sulphhydryl functionality and magnetism. This hydrogel-treated synthetic solution includes Cd^{2+}, Cu^{2+}, Ni^{2+}, and Pb^{2+}, ions (Hua and Li 2014).

The removal methods of ions were suggested to include a simultaneous ion exchange and chelation reaction between Fe_3O_4/P(Cys/HEA) hydrogel and metallic ions in the solution. A fixed-bed method was investigated using a polyampholyte hydrogel sorbent to treat industrial wastewater. The hydrogel demonstrated significant removal rates of 216.1 mg/g for Pb(II) and 153.8 mg/g for Cd(II). The hydrogel exhibited high effectiveness throughout a broad pH range of 3–6, thanks to its numerous active groups. The method by which heavy metals were absorbed may be ascribed to the combined effects of ion exchange and complexation.

12.7.3 Hydrophobic Interactions

The hydrophobic interaction in hydrogels offers many possibilities in hydrogel engineering since it is crucial in improving mechanical characteristics. During these interactions, copolymerization is achieved by incorporating hydrophobic sequences into hydrophilic polymers. During the beginning of the reaction, the hydrophobic microdomain develops physical cross-linking spots with the best mechanical stability. The process occurs by forming links between these cross-linking sites and additional polymer chains, forming a three-dimensional hydrogel network. Cheng et al. examined the governing mechanism of hydrophobic nanoparticle/ hydrogel composites (rGO-AgarBs) to remove organic contaminants from wastewater. The primary component for removing organic compounds is the hydrophobic interactions between the reduced graphene oxides (rGOs) inserted in the agarose hydrogel beads (AgarBs) and the functional groups. This research achieved a maximal uptake potential of 321.7 mg/g and 196.4 mg/g for RB and aspirin, respectively. Additionally, it demonstrated increased adsorption behavior due to salinity and a preference for removing organic pollutants with lesser solubility in water. The exceptional characteristics of hydrophobic forces may be used to improve the adsorption qualities of hydrogel for wastewater treatment (Cheng et al. 2018).

12.7.4 Chelation

Several investigators have exploited intramolecular metal chelation to enhance the mechanical characteristics of hydrogels. A polyethylenimine-functionalized ion-imprinting hydrogel was developed that selectively eliminates Cu(II) ions. The synthesized gel demonstrated exceptional adsorption efficacy and the ability to be regenerated. The primary interaction mechanism was the chelation between Cu(II) ions and functional groups. Hydrogels were produced using biopolymers derived from proteins and polysaccharides. The structure and physical characteristics of the material were determined by the alkaline conditions (resulting in spherical shapes) and acidic conditions (resulting in rod-like particles), as well as the concentration of the ionic solution. The pH of chitosan-based hydrogels is crucial since the formation of metal complexes (via ion exchange or chelation) relies on the protonation of the macromolecule. The chitosan polymer has a high concentration of amine sites and a strong affinity for binding metal ions via chelation. However, when exposed to acidic conditions, it becomes protonated and exhibits cationic polymer characteristics that promote electrostatic interactions (the strength of which depends on the degree of deacetylation). A chitosan-montmorillonite hydrogel was developed to remove cationic cupric and anionic nitrazine yellow (NY) dyes. (Ngwabebhoh et al. 2016) The primary removal process was revealed to be the interaction between eNH_2 groups and Cu^{2+} and NY dye. The research found that at pH levels below 5.6, the eNH_3^+ form is the most prevalent and interacts with the anionic NY dye. On the other hand, at elevated pH values, the NH_2 form prevails, where the N atoms share a lone electron pair with NY and Cu^{2+}, forming a composite.

12.7.5 *π–π Interactions*

Several organic pollutants, such as nitroaromatic compounds (NACs) and polycyclic aromatic hydrocarbons (PAHs), consist of aromatic rings with π-donor and π-acceptor characteristics that get charged under ambient pH conditions. The major sorption force in these hydrophobic compounds is believed to be the π–π electron donor–acceptor (EDA) interaction. Tiwari et al. synthesized three-dimensional rGO hydrogels that exhibited excellent adsorption properties for MB (100%) and RB (97%) owing to the presence of robust anion–cation and interactions (Tiwari et al. 2013).

These interactions are evident when electron-rich pollutants (such as simple monoaromatic chemicals, complex organic insecticides, polyphenols, and PAHs) are adsorbed onto electron-deficient sites in various adsorbents. The amine cations operate as π acceptors and generate π–π contacts with the π electron-rich, polyaromatic surface of manufactured hydrogels that consist of black carbon, biochar, and a graphene-rich matrix (Kumar and Gangawane 2022b). Significant quadrupole moments from charge polarization make these interactions more noticeable. A hydrogel composite consisting of GO and silver NPs was synthesized. These hydrogels exhibited exceptional mechanical strength due to the robust π–π stacking interactions between the graphene sheets and their derivatives. The hydrogel has extensive water/solid boundaries that allow for effortless absorption of organic dyes, thanks to many interconnected pores. Hu et al. discovered the vital function of the sp^2-conjugated carbon network of a graphene sheet in the process of separating organocations via π–π interactions. Therefore, the identification and characterization of these π–π stacking interactions in various heterogeneous materials will contribute to improving tensile strength and enhancing uptake potential for aromatic organic pollutants (Hu et al. 2017).

12.8 Conclusions and Future Perspectives

The environment, including aquatic life, human health, and the ecosystem, is adversely affected by the discharge of heavy metals and dyes by various industries, among the most prevalent pollutants in effluent. Consequently, it is essential to remediate effluent contaminants to ensure a secure environment. An efficacious wastewater remediation method is the adsorption procedure, which involves various varieties of adsorbents. Nevertheless, the majority of the adsorbents employed for treatment are nondegradable and artificial. The subsequent handling of the adsorbent presents an additional environmental issue that has prompted researchers to seek to use a naturally sourced and renewable material as an adsorbent. Hydrogels are the optimal prospects for this requirement, as they offer various additional advantages, such as cost-effectiveness, high removal efficacy, and simplified processing.

Polysaccharide-based hydrogels are highly sought after for wastewater treatment owing to their unique characteristics, cost-effectiveness, widespread availability, capacity to degrade naturally, renewable nature, lack of toxicity, and reusing ability. Cellulose, chitosan, and alginate-based hydrogels have great potential as adsorbents for removing various pollutants. The adsorption and physicochemical performance of the polysaccharide hydrogel were determined by internal factors such as functionalization, crosslinking, nanoparticles, magnetic particles, and polymers, as well as external factors such as temperature, ionic strength, and pH. Crosslinking is a widely used approach to improve the adsorption capabilities of polysaccharide hydrogels. Recently, scientists have embraced a novel technique called photo-crosslinking or crosslinking technology to enhance the dye adsorption capabilities of hydrogels. This approach is both cost-effective and ecologically sustainable. In addition, surface functionalization may enhance hydrogels by including active functional groups and improving their attraction to pollutants such as metals and charged dyes. The pollutant adsorption on a hydrogel is likely due to many interactions, including H bonding, electrostatic forces, π-π interactions, van der Waals forces, and electron donor–acceptor interactions.

The adsorption capacity of numerous CBHs is diminished following regeneration. According to certain studies, the hydrogels' adsorption capacity was only maintained when regeneration was performed with caustic soda. Additionally, the adsorption capacity of the CBHs can be somewhat improved by pretreatments and modifications (both physical and chemical) to cellulose. The gel characteristics and regeneration performances can be enhanced by experimenting with a variety of physical and chemical networking approaches, including the modification of graphene oxide (GO), NPs, carbon nanotubes (CNTs), and carbon quantum dots (CQDs), as well as the blending of other suitable synthetic or natural polymers. The robust structure and effective expansion of CBHs are essential for effluent remediation, particularly at elevated temperatures in the industrial waste stream. Certain CBHs tend to diminish and lose their mechanical stability after repeated enlargement. A chemically crosslinked network frequently enhances the adsorption performance and stability of hydrogels. Additionally, the swelling properties and adsorption capacity of the CBHs should be improved by investigating the integration of nanoparticles, magnetic particles, and various chemical catalysts. It is crucial to examine the potential of enhancing the mechanical stability of CBHs by enhancing their self-healing capabilities following a distended state. Their general morphology and physicochemical composition primarily define the CBHs' effectiveness in treating industrial wastewater. Developing a particular CBH that overcomes the difficulties and inadequacies of low turnover number, mechanical strength, and decreased resistivity depends critically on a suitable formulation and optimal synthesis circumstances. Addressing issues with pH sensitivity, thermal stability, and swelling ratio for complete implementation is also essential.

Because of their remarkable properties, reusability, excellent adsorption capacity, and fast kinetics, chitosan hydrogels have gained much interest in wastewater treatment. Apart from these qualities, chitosan hydrogels show poor stability and mechanical strength. Resolving these problems depends critically on the construction technique of these hydrogels. Emerging trends in chitosan-based hydrogels for wastewater treatment are presented in this chapter, and the hydrogels are categorized based on preparation techniques. Chitosan-based hydrogels improve mechanical qualities and removal efficiency using preparation and modification techniques. It is advised that using chitosan and chitosan-based hydrogels, attempts should be made to improve the adsorption capacity and removal rate of specific adsorbents. Investigating the effectiveness of chitosan-based hydrogels in treating industrial wastewater containing many contaminants is very necessary. It is advisable to develop innovative multifunctional chitosan-based hydrogels to remove diverse contaminants from effluent. The intricate interactions among the operational parameters of adsorption have received insufficient attention from researchers. Future investigations should incorporate optimization techniques, such as response surface methodology (RSM), to enhance the adsorption capacity of adsorbents. Additionally, conducting molecular simulations of contaminant adsorption on chitosan-based hydrogels is encouraged to deepen the understanding of the adsorption process. The reusability of hydrogels determines the economic efficacy of adsorbents. More comprehensive research should be conducted on the regeneration capabilities of chitosan hydrogels to ascertain their reusability in various cycles.

While environmental circumstances influence the adsorption behavior of hydrogels, it is uncommon to come across research that includes testing these hydrogels using actual water samples. It is crucial to emphasize the efficacy of hydrogels under difficult environmental circumstances. Authentic wastewater comprises various pollutants, such as medications, dyes, microplastics, heavy metals, lubricants, pesticides, detergents, and fertilizers. Hence, instead of just focusing on studying a single contaminant, future research on adsorbents made from polysaccharides should include the assessment of many pollutants. Similarly, a few papers discuss using hydrogels for selectively adsorbing colors from dye mixtures, separating oil and water, and removing pesticides. Hence, conducting cutting-edge research to eradicate these toxins is essential.

Additionally, it is essential to mention that carbon-based nanoparticles strengthen most CBH and alginate-based hydrogels. From an environmental standpoint, it is necessary to prioritize using natural biomass modifiers, including orange peel, cotton, and banana peel, to manufacture hydrogels. Hence, a thorough examination is needed in this domain. Furthermore, it is notable that the development of innovative smart polysaccharide hydrogels has the potential to fulfill future demands. Thus, it is essential to investigate the possibility of intelligent polysaccharide hydrogels for photocatalytic or sun degradation of pollutants. Furthermore, it is crucial to prioritize simulation studies to better understand the underlying process of hydrogel adsorption and to develop innovative hydrogels for various wastewater treatments. Ultimately, the economic feasibility of hydrogel adsorbents depends on their ability to be regenerated. Nevertheless, in other concerns, researchers prioritized examining the adsorption procedure over the material's ability to be regenerated. Further comprehensive investigations should be undertaken in this setting to understand the hydrogels' regenerative capacity throughout several cycles entirely. Finally, the research on hydrogel filtration membranes is in the nascent phase. Hence, it is anticipated that there will be several other occurrences, concepts, and discoveries to elucidate in the future.

References

Akhtar MF, Hanif M, Ranjha NM (2016) Methods of synthesis of hydrogels—a review. Saudi Pharm J 24:554–559. https://doi.org/10.1016/j.jsps.2015.03.022

Cheng C, Cai Y, Guan G, Yeo L, Wang D (2018) Hydrophobic-force-driven removal of organic compounds from water by reduced graphene oxides generated in agarose hydrogels. Angew Chemie Int Ed 57:11177–11181. https://doi.org/10.1002/anie.201803834

Çınar S, Kaynar ÜH, Aydemir T, Çam Kaynar S, Ayvacıklı M (2017) An efficient removal of RB5 from aqueous solution by adsorption onto nano-ZnO/Chitosan composite beads. Int J Biol Macromol 96:459–465. https://doi.org/10.1016/j.ijbiomac.2016.12.021

Cong HP, Qiu JH, Yu SH (2015) Thermoresponsive poly(N-isopropylacrylamide)/Graphene/Au nanocomposite hydrogel for water treatment by a laser-assisted approach. Small 11:1165–1170. https://doi.org/10.1002/smll.201401651

Deng Y, Tang L, Zeng G, Zhu Z, Yan M, Zhou Y, Wang J, Liu Y, Wang J (2017) Insight into highly efficient simultaneous photocatalytic removal of Cr(VI) and 2,4-diclorophenol under visible light irradiation by phosphorus doped porous ultrathin g-C_3N_4 nanosheets from aqueous media: performance and reaction mechanism. Appl Catal B Environ 203:343–354. https://doi.org/10.1016/j.apcatb.2016.10.046

Eldin MSM, Soliman EA, Elzatahry AAF, Elaassar MR, Eweida BY, Elkady MF, Rahman AMA, Yossef ME (2019) Carboxylated alginate hydrogel beads for methylene blue removal: formulation, kinetic and isothermal studies. Desalin Water Treat 168:308–323. https://doi.org/10.5004/dwt.2019.24628

Fang S, Wang G, Li P, Xing R, Liu S, Qin Y, Yu H, Chen X, Li K (2018) Synthesis of chitosan derivative graft acrylic acid superabsorbent polymers and its application as water retaining agent. Int J Biol Macromol 115:754–761. https://doi.org/10.1016/j.ijbiomac.2018.04.072

Farshidfar N, Iravani S, Varma RS (2023) Alginate-based biomaterials in tissue engineering and regenerative medicine. Mar Drugs 21:189. https://doi.org/10.3390/md21030189

Fosso-Kankeu E, Mittal H, Marx S, Ray SS (2017) Hydrogel-based bioflocculants for the removal of organic pollutants from biodiesel wastewater. J Polym Environ 25:844–853. https://doi.org/10.1007/s10924-016-0870-8

Gong Z, Zhang G, Zeng X, Li J, Li G, Huang W, Sun R, Wong C (2016) High-strength, tough, fatigue resistant, and self-healing hydrogel based on dual physically cross-linked network. ACS Appl Mater Interfaces 8:24030–24037. https://doi.org/10.1021/acsami.6b05627

Guo H, Jiao T, Zhang Q, Guo W, Peng Q, Yan X (2015) Preparation of graphene oxide-based hydrogels as efficient dye adsorbents for wastewater treatment. Nanoscale Res Lett 10:272. https://doi.org/10.1186/s11671-015-0931-2

Hennink WE, van Nostrum CF (2012) Novel crosslinking methods to design hydrogels. Adv Drug Deliv Rev 64:223–236. https://doi.org/10.1016/j.addr.2012.09.009

Hou N, Wang R, Wang F, Bai J, Jiao T, Bai Z, Zhang L, Zhou J, Peng Q (2019) Self-assembled hydrogels constructed via host-guest polymers with highly efficient dye removal capability for wastewater treatment. Colloids Surf A Physicochem Eng Asp 579:123670. https://doi.org/10.1016/j.colsurfa.2019.123670

Hu H, Chang M, Zhang M, Wang X, Chen D (2017) A new insight into PAM/graphene-based adsorption of water-soluble aromatic pollutants. J Mater Sci 52:8650–8664. https://doi.org/10.1007/s10853-017-1090-x

Hua R, Li Z (2014) Sulfhydryl functionalized hydrogel with magnetism: synthesis, characterization, and adsorption behavior study for heavy metal removal. Chem Eng J 249:189–200. https://doi.org/10.1016/j.cej.2014.03.097

Igberase E, Osifo P, Ofomaja A (2014) The adsorption of copper (II) ions by polyaniline graft chitosan beads from aqueous solution: Equilibrium, kinetic and desorption studies. J Environ Chem Eng 2:362–369. https://doi.org/10.1016/j.jece.2014.01.008

Kamoun EA, Kenawy E-RS, Tamer TM, El-Meligy MA, Mohy Eldin MS (2015) Poly (vinyl alcohol)-alginate physically crosslinked hydrogel membranes for wound dressing applications: characterization and bio-evaluation. Arab J Chem 8:38–47. https://doi.org/10.1016/j.arabjc.2013.12.003

Klinpituksa P, Kosaiyakanon P (2017) Superabsorbent polymer based on sodium carboxymethyl cellulose grafted polyacrylic acid by inverse suspension polymerization. Int J Polym Sci 2017:1–6. https://doi.org/10.1155/2017/3476921

Kolya H, Kang C-W (2023) Bio-based polymeric flocculants and adsorbents for wastewater treatment. Sustainability 15:9844. https://doi.org/10.3390/su15129844

Kumar A, Gangawane KM, Ramanjaneyulu B (2022) Ferrofluids for waste-water treatment. In: Advances in chemical, bio and environmental engineering. Springer, pp 723–744

Kumar A, Gangawane KM (2022a) Effect of precipitating agents on the magnetic and structural properties of the synthesized ferrimagnetic nanoparticles by co-precipitation method. Powder Technol 401:117298. https://doi.org/10.1016/j.powtec.2022.117298

Kumar A, Gangawane KM (2022b) Synthesis and effect on the surface morphology andamp; magnetic properties of ferrimagnetic nanoparticles by different wet chemical synthesis methods. Powder Technol 410:117867. https://doi.org/10.1016/j.powtec.2022.117867

Kumar A, Gangawane KM (2023a) Advanced computational approaches for water treatment. Advanced computational approaches for water treatment. CRC Press, New York, pp 1–12

Kumar A, Gangawane KM (2023b) Molecular dynamics for the membrane process. Advanced computational approaches for water treatment. CRC Press, New York, pp 13–38

Liu X, Zhang L (2015) Removal of phosphate anions using the modified chitosan beads: adsorption kinetic, isotherm and mechanism studies. Powder Technol 277:112–119. https://doi.org/10.1016/j.powtec.2015.02.055

Liu L, Gao ZY, Su XP, Chen X, Jiang L, Yao JM (2015) Adsorption removal of dyes from single and binary solutions using a cellulose-based bioadsorbent. ACS Sustain Chem Eng 3:432–442. https://doi.org/10.1021/sc500848m

Ma J, Zhang M, Ji M, Zhang L, Qin Z, Zhang Y, Gao L, Jiao T (2021) Magnetic graphene oxide-containing chitosan-sodium alginate hydrogel beads for highly efficient and sustainable removal

of cationic dyes. Int J Biol Macromol 193:2221–2231. https://doi.org/10.1016/j.ijbiomac.2021.11.054
Mahmoud GA, Hegazy EA, Badway NA, Salam KMM, Elbakery SM (2016) Radiation synthesis of imprinted hydrogels for selective metal ions adsorption. Desalin Water Treat 57:16540–16551. https://doi.org/10.1080/19443994.2015.1079256
Makhado E, Pandey S, Modibane KD, Kang M, Hato MJ (2020) Sequestration of methylene blue dye using sodium alginate poly(acrylic acid)@ZnO hydrogel nanocomposite: kinetic, isotherm, and thermodynamic investigations. Int J Biol Macromol 162:60–73. https://doi.org/10.1016/j.ijbiomac.2020.06.143
Matei E, Predescu AM, Râpă M, Ţurcanu AA, Mateş I, Constantin N, Predescu C (2022) Natural polymers and their nanocomposites used for environmental applications. Nanomaterials 12:1707. https://doi.org/10.3390/nano12101707
Mittal H, Alhassan SM, Ray SS (2018) Efficient organic dye removal from wastewater by magnetic carbonaceous adsorbent prepared from corn starch. J Environ Chem Eng 6:7119–7131. https://doi.org/10.1016/j.jece.2018.11.010
Ngwabebhoh FA, Erdem A, Yildiz U (2016) Synergistic removal of Cu(II) and nitrazine yellow dye using an eco-friendly chitosan-montmorillonite hydrogel: optimization by response surface methodology. J Appl Polym Sci 133:1–14. https://doi.org/10.1002/app.43664
Pal D, Sen S (2023a) Emerging petroleum pollutants and their adverse effects on the environment. In: Impact of petroleum waste on environmental pollution and its sustainable management through circular economy. Springer, pp 103–137
Pal D, Sen S (2023b) In-depth coverage of petroleum waste sources, characteristics, environmental impact, and sustainable remediation process. In: Impact of petroleum waste on environmental pollution and its sustainable management through circular economy. Springer, pp 1–38
Pal D, Sen S (2024a) Optimal synthesis of dolochar derived faujasite zeolite X for highly effective Cd(II) removal. Environ Res 240:117494. https://doi.org/10.1016/j.envres.2023.117494
Pal D, Sen S (2024b) Valorization of sponge iron industrial waste into iron-modified zeolite X for ciprofloxacin removal: a multi-parameter optimization study. Environ Sci Pollut Res 31:53381–53398. https://doi.org/10.1007/s11356-024-34730-8
Peng Q, Liu M, Zheng J, Zhou C (2015) Adsorption of dyes in aqueous solutions by chitosan–halloysite nanotubes composite hydrogel beads. Microporous Mesoporous Mater 201:190–201. https://doi.org/10.1016/j.micromeso.2014.09.003
Peng B, Yao Z, Wang X, Crombeen M, Sweeney DG, Tam KC (2020) Cellulose-based materials in wastewater treatment of petroleum industry. Green Energy Environ 5:37–49. https://doi.org/10.1016/j.gee.2019.09.003
Rodrigues F, Faria M, Mendonça I, Sousa E, Ferreira A, Cordeiro N (2024) Efficacy of bacterial cellulose hydrogel in microfiber removal from contaminated waters: a sustainable approach to wastewater treatment. Sci Total Environ 919:170846. https://doi.org/10.1016/j.scitotenv.2024.170846
Shan R, He Y, Zi T, Wang G, Liu X, Han Z, Zhang T, Zhu Y (2018) Immobilization of calcined layered double hydroxide into alginate hydrogel beads for PNP and PAP removal: kinetics, isotherms, thermodynamics, and mechanism. Water, Air, Soil Pollut 229:329. https://doi.org/10.1007/s11270-018-3976-x
Shi Y, Xue Z, Wang X, Wang L, Wang A (2013) Removal of methylene blue from aqueous solution by sorption on lignocellulose-g-poly(acrylic acid)/montmorillonite three-dimensional cross-linked polymeric network hydrogels. Polym Bull 70:1163–1179. https://doi.org/10.1007/s00289-012-0898-4
Sinha V, Chakma S (2019) Advances in the preparation of hydrogel for wastewater treatment: a concise review. J Environ Chem Eng 7:103295. https://doi.org/10.1016/j.jece.2019.103295
Soleimani K, Dadkhah Tehrani A, Adeli M (2018) Preparation of new GO-based slide ring hydrogel through a convenient one-pot approach as methylene blue absorbent. Carbohydr Polym 187:94–101. https://doi.org/10.1016/j.carbpol.2018.01.084

Soliman SMA, Ali AM, Sabaa MW (2017) Alginate-based hydrogel for water treatment. Desalin Water Treat 94:129–136. https://doi.org/10.5004/dwt.2017.21558

Tiwari JN, Mahesh K, Le NH, Kemp KC, Timilsina R, Tiwari RN, Kim KS (2013) Reduced graphene oxide-based hydrogels for the efficient capture of dye pollutants from aqueous solutions. Carbon N Y 56:173–182. https://doi.org/10.1016/j.carbon.2013.01.001

Tran TH, Okabe H, Hidaka Y, Hara K (2017) Removal of metal ions from aqueous solutions using carboxymethyl cellulose/sodium styrene sulfonate gels prepared by radiation grafting. Carbohydr Polym 157:335–343. https://doi.org/10.1016/j.carbpol.2016.09.049

Vakili M, Rafatullah M, Salamatinia B, Abdullah AZ, Ibrahim MH, Tan KB, Gholami Z, Amouzgar P (2014) Application of chitosan and its derivatives as adsorbents for dye removal from water and wastewater: a review. Carbohydr Polym 113:115–130. https://doi.org/10.1016/j.carbpol.2014.07.007

Varaprasad K, Jayaramudu T, Sadiku ER (2017) Removal of dye by carboxymethyl cellulose, acrylamide and graphene oxide via a free radical polymerization process. Carbohydr Polym 164:186–194. https://doi.org/10.1016/j.carbpol.2017.01.094

Voo W-P, Ooi C-W, Islam A, Tey B-T, Chan E-S (2016) Calcium alginate hydrogel beads with high stiffness and extended dissolution behaviour. Eur Polym J 75:343–353. https://doi.org/10.1016/j.eurpolymj.2015.12.029

Wang J, Li Z (2015) Enhanced selective removal of Cu(II) from aqueous solution by novel polyethylenimine-functionalized ion imprinted hydrogel: Behaviors and mechanisms. J Hazard Mater 300:18–28. https://doi.org/10.1016/j.jhazmat.2015.06.043

Wang Y, Zhao L, Peng H, Wu J, Liu Z, Guo X (2016) Removal of anionic dyes from aqueous solutions by cellulose-based adsorbents: equilibrium, kinetics, and thermodynamics. J Chem Eng Data 61:3266–3276. https://doi.org/10.1021/acs.jced.6b00340

Xie M, Liu X, Wang S (2022) Degradation of methylene blue through Fenton-like reaction catalyzed by MoS2-doped sodium alginate/Fe hydrogel. Colloids Surf B Biointerfaces 214:112443. https://doi.org/10.1016/j.colsurfb.2022.112443

Yu Z, Zhang X, Huang Y (2013) Magnetic Chitosan–iron(III) hydrogel as a fast and reusable adsorbent for chromium(VI) removal. Ind Eng Chem Res 52:11956–11966. https://doi.org/10.1021/ie400781n

Yu P, Wang H-Q, Bao R-Y, Liu Z, Yang W, Xie B-H, Yang M-B (2017) Self-assembled sponge-like chitosan/reduced graphene oxide/montmorillonite composite hydrogels without cross-linking of chitosan for effective Cr(VI) sorption. ACS Sustain Chem Eng 5:1557–1566. https://doi.org/10.1021/acssuschemeng.6b02254

Zainal SH, Mohd NH, Suhaili N, Anuar FH, Lazim AM, Othaman R (2021) Preparation of cellulose-based hydrogel: a review. J Mater Res Technol 10:935–952. https://doi.org/10.1016/j.jmrt.2020.12.012

Zhao D, Shen Z, Shen X (2021) Dual-functional calcium alginate hydrogel beads for disinfection control and removal of dyes in water. Int J Biol Macromol 188:253–262. https://doi.org/10.1016/j.ijbiomac.2021.07.177

Chapter 13
Nano-Hydrogel: A Cutting Edge Solution for Inorganic Wastewater Contaminants

Swati Dubey and Raje Sengar

Abstract One of the most common issues in the world is the lack of access to clean water for consumption due to the addition of inorganic contaminants owing to several anthropogenic activities combined with a fast pace of modernization. The inorganic contaminants have a severe influence on both flora and fauna along with their surroundings since these contaminants are non-biodegradable, poisonous, and carcinogenic in nature. As a result, there is an urgent need to recycle wastewater to save the ecosystem as a whole. Over recent years, hydrogels have been investigated as a promising material for the effective elimination of inorganic contaminants from water. The formulation of hydrogels into nano-range forms a multifunctional framework that facilitates the production of a hybrid system having a cumulative influence on innovative materials, due to the combined properties of hydrogels and nanomaterials. Nano-Hydrogels' hydrophilic 3D cross-linked network structure allows them to withstand large amounts of water or other water-based liquids without disintegrating. This demonstrates the potential for removing dangerous chemicals from water. This chapter summarizes current achievements in nano-based hydrogels, with a special emphasis on the removal of inorganic pollutants from water.

Keywords Wastewater treatment · Nano-hydrogels · Inorganic contaminants · Heavy metals · Adsorption · Nano-filtration

List of Abbreviations

•OH	Hydroxyl radicals
3D	Three-dimensional
AFGS	Amino-functional graphene sponge adsorbents
AM	Acrylamide
BIS	Bureau of Indian Standards

S. Dubey (✉) · R. Sengar
Department of Applied Chemistry, Samrat Ashok Technological Institute, Vidisha, M.P., India
e-mail: swatidubey82@gmail.com

S. Bardhan and M. P. Shah (eds.), *Applications of Hydrogels in Modern Wastewater Treatment*, Advances in Wastewater Research,
https://doi.org/10.1007/978-981-95-1598-1_13

BOD	Biological oxygen demand
CBHs	Cellulose-based hydrogels
CFN	Chitosan-functionalized nano-fiber
COD	Chemical oxygen demand
dc	Direct current
ED	Electro-dialysis
HG	Hydrogel
HN-AD	Heterotrophic nitrification–aerobic denitrification
HPMC	Hydroxypropyl methyl cellulose
LCN	Lignocellulose nano-fiber
MOF	Metal–organic framework
NF	Nanofiltration
NPs	Nanoparticles
O_2• −	Superoxide anions
PAMAM	Polyamidoamine
PNIPAM	Poly(N-isopropylacrylamide)
PVA	Polyvinyl alcohol
RO	Reverse osmosis
RSM	Response surface methodology
SPA	3-Sulfopropyl acrylate
TiO_2	Titanium dioxide
UV	Ultraviolet
WAC	Water absorption capacity
WHO	World Health Organization
κ-g-PHEAA/Fe3O4	Kappa carrageenan–graft–poly(N hydroxyethyl acrylamide)/ magnetite nanocomposite

13.1 Introduction

13.1.1 Background

Water is the most plentiful and vital element necessary for the survival of all forms of life on Earth. About 70% of Earth's surface is covered by water, earning it the nickname "blue planet." However, nearly 97% of this water is saltwater, which is neither usable nor drinkable (Falkenmark 1998). The unfortunate reality is that despite the vast amount of water available, only 3% is fresh, and of that, a mere 0.06% is readily accessible. Water is crucial for sustaining the diverse life forms on Earth. Freshwater is distributed through rivers, lakes, streams, dug wells, groundwater, and rainwater. 97% of the total available clean water on Earth comes from groundwater and is often the primary water source in both rural and urban areas, especially in developing countries like India. Historically, people have relied on groundwater to meet their drinking water needs. Consequently, maintaining the cleanliness of groundwater and protecting it from toxic contaminants remains a significant challenge in many countries worldwide. The usability of water in sectors such as agriculture, domestic use, and industry depends on its chemical composition. Both natural and human-made sources are significantly increasing inorganic contaminants, severely impacting water quality. Recently, it was reported that over 80 countries are experiencing water scarcity, with about 1.2 billion people worldwide consuming unclean water. WHO and UNICEF report that around 748 million people lack access to adequate and safe drinking water, relying instead on water contaminated with fecal matter. According to the UN, approximately 4.5 billion people will face water scarcity by 2040 (Baer 1996).

13.1.2 Overview of Water Contamination and Its Impact

Water contamination occurs when harmful substances pollute water sources, making it unsafe to consume and damaging the environment. Water contains diverse organic, inorganic and microbial pollutants. Numerous pollutants pose significant risks due to their high toxicity and carcinogenic nature. Impurities such as microorganisms, heavy metals, chemicals, plastics, and sediments can enter water through industrial processes, agricultural runoff, sewage, and waste mismanagement. This contamination poses significant health risks, causing waterborne diseases and chronic illnesses. One of today's key ecological issues is the presence of inorganic contaminants in both surface and groundwater. The emission of these contaminants remains disturbingly high in spite of many mitigation methods. Inorganic pollutants may be detected in water resources; however, their concentrations exceeding the allowable limit can result in various health complications. These pollutants may cause harm to the liver, kidneys, neurological system, cardiovascular system, digestive tract, cartilage, or epidermis, contingent upon the specific inorganic pollutant and the degree of exposure (Moore 2012). However, inorganic water toxins include cationic, anionic, or neutral forms of elements like arsenic, lead, cadmium, mercury, chromium, aluminum, nitrates, nitrites, and fluorides. Heavy metals make up a significant portion of inorganic pollutants. Their ubiquitous presence in the environment is mostly owing to numerous waste sources, including chemical fertilizers, industrial trash, mining debris, sewage, pesticides, and metallurgical operations (Singh et al. 2023). Because of their continued existence and lack of ability to degrade heavy metals pose a serious threat to living beings and sometimes cause carcinogenicity. While some heavy metals serve a crucial role in human metabolism, high doses can damage organs. These metals are water-soluble, allowing them to quickly infiltrate the food supply to aquatic organisms and subsequently impact humans, posing possible health hazards. (Uddin 2017). The Bureau of Indian Standards (BIS) and World Health Organization (WHO) have set permissible limits for inorganic contaminants, which are thoroughly detailed in Table 13.1. It also discusses the possible detrimental impacts these pollutants may have on living organisms and the surrounding environment (Hussain et al. 2021). Possible solutions involve better waste management, stricter regulations, sustainable farming practices, public education, and technological advancements in water purification. Combating water contamination necessitates a collaborative effort to provide safe and sustainable water supplies for future generations. To eliminate inorganic pollutants from wastewater, several treatment methods can be applied, such as reverse osmosis, precipitation, solvent (liquid) extraction, ion-exchange, filtration, coagulation, and adsorption. These techniques offer various mechanisms to effectively remove inorganic contaminants in both industrial and municipal wastewater treatment processes. (Ghosh et al. 2022).

Among the various techniques available, adsorption stands out as the most promising method for removing inorganic contaminants. Its appeal lies in its ease of operation, cost-effectiveness, high adsorption capacity, ability to undergo multiple cycles, potential for reuse, and feasibility of regeneration. These advantages have garnered significant interest from researchers, motivating them to explore new possibilities and the potential of different adsorbents. A range of affordable adsorbents have been employed to remove inorganic pollutants, including alumina, charcoal, activated carbon, brick powder, bone char, red clay, activated coconut shell, oxide ores, activated kaolinites, clays, modified chitosan, calcite, hydrogels and other polymeric material-based adsorbent (Giri and Badwaik 2022). These adsorbent materials have various drawbacks, such as difficulty in separating from water post-decontamination, high production costs, and limited economic viability for large-scale use, along with other concerns.

This necessitates the swift creation of a renewable, affordable, manageable, and environmentally friendly substance that effectively absorbs inorganic contaminants from water (Godiya et al. 2019). Hydrogels have sparked significant attention as adsorption materials for removing inorganic pollutants from wastewater due to their porous nature, robust mechanical qualities, tunable swelling actions, and simplicity of operation, which are essential for the reuse of adsorbent materials.

Hydrogels are unique three-dimensional, mesh-like polymer colloidal networks with high water compatibility and flexibility, akin to natural tissue, owing to their high moisture content (>90%), permeability, and soft texture. Hydrogels consist of hydrophilic groups like –OH, –COOH, $–NH_2$, $–O_3H$, $–CONH_2$, etc.,which enable them to swell in water and biological fluids. This swelling is reversible and highly reliant on the surrounding atmosphere. This feature characterizes them as "smart" or "intelligent" materials. They are also more stable in most polar solvents, and as time passes, they deteriorate, disintegrate, and eventually dissolve (Hossain et al. 2023). Hydrogels are consumable polymeric substances that are made primarily from polysaccharides, gelatin, methacrylate esters, integrated polyacrylamides, and other compounds. They can take numerous physical forms, including gels, nanoparticles (NPs), microparticles, sheets, films, and exterior coatings. Hydrogels and their composites are widely used in a variety of fields, including therapeutic regeneration, biomedical sciences, delivery of drugs, wound therapy, food and agricultural components, skincare products, and wastewater treatment, due to their potential to take on several forms while providing enhanced hydration reliability, permeability, and durability under diverse environments (Ho et al. 2022).

Table 13.1 Allowable limits for inorganic water and wastewater contaminants (Adapted from Sahu et al. 2024. Elsevier. License no: 6047560408786)

Contaminant	Permissible limits		Sources	Associated health impacts
	BIS limit (mg/L)	WHO limit (mg/L)		
Fluoride	1.0	1.5	Steel and aluminum industries, phosphate fertilizers, geological formations	Can result in dental and skeletal fluorosis; neurological impairments at high levels
Chloride	250	200–300	Soil runoff, marine aerosols, road de-icing agents	Causes gastrointestinal discomfort, water retention, and hypertension with prolonged exposure
Chlorite	–	0.7	Chlorination by-products in water treatment	Triggers eye, throat, and nasal irritation; possible blood-related disorders
Nitrite	0.02	3.0	Fertilizers, wastewater effluent, septic tanks, decaying organic matter	Leads to methemoglobinemia ("blue baby syndrome"), impacts blood oxygen transport
Bromide	0.5	–	Coal combustion, fertilizers, geological sources	Causes mucous membrane irritation, dizziness, and respiratory discomfort
Nitrate	45	50	Agricultural leachate, septic systems, drainage	Associated with headaches, elevated heart rate, and can induce hypoxia in infants
Phosphate	0.1	–	Fertilizer use, industrial discharges, poorly managed sewage	Leads to mineral imbalances, muscle spasms, and heart irregularities
Sulphate	200	–	Textile and tanning industries, sewage, detergents	Worsens symptoms of asthma, heart conditions; decreases lung function
Ammonia	0.5	1.5	Organic decomposition, fertilizers, corrosion inhibitors	Can cause irritation of eyes, lungs, and skin upon exposure

(continued)

Table 13.1 (continued)

Contaminant	Permissible limits		Sources	Associated health impacts
	BIS limit (mg/L)	WHO limit (mg/L)		
Lithium	0.01	–	Batteries, ceramics, mining activities	Triggers tremors, gastrointestinal issues, weight fluctuations
Molybdenum	0.07	0.07	Coal combustion, industrial effluents, sewage sludge	Linked to joint pain, elevated uric acid, liver and kidney dysfunction
Cyanide	0.05	0.07	Electroplating, photography, mining, steel plants	Interferes with cellular respiration; causes headaches, seizures, and cardiac arrest
Lead	0.01	0.01	Paint, mining, fuel combustion, old pipes	Neurotoxin, especially in children; causes anemia, hypertension, cognitive decline
Arsenic	0.01	0.01	Pesticides, smelters, natural groundwater deposits	Causes skin lesions, cancer, liver dysfunction, and immune suppression
Cadmium	0.003	0.003	Batteries, electroplating, phosphate fertilizers	Severe kidney and bone damage; anemia and lung irritation
Iron	0.3	–	Pipe corrosion, natural leaching	Linked to hemochromatosis, liver damage, and nausea
Mercury	0.001	0.006	Mining, industrial waste, dental amalgams	Highly toxic; affects the brain, kidneys, and can cause birth defects
Chromium	0.05	0.05	Leather tanneries, electroplating, dye industries	Carcinogenic; causes DNA damage, skin ulcers, and asthma
Copper	0.05	2.0	Pesticide residues, plumbing, mining	Results in nausea, liver damage, and CNS depression
Zinc	5.0	–	Plumbing, galvanization, paints	Causes stomach cramps, anemia, and can be phytotoxic in excess
Manganese	0.1	0.4	Ferromanganese alloys, fuel additives	Can lead to cognitive and motor impairments with prolonged exposure

(continued)

Table 13.1 (continued)

Contaminant	Permissible limits		Sources	Associated health impacts
	BIS limit (mg/L)	WHO limit (mg/L)		
Nickel	0.02	0.07	Batteries, alloy production, electroplating	Induces allergic reactions, skin eczema, and DNA damage
Dyes	0.01	–	Textile, dyeing, food coloring industries	Impacts aquatic photosynthesis, promotes mutagenicity and bioaccumulation

The findings suggest that hydrogel-based adsorbents are not particularly efficient at adsorbing inorganic pollutants. As a result, chemical and thermal modifications are needed to increase their adsorption capabilities. Furthermore, research has demonstrated that incorporating these adsorbents with different nanoparticles improves their selectivity and efficacy in adsorption. Several recent studies have described the use of nanohydrogels to purify and recover different inorganic contaminants from wastewater (Sethy et al. 2019). Nanohydrogels are multifaceted, swelling, interconnected frameworks of polymer chains with nanometer-sized particles. Nanogels, which have particle sizes ranging from 10–200 nm, and are economically viable adsorbents are extremely successful at removing inorganic pollutants from contaminated aquatic environments (Delgado-Pujol et al. 2025). Chemically interlinked nanohydrogels are formed by inner or outer covalent bridging of polymeric chains, using micro or mini emulsion methods, transplanting monomer molecules onto biopolymers, or self-assembling nanoaggregates.

This chapter explores the omission of inorganic toxins from wastewater using nanohydrogels, emphasizing the properties and synthesis techniques of nanohydrogels. It also covers various technologies used for eliminating inorganic pollutants from wastewater. Additionally, the chapter examines the applications and case studies of nanohydrogels in wastewater treatment.

13.2 Nano-Hydrogels: Properties and Synthesis

13.2.1 Properties of Nano-Hydrogel

13.2.1.1 Swelling Actions

Hydrogels composed of an interlinked (3D) framework of polymeric material, held together by physical and chemical bonds, exhibit excellent water-friendly and permeability behavior. When exposed to water, these gels enlarge due to osmosis, in which water molecules pass through the gel framework and cause it to dilate 1000 times its dry weight (Saha et al. 2023). This absorption process increases the gel's mechanical strength and elasticity, making it suitable for use in the administration of drugs, wound care, and environmental remediation.

The interlinking density of the gel influences swollen action, retention of water, pH, functional groups, and temperature. With this in mind, hydrogels are often described without molecular weight and referred to as infinitely macromolecular. The extent of water absorption in any hydrogel material is governed by the water absorbing functional groups on the polymeric chain, whereas the kind and extent of linkage influence its dissolving tolerance. Other elements, such as pH, variation in temperature, hydrolyzing enzymes, and electrically generated signals, can cause simultaneous changes in a hydrogel's physical characteristics, including retention of water, accumulation, and morphological changes. Volume variations occur when the number of mobile ions within the hydrogel molecules changes rapidly in comparison to the external solution because of pH fluctuations. Furthermore, pH influences the tendency to swell or the dimensions of the hydrogel by changing the electrostatic attraction of both basic and acidic functional units, as well as the degree of ionic bonding of the polymer backbone. Equation (13.1) is employed to determine the rate of water adsorption/swelling. (Alsaka et al. 2025).

$$\text{Swellingratio(SR\%)} = \frac{W_2 - W_1}{W_2} \times 100 \tag{13.1}$$

13.2.1.2 Mechanical Strength

Hydrogel polymers are substantially consumed in the healthcare and biomedical fields. Wound treatment, bone marrow scaffolding, medication delivery systems, muscle and ligament rehabilitation, tissue creation, and cartilage replacement are all applications that require an in-depth comprehension of their resistance to movement. The mechanical strength of a nano-hydrogel material is heavily dependent on the extent of inter-linking. Hydrogels with a higher level of inter-linking exhibit a reduced elongation percentage, a more brittle framework, and eventually link to a hydrogel structure that is more mechanically significant. Khan et al. explored the tensile properties of poly(acrylic acid) hydrogel electrolytes by conducting strain amplitude and frequency sweep tests. Their results demonstrated that the hydrogels exhibited viscoelastic behavior, characterized by a higher storage modulus than loss modulus, which signifies strong elasticity. This elastic nature was attributed to the presence of carboxylic acid functional groups. Additionally, the mechanical strength of the hydrogels was found to be influenced by the extent of coordination between carboxylate anions and metal cations (Khan et al. 2020).

13.2.1.3 Biological Properties

A biomaterial's cytotoxicity is an important element in determining its acceptability for use in biomedicine. Likewise, hydrogels must be non-toxic and environmentally friendly. Biocompatibility refers to a material's ability to perform its intended function while interacting safely and effectively with biological tissues. The biological safety and bioactivity are important aspects of biocompatibility. The use of hazardous compounds in hydrogel compositions can impair in vivo biologic compatibility. Polysaccharides, on the other hand, are commonly regarded as non-toxic, environmentally friendly polymers that are safe for consumption (Bashir et al. 2020). Tokura et al. performed in vivo cytotoxicity experiments on several biodegradable polysaccharides and discovered no serious adverse effects in rat blood circulation (Yang et al. 2012).

Biodegradable hydrogels are crucial in the biomedical field, as they break down into non-toxic materials inside the tissues. Degradation processes include hydrolysis and solubilization, with hydrogels being eliminated via bio-sorption or bio-erosion. Factors such as swelling nature, molecular weight, polymer-water association, and surrounding variables like pH and temperature influence the dissociation route (Fig. 13.1). Gao et al. (2013) investigated the in vitro biological breakdown of polyacrylic acid, its derived substances, and polyglutamic acid in buffer solutions with pH 1.2 and 7.4 when proteinase is present. They discovered that degradation was faster at pH 7.4 and slower at pH 1.2, which was most likely owing to enhanced water reactivity at pH 7.4 and higher hydrophobic behavior at pH 1.2.

13.2.2 Synthesis Methods of Nano-Hydrogel

Hydrogels made from natural polysaccharides offer advantages over synthetic materials, as they are biorenewable, readily available, and environmentally friendly. These gels contain various hydrophilic groups that absorb water upon contact, forming a three-dimensional gel structure.

13.2.2.1 Physical Crosslinking

The method, which relies on feeble interactions like H-bonding, ionic bonding, van der Waals forces, hydrophobic associations, and π–π attractions (Sharma et al. 2018), is often preferred for environment-friendly and non-toxic hydrogel preparation, since chemical crosslinking materials are not utilized. Though cellulose-based hydrogels (CBHs) formed through physical interlinking may have lower mechanical strength, their high porous nature increases their ability to adsorb pollutants. Additionally, they are less sensitive to pH, easily regenerated (since the process is reversible), and do not experience a loss in adsorption capacity from reactions with crosslinkers, as seen in chemical processes (Sinha and Chakma 2019).

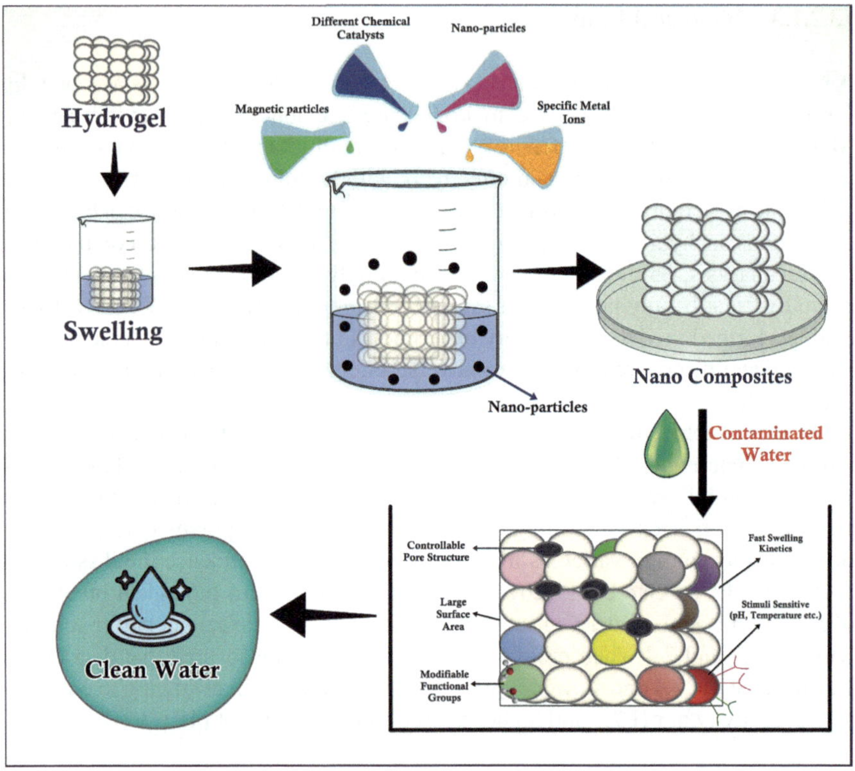

Fig. 13.1 Alterations in the techniques and properties of nano-hydrogels for inorganic contaminant removal from wastewater treatment

13.2.2.2 Chemical Crosslinking

Chemical crosslinking is favored for enhancing mechanical properties, as it involves agents that form robust molecular links like covalent or ionic attractions. This process can increase polymer crystallinity, leading to reduced swelling and rate of adsorption due to smaller pores and chain stiffness. Moreover, higher interlink density may block some active binding sites. Thus, it's essential to balance mechanical strength with optimal adsorption performance (Pakdel and Peighambardoust 2018).

13.3 Traditional Methods for Inorganic Contaminant Removal

Inorganic pollutants, like heavy metals (for example, lead, mercury, cadmium), salts, and other non-organic chemicals, are commonly found in industrial wastewater and pose significant environmental and health risks. The removal of these contaminants is crucial to ensure that processed wastewater is safely discharged into the environment or for its reuse. Several traditional ideas have been implemented to address the discharge of inorganic contaminants from wastewater. These methods are based on physical, chemical, and biological principles. Below is a detailed overview of the primary traditional methods used to process wastewater to remove impurities (Fig. 13.2).

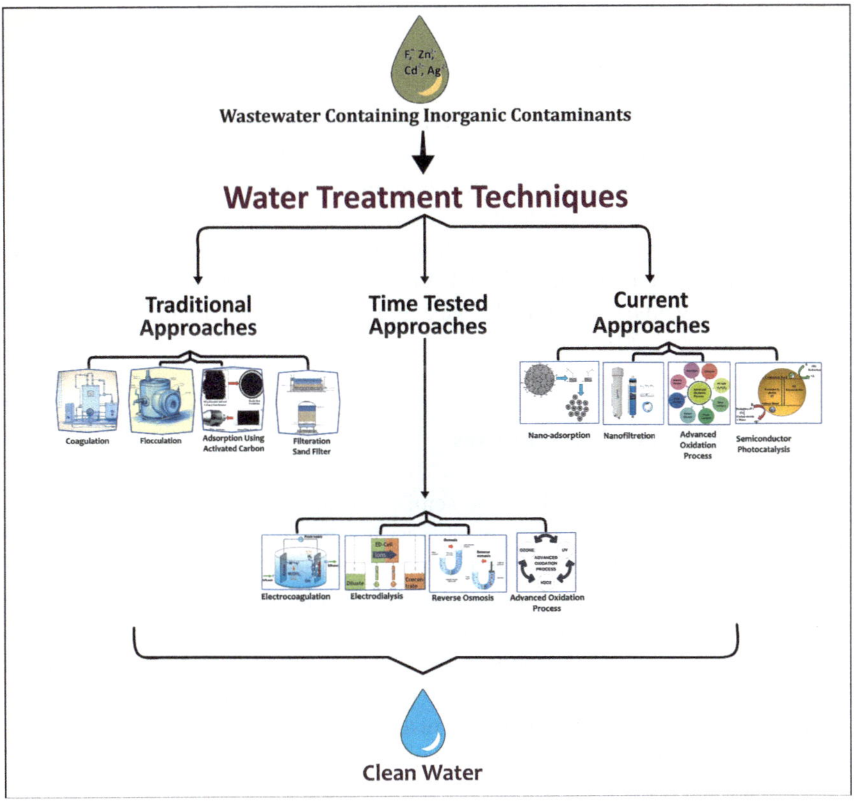

Fig. 13.2 Classification of water treatment techniques

13.3.1 Conventional Methods

13.3.1.1 Coagulation and Flocculation

Coagulation and flocculation are two chemical methods often utilized together to remove inorganic contaminants, particularly suspended particles, heavy metals, and colloidal materials. This procedure involves adding chemicals known as coagulants, such as iron chloride and alum, to wastewater in order to counteract the charges on the particles and cause them to combine. The coagulants destabilize the suspended particles, enabling them to form larger clumps or “flocs.” After coagulation, the water is slowly stirred that promotes the production of bigger, settleable clumps. This allows the flocs to separate from the water and settle out under gravity (Thines et al. 2017).

13.3.1.2 Ion Exchange Process

It is a process in which ions in the water are interchanged with other ions that are more readily removed. Water passes through a resin or other ion-exchange medium, which is loaded with exchangeable ions such as sodium or hydrogen. These ions replace the contaminant ions, such as heavy metal ions (e.g., lead, arsenic, and cadmium), as the water passes through the medium. Ion exchange is particularly useful for removing dissolved inorganic pollutants like heavy metals and other ions, especially in industrial effluents, where metals like copper, nickel, or chromium are commonly found (Thines et al. 2017).

13.3.1.3 Membrane-Based Process

This process is mainly based on the semi-permeability phenomenon. A selective filtration between the two adjacent phases depends upon the specific pore size, compound-specific permeability of the membrane and electrical charge on atoms for inorganic contaminant removal. Semi-permeable membranes act as a barrier for various pollutants, both organic and inorganic, suspended solids, pesticides and microorganisms. The membrane should be a thin, ceramic or metal material that must not break and disintegrate, made up of porous and nonporous polymeric film. Membrane-based processes, including reverse osmosis (RO), nanofiltration (NF), and electro-dialysis, have also been employed for the elimination of inorganic contaminants. (Thines et al. 2017).

Reverse Osmosis

This is a membrane-based separation technique that is capable of removing a wide range of inorganic pollutants, such as dissolved salts and heavy metals. Water is pushed across a semi-permeable barrier that blocks impurities like inorganic ions and only permits water molecules to go through. The purified water flows via the membrane, while the contaminants are retained and discarded as concentrate. RO is commonly used in desalination and for treating industrial wastewater that contains high levels of dissolved inorganic salts, heavy metals, and other dissolved contaminants (Thines et al. 2017).

Nanofiltration

Nanofiltration process is comparatively working on low-pressure and it primarily eliminates a large amount of dissolved solids, in contrast to the RO process. It is the new process that contains properties between the reverse osmosis and ultrafiltration. Contrarily, RO works at higher pressure by discarding all the dissolved solid impurities. Nanofiltration is appearing as a promising functional technology for upcoming problems related to the processing of industrial wastewater. The efficiency of the technique depends on various factors, including temperature, pressure, characteristics of raw water and regular monitoring and maintenance (Thines et al. 2017).

Electro-dialysis

Electro-dialysis is an emerging technique used for the separation of ionic contaminants from aqueous medium by using an ion-exchange membrane. This process is similar to RO, though ED works on applying potential electric current, direct current (DC), which acts as a driving force, in place of pressure. The global attention is diverting towards the ED processes for defluoridation due to their simplicity and capability to remove contaminants without using any chemical processes (Thines et al. 2017).

13.3.2 Mechanisms of Inorganic Contaminant Removal

Inorganic pollutants in wastewater, including heavy metals, salts, and other non-organic substances, pose significant environmental and health risks. Effective removal of these contaminants is essential to ensure safe wastewater treatment. Various mechanisms are employed for their removal, including **adsorption, ion**

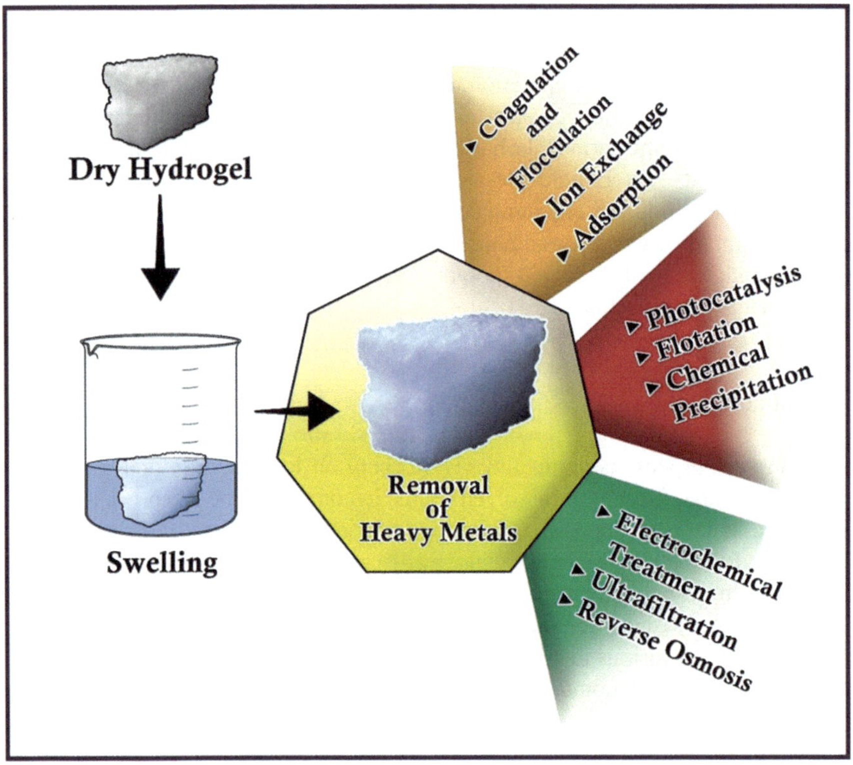

Fig. 13.3 Mechanism for inorganic contaminant removal through nano-hydrogel

exchange, and **photocatalytic degradation.** These processes rely on different principles, such as surface interactions, chemical reactions, and energy-driven transformations. Below is a detailed explanation of each mechanism involved in the removal of inorganic contaminants (Fig. 13.3).

13.3.2.1 Adsorption Mechanisms

This is one of the most widely used techniques for eliminating inorganic contaminants from wastewater. It refers to the process in which contaminants accumulate on the surface of a solid substance (adsorbent) from the liquid state (adsorbate). The kind of adsorbent substance, its surface area, porosity, and the type of pollutants all affect how well adsorption removes them.

There are two types of adsorption mechanisms: chemical and physical adsorption. The two mechanisms involve different forces and processes but share the common principle of attracting contaminants to the outer layer of the adsorbent material.

- **Surface area**: Larger the surface area of the adsorbent, the more contaminants it can attract.
- **Pore structure**: The size and structure of the pores in the adsorbent material determine the efficiency of contaminant removal.

Adsorbents commonly used for inorganic contaminant removal include activated carbon, zeolites, clay minerals, and metal oxides (Chai et al. 2021).

Physical Adsorption

Physical adsorption (physisorption) occurs when pollutants adhere to the surface of an adsorbent by weak forces between the molecules, such as van der Waals interactions. Since this kind of adsorption is usually reversible, it allows the adsorbed pollutants to be released by changing conditions like temperature or pressure. In this type of adsorption, the electrostatic attraction between the contamination and the sorbent causes the contaminating molecules to be drawn to the adsorbent surface. However, the attraction is weak and does not involve any significant chemical reaction.

Physical adsorption is characterized by several key features. It involves low energy compared to chemical adsorption, as the forces at play, such as van der Waals forces, are weaker and require less energy. Additionally, the process is reversible, meaning contaminants can be easily released from the adsorbent, making it ideal for applications where the adsorbent needs to be regenerated. However, physical adsorption is also dependent on environmental factors, such as temperature and pressure, which can influence its effectiveness. This method is commonly applied to remove volatile contaminants and inorganic pollutants, including toxic metals such as lead and cadmium, phosphates, and certain salts. Activated carbon is a common adsorbent utilized in physical adsorption because of its huge surface area and porosity, which makes it very effective for these kinds of contaminants. (Chai et al. 2021). Figure 13.3 depicts the mechanism for inorganic contaminant removal through nano-hydrogel.

Chemical Adsorption

When contaminants establish chemical interactions with the adsorbent's surface, this process is known as chemical adsorption, or chemisorption. Unlike physical adsorption, this process is typically stronger and more specific, as it involves the development of covalent or ionic bonds.

In chemisorption, the contaminant binds to the active sites on the adsorbent's surface through the formation of chemical bonds. This process is generally more complex and slower than physical adsorption, but results in stronger interactions between the contaminant and the adsorbent. Chemical adsorption is characterized by higher energy requirements compared to physical adsorption due to the formation of covalent or ionic bonds. Additionally, chemical adsorption tends to be irreversible under normal conditions, meaning once a contaminant is adsorbed, it is difficult to remove and requires more intense conditions for desorption. This process is also more selective, as it relies on the unique chemical makeup of the adsorbent and the adsorbate. Chemisorption is particularly useful for removing inorganic pollutants including toxic metals (e.g., mercury, arsenic, chromium) and other ionic pollutants from wastewater. Common adsorbents used in this process include metal oxides, typically iron and aluminum oxide, as well as functionalized materials specifically designed to form strong bonds with contaminants (Luo et al. 2018).

Ion Exchange Mechanisms

It is a popular and efficient technique for eliminating dissolved inorganic contaminants from wastewater. The procedure entails an interchange of ions within the ions on the outer layer of a solid resin or absorbent substance and the contaminating ions in the wastewater. Ion exchange resins contain charged sites that attract and bind contaminant ions, such as heavy metals like copper, nickel, and lead. These contaminant ions are replaced by other ions, typically sodium or hydrogen ions, which helps to remove the contaminants from the water. Ion exchange comes in two primary forms: cation exchange, which involves positively charged ions. (such as calcium, magnesium, or heavy metals) are replaced with other cations like sodium or potassium, and **anion exchange**, where negatively charged ions (like chloride, sulfate, or phosphate) are exchanged for other anions such as hydroxide or bicarbonate. Ion exchange is characterized by high selectivity, as resins can be tailored for specific contaminants. It is also regenerable, meaning the resins can be reused by washing them with a concentrated solution of the exchanged ion, such as sodium chloride or hydrochloric acid. Additionally, ion exchange is an efficient process, especially for removing low concentrations of contaminants. This method is commonly used in industrial effluent treatment, water softening, and applications that require ultrapure water, effectively removing inorganic contaminants like hardness ions and heavy metals from wastewater (Bashir et al. 2019).

Photocatalytic Degradation

Photocatalytic degradation is a sophisticated process of oxidation that breaks down pollutants in water by activating a photocatalyst with light, usually ultraviolet (UV) radiation. While traditionally applied to organic contaminants, photocatalysis has also proven effective for removing certain inorganic contaminants, particularly those that can undergo oxidative transformations. In this process, a semiconductor material, often titanium dioxide (TiO_2), serves as the photocatalyst. When exposed to UV light, the photocatalyst generates extremely active radicals like superoxide anions ($O_2\bullet -$) and hydroxyl radicals (•OH), which are capable of degrading contaminants by breaking their chemical bonds, leading to their removal from the water. For inorganic contaminants, photocatalysis can either directly degrade the contaminants or oxidize them into less harmful compounds, which can then be removed through methods like precipitation or adsorption. This process is characterized by energy efficiency, as it relies on light to activate the photocatalyst, and environmental friendliness, as it does not require harsh chemicals and produces minimal waste. Additionally, photocatalytic degradation can be tailored to selectively oxidize certain contaminants, including some heavy metals and other inorganic pollutants. While primarily used for removing organic pollutants, recent advancements have expanded its application to remove inorganic contaminants like arsenic and lead, making it valuable for wastewater treatment, groundwater cleanup, and ecological remediation (Khader et al. 2024).

The efficiency of contaminant removal from wastewater is influenced by several factors that affect the adsorption, ion exchange, or degradation processes. These factors include pH, temperature, concentration of contaminants, and contact time. Understanding how these factors interact can help optimize treatment methods for better removal of pollutants.

13.4 Factors Influencing Removal Efficiency

13.4.1 pH

The effectiveness of several pollutant removal procedures, particularly adsorption and ion exchange, is significantly impacted by the pH of the medium. The charge on the adsorbent material and the contaminating ions can be affected by pH, which can change the adsorption potential. For example, in the removal of heavy metals, the solubility of metal ions can be influenced by the pH, with certain metals becoming more soluble under acidic or basic conditions. For ion exchange processes, pH can affect the ionization of contaminants and the ion-exchange resin, which may lead to either enhanced or reduced ion exchange efficiency. Additionally, in processes like photocatalytic degradation, pH can impact the production of active species, which are responsible for contaminant breakdown. In general, the optimum pH range for effective contaminant removal varies depending on the nature of contaminants, the adsorbent material, and the process of treatment.

13.4.2 Temperature

The effectiveness of removing pollutants from wastewater is significantly influenced by temperature. It can impact the physical characteristics of the adsorbent, the kinetics of adsorption, and the reactivity of contaminants. For instance, higher temperatures can enhance the kinetic energy of molecular species, causing pollutants to diffuse onto the adsorbent surface more quickly, which may improve the adsorption process. However, excessive heat may cause desorption, reducing the efficiency of contaminant removal. In chemical adsorption processes, increased temperature can enhance the chemical reaction rates, leading to better adsorption and quicker removal of contaminants. In ion exchange processes, higher temperatures may reduce the viscosity of the solution, improving the rate of ion exchange. In photocatalytic degradation, temperature can affect the generation of reactive species, with higher temperatures potentially increasing the rate of photocatalysis. However, temperature must be optimized, as extremely high temperatures can lead to the degradation of adsorbent materials or the loss of their effectiveness.

13.4.3 Concentration of Contaminants

The concentration of contaminants in the wastewater is another important factor that directly influences the efficiency of removal. Generally, the higher the concentration of contaminants, the more adsorbent or treatment material is required to remove them. For adsorption processes, higher contaminant concentrations often lead to higher adsorption rates up to a certain limit, beyond which the adsorbent may become saturated, and the removal efficiency decreases. In ion exchange processes, the percentage of ions in the water influences the rate and extent of ion exchange, as higher concentrations of contaminants may lead to competition with other ions present in the solution. However, very high concentrations of contaminants may also saturate the ion exchange sites, reducing the effectiveness of the process. In photocatalytic degradation, a higher concentration of contaminants may reduce the efficiency due to the limited availability of reactive species to break down contaminants. Thus, optimal contaminant concentrations are crucial for maximizing treatment efficiency, and pre-treatment methods like dilution can sometimes be necessary.

13.4.4 Contact Time

Contact time, or the duration during which the contaminant is in contact with the adsorbent or treatment material, is a vital factor that influences the removal efficiency. For adsorption processes, longer contact times generally lead to higher contaminant removal rates because more time allows for increased interaction between the adsorbent and the contaminants. However, after a certain period, the adsorbent may become saturated, and the removal rate may level off or even decrease. In ion exchange processes, contact time is equally important; longer contact time allows for more complete ion exchange, particularly for low-concentration contaminants. In photocatalytic degradation, the effectiveness of the process often increases with extended exposure to UV light, as the reactive species generated by the photocatalyst can continue breaking down contaminants over time. However, excessively long contact times can lead to unnecessary energy consumption and potential loss of efficiency, so it is essential to balance this factor according to the system's needs.

13.5 Case Studies of Nano-Hydrogels in Wastewater Treatment

Nano-hydrogels have attracted considerable attention in the wastewater treatment sector owing to their unique properties, including versatility, selective pollutant removal, superior purification capabilities, and enhanced water recovery efficiency. Their ability to tune physicochemical characteristics enables targeted capture and controlled release of contaminants, allowing them to perform effectively under a variety of environmental conditions. As a result, they are highly efficient in eliminating a broad spectrum of pollutants, including heavy metals, nitrates, phosphates, and other inorganic substances (Table 13.2).

13.5.1 Heavy Metal Removal

To preserve the ecosystem and public health, it is crucial to eliminate harmful heavy metal ions from wastewater. Nano-hydrogels are effective in removing toxic metals like Pb, Cd, Cu, Cr, Zn, Hg, and Cr from industrial wastewater due to their high adsorption capacity. An ideal hydrogel should be cost-effective, highly adsorptive, fast-acting, biodegradable, stable, non-toxic, pH-neutral, and capable of swelling-deswelling, etc. They contain functional groups like amine, amide, hydroxyl, carboxyl, thiol, etc., which facilitate ionic interchange and complex formation, thereby increasing their ability to bind and remove toxic metals. An ideal hydrogel for eliminating toxic metals from wastewater should be cost-effective, highly adsorptive, fast-acting, biodegradable, easily modifiable, and low in residual monomers. It must also be stable, non-toxic, pH-neutral, and capable of deswelling and rehydration (Visan and Negut 2025). In another investigation, Zhao et al. (2023) suggested that Hydrogels are recognized as efficient adsorbents, especially those derived from biopolymers, as they are environmentally friendly, biodegradable, non-toxic, safe to use, and are being increasingly applied for the adsorption of toxic metal ions.

A recyclable, insoluble PGA-based hydrogel was created by Chen et al. (2024) to solve the problem of removing PGA, which accumulates heavy metals, from water. The study assessed the hydrogel's effectiveness in removing heavy metals by incorporating a third-generation polyamidoamine (PAMAM) dendrimer, rich in terminal amino groups, as a cross-linking agent. Copper ions (Cu^{2+}) were used as the representative heavy metal contaminant. The hydrogel underwent characterization, and the influence of different adsorption factors was analyzed. Furthermore, adsorption isotherms, kinetic and thermodynamic models, along with the potential mechanism of Cu^{2+} adsorption, were examined. In order to remove inorganic pollutants from water, a recent study created hydrogels that are incredibly swelling and sensitive to temperature and pH. Acrylamide (AM) and 3-sulfopropyl acrylate (SPA)

Table 13.2 Removal of inorganic contaminant using nano-hydrogel and its composites (Adapted from Alsaka et al. 2025. Open access)

S. No.	Nano-hydrogel	Removal mechanism	Target pollutants	Advantages	Limitations	Adsorption efficiency/ capacity	Refs.
1	Carbon-based nanofiber membranes	Adsorption via electrostatic forces, hydrogen bonding, and surface complexation	Pb^{2+}, Cr^{3+}, methylene blue	High surface area, rapid adsorption kinetics, efficient even at high flux, reusable and regenerable	Efficiency may drop after repeated use cycles, complex fabrication	Methylene blue: up to 818.6 mg/g; Pb^{2+}: 423.7 mg/g; Cr(VI): 221.3 mg/g	Liang et al. (2011)
2	Kappa carrageenan–graft–poly(N-hydroxyethyl acrylamide)/magnetite (κ-g-PHEAA/ Fe_3O_4) hydrogel	Adsorption and magnetic separation	Cationic dyes (MB, R6G), Cu^{2+}, Hg^{2+}	High removal efficiency, easy recovery due to magnetic properties, reusability	Complex synthesis route, potential long-term instability	MB: 196.1 mg/g; R6G: 307.8 mg/g; Cu^{2+}: 243.4 mg/g; Hg^{2+}: 229.9 mg/g	Kunal and Badalamoole (2020)
3	Poly (NIPAAm/DAPB/AA) amphoteric nanohydrogel	Adsorption	Pb^{2+}, Hg^{2+}	Amphoteric nature allows flexibility for multiple ions	Not explicitly mentioned	Pb^{2+}: 84%; Hg^{2+}: 64%	Mahida and Patel (2016)
4	Amino-functionalized graphene sponge (AFGS) with graphene nanosheets	Adsorption	Cr(VI)	Excellent mechanical strength, durable, good cycling stability	Not specified	166.46 mg/g	Sun et al. (2021)

(continued)

Table 13.2 (continued)

S. No.	Nano-hydrogel	Removal mechanism	Target pollutants	Advantages	Limitations	Adsorption efficiency/ capacity	Refs.
5	Poly(N-isopropylacrylamide) (PNIPAM)/Chitosan nanohydrogel	Froth flotation and electrostatic adsorption	Ni^{2+}, Cr(VI)	Eco-friendly and biodegradable, functional below 305 K (LCST)	Not detailed	Ni^{2+}: 86.1%; Cr(VI): 91.5%	Zheng et al. (2021)
6	Chitosan- and lignocellulose-based nano-fiber hydrogels (CFN, LCN)	Adsorption	NO_2^-	Biodegradable, compatible with biological systems	Not stated	CFN: 93.1%; LCN: 68.96%	Darabitabar et al., 2020
7	Aminated-nanocellulose/MOF hydrogel composite	Adsorption	Cr(VI)	Green and biodegradable with highly efficient metal ion capture	Not listed	98.6%	Sun et al. (2021)

were grafted onto hydroxypropylmethyl cellulose (HPMC) by radical polymerization, and a divinyl comonomer was used for crosslinking. Although it had a reduced mechanical strength, the hydrogel with the highest SPA content (AM/SPA = 0.5) showed a 12,100% swelling ratio, quick kinetics, and strong responsiveness. Higher AM/SPA ratio hydrogels were less sensitive but had superior mechanical qualities. Cr (VI) adsorption tests showed 90–96% removal efficiency. Hydrogels with AM/SPA ratios of 0.5 and 1 were suitable for repeated use with pH regeneration. These hydrogels offer an effective approach to water purification ions on the PGA hydrogel were analyzed. The study also examined the hydrogel's reusability (Tohamy et al. 2023).

In a separate case study, polyacrylamide-based nanohydrogels functionalized with thiol groups and embedded with Fe_3O_4-coated magnetic nanoparticles (PHPAm/Fe_3O_4@SiO_2-SH) showed outstanding efficiency in removing mercury (Hg) and lead (Pb) ions from wastewater. The thiol groups on the hydrogel surface formed stable complexes with the heavy metal ions, enabling easy separation from the water. Furthermore, the cross-linked structure of the hydrogels ensures that they maintain their structural integrity during the adsorption process, improving their stability and reusability (Ebrahimpour and Kazemi 2023).

13.5.2 Nitrate and Phosphate Removal

Excessive concentrations of nitrates and phosphates in wastewater, often from agricultural runoff and industrial discharges, result in eutrophication and the degradation of water quality. Nano-hydrogels offer a promising solution for removing these nutrients from water due to their ability to be functionalized with specific groups capable of adsorbing or removing nitrate and phosphate ions. For instance, a case study involving polyvinyl alcohol (PVA)-based nano-hydrogels functionalized with anion-exchange groups showed effective elimination of nitrate ions from synthetic wastewater. The ion-exchange process allows the hydrogels to replace the nitrate ions with other less toxic ions, decreasing the overall nitrate concentration. Similarly, phosphate removal can be achieved by incorporating phosphate-binding materials such as iron or aluminum salts into the nano-hydrogel matrix. The material's high swelling capacity and large surface area further enhance its ability to adsorb these ions, making it a viable option for reducing nutrient pollution in wastewater.

Orbuleţ et al. (2023) successfully synthesized a polymeric substance (A400-nZVI) for nitrate elimination from simulated groundwater. Characterization techniques (FTIR, SEM, EDAX, and XRD) confirmed the successful incorporation of nano zerovalent iron (nZVI) into the A400 polymer matrix, with nZVI particles measuring 75–100 nm. Kinetic studies revealed that nitrate reduction followed a pseudo-second-order model, with optimal removal (>80%) under acidic conditions due to enhanced H_2 production. Thermodynamic analysis indicated that nitrate adsorption fit the Freundlich isotherm model. The material's high efficiency suggests its potential for removing other pollutants, such as dyes and heavy metal ions, from groundwater and wastewater. In another study, Asadzadeh and Pirsa (2020) developed a biodegradable biohydrogel (ISP/TG/MPS/Lyc) using isolated soy protein, tragacanth gum, mesoporous silica nanoparticles, and lycopene pigment. Characterization techniques (SEM, FTIR, XRD, and TGA) confirmed its structure and properties. The bio-hydrogel demonstrated enhanced mechanical strength, antioxidant activity, water absorption capacity (WAC), and thermal stability. When applied to Lake Urmia sediments, it significantly reduced chemical oxygen demand (COD), biochemical oxygen demand (BOD), and nitrite levels by up to 90%, while lowering nitrate content by 30%. The study highlights the biohydrogel's potential for selective nitrite removal and sediment purification. Similarly, Nishida et al. (2021) studied nitrate-nitrogen (NO_3^--N) removal from groundwater in Kumamoto City, Japan, where nitrate levels are rising due to agricultural and domestic waste. They used a heterotrophic nitrification–aerobic denitrification (HN-AD) system with *Rhodotorula graminis* NBRC0190, a red yeast immobilized on calcium alginate hydrogel beads. The system achieved over 98% NO_3^--N removal for concentrations less than 10 mg/L (Nishida et al. 2021) in model underground water and maintained its efficiency over multiple treatments. When applied to actual groundwater, the method further improved water quality, demonstrating its effectiveness for nitrate removal.

13.5.3 Fluoride Removal

Fluoride pollution in water poses a significant issue, especially in areas where natural water sources contain elevated fluoride concentrations. Excessive fluoride exposure can cause serious health-related problems, including dental and skeletal fluorosis. Nano-hydrogels have been investigated for their potential to eliminate fluoride from contaminated water. A study focusing on the use of chitosan-based nano-hydrogels functionalized with aluminum oxide nanoparticles showed promising results for fluoride removal. The aluminum oxide nanoparticles act as efficient adsorbents, capturing fluoride ions through electrostatic interactions and ionic bonds. These nano-hydrogels demonstrated not only higher adsorption potential but also high selectivity for fluoride, making them highly effective in water treatment. Moreover, the hydrogels' biocompatibility and ease of regeneration make them an attractive option for continuous use in water purification systems.

Kavisri et al. (2023) investigated the potential of marine microalgae *Dunaliella salina* for fluoride removal from aqueous solutions using phytoremediation. The study involved collecting, identifying, and mass-culturing the microalgae, followed by characterization through FT-IR, NMR, XRD, and SEM analysis before and after fluoride adsorption. Fluoride removal was optimized using response surface methodology (RSM) and kinetic modeling, revealing the influence of temperature, pH, dosage, and time on efficiency. The process followed isotherms, like Langmuir, Freundlich, and Temkin, and also pseudo-first-order, pseudo-second-order, and intra-particle diffusion models. The results demonstrated that *D. salina* is a low-cost, sustainable, and efficient biosorbent for fluoride eradication in water and industrial wastewater. In another article, Arab et al. (2024) highlight the global issue of groundwater contamination with fluoride, emphasizing its serious health risks at high concentrations. The study explores multiple methods for fluoride removal, including electrocoagulation, adsorption, photocatalysis, membrane filtration, and ion exchange, with a focus on nanotechnology-based solutions. Nanoparticles are highly effective owing to their fast reactivity and strong oxidative properties. The study explores optimized conditions for fluoride removal and compares nanotechnology to traditional methods. Ultimately, advanced nanotechnologies offer an effective approach to improving water quality and protecting public health.

13.6 Conclusion and Future Prospects

Hydrogel-based materials have emerged as remarkable and versatile components for nano-composite formation, particularly useful in wastewater treatment. They have strategically secured a leading position in today's composite market. This success is largely due to their unique properties, including three-dimensional, cross-linked polymeric colloidal networks with high water compatibility and flexibility, akin to natural tissue, owing to their high moisture content (>90%), permeability, and soft texture. As a result, hydrogels have become a preferred raw material for creating lightweight yet strong composites, which enable them to swell in water and organic solvents for the elimination of heavy metal ions.

This chapter explored the removal of inorganic contaminants from wastewater using nanohydrogels, emphasizing their properties and synthesis techniques. Due to excellent water absorption capacity and high crosslinking framework, these nanohydrogels show enhanced mechanical strength and flexibility, making them a promising candidate for environmental remediation. It also covered various technologies used for eliminating inorganic pollutants from wastewater, including traditional, time-tested approaches and latest approaches such as coagulation, flocculation, adsorption, nano-filtration, etc. Additionally, the chapter examines the applications and case studies of nanohydrogels in wastewater treatment, which revealed that the use nanohydrogels as an adsorbent sparked significant attention in water treatment technologies.

The future of nano-hydrogels in wastewater treatment will likely focus on developing sustainable and cost-effective synthesis methods. Research is expected to prioritize the use of low-cost, abundant raw materials and environmentally friendly, scalable production techniques. Advances in green chemistry and the utilization of biopolymers could lead to more affordable and sustainable nano-hydrogels, making them viable for large-scale applications. In addition, there will be a push towards enhancing the functionalization of nano-hydrogels to improve their selectivity and potential in removing specific inorganic pollutants like toxic metals, nitrates, phosphates, and fluoride. As understanding of nano-hydrogel chemistry advances, tailored surface chemistries could be designed to target these pollutants with greater specificity. Another promising area of research is the development of advanced regeneration techniques, such as using electrochemical, thermal, or magnetic fields, to restore the hydrogel's adsorption capacity in a cost-effective manner. Furthermore, the integration of nano-hydrogels with other nanomaterials, such as carbon nanotubes, magnetic nanoparticles, or metal oxides, could enhance their adsorption capacity, selectivity, and stability, addressing current limitations. The emergence of smart and stimuli-responsive hydrogels, which adapt their properties in response to changes in pH, temperature, or contaminant presence, is also an exciting development that could improve real-time control and efficiency in wastewater treatment. Additionally, integrating nano-hydrogels with other treatment technologies like photocatalysis, membrane filtration, or electrochemical processes could create hybrid systems that offer superior performance in removing a broader range of contaminants. Lastly, to facilitate the widespread adoption of nano-hydrogels, further ecological and toxicological studies are needed to assess their life cycle, degradation, and potential hazards, ensuring their safety and suitability for real-world applications.

References

Alsaka L, Alsaka L, Altaee A, Zaidi SJ, Zhou J, Kazwini T (2025) A review of hydrogel application in wastewater purification. Separations 12(2):51

Arab N, Derakhshani R, Sayadi MH (2024) Approaches for the efficient removal of fluoride from groundwater: a comprehensive review. Toxics 12(5):306

Asadzadeh F, Pirsa S (2020) Specific removal of nitrite from Lake Urmia sediments by biohydrogel based on isolated soy protein/tragacanth/mesoporous silica nanoparticles/lycopene. Glob Challenges 4(12):2000061

Baer A (1996) Not enough water to go round? Int Soc Sci J 48(148):277–292

Bashir A, Malik LA, Ahad S, Manzoor T, Bhat MA, Dar GN, Pandith AH (2019) Removal of heavy metal ions from aqueous system by ion-exchange and biosorption methods. Environ Chem Lett 17:729–754

Bashir S, Hina M, Iqbal J, Rajpar AH, Mujtaba MA, Alghamdi NA, Ramesh S (2020) Fundamental concepts of hydrogels: synthesis, properties, and their applications. Polymers 12(11):2702

Chai WS, Cheun JY, Kumar PS, Mubashir M, Majeed Z, Banat F, Show PL (2021) A review on conventional and novel materials towards heavy metal adsorption in wastewater treatment application. J Clean Prod 296:126589

Chen F, Zhao Y, Zhao H, Zhou X, Liu X (2024) Heavy metal removal from wastewater using poly(gamma-glutamic acid)-based hydrogel. Gels 10(4):259

Darabitabar F, Yavari V, Hedayati A, Zakeri M, Yousefi H (2020) Comparison of the performance of plant nano-hydrogels as bio-filters for nitrite uptake from effluent of fish farms. Iranian Sci Fish J 29(2):43–52

Delgado-Pujol EJ, Martínez G, Casado-Jurado D, Vázquez J, León-Barberena J, Rodríguez-Lucena D, Begines B (2025) Hydrogels and nanogels: pioneering the future of advanced drug delivery systems. Pharmaceutics 17(2):215

Ebrahimpour E, Kazemi A (2023) Mercury (II) and lead (II) ions removal using a novel thiol-rich hydrogel adsorbent; PHPAm/Fe3O4@ SiO2-SH polymer nanocomposite. Environ Sci Pollut Res 30(5):13605–13623

Falkenmark M (1998) Dilemma when entering the 21st century-rapid change but lack of a sense of urgency. Water Policy 1:421–436

Gao X, He C, Xiao C, Zhuang X, Chen X (2013) Biodegradable pH-responsive polyacrylic acid derivative hydrogels with tunable swelling behavior for oral delivery of insulin. Polymer 54(7):1786–1793

Ghosh S, Othmani A, Malloum A, Christ OK, Onyeaka H, AlKafaas SS, Koduru JR (2022) Removal of mercury from industrial effluents by adsorption and advanced oxidation processes: a comprehensive review. J Mol Liq 367:120491

Giri TK, Badwaik H (2022) Understanding the application of gum ghatti based biodegradable hydrogel for wastewater treatment. Environ Nanotechnol Monitor Manag 17:100668

Godiya CB, Cheng X, Li D, Chen Z, Lu X (2019) Carboxymethyl cellulose/polyacrylamide composite hydrogel for cascaded treatment/reuse of heavy metal ions in wastewater. J Hazard Mater 364:28–38

Ho TC, Chang CC, Chan HP, Chung TW, Shu CW, Chuang KP, Tyan YC (2022) Hydrogels: properties and applications in biomedicine. Molecules 27(9):2902

Hossain MS, Hossain MM, Khatun MK, Hossain KR (2023) Hydrogel-based superadsorbents for efficient removal of heavy metals in industrial wastewater treatment and environmental conservation. Environ Funct Mater 2(2):142–158

Hussain S, Abid MA, Munawar KS, Saddiqa A, Iqbal M, Suleman M, Amjad M (2021) Choice of suitable economic adsorbents for the reduction of heavy metal pollution load. Polish J Environ Stud 30(3)

Kavisri M, Abraham M, Moovendhan M (2023) Effective removal of fluoride ions from aqueous solution by marine microalgae as natural biosorbent. Chemosphere 313:137312

Khader EH, Muslim SA, Saady NMC, Ali NS, Salih IK, Mohammed TJ, Zendehboudi S (2024) Recent advances in photocatalytic advanced oxidation processes for organic compound degradation: a review. Desalin Water Treat 100384

Khan Y, Bashir S, Hina M, Ramesh S, Ramesh K, Lahiri I (2020) Effect of salt concentration on poly (acrylic acid) hydrogel electrolytes and their applications in supercapacitor. J Electrochem Soc 167(10):100524

Kulal P, Badalamoole V (2020) Hybrid nanocomposite of kappa-carrageenan and magnetite as adsorbent material for water purification. Int J Biol Macromol 165:542–553

Liang HW, Cao X, Zhang WJ, Lin HT, Zhou F, Chen LF, Yu SH (2011) Robust and highly efficient free-standing carbonaceous nanofiber membranes for water purification. Adv Func Mater 21(20):3851–3858

Luo T, Abdu S, Wessling M (2018) Selectivity of ion exchange membranes: a review. J Membr Sci 555:429–454

Mahida VP, Patel MP (2016) Removal of heavy metal ions from aqueous solution by superabsorbent poly (NIPAAm/DAPB/AA) amphoteric nanohydrogel. Desalin Water Treat 57(29):13733–13746

Moore JW (2012) Inorganic contaminants of surface water: research and monitoring priorities. Springer Science and Business Media

Nishida M, Matsuo S, Yamanari K, Iwahara M, Kusakabe K (2021) Removal of nitrate nitrogen by *Rhodotorula graminis* immobilized in alginate gel for groundwater treatment. Processes 9(9):1657

Orbuleţ OD, Dăncilă AM, Căprărescu S, Modrogan C, Purcar V (2023) Nitrates removal from simulated groundwater using nano zerovalent iron supported by polystyrenic gel. Polymers 15(1):61

Pakdel PM, Peighambardoust SJ (2018) A review on acrylic based hydrogels and their applications in wastewater treatment. J Environ Manage 217:123–143

Saha A, Bardhan S, Roy S, Dutta S, Das S (2023) Biopolymeric hydrogels: a new era in combating heavy metal pollution in industrial wastewater. In: Membranes for water treatment and remediation. pp 209–226

Sahu D, Pervez S, Karbhal I, Tamrakar A, Mishra A, Verma SR, Satnami ML (2024) Applications of different adsorbent materials for the removal of organic and inorganic contaminants from water and wastewater–a review. Desal Water Treat 100253

Sethy TR, Pradhan AK, Sahoo PK (2019) Simultaneous studies on kinetics, bio-adsorption behaviour of chitosan grafted thin film nanohydrogel for removal of hazardous metal ion from water. Environ Nanotechnol Monit Manag 12:100262

Sharma G, Kumar A, Naushad M, García-Peñas A, Al-Muhtaseb AAH, Ghfar AA, Stadler FJ (2018) Fabrication and characterization of Gum arabic-cl-poly (acrylamide) nanohydrogel for effective adsorption of crystal violet dye. Carbohyd Polym 202:444–453

Singh V, Singh N, Rai SN, Kumar A, Singh AK, Singh MP, Mishra V (2023) Heavy metal contamination in the aquatic ecosystem: toxicity and its remediation using eco-friendly approaches. Toxics 11(2):147

Sinha V, Chakma S (2019) Advances in the preparation of hydrogel for wastewater treatment: a concise review. J Environ Chem Eng 7(5):103295

Sun Y, Yu F, Li L, Ma J (2021) Adsorption-reduction synergistic effect for rapid removal of Cr (VI) ions on superelastic NH_2-graphene sponge. Chem Eng J 421:129933

Thines RK, Mubarak NM, Nizamuddin S, Sahu JN, Abdullah EC, Ganesan P (2017) Application potential of carbon nanomaterials in water and wastewater treatment: a review. J Taiwan Inst Chem Eng 72:116–133

Tohamy HAS, El-Sakhawy M, Strachota B, Strachota A, Pavlova E, Mares Barbosa S, Kamel S (2023) Temperature-and pH-responsive super-absorbent hydrogel based on grafted cellulose and capable of heavy metal removal from aqueous solutions. Gels 9(4):296

Uddin MK (2017) A review on the adsorption of heavy metals by clay minerals, with special focus on the past decade. Chem Eng J 308(438):462

Visan AI, Negut I (2025) Environmental and wastewater treatment applications of stimulus-responsive hydrogels. Gels 11(1):72

Yang Z, Han B, Fu D, Liu W (2012) Acute toxicity of high dosage carboxymethyl chitosan and its effect on the blood parameters in rats. J Mater Sci Mater Med 23:457–462

Zhao C, Liu G, Tan Q, Gao M, Chen G, Huang X, Xu D (2023) Polysaccharide-based biopolymer hydrogels for heavy metal detection and adsorption. J Adv Res 44:53–70

Zheng H, Zhang S, Yang C, Yin H, Liu W, Lu K (2021) Simultaneous removal of Ni (II) and Cr (VI) from aqueous solution by froth flotation using PNIPAM-CS intelligent nano-hydrogels as collector. J Mol Liq 342:117551

Chapter 14
Policy Considerations and Regulatory Framework for Hydrogel Application in Wastewater Treatment to Advance Eco-Friendly and Regenerative Practices

Sumanta Bhattacharya and Bhavneet Kaur Sachdev

Abstract The application of hydrogels in wastewater treatment has significant promise for addressing environmental concerns and promoting the principles of sustainability and the circular economy. This paper analyses the legal framework and policy consequences of using hydrogel technology, focusing on its advantages, difficulties, and the required support systems for advancing environmental sustainability and circular economy concepts. Regulatory frameworks are crucial in promoting the acceptance of hydrogel technology in wastewater treatment. The safety and effectiveness of hydrogels are ensured by adhering to standards and guidelines for their manufacturing, usage, and disposal. The policy implications for increasing the use of hydrogels in wastewater treatment involve providing incentives for research and development, such as funding possibilities and tax incentives, to encourage innovation and technological progress. By implementing policies that encourage the use of hydrogel-based wastewater treatment systems, companies can be motivated to adopt these systems, leading to the closure of the resource loop and contributing to a more sustainable and circular economy. The establishment of regulations and policies regarding the use of hydrogels in wastewater treatment is crucial for advancing environmental sustainability and supporting the concepts of a circular economy.

Keywords Water resource management · Circular economy · Hydrogel · Sustainable development · Innovative policy development

S. Bhattacharya (✉)
Asian International University, Manipur, India
e-mail: sumanta.21394@gmail.com

B. K. Sachdev
Suresh Gyan Vihar University, Jaipur, India

S. Bardhan and M. P. Shah (eds.), *Applications of Hydrogels in Modern Wastewater Treatment*, Advances in Wastewater Research,
https://doi.org/10.1007/978-981-95-1598-1_14

List of Abbreviations

WWTP	Waste Water Treatment Plant
UNEP	United Nations Environment Program
WHO	World Health Organization
UWWTD	Urban Waste Water Treatment Directive
WFD	Water Framework Directive
EU	European Union
WWT	Waste Water Treatment
CWA	Clean Water Act
NPDES	National Pollutant Discharge Elimination System
BAT	Best Available Techniques
EPA	Environmental Protection Agency
NGT	National Green Tribunal
CPCB	Central Pollution Control Board
LCA	Life Cycle Assessment
R & D	Research and Development
SME	Small and Medium Enterprises
PPP	Public–Private Partnership
NGO	Non-Government Organization
MBR	Membrane bioreactor
ISO	International Organization for Standardization

14.1 Introduction

Hydrogels are a family of smart materials that belong to the network polymers, which consist of three-dimensional cross-linked macromolecular structures having a capacity to absorb a large amount of water or biological fluids. This feature of structure allows for hydrogels to serve as super adsorbents for most pollutants in cases of wastewater treatment. The ability of hydrogels to adsorb the pollutants is due to their porous nature, high specific surface area and the polar groups which provide an active site to interact with the pollutants containing heavy metals, dye and organic components. In wastewater treatment, hydrogels work using one of the methods, including the absorption of pollutants by physique adsorption, ion exchange or chemical attachment. Due to which, they can tackle a huge variety of pollutants, including recalcitrant ones, which are difficult to treat with other means. For instance, the interaction of hydrogels can efficiently remove heavy metal ions such as lead, cadmium and mercury ions from industrial wastewater, which is highly dangerous and hazardous to the environment. In addition, the hydrogels can also be modified for the improvement of the selectivity as well as the adsorption abilities for certain pollutants. This is because through altering the polymer type, adding or changing functional groups or integrating nanoparticles, scientists are able to design specific types of hydrogels to meet the specific pollution threats. This versatility for use in combination with their high performance makes the hydrogels suitable for applications in wastewater treatment across the board, from industrial discharge to agricultural leachate. Besides cleaning up the contaminated water, hydrogels are also appreciated for their ability to cooperate with other processes, including filtration and biological degradation and so on. As science expands further, hydrogels will remain at the center of novel wastewater treatment technologies and methods (Chang et al. 2023).

It is necessary to note that hydrogels are rather astounding materials with applicable characteristics, which makes them very useful in different scientific and medical fields. These 3D polymer networks consist mostly of water, while they are capable of swelling and holding a lot of water without compromising their mechanical characteristics. They have many advantages and hence their applicability is vast, especially when it comes to biodegradability, high water retention capacity, and regenerative medicine applications; they are of significant interest to the advanced research and development (Bahram et al. 2016).

An important feature of hydrogels is their biodegradability, which has been considered as one of their strongest sides. Biodegradable hydrogels, once they are done with their jobs, are expected to either degrade in the body or in the environment, producing non-hazardous byproducts. Biodegradable hydrogels have benefits wherever there is no need for substances to linger in the body or natural environment. This property makes the application of this property very crucial, especially in areas such as controlled drug release, wound healing and tissue engineering. For instance, in drug delivery, hydrogels can be bioactive, that is, capable of being designed to release drugs in specific quantities and then disintegrate safely in the body. This is less invasive as opposed to surgical removal, hence causing less discomfort to the patient as well as having fewer complications. Also, they are biodegradable, which makes their use eco-friendly, as seen in use in agriculture to reduce waste by using such hydrogels to enhance moisture retention in the soil (Sannino et al. 2009).

Hydrogels also have high water retention as one of their major properties. As seen above, these materials can absorb water content in terms of their dry weight and are compared to natural tissues based on their mechanical and physical properties. Due to its water retention properties, hydrogels are well suited for use in applications where there is a need to moderate moisture levels. Hydrogels are common in the medical profession, especially as dressings for wounds, because they help to keep the wound moist and, at the same time, can also take up the fluid that the wound excretes. This has the added advantage of compounding the healing process while at the same time decreasing the chances of contracting an infection. In agriculture, hydrogels are used to enhance water availability in soil, and this is of more advantage in regions where water is valuable. Stimulating the growth of plants and decreasing the frequency of watering is possible by using hydrogels, which swell during irrigation or rain and slowly release water (Abdallah 2019).

It is said that one of the most useful fields associated with hydrogels is the medicare industry. Hydrogels are suitable for tissue engineering applications in that they are biocompatible, possess controllable mechanical characteristics and have similarities to the extracellular matrix that surrounds cells in real tissues. These cell components supply all the necessary nutrients, which are helpful for cell division and specialization, necessary for the healing of tissues. Therefore, it is possible that hydrogels of tissue-specific composition can be designed to encompass growth factors, cells or bioactive molecules that may enhance tissue formation. For instance, hydrogels have been applied in the construct of artificial cartilage, in which they promote the attachment and growth of chondrocytes and provide mechanical properties that are similar to natural cartilage samples. In addition, injectable hydrogels have been considered a means of delivering stem cells or therapeutic factors to definite target areas within the body with potential applications in diseases such as myocardial infarction, spinal cord injuries and osteoarthritis (Mantha et al. 2019).

Therefore, hydrogels are all-purpose materials given the facts that they are biocompatible have high water absorbing capacity, and therapeutic relevance in regenerative processes. Due to these properties, they became popular and are widely used in medicine, agriculture, as well as in environmental studies, as they provide future-oriented solutions which are not damaging the environment. As future research emerges, the creation of sophisticated hydrogel systems with advanced applications will be unveiled, specifically in the field of individualized medicine and biotechnology. (Revete et al. 2022).

14.2 Current Applications of Hydrogels in Water Resource Management

Today, hydrogels are regarded as potential candidates in water treatment because of their properties like high water retention capacity, tunable pore structure, and remarkable chemical compatibility that allows for the loading of some functional groups that are capable of reacting with pollutants. These properties make it possible for hydrogels to be used in the removal of almost all common contaminants in wastewater, such as heavy metals, dyes, organic pollutants and pathogenic pollutants.

It is established that one of the applications of hydrogels in wastewater treatment is through the process of adsorption of heavy metals. Hydrogels can be synthesized to incorporate functional groups such as carboxyl, hydroxyl or amino groups that have consideration to ion like lead, cadmium, mercury and chromium. They entrap these metals and immobilize them, thus they can easily be separated from water bodies, reduce direct interaction of toxic species with…" For example, chitosan and alginate hydrogel beads are commonly used for this purpose by taking advantage of biopolymers' ability to selectively adhere to metal ions and their biodegradability (Hossain et al. 2023).

Another major use is the decolourization of industrial wastes, especially the effluents from textile industries. Dyes are non-biodegradable substances and are toxic to aquatic life because they decompose very slowly. It was established that hydrogels derived from polymers such as polyacrylamide or polyvinyl alcohol are capable

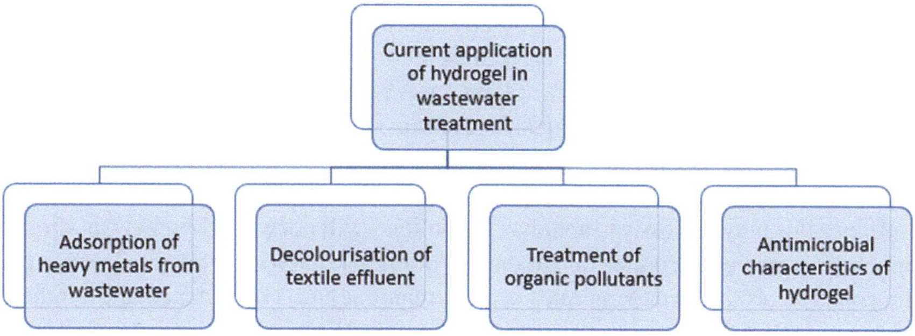

Fig. 14.1 Current application of hydrogel in wastewater treatment

of adsorbing different dyes from wastewater. Hydrogels have a large surface area and are porous in nature and can be functionalized with certain chemical functional groups that make it easy to capture dye molecules and thereafter remove them from the wastewater stream (Flores-Valenzuela et al. 2023).

Hydrogels in the treatment of organic pollutants, such as pesticides and pharmaceuticals, are also employed. These pollutants are of increasing concern as they are known to persist in the environment while raising concern over the ability of human and Ecological systems to cope with their impacts. Other organic contaminants can be incorporated or designed in hydrogels that have specific enzymes or catalysts which break down these contaminants into less hazardous compounds. For instance, hydrogels incorporating titanium dioxide nanoparticles can utilize photocatalytic processes for the elimination of organic matter in water, provided adequate illumination; therefore, the use of hydrogels can be employed as a solution to water pollution. Furthermore, hydrogels are employed in the process of pathogen elimination in water treatment plants from wastewater. Hydrogels have a highly permeable structure of their network, so they are able to capture bacteria, viruses, and other microbes and stop their spread through waterworks. Some hydrogels are also surface modified with antimicrobial molecules or loaded with metal such as silver, which has the capacity to destroy pathogenic microorganisms on contact, making them suitable in water treatment to meet health requirements (Ahmaruzzaman et al. 2023) (Fig. 14.1).

14.3 Challenges of the Application of Hydrogels in Wastewater Treatment

Scalability is one of the biggest issues in organizations as they require a large amount of resources to function. Though hydrogels work at the laboratory scale, it is not easy to employ them at the industrial level required for wastewater treatment. Hydrogel production can be very challenging, particularly when one is tasked with the responsibility of preparing large volumes of the material while at the same time ensuring that the final product is of the right quality and has excellent performance. Also, the structures that would need to be erected to incorporate the hydrogel-based systems into existing wastewater treatment plants may be capital intensive, which makes it unprofitable for industries seeking an economic system to implement (Persano et al. 2024).

The spending tends to play a major role in the formulation of favourable policies, and since this may be costly to achieve, the policies will not work. Synthetic hydrogels may be expensive to produce due to the use of higher quality raw materials as well as the complex production techniques used in manufacture. For instance, the synthesis of functionalized hydrogels that are able to selectively adsorb specific pollutants will need expensive monomers, cross-linkers or catalysts. However, the regeneration and reuse of hydrogels, which is crucial in maintaining cost effectiveness, may turn into further expenses in the operational costs due to the use of severe chemicals or high-energy-consuming procedure in the regeneration process. This makes hydrogel-based treatments possibly less competitive to other currently used methods, such as activated carbon adsorption or chemical coagulation (Ahmed 2015).

One more disadvantage that should be mentioned is that it is still questionable how synthetic hydrogels affect the environment. Several synthetic hydrogels result from petroleum polymeric materials and are non-biodegradable if not disposed of correctly may have long-term effects on the environment. Hydrogels disposal, especially in their spent form and when laden with adsorbed contaminants, presents a major environmental concern. It is also important to consider that breakdown products of synthetic hydrogels released into the ecosystem may include microplastics or other carcinogenic products. Hydrogels' environmental implications are, thus, presented as more complex than the often cited advantages of this technology concerning its sustainability. There are also some regulatory issues making it hard for hydrogels to be adopted in wastewater treatment. It may easily be noted that there is a paucity of universal standards and protocols restricting the practice of using hydrogels, particularly synthetic ones, for environmental purposes. This lack of clarity with regulatory bodies creates confusion and raises questions of misapplication and use that adversely affect the efficacy and safety of hydrogel-based treatments. Thus, in the absence of comprehensive guidelines and requirements for screening and approval of technologies, many companies may face difficulties the understanding the multiple stages that are to be completed to introduce new technologies to WWTPs. This could create scope for uncertainty and risk, which in turn will discourage investors and technicians from exploring more on the development of enhanced hydrogel technologies to fit the large-scale usage (Nikolić et al. 2018).

Consequently, although functionalized hydrogels represent promising new developments in the retrieval of wastewater, several concerns regarding the practical difficulty, cost techniques, effects on the ecosystem, and legal issues ought to be clarified. Mitigating these challenges will be fundamental in enhancing the uptake and use of hydrogel technology for wastewater management and to support the greater contribution towards environmental management (Fig. 14.2).

Fig. 14.2 Challenges of the use of hydrogel

14.4 Policy and Regulatory Landscape for Effective Application of Hydrogel in Wastewater Treatment

The overall laws that guide the use of Wastewater treatment technologies are policies, guidelines and standards meant to ensure people and the environment are protected. The measures applied also differ considerably and are preceded by certain socio-economic conditions with effective resources, advanced technologies, or particular environmental problems of each territory, among others. Overall, the wastewater treatment standards aim at regulating the discharge of pollutants to water sources and provide maximum acceptable levels of pollutants in the treated effluent that will be safe to aquatic life and human beings. At the international level of regulation, there are general guiding frameworks endorsed, for example, by the United Nations and World Health Organization that act as preparatory requirements towards the country-level regulations. For instance, UNEP has provided guidelines that focus on integrated water management and sustainable wastewater treatment disciplines. The guidelines for using wastewater for Agricultural and aquacultural purposes are given by WHO, which includes precautionary measures that help reduce health risks in the case of using safely treated water. These international frameworks are normally adopted by individual countries but complied with in a way that suits individual countries best. Within the European Union, the Urban Waste Water Treatment Directive (UWWTD) is one among the principal measures that lays down objectives for the collection and treatment of urban wastewaters and industrial effluents. The directive has a provision that means that all member states have to provide proper capacity of wastewater treatment for all agglomerations with a population of more than twenty thousand people. It also fixes the standard concentrations of some of the major pollutants, like nitrogen and phosphorus, during treatment of the effluents for the prevention of eutrophication. Also, there is the Water Framework Directive (WFD), which forms the legal foundation of water management in the EU, and aims at the protection and recovery of water bodies through, among other activities, control of pollutants from the WWT discharges. The most important act regulating the discharge of pollutants into waters to protect the quality of the nation's waters, in the United States, is the Clean Water Act (CWA). Accordingly, the CWA lays the framework for the federal government's regulation of WWT by creating the NPDES in this section to prevent the discharge of pollutants into the U. S. waters without permits from certified firms. Standing permits specify legal concentrations of certain pollutants that can be released and require the use of BAT to reduce emissions. It is important to note that the EPA is charged with the responsibility of implementing these regulations with effluent guidelines as well as standards being reviewed occasionally, with a view to implementing technological improvements, as well as environmental surveys being made. For the latter, a number of factors can influence the regulation of wastewater treatment all the more in developing countries, such as material and capital constraints, insufficient presence of infrastructure, and various socio-economic considerations. But at the same time, they are trying to enhance their legislation due to the increase in the level of concern for the environment and the requirement of globalization. For instance, India has adopted the National Green Tribunal (NGT), which is responsible for the implementation of laws regarding environmental issues such as wastewater management. The CPCB in India also recommends guidelines for effluent quality as well as checks on the compliance of the recommended standards in industries. Likewise, the People's Republic of China has also reinforced its rules in wastewater treatment in the last years as part of the country's commitment to environmental conservation. The following are the measures that the country has put in place in its Water Pollution Prevention and Control Action Plan. The country targets the following water pollution prevention measures. The country has set tough goals with regard to the improvement of water pollution at all levels and enhanced wastewater treatment facilities. This plan requires that all the wastewater treatment plants have standard discharges and encourages the use of higher level technologies in the treatment of wastewater. In general, the regulation guidelines of the WWTs aim at the achievement of quality of the treated effluent to ensure the public health and environmental safety. These regulations are also geographical-specific, but in broad terms, they cover the application of BAT(s), management of significant pollutants, and safe water management. However, the applicability of these regulations depends on several factors, including compliance, technological advancement and the ability of countries to finance and develop the infrastructure for wastewater treatment (Catoira et al. 2020).

It can be stated that the current regulations allowing the use of technologies in wastewater treatment industries are somewhat comprehensive in most aspects; however, the same cannot be said with regards to hydrogels. Thus, hydrogels as materials with an exceptional porosity can serve as an example of the opportunities and threats that are not covered by modern legislation yet. One of the major problems is that there are no individual provisions or norms regulating the application of hydrogels in wastewater treatment. Many regulatory guidelines and recommendations address traditional treatments, including biological treatment, chemical coagulation, and physical removal, and very few, if any, make reference to hydrogels. This means that there can be significant ambiguities within the application of these formulated regulations, especially on the effects of the hydrogel-based treatments on the environment and the health of the individuals who engage in the treatment. The lack of specific guidelines could discourage industries from engaging in the use of hydrogels and at the same time, it may be difficult for the regulators to sanction the safe and effective use of hydrogels. The last set of gaps pertains to the categorization of hydrogels when it comes to the regulatory authorities and frameworks. The hydrogels are mostly classified with other polymers or materials without taking into account the unique properties of the hydrogels. This lack of differentiation can lead to regulations that are over- or under-prescribed when applied to hydrogels and their interactions in wastewater treatment processes. For instance, the rules that would be set for non-biodegradable synthetic polymers would not be fit for biodegradable hydrogels and this will create a situation whereby the hydrogels are either over-regulated or under-regulated. There is also a general lack of adequate current regulation regarding the effects of synthetic hydrogels on the environment. Although numerous synthetic hydrogels have been reported to have several advantages in wastewater treatment, the environmental concerns, including biodegradability and formation of microplastics, are not well managed in the current polices. Existing laws usually prescribe the efficiency of the treatment technologies without taking into account the impacts that such technologies will have on the environment in the future. This seems rather odd given the increasingly growing alarm about the presence of microplastics in water bodies and the possible negative impacts on marine life and humans. There are also no well-rounded lifecycle assessments or LCAs for the hydrogel-based technologies within the regulatory measurements that have been established. LCAs are necessary to estimate the environmental effect of a product or a technology during its life cycle, that is, from manufacturing up to the stage of its discharge. However, the current regulatory framework does not insist on elaborate LCAs in the application of hydrogels in wastewater treatment and, as a consequence, a number of aspects concerning the overall sustainability of the process remain unidentified. This could result in the adoption of hydrogel technologies without taking full compliance with the environmental and economic ramifications into account due to this gap in the regulatory requirements. However, other issues such as regulatory disparities on cost and achievement ability of hydrogel-based technology have not been left out. Some regulations that have been put in place encourage the use of BAT and yet do not offer guidance as to how to assess the feasibility and potentially wide-scale integration of new technologies such as hydrogels for cost-effective management of hospital-acquired infections. This may cause inequality in the assessment of hydrogel technologies, together with their application in the various parts of the world, or in various sectors of the economy. At other times, developments like hydrogel-based treatments may not go through despite the benefits that they offer when applied because they are costly or cannot easily be scaled up, given current regulatory structures, since the current regulations are not flexible enough to allow for the use of newer technologies. The absence of international legislation harmonization also makes the problem of gaps in regulation even more acute concerning standards and guidelines for hydrogels. There are countries or regions that have more developed or specialized regulation of hydrogel application, while in others, there are no regulations at all, which combines the uneven development, testing and application of these technologies. Such a lack of harmonization can lead to instances in which companies and researchers operating across different jurisdictions face lots of barriers because different jurisdictions have different rules, which would be different from the requirements in another country. This is because there is also a lack of clear channels for the certification and application of hydrogels in wastewater treatment, which in turn will slow down growth. Hydrogel technology startups have to face the problem of obtaining lengthy approvals and unpredictable outcomes, making the development of new technologies expensive and time-consuming. All in all, it can be said that the possibilities of hydrogels' application for enhancing the process of wastewater treatment are vast; however, the contemporary legislation does not offer sufficient tools for handling them. Some of the issues which are still uncovered in the existing regulations include environmental impacts, life cycle assessment, cost analysis, scalability and harmonization between countries for the use of hydrogels in managing wastewater to the required set standards. Overcoming these regulatory deficits will involve engagement of policymakers, researchers and industry players to formulate regulations and policies based on the peculiarities of hydrogel technologies (Mohapatra et al. 2021).

14.5 Environmental and Health Safety Standards for the Application of Hydrogel in Wastewater Treatment

The present use of hydrogels is widely attributed to their useful characteristics such as high water stability, biocompatibility, and controllable mechanical strength, which makes them relevant in wastewater treatment, agriculture, and in biomedical science. But the extensive application of hydrogels requires the strict regulation of the environment and health effects in the process of their production, application and recycling. It is for such reasons that the need to develop such standards can be underscored in the following areas. Second, issues related to the degradation of synthetic hydrogels and their subsequent hazard for ecosystems, especially those based on petroleum products. Biodegradable hydrogels degrade and can therefore pose temporary problems once they're released into the environment, while non-biodegradable hydrogels remain in the ecosystem for long and cause more severe effects, such as pollution of soil and water, effects that are lethal to plant and animal life. Furthermore, the breakdown of some synthetic hydrogels may involve the release of hazardous by-products, which would be injurious to the environment and human beings. Second, this book points out that the immunogenicity of hydrogels in biomedical applications, ranging from drug delivery systems, wound dressings, to tissue-engineering scaffolds, must be critically examined. Hydrogels should be toxicologically inactive, immunologically inactive and rid of any biologically active leachable substances that can potentially induce harm. Due to the fact that hydrogels come into contact with biological tissues and fluids, their biocompatibility must be strongly assessed to avoid inflammatory reactions or other undesirable health outcomes. Furthermore, in the use of hydrogels to improve the effectiveness of agricultural production through their incorporation in the soil or as water-absorbing agents, the use of the hydrogels has to be well controlled so as not to harm the productivity of the crops and the general safety of the food supply. The last major aspect is the life cycle of hydrogel materials which must also be incorporated into environmental protection. This ranges from establishing the environmental effects of hydrogels in the extraction and synthesis of their components and constituents to their application as well as disposal and recycling. Thus, when guidelines for the life cycle of hydrogels are clearly defined, it means that the impact of these products on the environment can be reduced and the creation of the biodegradable hydrogels is encouraged. Therefore, there is a need for adaptive regulatory mechanisms, especially because of the continuous creation of new hydrogel formulations and technologies. These frameworks should be capable of capturing the dynamic nature of hydrogel materials, safety standards should be able to grow with the new technologies. In addition, more information provided to the public and people in order to avoid ineffective usage and disposal of hydrogel products helps prevent the occurrence of events associated with improper handling of these materials. In conclusion, there is a need for the hydrogel materials to have increased measures of environmental and health safety as they are used in various industrial applications. These standards for hydrogels may help to minimize the risks of negative impacts on the environment and citizens' health, thus contributing to the protection of the latter and sustainable development of hydrogel technologies (Todman and Mauvais 2021).

Several safety measures and guidelines need to be complied with when dealing with hydrogel materials, and therefore they require to undergo rigorous testing to determine their safety implication on the environment and humans before they can be used in the different industries. The first recommendation is therefore the development of normative procedures for determining the biodegradability and the environmental footprint of the hydrogel materials. This should comprise the leaching rate test, toxicity test and the ecotoxicity of the material, and test on the emission of noxious substances. For example, concerns with long-term effects and the interaction of hydrogels with the natural environment, such as soil and liquid, could be investigated by exposing hydrogels to water and soil for some time. More so, the effect of temperature, pH, and microbial growth of biodegradable products must also be studied. Such tests will offer a better idea of the actual performance of hydrogels and make sure only safe materials are given certification for release into the market. The second important suggestion is the subjecting of hydrogels for biomedical purposes to strict tests on biocompatibility. There should be cytotoxicity tests, tests on hemocompatibility; and an evaluation of whether the sample may prove immunogenic. Moreover, the leachables and extractables study also needs to be carried out to determine if there are toxic and potentially deleterious substances that can be leached out from the hydrogel. Further, to evaluate the long-term safety concerns and utilities of hydrogels, in vivo evaluation should also be done using appropriate animal models. When assessing hydrogels for agricultural applications, emphasis should be placed on how they influence the soil and plant germination and the safety of foods produced on such soil. This involves evaluating the capacity of hydrogels to change any of the characteristics, such as pH, availability of nutrients as well and microbial population in the soil. Further, special experiments should be carried out to reveal if hydrogels cause any toxic substances to leach to the ground or water level that may be dangerous to crops and thus the food chain. To enhance the coverage of safety assessment, the lifecycle assessment (LCA) should be included in the certification procedure. LCA also entails the assessment of the effects of hydrogel materials on the environment during their life cycle, starting from the extraction of raw material to production and utilization, end utilization, whether by disposal or recycling. Including LCA as a part of certification, it is possible to analyze possible environmental impacts of hydrogels at all stages of their life cycle and minimize these impacts. The other recommendation that I would like to put forward is the creation of a certification body for hydrogel materials. This body should be in charge of preparing the safety standards and guidelines, carrying out tests independent of the manufacturer and certifying hydrogel products to premium safety standards. It was also recommended that the certification body cooperate with regulatory agencies, industry players and research institutes in order to develop credible and scientifically rigorous testing and certification procedures. Furthermore, the process of certification should be flexible, which means that it will accommodate modifications in the new technologies and formulations of hydrogels in the market. Last but not least; public awareness and information towards proper usage and/or safely removing hydrogel products is the final means towards conservation. This can be done by provisions of labelling requirements, preparation of standard safety data sheets, and necessary public awareness campaigns that would give necessary specific directions on the right handling and disposal of hydrogel materials. Should these recommendations be embraced, it will be possible to reduce any harm that these hydrogel materials may cause to the environment or the health of people as they are used in various industries safely (García-Villén et al. 2020) (Fig. 14.3).

14.6 Encouraging Innovation and Sustainable Practices in the Application of Hydrogel in Wastewater Treatment

Promoting and supporting research and development of green hydrogel chemistry could help pharmaceutical, agriculture, biomedical and wastewater industries advance their innovative sustainability tactics. The governments can be relevant to develop the hydrogel materials that are environmentally friendly through putting forward sound policy measures. Any incentives that will need to be offered should be

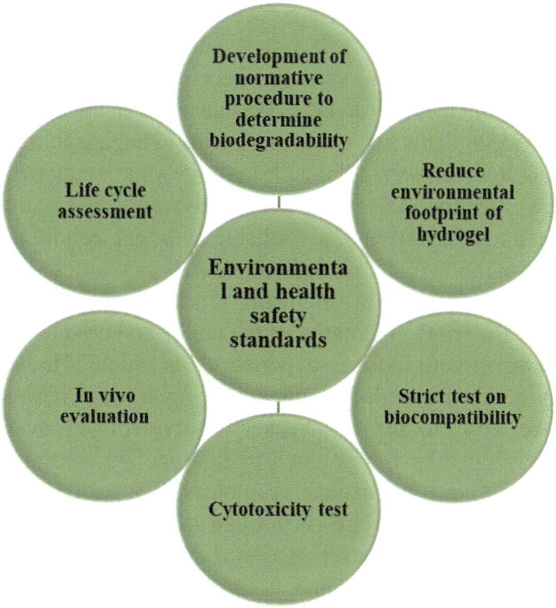

Fig. 14.3 Environmental and health safety standards

for the purpose of decreasing the negative effects that hydrogel production, as well as utilization, has on the environment, besides supporting biodegradable and non-toxic hydrogel products. The following is the policy incentives: Promoting the establishment of R&D subsidies and funding programs aimed at developing green hydrogel formulations. Public bodies can grant money for the research towards the development of hydrogels from renewable sources, for example, plant polymers or any other biodegradable material. These grants can be provided to university, research institutes and other business entities working in the field of developing sustainable hydrogel technologies. The benefit of such programs is that they can address and support early-stage research and development activities that are required for the discovery and commercialization of environmentally friendly hydrogel formulations. Apart from direct subsidies, governments may facilitate financial support through taxation measures that would facilitate the companies' commitment to research and development of environmentally friendly hydrogels. For instance, when formulating policies that encourage companies' innovation on sustainable hydrogel materials, R&D tax credit incentives can be granted to those firms that spend some preset percentage or proportion of their total revenues on such research and development initiatives. The above-mentioned tax credits may be claimed in relation to research expenditure, including lab expenses, people as well as equipment. Now, tax incentives make it possible to minimize the costs of the R&D required for developing novel environmentally safe hydrogel technologies rather than using the traditional approaches that are not sustainable. Moreover, the governments should enable fast-track regulatory processes, especially for the eco-friendly hydrogel products. The process of commercialization of new materials is always obstructed by regulations, which is a big problem for SMEs. In addition, by expediting the approval process on those hydrogel formulations that are environmentally friendly, governments are able to cut down on the time and the costs involved in the development of sustainable hydrogel products. This may encourage firms to engage in research with a view to developing environmentally sustainable hydrogels since they will expect a reduced level of policy intervention. A second policy motivator is the creation of PPPs aimed at the creation and marketization of sustainable hydrogel technologies. Funding of projects for the preparation of eco-friendly hydrogels can be achieved through cooperation between government and private firms, research centers and NGOs. These partnerships can use the complementarity of the public and private sectors by promoting government support with industry experiences and funds. However, PPPs can also assist in the process of knowledge and technology transfer between the research institutions and the private sector, so as to ensure that existing innovative hydrogel formulations often developed by the research institutions are commercialized. Another way through which governments can promote the formation of innovation and proper usage of eco-friendly hydrogels is through green procurement policies. In the same light, guaranteeing the public sector as the key consumer of hydrogels composed of sustainable raw materials, governments can obtain a stable and regular demand for ecologically friendly products. This may just be the push that companies need to invest in the research and synthesis of biodegradable hydrogels, since they know that they will have access to the market. Hydrogels find use in various sectors such as health care, agriculture and environmental management, hence green procurement policies can be used across sectors. Consequently, incentives like the R&D grants, tax incentives, fast-track approval for regulatory purposes, PPP, as well as green procurement policies are inevitable for the development of eco-friendly hydrogel formulations. Via these incentives, a favorable climate for innovation is fostered, and the development and commercialization of sustainable hydrogel technologies will be promoted as a way of advocating for environmental conservation and increases in public health (Sinha and Chakma 2019).

It is imperative to use the regenerative wastewater treatment technologies in order to conserve water and minimize pollution of the environment. As a result, the governments should provide the companies with subsidies or tax exemptions since these technologies are costly to implement. It means that such incentives can provide money-back guarantees for embracing regenerative wastewater treatment systems, thus making the implementation of sustainable production practices financially feasible. Subsidies refer to another instrument that governments can use to support companies that have adopted regenerative wastewater treatment technology in a unique way. These subsidies can be provided in the form of grants, low-interest loans and direct cash incentives to meet the capital outlay required for acquiring and installing improved wastewater treatment facilities. For instance, governments can provide subsidies on funds to firms that embrace technologies like the MBRs, constructed wetlands, or advanced oxidation processes in order to conserve water as well as reducing on the impacts on the environment. The implication, therefore, is that subsidies effectively reduce the costs associated with the implementation of the improvement technologies and more firms will embrace sustainable wastewater treatment systems. Besides subsidies policy, there are other policy instruments, such as tax credit policies, which can be used by governments to encourage the use of regenerative wastewater treatment technologies. Possible tax incentives include the deductions, credits or exemptions on the overall tax burden of the firms that opt to use environmentally friendly wastewater treatment technologies. For example, there could be tax credits for a certain percentage of expenditures made on enhancing the wastewater treatment facilities among the companies for regenerative technologies. These tax incentives can be used to provide relative profitability for the utilization of sustainable wastewater treatment in industries, where the cost of following the environmental legislation is rather high. Another possible change could be to allow expensing of, or to implement first-year expensing of, investment in regenerative wastewater treatment technologies at an accelerated rate. Essentially, it allows the companies concerned to depreciate the money spent on subsidizing facilities to treat wastewater at a faster rate than the latter. This accelerated depreciation can some time offer instant tax savings for the companies and increase their cash flow, making it more worthwhile to invest in the advanced treatment systems. Another way through which governments can encourage the use of regenerative wastewater treatment technology among companies is by providing property tax reliefs for the purposes of implementing the technology. This incentive would also be especially helpful for large organizations that own property and where the integration of such technologies would help in invaluable ways in lessening harm to the environment. Lowering the property tax as a measure can go a long way to improving the company's sustainable wastewater treatment practices. Furthermore, governments can also offer contracts and performance-based specifications whereby companies are compelled to implement certain environmental objectives and goals, such as a regenerative wastewater treatment system. For instance, corporations that show marked improvement in water use efficiency, emissions to water, or greenhouse gas emissions in their WWT processes may be further exempted from taxes or receive subsidization funds. It can, in its turn, help to set and reach goals that would guarantee the best possible outcomes in terms of environmental improvement and thus also stimulate the implementation of regenerative technologies and the necessary organizational and managerial changes. Moreover, it is discovered that governments analyse, industry-specific incentive programs for the implementation of regenerative wastewater treatment technologies. There is a significant degree of heterogeneity in the water consumption and the generation of wastewater across industries; therefore, it is sensible that specific incentives should be of more utility to each industry. For instance, industries such as textile, pharmaceutical and food processing industries require a lot of water and discharge high quantities of wastewater; thus, aspects like subsidies and tax incentives should foster regenerative technologies. Thus, subsidies and tax benefits reveal themselves as leverage that Governments can utilize in order to make companies switch to regenerative wastewater treatment technologies. Through subsidizing the costs of implementation of such complex systems, these incentives could encourage the use of sustainable water management practices, therefore conserving the environment as well as using resources efficiently (The Economic Times 2024).

14.7 Implementation of Regenerative Practices and Principles of Circular Economy Through Hydrogel-Based Wastewater Treatment

A hydrogel-based treatment system for wastewater in line with the existing circular economy concepts has vast possibilities to increase sustainability measures across industries. This means that instead of the conventional model of Take-Make-Disposing, where the raw materials are taken, used, and disposed of, the circular economy model takes the reuse, recycling and reutilizing option. Conventional wastewater treatment structures can be based on hydrogel substances that are capable of water absorption selectivity and encapsulation of pollutants, hence taking a central role in the execution of circular economy principles. Resource efficiency is thus the first way in how hydrogel-based wastewater treatment correlates to the circular economy—through recovery and recycling of as many resources as possible after the wastewater has been treated. Conventional wastewater treatment processes generally lead to loss of good nutrients, chemicals, and water that are usually dumped to the drain. Hydrogel technologies are, however, designed to have the ability to selectively adsorb and desorb particular contaminants without compromising the valuable material, which can be retrieved and recycled within production processes. For example, hydrogels can be made to selectively remove heavy metals, phosphates or nitrates from wastewater for their recovery and re-utilization in agriculture as fertilizer or soil conditioners. This not only contributes to using fewer virgin resources but also helps to solve the problem of the ecological footprint of new materials' extraction and processing. In addition, the implementation of hydrogel-based systems in wastewater treatment will intensify the progress of water recycling processes. Because of the high water absorption capacity of hydrogels, they can swell and filter water, ensuring that the water that they release is free from contaminants and can therefore be recycled for use in industrial processes, to irrigate crops or for drinking purposes. It also helps to drive the circular economy through its reduction of the demand for freshwater resources during production processes, which in turn decreases the water footprint of establishments. The application of hydrogel-based treatment systems can also result in the development of new value recovery streams in the circular economy. For instance, these derived by-products from hydrogel-based treatment processes include nutrient-rich sludges or metal-laden hydrogels, which can be later on utilized or sold as value-added products. The by-products can be employed in other sectors like in agriculture, construction or the manufacturing sector. This opens greater economic value and optimizes the usage of available resources. In addition, due to the modularity and flexibility of hydrogel-based treatment systems, they fit well into decentralized wastewater treatment solutions, which are steadily being offered as part of circular economy approaches. On-site systems, such as wastewater disposal at an individual source, centralized systems on a farm, can be cheaper and faster than other centralized systems in that they are installed close to the source of wastewater. Some of the hydrogel-based technologies may be adapted to various functions depending on the operational requirement and thus suitable for decentralized use in industries, municipalities or even rural settings. Further, the replacement of the used hydrogels with sustainable and biodegradable ones in wastewater treatment can make the concept even more applicable to the circular economy model. It is therefore possible to reduce the environmental burden of hydrogel manufacturing and disposal if biodegradable and bio-based hydrogels substitute the synthetic hydrogels. This eliminates the creation of waste and also puts it in a position that the products used in the treatment of wastewater can be recycled and returned to the ecosystem after their usefulness has been completed, in line with the circular economy. In conclusion, through the evaluation of the materials and methods used in developing the hydrogel-based wastewater treatment, it can be viewed that the inclusion of the hydrogel in the circular economy framework provides a practical application for resource recovery, that is, water reuse with minimal waste disposal. Through the utilization of hydrogels, industries are able to come up with systems that conform and are sustainable in a circular economy, hence conserving the environment while at the same time improving the economic stability (Singh et al. 2023).

The use of hydrogels has a vast potential to enhance the use of regenerative practices in wastewater treatment, especially concerning nutrients and water recycling. Sustainable practices involve fairly using the natural resources in order to rebuild and replenish the natural resources that are in use, in order to make informed decisions that work in the best interest of society and the earth. Some of the features that make hydrogels particularly suitable with regard to these goals are high water retention capacity, selectivity of absorption and biocompatibility. From the point of view of nutrient recovery, hydrogels can be designed to have the ability to absorb and release nutrient elements, for instance nitrogen and phosphorus that can be found in wastewaters. These nutrients, the very compounds which are injurious in higher concentrations, are useful plant nutrients for agricultural utilization. In the case of nutrient pollution, it is possible to capture these nutrients and recover it from the wastewater streams for the purpose of fertilizers or modifying the soil properties—nutrient recovery. This approach also helps in minimizing the problem and effects of nutrient pollution in water bodies, including those that cause eutrophication. It also works towards minimizing the use of synthetic fertilizers, which are produced using a lot of energy and are derived from non-renewable sources. Thus, the use of hydrogels for encapsulation and nutrient delivery control is well suited for this purpose. For instance, hydrogels can be engineered to deliver nutrients gradually to the system and this enables a plant to receive necessary elements in a controlled manner without polluting the environment. It helps the restoration of the regenerative agricultural system, as it enables the improvement of the fertility of the soil and expansion of the productivity of crops. Besides nutrient recovery, it is found that hydrogels have a great potential for enhancing water reuse processes. Hydrogels have the capacity for water absorption and purification of high volumes of water, which makes them ideal for use in wastewater through wastewater recycling. Through incorporating hydrogels in water treatment systems, Industries and municipalities are helped to lower the consumption of fresh water, hence recycling water after the treatment process for purposes such as irrigation purposes, some industrial usage, or sometimes even for human consumption. This not only helps in saving water but also works in saving the environment around, bearing the cost of water extraction and distribution. It is also important to admit that hydrogels can be implemented in combination with other approaches for promoting regenerative processes for increasing the efficiency and effectiveness of water reuse practices. For instance, hydrogels can be integrated into more sustainable systems commonly known as the green infrastructure technologies, like constructed wetlands or green roofs, whereby the hydrogels assist in absorbing and purifying stormwater gradients, lowering the pressure on the traditional wastewater treatment plants. Hem water in these systems can be released slowly to plant growth or funneled into the soil, which will enhance urban plant growth and also help the urban green space. Also, the ability to build hydrogels from bio-based and biodegradable materials extends their application to regeneration activities. While synthetic hydrogels are notoriously known to be ecologically unfriendly and to contribute to the growth of microplastic pollution, bio-based hydrogels can be naturally decomposed at the end of their service life. This corresponds with the principles of regenerative design, extending to the intent utilization of materials that can be safely returned to the biological communities. In the case of bio-based hydrogels in WWTP and WRR, bio-based hydrogels offer the opportunity to develop closed-loop life cycles contributing to natural resource replenishment. Thus, hydrogels have substantial opportunities to contribute to the development of initiatives that enhance the regenerative potential of wastewater treatment. Due to their specific properties, these objects are significant in achieving resource efficiency, improving the circular economy and supporting the rehabilitation of native ecosystems. The application of hydrogels in regenerative wastewater treatment approaches enhances the efficiency of the systems towards environmental conservation efforts as well as resource utilization (Liu et al. 2022) (Fig. 14.4).

14.8 Future Perspectives

With the increasing identification of hydrogels towards wastewater remedy, therefore, it is crucial to design suitable policy and regulation mechanisms for the comprehensive deployment that will uphold ecological, sustainable and regenerative principles. Future use of hydrogel in wastewater treatment will involve a delicate blend of research and development to ensure advancement of methods without having to compromise the health of the environment and the people.

Another major policy factor is the need to develop stringent measures for environmental safety, especially in relation to hydrogel materials. Since the development of hydrogels covers a wide variety of compositions and functional properties, the

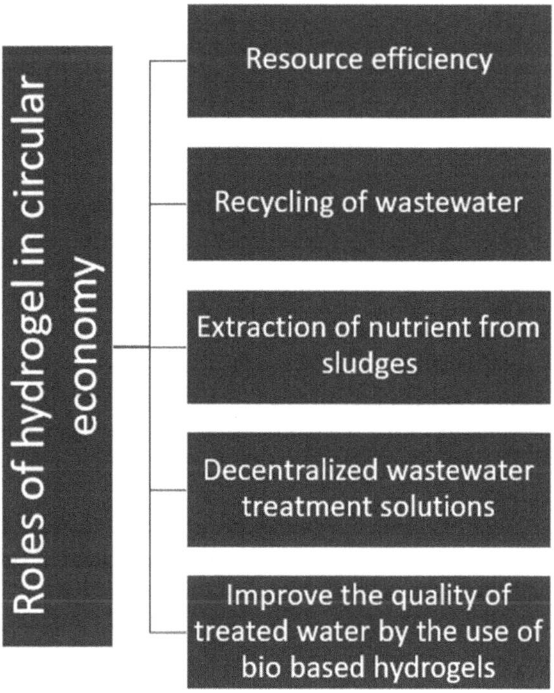

Fig. 14.4 Roles of hydrogel in circular economy

principles of their regulation should include statements that position these materials as non-toxic, biodegradable, and environmentally friendly. This may mean implementing standards for the degree of biodegradability, especially of hydrogels with high density in wastewater treatment plants for mass use and indefinite disposal. In the same regard, measures on proper disposal or recycling of used hydrogels should also be developed so as not to harm the environment.

The regulation should also take into consideration the Lifecycle analysis LCA of hydrogel products from the purchase of raw materials up to the disposal of products. To this effect, policymakers should deem LCA necessary steps in approving new hydrogel formulations for use to allow only those products with the least impact on the environment. Such a policy would encourage the creation of bio-based and biodegradable hydrogels that are in tandem with the concept of regeneration. In addition, it would be possible to note that the installation of LCA requirements would allow for the identification of potential environmental exchange and ensure that the advantages of the hydrogels' usage in WWT exceed the possible problematic effects.

To encourage the use of hydrogel in wastewater treatment, the following should be recommended to the policy makers: The use of incentives which include subsidies or provisions such as tax exemptions to companies that use hydrogel technology in its environmentally friendly applications. These could be made contingent on the actual accomplishment of particular results within the environment, for instance, a decreased content of water pollution or improved nutrient productivity. Since governments can attach conditions to their financial support towards secure organizations, it becomes possible to entrench regenerative practices in the way communities deal with wastewater.

Another essential element of the regulatory framework is that the latter should also introduce requirements for the safety and efficiency of hydrogel usage in different cases of wastewater treatment. This entails defining goals on the efficiency of the removal process of certain pollutants such as heavy metals or organic compounds and guidelines on the assessment of the efficacy of hydrogel-based systems. Other issues that regulations should cover include risk of emission or release of hydrogels chemistries; this could occur through accidental leaching of hydrogels with chemistries likely to contaminate sensitive ecosystems such as agricultural land or natural water sources. Thus, by defining clear standards for the use of hydrogels in various areas, the regulators would be in a position to make sure that the procedure is very effective, not forgetting the fact that it needs to be relatively safe for purposes of use.

Hydrogel-based wastewater treatment technologies it is considered that the involvement of PPPs might help improve their development of these technologies. Industry players, research institutions and NGO can work with the government to sponsor and encourage more R&D in the formulation and application of hydrogels. These technologies can also be enhanced by the PPPs to assist in knowledge transfer and capacity building to remove knowledge and cost barriers to these technologies, especially in the developing world.

Moreover, the specific regulatory structure should be able to be changed and dynamic due to the position of new hydrogel applications. Due to new formulations and applications becoming more frequent, the same must be aligned with the current scientific knowledge and technology. Such a body could consist of a regulatory head or an advisory committee to monitor the application of hydrogels in the context of wastewater treatment while making sure that the existing policies are up to date and efficient.

Lastly, the cooperation of the international community and a common uniform set of regulations will be crucial to implement hydrogel-based WWTPs around the world. Since most of the environmental issues cut across borders for instance, water pollution, it is important that the approach to regulation should also be in harmony. This could be done via the United Nations Environment Program (UNEP), in collaboration with ISO, which encourages the swapping of practices from different countries that can be effective in the promotion and utilization of the eco-friendly technologies on hydrogels (Fig. 14.5).

Accordingly, the future of hydrogel application in wastewater treatment depends on the presence of an adequate and visionary policy and legal environment. In addressing environmental safety, encouraging inventiveness, and encouraging intergovernmental cooperation, policy-makers can facilitate the application of hydrogels and positively impact the development of eco-sustainability and regeneration of the industry dealing with wastewater management.

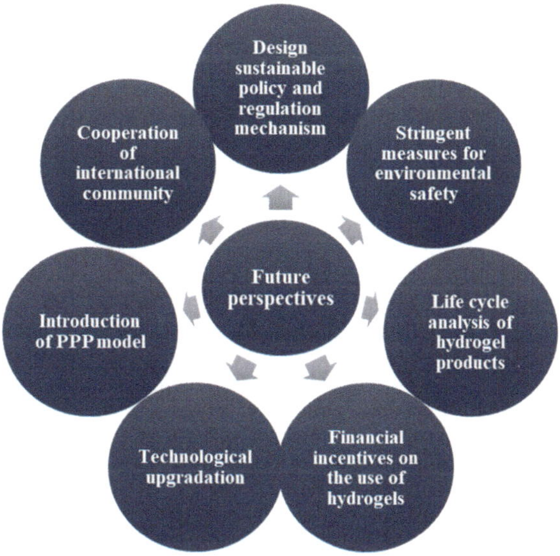

Fig. 14.5 Future perspectives

14.9 Conclusion

The use of hydrogels in wastewater treatment can be one of the sustainable techniques for water purification and regeneration in the future. With these innovative materials continuing to grow in popularity, it is important to develop a clear policy and legal framework for these materials, which will regulate their safety and efficiency of usage as well as stimulate the creation of environmentally friendly formulations. Discussed below are some of the ways through which governments can encourage the use of hydrogels in a sustainable manner: Governments can specify environmental safety requirements towards encouraging research and development of hydrogels, as well as involve the private sector in the production of hydrogels. The fact that the technology is dynamic also means that the regulation must be dynamic in that it has to be capable of being adjusted in line with new scientific discoveries and technological developments. Just to sum up, much attention will be paid to the international cooperation in standardization and promotion of new hydrogel-based wastewater treatment technologies. If policymakers want to maximize the positive potential of hydrogels, the must take a proactive approach in order to guarantee that these materials will significantly shift towards the circular economy, where resource usage is responsible and environmental footprint is minimized.

References

Abdallah AM (2019) The effect of hydrogel particle size on water retention properties and availability under water stress. Int Soil Water Conserv Res 7(3):275–285. https://doi.org/10.1016/j.iswcr.2019.05.001

Ahmaruzzaman M, Roy P, Bonilla-Petriciolet A, Badawi M, Ganachari SV, Shetti NP, Aminabhavi TM (2023) Polymeric hydrogels-based materials for wastewater treatment. Chemosphere 331:138743. https://doi.org/10.1016/j.chemosphere.2023.138743

Ahmed ED (2015) Hydrogel: preparation, characterization, and applications: a review. J Adv Res 6(2):105–121. https://doi.org/10.1016/j.jare.2013.07.006

Bahram M, Mohseni N, Moghtader M (2016) An introduction to hydrogels and some recent applications. Emerg Concepts Anal Appl Hydro. https://doi.org/10.5772/64301

Catoira MC, González-Payo J, Fusaro L, Ramella M, Boccafoschi F (2020) Natural hydrogels R&D process: technical and regulatory aspects for industrial implementation. J o Mater Sci Mater Med 31(64). 10.1007%2Fs10856–020–06401-w

Chang H, Zhao H, Qu F, Yan Z, Liu N, Lu M, Liang Y, Lai B, Liang H (2023) State-of-the-art insights on applications of hydrogel membranes in water and wastewater treatment. Separat Purif Technol 308:122948. https://doi.org/10.1016/j.seppur.2022.122948

Flores-Valenzuela LE, González-Fernández JV, Carranza-Oropeza MV (2023) Hydrogel applicability for the industrial effluent treatment: a systematic review and bibliometric analysis. MDPI Polym 15(11):2417. https://doi.org/10.3390/polym15112417

García-Villén F, Sánchez-Espejo R, Borrego-Sánchez A, Cerezo P, Perioli L, Viseras C (2020) Safety of nanoclay/spring water hydrogels: assessment and mobility of hazardous elements. MDPI Pharmaceut 12(8):764. https://doi.org/10.3390/pharmaceutics12080764

Hossain MS, Hossain MM, Khatun MK, Hossain KR (2023) Hydrogel-based superadsorbents for efficient removal of heavy metals in industrial wastewater treatment and environmental conservation. Environ Funct Mater 2(2):142–158. https://doi.org/10.1016/j.efmat.2024.01.001

Liu Y, Wang J, Chen H, Cheng D (2022) Environmentally friendly hydrogel: a review of classification, preparation and application in agriculture. Sci Total Environ 846:157303. https://doi.org/10.1016/j.scitotenv.2022.157303

Mantha S, Pillai S, Khayambashi P, Upadhyay A, Zhang Y, Tao O, Pham HM, Tran SD (2019) Smart hydrogels in tissue engineering and regenerative medicine. MDPI Mater 12(20):3323. https://doi.org/10.3390/ma12203323

Mohapatra S, Mirza MA, Hilles AR, Zakir F, Gomes AC, Ansari MJ, Iqbal Z, Mahmood S (2021) Biomedical application, patent repository, clinical trial and regulatory updates on hydrogel: an extensive review. MDPI Gels 7(4):207. https://doi.org/10.3390/gels7040207

Nikolić LB, Zdravković AS, Nikolić VD, Ilić-Stojanović SS (2018) Synthetic hydrogels and their impact on health and environment. In: Cellulose-based superabsorbent hydrogels. pp 1–29

Persano F, Malitesta C, Mazzotta E (2024) Cellulose-based hydrogels for wastewater treatment: a focus on metal ions removal. MDPI Polym 16(9):1292. https://doi.org/10.3390/polym16091292

Revete A, Aparicio A, Cisterna BA, Revete J, Luis L, Ibarra E, González EAS, Molino J, Diego R (2022) Advancements in the use of hydrogels for regenerative medicine: properties and biomedical applications. Int J Biomater. https://doi.org/10.1155/2022/3606765

Sannino A, Demitri C, Madaghiele M (2009) Biodegradable cellulose-based hydrogels: design and applications. MDPI Mater 2(2):353–373. https://doi.org/10.3390/ma2020353

Singh AK, Itkor P, Lee YS (2023) State-of-the-art insights and potential applications of cellulose-based hydrogels in food packaging: advances towards sustainable trends. MDPI Gels 9(6):433. https://doi.org/10.3390/gels9060433

Sinha V, Chakma S (2019) Advances in the preparation of hydrogel for wastewater treatment: a concise review. J Environ Chem Eng 7(5):103295. https://doi.org/10.1016/j.jece.2019.103295
The Economic Times (2024) Tax incentives can help enhance adoption of sustainability manufacturing: experts. https://economictimes.indiatimes.com/news/economy/policy/tax-incentives-can-help-enhance-adoption-of-sustainability-manufacturing-experts/articleshow/110115593.cms?from=mdr
Todman W, Mauvais L (2021) Hydrogel: a promising solution for water scarcity in the middle east? Centre for Strategic & International Studies. https://www.csis.org/analysis/hydrogel-promising-solution-water-scarcity-middle-east

Chapter 15
Challenges and Future Perspectives of Hydrogels in Wastewater Treatment

Kshama G. Dev and Mamta Pal

Abstract Hydrogels are three-dimensional polymeric networks, effective in trapping pollutants due to their greater surface area and porous structure. Hydrogels with excellent absorbing and water-retaining properties are innovative in the discipline of wastewater treatment. Introduction of functionalizing groups increases affinity for various pollutants, like heavy metals, and organic pollutants, along with selective nutrient adsorption. Although hydrogels possess vast potential, there are enormous challenges as well. For instance, for large-scale applications, hydrogels degrade under the dynamic environment of wastewater treatment plants. Hence, the mechanical stability of hydrogels needs to be increased. Furthermore, effective regeneration and reuse are required for the process to be economically viable. Besides, scaling up is a major concern for effective performance at laboratory scale, which may not be valid for industrial scale. Therefore, the present study focuses on novel functionalization methods like nanomaterial loading and composite hydrogel designing, hybrid technologies to enhance their efficiencies and make them applicable for a broad spectrum of contaminants. Application of hydrogels reduces environmental impact due to fewer adverse effects with the use of alternative raw materials and energy-efficient processes for cost reduction. Conclusively, cross-disciplinary research, their environmental impact studies, and sustainable practices can promote the importance of hydrogel technologies to meet the challenges of global water pollution.

Keywords Hydrogels · Wastewater treatment · Functionalization of hydrogels · Mechanically stable hydrogels · Scalability of hydrogels

---

K. G. Dev
RJ Forsec Solutions, Village Chhatarpur, New Delhi, India

M. Pal (✉)
Department of Chemistry, Biochemistry and Forensic Science, Amity School of Applied Science, Amity University Haryana, Haryana, India
e-mail: mamtapal1717@gmail.com

S. Bardhan and M. P. Shah (eds.), *Applications of Hydrogels in Modern Wastewater Treatment*, Advances in Wastewater Research,
https://doi.org/10.1007/978-981-95-1598-1_15

List of Abbreviations

TiO_2	Titanium dioxide
UV	Ultra violet
IPNs	Interpenetrating polymer networks
EPA	Environmental Protection Agency
EIA	Environmental impact assessment

15.1 Introduction

The presence of water in various aquatic bodies is crucial for sustaining diversified ecosystems and in supporting human life in multiple ways, like drinking water supply, irrigation, in facilitating industrial and laboratory processes, etc.

However, the various pollutants from industrial waste, agricultural runoff, municipal wastewater, and improper waste disposal can enter these water bodies and can have detrimental effects on the water quality and ecosystem health (Weldeslassie et al. 2018). For example, nutrients from the agricultural runoff can cause eutrophication in lakes and rivers, leading to algal blooms that deplete oxygen and harm aquatic life. Industrial waste can introduce heavy metals and toxic chemicals that accumulate in the food chain, posing risks to both wildlife and humans. Besides, in industrial and laboratory settings, water purification is often necessary to ensure that water is free from contaminants that could interfere with desired processes and experiments carried out there.

To provide a more quantitative perspective, it is important to consider the percentage of water available for different uses along with the percentage of pollutants present in water bodies. Globally, the distribution of water is such that the freshwater approximately constitutes 2.5% of the earth's water, with the majority locked in glaciers and ice caps. Only a small fraction (around 0.3%) is readily available for human use in rivers, lakes, and groundwater. The remaining 97.5% of the Earth's water is saltwater, primarily found in oceans.

Regarding pollutants, it is challenging to provide a single percentage due to the varying levels of contamination in different water bodies and regions. Table 15.1 shows the major contributors of water pollutants along with the impacts on the environment and living organisms.

Therefore, in the view of wastewater treatment, hydrogels, a class of materials with unique properties, offer promising potential for water purification as hydrogels can selectively absorb and retain the pollutants (Van Tran et al. 2018). These have emerged as versatile materials with applications spanning biomedicine, agriculture, and environmental engineering. Apart from these, it's also evident that hydrogels can play a critical role in ensuring to make the clean water, promoting sustainable development and thus, protecting public health. Since rapid industrialization and urbanization worldwide have resulted in increased discharge of pollutants into

Table 15.1 Major water pollutants, their sources and impacts on environment and living organisms

Types of contaminants in water	Sources	Impacts on environment and living organisms	References
Heavy metals	Mining, mineral erosion, weathering, etc.	Biomagnification, affects aquatic environment, unsafe drinking water, etc.	Obasi and Akudinobi (2020), Khayatzadeh and Abbasi (2010), Fu and Xi (2020)
Dyes	Textiles, tannery, cosmetic industry, etc.	Depletion of water quality, carcinogenicity, mutagenicity, teratogenicity, etc.	Gita et al. (2017), Borgerding and Hites (1994), de Lima et al. (2007)
Organic pollutants	Agricultural fields, industries, etc.	Biomagnification, depletion of oxygen in water, causing disease	Rashed (2013), Saha et al. (2017)
Biological waste	Medical waste, human and animal waste, etc.	Water-borne diseases, depletion of water quality	Janagi et al. 2015, Akinluyi and Odeyemi (1984)
Sewage waste	Inadequate sewage treatment	Water-borne diseases	Rajasulochana and Preethy (2016), Burge and Marsh (1978)
Suspended solids	Anthropological acts, plastic accumulation, ship wreckage, etc.	Depletion of oxygen in water, causes threatening to life, etc.	Ward (2003), Wan et al. (2022)
Nutrients	Daily domestic activities, agricultural activities, soaps, detergents, etc.	Eutrophication, causes threatening to life, etc.	Ammary 2004, Preisner et al. (2020), Ziemba et al. (2018), Shannon (1975)
Microbes	Failing sewage disposal, disoriented environmental, etc.	Serious risk to life, especially to humans, causes various water-borne diseases	Bohra et al. 2012, Leclerc et al. (2002)
Fats, oils and grease	Restaurants, domestic, industrial waste, etc.	Harmful to life, results in poor water quality, etc.	Mohana et al. 2023, Yau et al. (2021), Ghaly et al. (2006)

water bodies, posing substantial risks to both human health and the environment, the conventional wastewater treatment methods are effective to a certain extent only, which often struggle with the removal of specific contaminants such as heavy metals, dyes, and persistent organic pollutants. Therefore, there is an urge to develop more efficient and sustainable treatment technologies, and hydrogels are emerging as a promising candidate.

Despite their promising capabilities, several challenges hinder the widespread adoption of hydrogels in wastewater treatment. The cost associated with the production and functionalization of hydrogels remains a significant barrier, even though many high-performance materials require expensive raw materials and complex synthesis processes; hydrogels are much more cost-efficient. Ensuring the stability and durability of hydrogels under varying pH, temperature, and chemical conditions in real-world wastewater scenarios is another critical challenge. Moreover, the environmental impact of hydrogel degradation products must be carefully assessed to prevent secondary pollution and ensure the overall environmental sustainability. Thus, to overcome these challenges, ongoing research focuses on technological innovations aimed at enhancing the cost-effectiveness, stability, and sustainability of hydrogels in wastewater treatment.

Therefore, this study explores the application of hydrogels in wastewater treatment, including their properties, types, challenges, innovations, and efficacy in removing pollutants from water. It will also delve into the current state of water pollution, sources of pollutants, and the need for sustainable solutions like hydrogels. While our aim is to highlight the challenges faced by hydrogels in water treatment along with innovative advancements that can be brought about in the future to counter the challenges of hydrogels and in addressing the global water crisis and promoting a cleaner, healthier, and sustainable future.

15.2 Hydrogel Technology in Wastewater Treatment

Hydrogels in wastewater treatment are considered an advanced method in making sustainable and efficient water for the current and upcoming generations. Even though greater changes and advancements have been brought forward in hydrogel structure and its selectiveness, however, the entire process of their manufacturing as well as their implementation remains the same. The treatment process begins with the wastewater inlet, where water enters the system and is initially characterized for parameters such as pH, temperature, turbidity, and pollutant concentrations. This step is crucial for understanding the nature of the wastewater and tailoring the treatment process accordingly.

Pre-treatment follows the removal of large objects and debris, then grinds it to reduce its particle size, followed by sedimentation to settle out the suspended solids. These steps are essential for preventing damage to equipment and ensuring that the wastewater is suitable for the subsequent treatment stages. Subsequently, the core of the process of hydrogel addition comes, where the hydrogels, such as polyacrylamide or polyethylene glycol, are introduced to the wastewater. The amount of hydrogel added is typically between 0.1 and 1.0% w/w and it is mixed thoroughly with wastewater to ensure maximum contact. The mixing and contact time can vary from 30 min to 2 h for mixing and 2–24 h for contact, depending on the specific requirements of the treatment process as well as on the properties of the hydrogel used. The mixing and contact time create an environment for the pollutants to interact with the hydrogels, and is the time taken for the absorption of pollutants. During that time, the hydrogel interacts with the wastewater, resulting in the binding with heavy metals like lead, mercury, and arsenic, as well as with organic compounds such as pesticides and pharmaceuticals. The absorption capacity of the hydrogel can vary, making it an efficient method for removing these pollutants. Once the absorption is completed, the hydrogel is separated from the treated water through various methods such as filtration, centrifugation, or sedimentation. The choice of separation technique depends on the requirements of the type of hydrogel to be employed and the quality of water to be treated.

Since the hydrogel with the absorbed pollutants needs to be ready for the next batch of treatment, the contaminants attached to the used hydrogels need to be removed, and thus, the hydrogel recharging and regeneration are carried out. The regeneration step is vital for the sustainability of the process. The washed and subsequently recharged hydrogel will be allowed to be reused for wastewater treatment. The frequency of regeneration can vary, depending on the hydrogel's performance and the pollutant load in the wastewater. Thereafter, the treatment process, the treated water outlet releases water which is then treated and characterized again for the pH, turbidity, pollutant concentrations, etc. (Mu et al. 2020). The water can then be discharged into the environment and is reused for the numerous applications, depending on the quality obtained.

Sludge management is also an important aspect of the process; the byproduct or the absorbed pollutants need to be removed and discharged safely, considering human health and its environmental aspect. Sludge handling includes dewatering, drying, and disposal. The sludge needed to be monitored in terms of moisture content and concentration of various pollutants to ensure its proper management.

Finally, monitoring and controlling are continuous processes throughout the treatment system. Proper handling and supervision are immensely important in this process. Water quality parameters need to be regularly checked to ensure compliance with standards, and the performance of the hydrogel is monitored to optimize its absorption capacity and regeneration frequency. System control can be automated or manual, allowing for adjustments to be made in real-time to maintain the efficiency and effectiveness of the treatment process. The entire wastewater treatment methodology has been represented in Fig. 15.1.

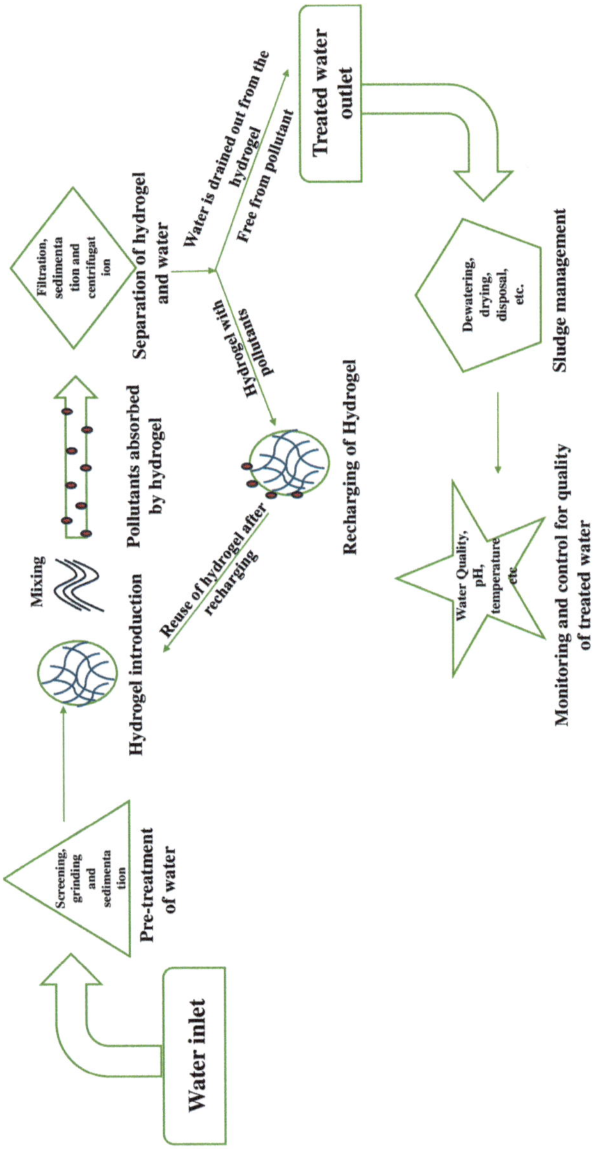

Fig. 15.1 Illustration of hydrogel technology in wastewater treatment

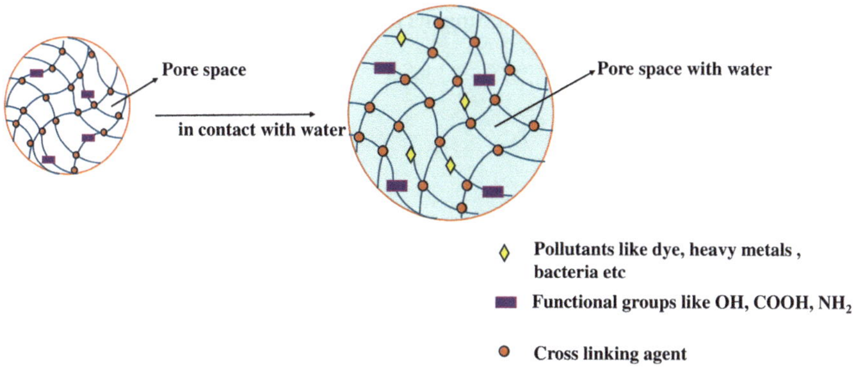

Fig. 15.2 Diagrammatic representation of a hydrogel on the absorption of pollutants-contaminated water

15.3 Mechanism Behind Hydrogel Technology

Hydrogels are versatile materials in wastewater treatment due to their unique structural and chemical properties. The mechanisms by which they act can be deeply understood by examining each process in detail as follows.

15.3.1 Adsorption

Hydrogel's ability to adsorb pollutants relies on its high surface area and the presence of functional groups within the polymer network. These functional groups (such as carboxyl, hydroxyl, and amine groups) interact with contaminants through the van der Waals forces, hydrogen bonding, and other intermolecular interactions (Seida and Tokuyama 2022). The effectiveness of adsorption is influenced by the hydrogel's surface chemistry, pore size, and crosslinking density. For instance, hydrogels with higher porosity provide more active sites for pollutant binding (Van Tran et al. 2018). The absorption of contaminated water, constituting various pollutants through the hydrogel, has been depicted in Fig. 15.2.

15.3.2 Absorption

Hydrogels can swell significantly by absorbing water due to their hydrophilic nature. This property allows them to retain large volumes of water within their network. Absorption is primarily driven by the osmotic pressure difference between the hydrogel and the surrounding wastewater. This ability is beneficial in reducing the volume of wastewater and concentrating the pollutants, making subsequent treatment processes more efficient.

15.3.3 Encapsulation

Through encapsulation, hydrogels trap pollutants within their 3D network. This entrapment prevents the release of contaminants back into the water, ensuring the permanent removal. Encapsulation is particularly useful for handling toxic substances, as it immobilizes them and reduces their mobility in such an environment. The encapsulation efficiency depends on the hydrogel's mesh size and the nature of the pollutants (Van Tran et al. 2018).

15.3.4 Ion Exchange

Ion exchange refers to the process of exchanging ions between the wastewater and the hydrogel. Hydrogels containing ion exchange groups (e.g., sulfonate, carboxylate) can effectively remove heavy metals and other ionic pollutants. The process works by exchanging undesirable ions in the water with benign ions within the hydrogel. Factors affecting ion exchange include the hydrogel's ion-exchange capacity, the concentration of ions in the wastewater, and the affinity of the hydrogel for specific ions.

15.3.5 Catalysis

Some hydrogels are engineered to possess catalytic properties, either inherently or by embedding catalytic nanoparticles. These catalytic hydrogels can accelerate chemical reactions that break down pollutants into less harmful substances. For example, hydrogels incorporating metal nanoparticles can catalyze the degradation of organic dyes through redox reactions. The catalytic activity is influenced by the distribution and stability of the catalytic sites within the hydrogel (Lu et al. 2007).

15.3.6 Biodegradation

Hydrogels can support microbial communities that degrade organic pollutants; therefore, microbes are incorporated into the hydrogels, which provide a conducive environment for microorganisms by maintaining moisture and nutrients. Since hydrogels gradually decompose, they release the encapsulated microorganisms into the wastewater, this process enhances the natural biodegradation, treating organic contaminants like oils and greases. The rate of biodegradation depends on the microbial load, the hydrogel's biodegradability, and the nature of the organic pollutants (Shalla et al. 2018).

15.3.7 Flocculation

When suspended particles and colloids combine to form bigger flocs that are easily removed by sedimentation or filtration, hydrogels can cause flocculation. This mechanism involves the bridging of particles by the hydrogel chains, leading to the formation of larger and heavier aggregates. The effectiveness of flocculation is influenced by the hydrogel's charge density, molecular weight, and the nature of the suspended particles (Mittal et al. 2015).

15.3.8 Emulsification

Hydrogels can emulsify hydrophobic contaminants such as oils, enabling their dispersion in the water (Seida and Tokuyama 2022). This emulsification process increases the surface area of the contaminants, making them more accessible to treatment processes. Emulsifying hydrogels often contain amphiphilic groups that stabilize the oil–water interface. The efficiency of emulsification depends on the hydrogel's amphiphilicity and the nature of the hydrophobic contaminants.

15.4 Application of Hydrogel in Wastewater Treatment

Specifically discussing the hydrogel's properties, these offer promising solutions for various challenges in wastewater treatment. The major applications of hydrogel include the removal of heavy metals, adsorption of dye, degradation of organic pollutants, and recovery of nutrient. Such versatile materials can address multiple contaminants due to their unique properties and functional capabilities, which have been intricate under the following heads.

15.4.1 Removal of Heavy Metals

Heavy metals, such as lead, cadmium, and mercury, pose significant environmental as well as health risks. Hydrogels can effectively remove heavy metals from wastewater through adsorption, ion exchange, and chelation mechanisms.

The adsorption of heavy metals onto hydrogels occurs through various mechanisms, such as electrostatic interaction, coordination bonding, and ion exchange. The efficiency of removal of metals depends on the hydrogel's surface area, functional groups, and the presence of competing ions in the water. For example, functional groups like carboxyl, hydroxyl, and amine groups enhance the hydrogel's ability to bind heavy metals. A study by Wu et al. (2019) detailed the mechanisms of heavy metal removal by hydrogels, highlighting the role of functional groups in enhancing the adsorption capacity.

Hydrogels can also be designed to target specific metals. For instance, thiol-functionalized hydrogels show high affinity for mercury, while hydrogels with iminodiacetic acid groups are effective for chelating cadmium and lead. The porous structure of hydrogels further increases their surface area, hence allowing for greater contact with contaminants and improving the removal efficiency.

15.4.2 Removal of Dye Through Adsorption

Dyes from the textile and industrial effluents are major pollutants that require effective treatment methods. Hydrogels can adsorb these dyes through physical entrapment, electrostatic interactions, and hydrogen bonding.

The adsorption capacity of hydrogels for dyes depends on their porosity, surface charge, and the presence of specific functional groups. Anionic and cationic dyes, for instance, can be efficiently adsorbed by a hydrogel containing carboxyl and amine groups, respectively. Zhang and Khademhosseini's (2017) study showed that functionalized hydrogels are very effective at removing dye from wastewater.

Hydrogels can also be tailored to target specific types of dyes. Cationic hydrogels are effective for anionic dyes, while anionic hydrogels work well for cationic dyes. The adsorption process often follows a pseudo-second-order kinetic model, indicating that chemisorption is the primary mechanism. This involves the formation of chemical bonds between the dye molecules and the functional groups on the hydrogel.

15.4.2.1 Degradation of Organic Pollutants

Organic pollutants, such as pesticides, pharmaceuticals, and industrial chemicals (dyes), are persistent and toxic. Hydrogels can be designed to degrade these pollutants through the catalytic and enzymatic processes as follows.

Enzyme-Embedded Hydrogels

Incorporating enzymes into hydrogels enables the degradation of specific organic pollutants. For example, a hydrogel embedded with laccase enzyme can degrade phenolic compounds in wastewater. Laccase catalyzes the oxidation of phenols, resulting in less toxic compounds. Wang et al. (2019a, b) demonstrated that laccase-embedded hydrogels effectively degraded phenolic pollutants, showcasing the potential of biocatalysis in wastewater treatment.

Catalytic Degradation

Hydrogels can also contain catalytic nanoparticles, such as titanium dioxide (TiO_2) or zinc oxide, which facilitate the breakdown of organic pollutants under light exposure. This photocatalytic process is highly effective for degrading persistent organic contaminants. A study by Wang et al. (2019a, b) demonstrated the use of TiO_2-hydrogel composites for the photocatalytic degradation of pharmaceuticals. When exposed to ultraviolet (UV) light, TiO_2 generates reactive oxygen species that attack and decompose organic molecules, making the process efficient and sustainable.

15.4.3 Recovery of Nutrient

Nutrient recovery from contaminated water should be carried out for the sustainable management of resources. Hydrogels can capture and release valuable nutrients, such as phosphates and nitrates are released as industrial effluent from the leather, fertilizer and pharmaceutical industries, for their reuse in agriculture.

Hydrogels can be functionalized with ion exchange groups that selectively bind phosphates and nitrates. For instance, a hydrogel with quaternary ammonium groups can effectively remove nitrates from wastewater. Research by Xiao et al. (2018) explored the use of functionalized hydrogels for nutrient recovery, highlighting their efficiency in phosphate and nitrate removal and found that incorporating these functional groups into the hydrogels can capture the nutrients from wastewater streams, preventing eutrophication and enables their reuse in agriculture.

15.5 Challenges in Using Hydrogel for Wastewater Treatment

Despite hydrogels' potential, several challenges hinder the widespread adoption of hydrogels in wastewater treatment processes. A few primary issues are mechanical stability and durability, the regeneration and their reuse, cost and scalability, selectivity and efficiency and its environmental impact and disposal, etc. All together, these challenges have been summarized in Fig. 15.3. This section will delve into these challenges, examination and their implications, to explore the potential solutions. Besides, how various factors like the chemical structure, molecular weight of the polymer, crystallization, network topology, and crosslinking density affect the mechanical properties of hydrogels have also been discussed.

15.5.1 Mechanical Stability and Durability

15.5.1.1 Mechanical Strength of Hydrogels

Ensuring the mechanical stability and durability of hydrogels is formidable to acquire due to their high-water content, which makes them prone to deformation, swelling, and degradation under mechanical stress. Thus, enhancing crosslinking density and incorporating reinforcing agents can improve resilience; however, higher

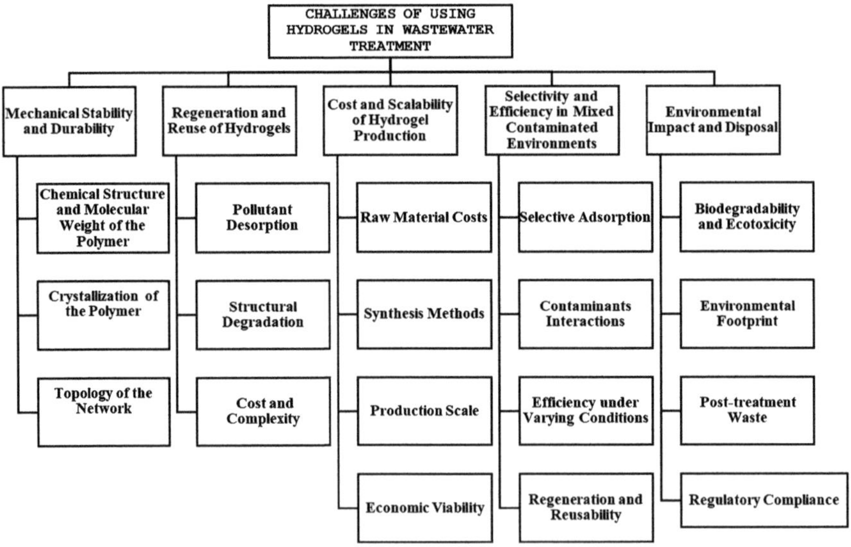

Fig. 15.3 Challenges of using hydrogels in wastewater treatment

crosslinking density and added reinforcing agents can also reduce the hydrogel's flexibility, potentially hindering its ability to adapt to varying contaminants present and to interact efficiently with different pollutants. Moreover, these modifications may also affect the hydrogel's capacity for the selective absorption of contaminants and its overall efficiency in treating wastewater. Therefore, achieving an optimal balance between the mechanical strength, flexibility, and contaminant removal efficiency is complex (Seida and Tokuyama 2022) and remains a significant challenge in the development of hydrogels for effective and sustainable wastewater treatment.

Factors Affecting Mechanical Properties of Hydrogels

Hydrogels are complex biomaterials whose mechanical properties are influenced by a combination of factors. Understanding these factors is crucial for designing and engineering hydrogels with specific properties for their applications. The factors affecting the mechanical strength of hydrogels include:

Chemical Structure and Molecular Weight of the Polymer

- **Chemical Structure**: The backbone structure of the polymer influences the rigidity and flexibility of the hydrogel. Polymers with more rigid backbones tend to form stronger and more stable hydrogels.
- **Molecular Weight**: Higher molecular weight polymers generally result in hydrogels with greater mechanical strength due to the increased entanglement of polymer chains.

Crystallization of the Polymer

- **Degree of Crystallinity**: The presence of crystalline regions within the polymer matrix can enhance the mechanical strength but may also reduce elasticity (Zahib et al. 2021). Thus, to balance the crystallinity and the amorphous region in a polymer is critical for optimizing performance.

Topology of the Network

The arrangement and connectivity of polymer chains within the hydrogel network significantly impact mechanical properties. More interconnected networks typically exhibit higher mechanical strength.

- **Nature and Density of Crosslinking**: Type of crosslinking, like covalent crosslinks, provides permanent bonding between polymer chains and results in robust hydrogels, while physical crosslinking (such as hydrogen bonds or ionic interactions) offers reversible and tunable properties like swelling behaviour, self-healing ability, degradation rate, etc. (Mayumi et al. 2021),

- **Crosslinking Density:** Higher crosslinking density generally increases mechanical strength; however, it can also decrease the swelling capacity. Thus, the optimal crosslinking density needs to be determined for each specific application.

Enhancement of Mechanical Stability

Even though the hydrogels have great capability in absorption and retention of water, it has certain difficulty in maintaining their mechanical stability, which can eventually limit their applications in various fields. The mechanical stability usually helps in withstanding external forces, such as compression, shear and tension forces without undergoing major deformity and destruction to the hydrogel, that are compromised due to certain factors that need to be improved. To address these issues, researchers have been exploring various strategies to enhance the mechanical stability of hydrogels in the following ways:

Optimization of Crosslinking Density

Adjusting the crosslinking density within the hydrogel network can improve mechanical strength while maintaining adequate porosity for water absorption (Li et al. 2022).

Use of Composite Hydrogels

Incorporation of reinforcing materials such as nanoparticles, carbon nanotubes, or natural fibres in the hydrogels can enhance their mechanical properties (Bratovcic 2020). Thereafter, these composites can provide additional strength and resistance to undergo deformation.

Interpenetrating Polymer Networks (IPNs)

Formation of IPNs, where two or more polymer networks are physically entangled, however, not covalently bonded, can improve the mechanical properties without significantly compromising the swelling capacity of the hydrogel (Crosby et al. 2022).

15.5.1.2 Durability of Hydrogels

Durability is a critical factor for the practical application of hydrogels in wastewater treatment for their long-term performance. Degradation is one important factor that affects the durability of the hydrogel. The susceptibility to degradation majorly occurs through several mechanisms, such as hydrolysis, enzymatic degradation, oxidative degradation, etc. As the hydrogels are exposed to degradation, their durability starts to deteriorate and leads to a compromise of certain properties of the hydrogels, which include loss in their mechanical properties such as strength, stiffness, and toughness, changes in the manner of swelling in the hydrogel, which can later alter the absorption and retention of water by the hydrogel. These factors will reduce the functionality of the hydrogel to a great extent. However, hydrogel must function effectively over extended periods to be economically viable. Apart from the impacts from degradation, the durability of the hydrogel is greatly compromised by its prolonged exposure to wastewater, which leads to:

- **Fouling:** Accumulation of organic and inorganic waste matter on the hydrogel surface can hinder its performance in water bodies during the treatment process. Fouling can block pores, reduce water permeability, and impair the hydrogel's ability to absorb the pollutants (Zulkefli et al. 2021).
- **Chemical and Physical Wear:** Continuous exposure of hydrogels to the fluctuation in the pH levels and variation in temperatures can lead to the degradation of the hydrogel material, consequently, reducing the efficiency and eventually resulting in failure (Bonyadi and Dunn 2020).

Improvement of Durability

To improve the durability of hydrogels, several approaches are involved as follows:

- **Outer Structure Modification**: Modifying the surface of hydrogels with antifouling agents or hydrophilic coatings can help to prevent the buildup of contaminants on the hydrogel's surface, thereby protecting it from fouling and degradation that could result from contaminant deposition (Zulkefli et al. 2021). This reduction in degradation and extension of lifespan can significantly enhance the hydrogel's durability.
- **Stabilizing Additives:** Incorporation of additives that stabilize the hydrogel network against chemical and thermal degradation can again improve their durability. These additives may include antioxidants, UV stabilizers, and other protective agents (Zafar et al. 2019).
- **Optimization of Formulation:** Development of such hydrogel formulations that are specifically tailored to withstand the conditions present in wastewater can enhance their longevity (Lefnaoui and Moulai-Mostefa 2014). Therefore, the optimization process involves the selection of such polymers and crosslinking agents that resist chemical and biological degradation.

15.5.2 Regeneration and Reuse of Hydrogels

For hydrogels to be sustainable and cost-effective in wastewater treatment, they must be regenerable and reusable. This involves the restoration of the hydrogel's capacity to absorb pollutants after each treatment cycle without significant loss of performance. Regeneration and reuse of hydrogels will provide effective benefits in the field of wastewater treatment. The advantages that are involved include cost saving, which can significantly help in reducing the cost associated with the production of hydrogel and wastewater treatment, the development of sustainable and reused water treatment technologies is very much essential for addressing the global water crisis, resulting in increased efficiency and minimizing the amount of waste deposition in the water streams. Regeneration is particularly challenging due to the complex and variable composition of wastewater containing chemical waste like heavy metals, dyes, human and animal waste, etc. Though there are major advantages associated with the regeneration and reuse of the hydrogels, there are certain challenges as well, which are faced in the development of hydrogels with such interest, which include:

15.5.2.1 Challenges in Regeneration

Pollutant Desorption

Effective removal of adsorbed pollutants from the hydrogel matrix can be difficult as pollutants may bind strongly to the hydrogel; consequently, desorption becomes inefficient and incomplete.

Structural Degradation

Regeneration processes, such as washing with solvents or applying thermal treatments, can damage the hydrogel's structure, which reduces its capacity and effectiveness in subsequent cycles.

Cost and Complexity

Besides, regeneration methods need to be economically viable and straightforward to implement, as complex or expensive regeneration processes can offset the benefits of using hydrogels.

15.5.2.2 Strategies for Effective Regeneration

Several strategies are being explored to improve the regeneration and reuse of hydrogels, like use of chemical components, introduction of physical factors and the implementation of biological treatments can lead to effective regeneration of hydrogels as explained below.

Chemical Regeneration

Use of chemicals to desorb the pollutants from the hydrogels can be effective, but needs to be optimized to prevent damage to it. Solutions such as acids, bases, or specific desorbing agents can be used, depending on the nature of the pollutants to be removed.

Physical Regeneration

Methods that involve physical factors like thermal treatment, introduction of ultrasound, or UV irradiation can be used to regenerate the hydrogels. These methods should be carefully controlled to avoid the degradation of the structure of the hydrogel (Zhang et al. 2013).

Biological Regeneration

Incorporating the microbial or enzymatic treatments to degrade or transform the adsorbed pollutants from the hydrogels into less harmful forms, as it makes them easy to remove from the hydrogel (Seida and Tokuyama 2022).

15.5.3 Cost and Scalability of Hydrogel Production

One of the primary challenges of using hydrogels in wastewater treatment is the cost and scalability of their production (Guo et al. 2020). Hydrogels can be synthesized using various methods, including chemical crosslinking, radiation crosslinking, and physical crosslinking. Each method has its advantages and disadvantages in terms of cost, efficiency, and scalability. The factors affecting are the raw material used, the synthesis methods, like microfluidics and the application of purification and further processing of hydrogels also drastically increase the cost. The recent advancement in technology with nanoparticle materials and crosslinking methods has created a surge in the cost in the market again. Thus, the adaptation of developing of optimized formulation is required in raw material selection and processing.

15.5.3.1 Raw Material Costs

The production of hydrogel is expensive due to the high cost of raw materials. Many hydrogels are synthesized using polymers such as polyacrylamide, polyvinyl alcohol, and natural biopolymers like chitosan and alginate. Synthetic polymers can be expensive due to their complex synthesis processes, which involve the costs associated with petrochemical feedstocks (Ahmad et al. 2022), while the natural polymers may offer a comparatively low cost advantage; however, it may require extensive purification methods and modification, which again increases the cost of production.

15.5.3.2 Synthesis Methods

A range of techniques, including chemical, radiation, and physical crosslinking, can be used to create hydrogels. Among them, chemical crosslinking often requires the use of toxic chemicals and can involve complex reactions that are not easily scalable. On the other hand, radiation crosslinking avoids the use of harmful chemicals; however, it requires sophisticated equipment that can produce the intense energies in the environment. While the physical crosslinking methods, often simpler and cost-effective, but may result in hydrogels with inferior mechanical properties and stability, hence, limiting their effectiveness in the applications of wastewater treatment (Kaczmarek et al. 2020). Thus, each method has its own merit and demerit.

15.5.3.3 Production Scale

Scaling up the production of hydrogels from laboratory settings to industrial scales presents significant challenges. Uniformity in hydrogel properties, such as mechanical strength, swelling capacity, and porosity must be maintained across the large batches, which is difficult to achieve. Additionally, large-scale production facilities need substantial capital investment, infrastructure, and energy, which contribute to the overall cost (Sorour et al. 2020).

15.5.3.4 Economic Viability

For hydrogels to be economically viable in wastewater treatment, the cost of production must be comparable to or lower than existing treatment technologies. This includes not only the initial synthesis but also increases the operational and maintenance costs associated with hydrogel-based treatment systems. Therefore, the development of cost-effective and scalable production methods remains a significant hurdle.

15.5.4 Selectivity and Efficiency in Mixed Contaminated Environments

The performance of the hydrogels in wastewater treatment applications is majorly challenged by the complexity and variable composition of wastewater. Wastewater often contains a diverse range of pollutants, including organic compounds, heavy metals, nutrients, and pathogens, which can compete with each other for binding sites on the hydrogel. In wastewater treatment, the ability of hydrogels to selectively adsorb and remove specific contaminants from complex mixtures is crucial and reduces overall performance. (Seida and Tokuyama 2022). Furthermore, the presence of multiple pollutants in wastewater can lead to a synergetic effect, where the combined effect of the pollutants creates difficulty in the removal of individual pollutants. Thus, the performance of hydrogels in such environments is challenged by several factors, which have been described in the subsequent sections.

15.5.4.1 Selective Adsorption

Hydrogels must be designed to selectively adsorb the targeted contaminants, excluding the non-targeted species. This requires precise control over the functional groups and surface properties like surface charge, porosity, roughness, etc. of the hydrogel (Zhu et al. 2023). To achieve this selectivity in the mixed contaminated environments is complex, as the presence of multiple pollutants can lead to competitive adsorption and hence, subsequently reduce the efficiency of the hydrogels.

15.5.4.2 Contaminants Interactions

The interactions between different contaminants in wastewater can complicate the adsorption process as discussed earlier. For example, the presence of certain ions or organic compounds can hinder the binding of other contaminants to the hydrogel, decreasing its overall effectiveness. This necessitates the development of hydrogels with multi-functional properties capable of handling a broad spectrum of pollutants simultaneously.

15.5.4.3 Efficiency Under Varying Conditions

The hydrogel's performance can be affected by a number of environmental changes, like variation in pH, temperature, and ionic strength (Dragan and Cocarta 2016). Since wastewater characteristics can vary widely depending on the source and stage of treatment, hydrogels must maintain high efficiency across these varying conditions in practical applications. It means the hydrogel's adsorption capacity and selectivity should remain effective despite fluctuations in environmental parameters. Thus, designing the hydrogels that are resilient and adaptable to these changes is crucial for their successful use in the diverse wastewater treatment scenarios.

15.5.4.4 Regeneration and Reusability

For hydrogels to be practical and sustainable in wastewater treatment, they must be easily regenerable and reusable without significant loss of efficiency. The regeneration process should effectively desorb the captured contaminants to again restore the adsorption capacity of the hydrogel. However, repeated use of hydrogels and their regeneration cycles can degrade the hydrogel structure and reduce its performance over time (Van Tran et al. 2018).

15.5.5 Environmental Impact and Disposal

The environmental impact and disposal of hydrogels used in wastewater treatment are critical concerns that need to be addressed to ensure their sustainability and compliance with environmental regulations. The hydrogels potentially release chemicals during their production, use, and disposal, which can contribute to increased chemical and soil pollution. Soil pollution can further lead to the pollution of groundwater. Furthermore, the unique properties of the hydrogel create significant challenges in the disposal of hydrogel, such as their ability to absorb and retain water, in incineration, chemical treatment, making it difficult for landfill disposal and recycling. To address these challenges, sustainable solutions need to be developed, including the design of biodegradable hydrogels, the development of recycling technologies, and the implementation of sustainable production methods. Some of them have been described under the following heads.

15.5.5.1 Biodegradability and Ecotoxicity

Many synthetic hydrogels are not biodegradable, posing a disposal challenge after their optimal life. Non-biodegradable hydrogels can persist in the environment, leading to potential long-term ecological impacts (Qin et al. 2024). Though the biodegradable hydrogels also need to be assessed for their breakdown products, which should be non-toxic and environmentally benign.

15.5.5.2 Environmental Footprint

The environmental footprint of hydrogel production includes energy consumption, greenhouse gas emissions, and the use of hazardous chemicals. Thus, to reduce the environmental impact of the synthesis of hydrogel requires the development of greener production technologies, such as the use of renewable raw materials, energy-efficient processes, and safer chemicals (Madani et al. 2022).

15.5.5.3 Post-Treatment Waste

After being used for contaminant removal, hydrogels may become loaded with concentrated pollutants, creating the challenge of proper disposal. To prevent secondary pollution, the safe disposal or treatment of spent hydrogels must be ensured. Strategies for this include incineration, landfilling under controlled conditions or further treatment to recover and recycle the valuable materials. Recovery and recycling involve the extraction of the adsorbed contaminants for reuse of the hydrogel material, thereby reducing waste and promoting sustainability in pollutant management (Devre and Gore 2023).

15.5.5.4 Regulatory Compliance

Hydrogels used in wastewater treatment must comply with environmental regulations regarding the release of chemicals and materials into the environment (Van Tran et al., 2018). This includes ensuring that any leachates or by-products obtained from the degradation of hydrogel do not exceed the permissible limits for environmental discharge as per the regulatory bodies like World Health Organization (Bashir et al. 2012).

15.6 Technological Innovations and Future Directions in Wastewater Treatment

Hydrogels, known for their exceptional water-absorbing capabilities and have found extensive applications in various fields, including medicine, agriculture, environmental engineering, etc. The utility of hydrogels can be attributed to their unique structural properties, which allow them to retain significant water retention and their interaction with a variety of substances. The technological innovations for extending the usage of hydrogels for wastewater treatment, majorly concentrating on the advanced functionalization techniques, integration with other treatment technologies, development of biodegradable and sustainable hydrogels, smart hydrogels with stimuli-responsive properties, cost-effective production methods, importance of field trials and pilot studies, which all have been illustrated in Fig. 15.4 as well.

15.6.1 Advanced Functionalization Techniques

Functionalization of hydrogels involves modifying their chemical structure to enhance their specific properties, such as mechanical strength, responsiveness to environmental stimuli, and ability to interact with specific pollutants. The very recent

Fig. 15.4 Technological innovations and future directions for the improvement of usage of hydrogel in wastewater treatment

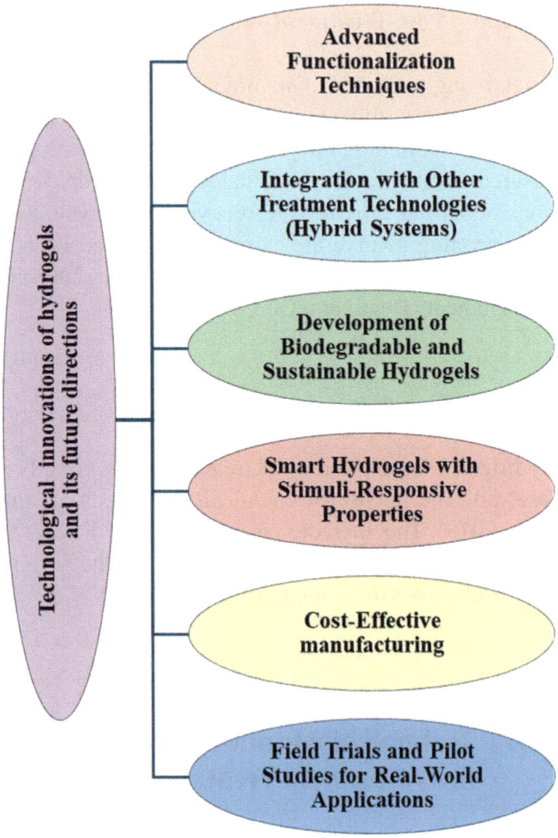

advancements in functionalization techniques include the incorporation of nanoparticles, grafting of functional groups, and the use of molecular imprinting, which have been discussed as follows.

15.6.1.1 Nanoparticle Incorporation

Integrating nanoparticles into the hydrogels has a significant effect on enhancing specific properties. For example, adding silver nanoparticles imparts antimicrobial properties, while magnetic nanoparticles can facilitate the removal of pollutants via magnetic separation. A study by Li et al. (2018) demonstrated the use of silver nanoparticle-loaded hydrogels for the effective microbial inhibition and heavy metal removal.

15.6.1.2 Grafting Functional Groups

By attaching specific functional groups to the polymer chains of hydrogels, researchers can tailor their interaction with targeted pollutants. Functional groups such as carboxyl, amine, and thiol groups can improve the hydrogel's affinity for heavy metals and organic pollutants. For instance, a study by Zheng et al. (2023) showed that thiol-functionalized hydrogels exhibited high affinity for strontium II ions in wastewater.

15.6.1.3 Molecular Imprinting

This technique involves creating specific binding sites within the hydrogel that are complementary in shape and functional groups to the targeted molecules. Molecularly imprinted hydrogels are highly selective and can be designed and modified to capture specific contaminants from wastewater.

A recent study by Ahmad et al. (2022) developed an amino-ethyl carboxymethyl cellulose crosslinked ampholyte hydrogel for the removal of methyl orange from wastewater, which increased the efficiency for the removal of certain highly contagious and difficult dyes that are polluting the water sources.

15.6.2 Hybrid Systems—an Integration with Other Treatment Technologies

Combining hydrogels with other treatment technologies can create hybrid systems that pull the strengths of each component for more effective and efficient wastewater treatment. Such integrations can include:

15.6.2.1 Membrane Filtration

Embedding hydrogels into membrane filtration systems can enhance pollutant removal and also reduce the fouling effect. Thereby, hydrogels can adsorb contaminants, allowing cleaner water to pass through the membrane filtration system. A study by Ma et al. (2017a, b) demonstrated that hydrogel-coated membranes effectively reduced fouling and enhanced the removal of organic pollutants in wastewater.

15.6.2.2 Photocatalysis

The incorporation of photocatalytic materials, such as TiO_2, into hydrogels enables the degradation of organic pollutants under light exposure. This hybrid system combines adsorption and degradation, making it highly effective for removing persistent organic contaminants.

15.6.2.3 Electrochemical Treatment

Integrating the hydrogels with electrochemical cells can improve the electrochemical degradation of pollutants. The hydrogel can serve as an electrolytic medium that enhances the contact between the pollutant and the electrode surface. Liu et al. (2020) explored the application of electrochemical hydrogel systems for the degradation of dye pollutants, showing significant improvement in their efficiency.

15.6.3 Development of Biodegradable and Sustainable Hydrogels

The environmental impact of hydrogels is a critical consideration in the development and application. Biodegradable and sustainable hydrogels are designed to minimize the ecological footprint while maintaining their functional properties. Thus, the key approaches for development of hydrogels in such area include the following, to be considered:

15.6.3.1 Natural Polymers

Application of natural polymers can be used to produce biodegradable hydrogels, such as components using cellulose, chitosan, and alginate. These materials are renewable, non-toxic, and decompose into harmless byproducts. A review by Gohil and Nimmagadda (2020) highlighted the use of natural polymer-based hydrogels for sustainable environmental applications.

15.6.3.2 Green Synthesis Methods

Emerging new trends in green synthesis and sustainable practice have gained a great interest in this current decade. A biodegradable hydrogel that can be disposed of easily onto the earth's surface without any harmful effects has imposed a higher user-friendly nature of the material. Hydrogel's basic structure is composed mainly of water, providing non-toxic and non-chemical structural advantages. The green synthesis hydrogel has further made the removal of diclofenac sodium from water easier and effective. Developing an environmentally friendly synthesis method, such as using water as a solvent and avoiding any toxic chemicals, can reduce the environmental impact of hydrogel production. Dhavade et al. (2024) used and studied silver nanoparticles for the Azo dye degradation, which effectively showed the enhanced method for the removal of dyes with minimal waste contaminants left in the water. Further, research by Xiao et al. (2018) demonstrated a green synthesis route for hydrogels using aqueous media and biodegradable crosslinkers.

15.6.3.3 Recyclability and Reusability Criterion

Designing hydrogels that can be easily regenerated and reused for multiple cycles can enhance their sustainability. Wu et al. (2019) developed a recyclable hydrogel for heavy metal removal that could be regenerated and reused without significant loss in performance.

15.6.4 Smart Hydrogels with Stimuli-Responsive Properties

Smart hydrogels can respond to external stimuli such as pH, temperature, light, and magnetic fields. These properties enable precise control over their behaviour and functionality, making them highly adaptable for various applications.

15.6.4.1 PH-Responsive Hydrogels

These hydrogels can swell or shrink in response to changes in pH. They are particularly useful in wastewater treatment for the targeted release of adsorbed contaminants or nutrients in response to environmental pH changes. Wang et al. (2018) demonstrated the use of pH-responsive hydrogels for controlled drug delivery and pollutant removal from contaminated water.

15.6.4.2 Temperature-Responsive Hydrogels

Temperature-Responsive Hydrogels are also known as thermoresponsive hydrogels; they undergo phase transitions at specific temperatures. This property can be exploited for the controlled release of substances or for self-cleaning applications in wastewater treatment systems. Zhang et al. (2017a, b) reported the application of temperature-responsive hydrogels for the controlled release of therapeutic agents.

15.6.4.3 Light-Responsive Hydrogels

Incorporating photochromic or photosensitive groups into the hydrogels enables them to change properties under light exposure. This can be useful for controlled drug delivery or for the degradation of pollutants through photocatalytic reactions. Liu et al. (2018) explored the use of light-responsive hydrogels for targeted drug delivery applications.

15.6.4.4 Magnetic-Responsive Hydrogels

Embedding of magnetic nanoparticles allows hydrogels to be manipulated using external magnetic fields. This facilitates easy separation and recovery of the hydrogel from the treated water. One such study was reported by Li et al. (2018), which implicate the use of magnetic hydrogels for the removal of heavy metals from water, demonstrating high efficiency and ease of recovery.

15.6.5 Cost-Effective Manufacturing

The economic viability of hydrogels for large-scale applications depends on the development of cost-effective production methods. Strategies to reduce production costs include:

15.6.5.1 Bulk Synthesis

Scaling up the synthesis process to produce hydrogels in large quantities can reduce per-unit cost. Continuous production methods and automation can further enhance efficiency. Chen et al. (2020) discussed the industrial-scale production of hydrogels using bulk polymerization techniques with less raw materials and cost-effectiveness.

15.6.5.2 Low-Cost Raw Materials

Using inexpensive and readily available raw materials, such as agricultural byproducts or industrial waste, can significantly lower the production costs. The use of agricultural waste-derived cellulose for hydrogel production, highlighting their cost-effectiveness and sustainability by Zhang and co-workers in 2019 (Zhang et al. 2019).

15.6.5.3 Energy-Efficient Processes

Implementing energy-efficient synthesis techniques, such as microwave-assisted polymerization or UV curing, can reduce energy consumption and, hence, the production costs. Luo et al. (2018) explored the use of UV curing for the rapid and energy-efficient synthesis of hydrogel.

15.6.6 Field Trials and Pilot Studies for Real-World Applications

Before hydrogels can be widely adopted for wastewater treatment, they must be tested in real-world conditions to evaluate their performance, durability, and scalability. Field trials and pilot studies are essential steps in this process and can be assessed by adopting the following methods.

15.6.6.1 Performance Evaluation

Field trials provide valuable data on the effectiveness of hydrogels in removing specific pollutants under varying environmental conditions. This helps in optimizing hydrogel formulations and treatment processes. A pilot study was conducted by Singh et al. (2020), which evaluated the performance of hydrogels in removing heavy metals from industrial effluents, providing insights into the practical application, challenges, and solutions.

15.6.6.2 Durability Testing

Long-term studies assess the stability and reusability of hydrogels in real-world applications. This is crucial for determining their lifespan and maintenance requirements. Kim et al. (2019) research focused on the long-term stability and reusability of hydrogels for wastewater treatment, highlighting their potential for sustainable use.

15.6.6.3 Scalability Assessment

Pilot studies help to identify the challenges associated with scaling up hydrogel production and applications. This includes evaluating the logistics of deployment, the cost of large-scale operations, and potential environmental impacts. Dharmapriya et al. (2021) comprehensively reviewed the scalability of hydrogel technologies for environmental applications, addressing key considerations and future directions on the need for advancement for wastewater treatment and for the effectiveness of hydrogel use.

15.7 Case Studies Based on the Efficacy of Hydrogels in Wastewater Treatment

Numerous studies have been carried out to explore the efficacy of hydrogels in different environments of wastewater, constituting the various pollutants during the treatment process, which have been described below.

15.7.1 Removal of Heavy Metals

A study was conducted by Gohil and Nimmagadda (2020) to investigate the effectiveness of a chitosan-based hydrogel including amino and hydroxyl groups in removing lead ions from contaminated water and the hydrogels were found to remove more than 90% of lead ions, demonstrating their high adsorption capacity and potential for industrial effluent treatment. The chitosan hydrogel's functional groups, including amino and hydroxyl groups, were also found to play a crucial role in binding the lead ions through coordination bonding and ion exchange mechanisms.

15.7.2 Removal of Dye Based on Adsorption and Recovery of Nutrients

Ma et al. (2017a, b) examined the use of a polyacrylamide-based hydrogel in removing the methylene blue dye from contaminated water. The hydrogel successfully removed over 95% of the dye by adsorption following a pseudo-second-order kinetic model. This indicated strong interactions between the dye molecules and the hydrogel, driven by electrostatic interactions and hydrogen bonding.

Zhang et al. (2017a, b) highlighted the efficiency of functionalized hydrogels for removing both anionic and cationic dyes from wastewater. The hydrogel's carboxyl groups effectively adsorbed anionic dyes, while amine groups targeted the cationic dyes, and its dual functionality allowed for the simultaneous removal of multiple dye types from industrial effluents.

Research by Dhavade et al. (2024) provided supporting evidence to demonstrate an effective pathway to remove the Azo dyes in wastewater using hydrogels incorporated with silver nanoparticles. Xiao et al. (2018) investigated the role of a functionalized hydrogel with quaternary ammonium groups for the removal of nitrate from the wastewater. The hydrogel showed high selectivity for nitrates, effectively captured them from the complex wastewater matrices. The study highlighted the hydrogel's potential for nutrient recovery and its application in reducing nutrient pollution. The alginate-based hydrogel has been used for removing and recovering the phosphates from agricultural runoff. The hydrogel was found to efficiently capture the phosphate ions through ion exchange mechanisms, and the recovered phosphates were used as fertilizers. This closed-loop system showed the potential of hydrogels to contribute to sustainable nutrient management in the agriculture domain.

15.7.3 Degradation of Organic Pollutants

Wang et al. (2019a, b) demonstrated the degradation of phenolic compounds in wastewater using a hydrogel embedded with the laccase enzyme. The enzyme catalyzed the oxidation of phenols, breaking them down into less toxic substances. This approach utilized the biocatalytic properties of laccase, showing its potential for the enzyme-embedded hydrogels in treating organic pollutants.

Liu et al. (2020) explored the use of TiO_2-hydrogel composites in the photocatalytic degradation of pharmaceuticals. The hydrogel matrix was found to support the dispersion of TiO_2 nanoparticles and enhance their exposure to UV light and promote the generation of reactive oxygen species. This resulted in the efficient breakdown of pharmaceutical contaminants, demonstrating the effectiveness of photocatalytic hydrogels in treating complex organic pollutants.

15.8 Regulatory and Environmental Considerations

The regulatory frameworks govern the safety, efficacy, and environmental sustainability of technologies introduced. The environmental impact assessments can evaluate hydrogels effects on ecosystems, while the safety and toxicity evaluations can address potential human health risks. The regulatory bodies review hydrogel materials, manufacturing processes, and disposal methods to mitigate the environmental harm and to ensure their safe use, while driving innovation towards more eco-friendly hydrogel treatment for wastewater.

15.8.1 Regulatory Frameworks for Wastewater Treatment

The implementation of hydrogel-based wastewater treatment technologies requires compliance with regulatory frameworks established by environmental protection agencies (EPA). These frameworks define the permissible levels of contaminants in wastewater after being treated, the safety of treatment materials, and disposal methods for spent hydrogels. Regulations also address the performance standards for treatment systems and monitoring requirements to ensure consistent compliance. Monitoring by regulatory bodies can regulate the technologies of wastewater treatment at the early stage, which could facilitate the proper adoption of innovative methodologies.

15.8.2 Environmental Impact Assessment of Hydrogel Use

Conducting an environmental impact assessment (EIA) is essential to evaluate the potential effects of hydrogel use in wastewater treatment. For this, the EIA considers the factors such as the lifecycle of hydrogels, their degradation products, and the potential of secondary pollutants generated. It also assesses the ecological footprint of hydrogel production, including energy consumption and raw material sourcing. EIA comprehensively ensures that hydrogel-based technologies provide environmental benefits without introducing new risks or unintended consequences.

A recent study by Amornpitoksuk et al. (2024) used a heterogenous Fenton-like catalyst using the polyacrylamide hydrogel, which is immobilized with the iron. This enhanced the pollutant degradation efficiency, and minimized iron leaching along with the promising use of the substance for wastewater treatment.

15.8.3 Safety and Toxicity Issues Pertaining to Hydrogel Materials

Safety and toxicity issues are the most critical considerations when deploying hydrogels for wastewater treatment. The materials used in hydrogel synthesis should be non-toxic and environmentally benign. Additionally, the degradation products of hydrogels should not pose any harm to the aquatic organisms and human health. Ensuring the biocompatibility and safety of hydrogels involves rigorous testing and validation of the potential leachates, including acute and chronic toxicity studies and biodegradability assessments, and other relevant analyses. Hence, the overall safety data generated pertaining to the toxicity studies makes a transparent communication and can build trust and acceptance among the stakeholders and regulatory authorities.

15.9 Conclusion

The chapter underscores the effective potentials of hydrogels in wastewater treatment, driven by their unique physicochemical properties. Hydrogels, with various properties, offer a versatile platform for targeting a wide range of contaminants. Despite these advantages, several challenges have also been addressed for harnessing their usage in practical applications. Key issues should be emphasized to improve their mechanical stability to withstand the operational stresses of large-scale water treatment processes, developing cost-effective and sustainable synthesis methods along with enhancing their reusability and regeneration capabilities to reduce the waste as well as the operational costs. Additionally, the environmental impact concerning their biodegradability and release of secondary pollutants should also be carefully ly considered. Advancements in development of hybrid hydrogels and the integration of nanotechnology to enhance pollutant adsorption and degradation have also been reflected, which could improve their performance. The exploration of smart hydrogels with stimuli-responsive properties, which can adapt to different environmental conditions, would also improve their performance under varied environmental factors. In the future, the proper monitoring and implementation of rules governed by the EPA may also overcome the issues associated with the disposal and degradation of used hydrogels. Moreover, the research should focus on optimizing hydrogel formulations to balance the performance, cost, environmental sustainability, etc. By addressing all these challenges together, hydrogels can prove to be potential solution in the revolutionization of wastewater treatment, which can provide a sustainable and efficient solution that contributes to environmental protection, conservation of water resources, and recovery of essential nutrients from wastewater.

References

Ahmad Z, Salman S, Khan SA, Amin A, Rahman ZU, Al-Ghamdi YO, Khan SB (2022) Versatility of hydrogels: from synthetic strategies, classification, and properties to biomedical applications. Gels 8(3):167

Akinluyi TO, Odeyemi O (1984) Human waste disposal and the faecal pollution of the River Niger Delta waters. Water Int 9(1):37–41

Ammary BY (2004) Nutrients requirements in biological industrial wastewater treatment. Afr J Biotech 3(4):236–238

Amornpitoksuk P, Kongseng P, Chantarak S, Suwanboon S (2024) Degradation of organic pollutants by floatable Fe-PAM hydrogel. Desalination 571:117103

Bashir MJK, Aziz HA, Yusoff MS, Aziz SQ (2012) Investigation of color and COD eliminations from mature semi-aerobic landfill leachate using anion-exchange resin: equilibrium and kinetic study. Environ Eng Sci 29:297–305

Bohra DL, Modasiya V, Bahura CK (2012) The distribution of coliform bacteria in waste water. Microbiol Res 3(1):e2

Bonyadi SZ, Dunn AC (2020) Compositional dependence of polyacrylamide hydrogel abrasive wear resistance. ACS Appl Polym Mater 2(12):5444–5451

Borgerding AJ, Hites RA (1994) Identification and measurement of food and cosmetic dyes in a municipal wastewater treatment plant. Environ Sci Technol 28(7):1278–1284

Bratovcic A (2020) Nanocomposite hydrogels reinforced by carbon nanotubes. Int J Eng Res Appl 10(5):30–41

Burge WD, Marsh PB (1978) Infectious disease hazards of landspreading sewage wastes. J Environ Qual 7(1):1–9

Chen G, Chen S, Zhang X (2020) Industrial-scale production of hydrogels using bulk polymerization techniques. Polym Chem 11(14):2275–2284

Crosby CO, Stern B, Kalkunte N, Pedahzur S, Ramesh S, Zoldan J (2022) Interpenetrating polymer network hydrogels as bioactive scaffolds for tissue engineering. Rev Chem Eng 38(3):347–361

de Lima ROA, Bazo AP, Salvadori DMF, Rech CM, de Palma Oliveira D, de Aragão Umbuzeiro G (2007) Mutagenic and carcinogenic potential of a textile azo dye processing plant effluent that impacts a drinking water source. Mutation Res/Genet Toxicol Environ Mutagen 626(1–2):53–60

Devre PV, Gore AH (2023) Agro-waste valorization into carbonaceous eco-hydrogel: a circular economy and zero waste tactic for doxorubicin removal in water/wastewater. Langmuir 40(1):141–158

Dharmapriya TN, Lee DY, Huang PJ (2021) Novel reusable hydrogel adsorbents for precious metal recycle. Sci Rep 11(1):19577

Dhavade AK, Hinge S, Sabharwal SG (2024) Green synthesized AgNPs entrapped in polymer hydrogels for azo dye degradation. J Mater Sci Mater Electron 35(19):1368

Dragan ES, Cocarta AI (2016) Smart macroporous IPN hydrogels responsive to pH, temperature, and ionic strength: synthesis, characterization, and evaluation of controlled release of drugs. ACS Appl Mater Interfaces 8(19):12018–12030

Fu Z, Xi S (2020) The effects of heavy metals on human metabolism. Toxicol Mech Methods 30(3):167–176

Ghaly AE, Snow A, Faber BE (2006) Treatment of grease filter washwater by chemical coagulation. Can Biosyst Eng 48:6

Gita S, Hussan A, Choudhury TG (2017) Impact of textile dyes waste on aquatic environments and its treatment. Environ Ecol 35(3C):2349–2353

Gohil JM, Nimmagadda A (2020) Natural polymer-based hydrogels for sustainable environmental applications: a review. J Environ Chem Eng 8(5):104217

Guo Y, Lu H, Zhao F, Zhou X, Shi W, Yu G (2020) Biomass-derived hybrid hydrogel evaporators for cost-effective solar water purification. Adv Mater 32(11):1907061

Janagi R, Shah J, Maheshwari D (2015) Scenario of management of medical waste in US and UK: a review. J Glob Trends Pharm Sci 6(1):2328–2339

Kaczmarek B, Nadolna K, Owczarek A (2020) The physical and chemical properties of hydrogels based on natural polymers. In: Hydrogels based on natural polymers. pp 151–172

Khayatzadeh J, Abbasi E (2010) The effects of heavy metals on aquatic animals. In: The 1st International applied geological congress, vol 1. Department of Geology, Islamic Azad University–Mashad Branch, Iran, pp 26–28

Kim H, Choi Y, Park J (2019) Long-term stability and reusability of hydrogels for wastewater treatment. Environ Sci Technol 53(3):1505–1512

Leclerc H, Schwartzbrod L, Dei-Cas E (2002) Microbial agents associated with waterborne diseases. Crit Rev Microbiol 28(4):371–409

Lefnaoui S, Moulai-Mostefa N (2014) Investigation and optimization of formulation factors of a hydrogel network based on kappa carrageenan–pregelatinized starch blend using an experimental design. Colloids Surf A 458:117–125

Li J, Li Y, Ye Z (2018) Silver nanoparticle-loaded hydrogels for microbial inhibition and heavy metal removal. J Hazard Mater 359:1–10

Li D, Zhan W, Zuo W, Li L, Zhang J, Cai G, Tian Y (2022) Elastic, tough and switchable swelling hydrogels with high entanglements and low crosslinks for water remediation. Chem Eng J 450:138417

Liu X, Li J, Chen Y (2018) Light-responsive hydrogels for targeted drug delivery applications. Adv Healthcare Mater 7(1):1701125

Liu Y, Wang L, Chen F (2020) Electrochemical hydrogel systems for the degradation of dye pollutants. Chem Eng J 387:124057

Lu Y, Spyra P, Mei Y, Ballauff M, Pich A (2007) Composite hydrogels: robust carriers for catalytic nanoparticles. Macromol Chem Phys 208(3):254–261

Luo Y, Wang Z, Zhang Z (2018) UV curing for rapid and energy-efficient hydrogel synthesis. Polymer 140:144–153

Ma J, Yang Z, Tang C (2017a) Polyacrylamide-based hydrogel for the adsorption of methylene blue dye: Kinetic and isotherm studies. Desalin Water Treat 67:234–242

Ma Y, Zhang X, Wang P (2017b) Hydrogel-coated membranes for fouling reduction and enhanced pollutant removal. J Membr Sci 541:160–169

Madani M, Hosny S, Alshangiti DM, Nady N, Alkhursani SA, Alkhaldi H, Gaber GA (2022) Green synthesis of nanoparticles for varied applications: green renewable resources and energy-efficient synthetic routes. Nanotechnol Rev 11(1):731–759

Mayumi K, Liu C, Yasuda Y, Ito K (2021) Softness, elasticity, and toughness of polymer networks with slide-ring cross-links. Gels 7(3):91

Mittal H, Jindal R, Kaith BS, Maity A, Ray SS (2015) Flocculation and adsorption properties of biodegradable gum-ghatti-grafted poly (acrylamide-co-methacrylic acid) hydrogels. Carbohyd Polym 115:617–628
Mohana AA, Roddick F, Maniam S, Gao L, Pramanik BK (2023) Component analysis of fat, oil and grease in wastewater: challenges and opportunities. Anal Methods 15(39):5112–5128
Mu R, Liu B, Chen X, Wang N, Yang J (2020) Hydrogel adsorbent in industrial wastewater treatment and ecological environment protection. Environ Technol Innov 20:101107
Obasi PN, Akudinobi BB (2020) Potential health risk and levels of heavy metals in water resources of lead–zinc mining communities of Abakaliki, southeast Nigeria. Appl Water Sci 10(7):1–23
Preisner M, Neverova-Dziopak E, Kowalewski Z (2020) Analysis of eutrophication potential of municipal wastewater. Water Sci Technol 81(9):1994–2003
Qin C, Wang H, Zhao Y, Qi Y, Wu N, Zhang S, Xu W (2024) Recent advances of hydrogel in agriculture: Synthesis, mechanism, properties and applications. Europ Polym J 113376
Rajasulochana P, Preethy V (2016) Comparison on efficiency of various techniques in treatment of waste and sewage water–a comprehensive review. Resour-Effic Technol 2(4):175–184
Rashed MN (2013) Adsorption technique for the removal of organic pollutants from water and wastewater. In: Organic pollutants-monitoring, risk and treatment. IntechOpen
Saha JK, Selladurai R, Coumar MV, Dotaniya ML, Kundu S, Patra AK, Patra AK (2017) Organic pollutants. In: Soil pollution-an emerging threat to agriculture. pp 105–135
Seida Y, Tokuyama H (2022) Hydrogel adsorbents for the removal of hazardous pollutants—requirements and available functions as adsorbent. Gels 8(4):220
Shalla AH, Bhat MA, Yaseen Z (2018) Hydrogels for removal of recalcitrant organic dyes: a conceptual overview. J Environ Chem Eng 6(5):5938–5949
Shannon EE (1975) Effects of detergent formulation on wastewater characteristics and treatment. J Water Pollut Control Federat 2371–2383
Singh J, Sharma P, Yadav A (2020) Performance evaluation of hydrogels in removing heavy metals from industrial effluents: a pilot study. Environ Sci Pollut Res 27(2):1857–1868
Sorour MH, Tewfik SR, Abulnour AG, Shaalan HF, Hani HA, Al-Bazedi GA (2020) Development and assessment of adsorptive hydrogel: process engineering and economic considerations. Eng Rep 2(4):e12140
Van Tran V, Park D, Lee YC (2018) Hydrogel applications for adsorption of contaminants in water and wastewater treatment. Environ Sci Pollut Res 25:24569–24599
Wan S, Yang X, Chen X, Qu Z, An C, Zhang B, Bi H (2022) Emerging marine pollution from container ship accidents: risk characteristics, response strategies, and regulation advancements. J Clean Prod 376:134266
Wang H, Chen Q, Wang X (2018) PH-responsive hydrogels for controlled drug delivery and pollutant removal. J Mater Chem B 6(2):182–192
Wang J, Xu X, Zhang P (2019a) Enzyme-embedded hydrogels for the degradation of phenolic compounds in wastewater. Chemosphere 220:679–686
Wang X, Zhang Y, Zhang X (2019b) TiO_2-hydrogel composites for the photocatalytic degradation of pharmaceuticals in wastewater. Appl Catal B 243:195–203
Ward J (2003) A plastic solar water purifier with high output. Sol Energy 75(5):433–437
Weldeslassie T, Naz H, Singh B, Oves M (2018) Chemical contaminants for soil, air and aquatic ecosystem. In: Modern age environmental problems and their remediation. pp 1–22
Wu D, Li H, Zhang X (2019) Mechanisms of heavy metal removal by hydrogels: a review. J Environ Manage 234:173–182
Xiao L, Ling Q, Chen Y (2018) Green synthesis of hydrogels using aqueous media and biodegradable crosslinkers. Green Chem 20(12):2884–2891
Yau YH, Rudolph V, Lo CCM, Wu KC (2021) Restaurant oil and grease management in Hong Kong. Environ Sci Pollut Res 28:40735–40745
Zafar MS, Quarta A, Marradi M, Ragusa A (2019) Recent developments in the reduction of oxidative stress through antioxidant polymeric formulations. Pharmaceutics 11(10):505

Zahib IR, Tahir PM, Talib M, Mohamad R, Alias AH, Lee SH (2021) Effects of degree of substitution and irradiation doses on the properties of hydrogel prepared from carboxymethyl-sago starch and polyethylene glycol. Carbohyd Polym 252:117224

Zhang Z, Xiao F, Guo Y, Wang S, Liu Y (2013) One–pot self–assembled three–dimensional TiO_2–graphene hydrogel with improved adsorption capacities and photocatalytic and electrochemical activities. ACS Appl Mater Interfaces 5(6):2227–2233. https://doi.org/10.1021/am303299r

Zhang H, Liu Y, Jin Y (2017a) Functionalized hydrogels for dye removal from wastewater: synthesis, performance, and mechanisms. J Water Proc Eng 18:128–136

Zhang Q, Wang J, Lu S (2017b) Temperature-responsive hydrogels for the controlled release of therapeutic agents. Acta Biomater 54:129–138

Zhang Y, Zhang X, Chen Q (2019) Agricultural waste-derived cellulose for hydrogel production: cost-effectiveness and sustainability. Biores Technol 282:16–23

Zhang YS, Khademhosseini A (2017) Advances in engineering hydrogels. Science 356(6337):eaaf3627

Zheng B, Yin J, Zhu L, Zhou B, Shen H, Harbottle D, Zhang H (2023) Thiol-rich and ion-imprinted alginate hydrogel as a highly adsorptive and recyclable filtration membrane for rapid and selective Sr (II) removal. Chem Eng J 465:142752

Zhu T, Ni Y, Biesold GM, Cheng Y, Ge M, Li H, Lai Y (2023) Recent advances in conductive hydrogels: classifications, properties, and applications. Chem Soc Rev 52(2):473–509

Ziemba C, Larivé O, Reynaert E, Morgenroth E (2018) Chemical composition, nutrient-balancing and biological treatment of hand washing greywater. Water Res 144:752–762

Zulkefli NF, Alias NH, Jamaluddin NS, Abdullah N, Abdul Manaf SF, Othman NH, Kusworo TD (2021) Recent mitigation strategies on membrane fouling for oily wastewater treatment. Membranes 12(1):t26